Introduction to Theory of Computation

Bikash Kanti Sarkar

MSc (Mathematics), MCA, MPhil (Computer Science), PhD (Computer Science)

Member of Faculty

Department of Computer Science and Engineering

BIT Mesra, Ranchi

Ambuj Kumar

BE (Computer Science), MTech (Computer Science)

Senior Software Professional, Hyderabad

Universities Press

INTRODUCTION TO THEORY OF COMPUTATION

UNIVERSITIES PRESS (INDIA) PRIVATE LIMITED

Registered Office
3-6-747/1/A & 3-6-754/1, Himayatnagar, Hyderabad 500 029, Telangana, India
info@universitiespress.com; www.universitiespress.com

Distributed by
Orient Blackswan Private Limited

Registered Office
3-6-752 Himayatnagar, Hyderabad 500 029, Telangana, India

Other Offices
Bengaluru / Bhopal / Chennai / Guwahati / Hyderabad / Jaipur
Kolkata / Lucknow / Mumbai / New Delhi / Noida / Patna / Visakhapatnam

ISBN 978-93-86235-75-6

Typeset in Times LT Std 10 points *by*
MacroTex Solutions, Chennai 600 091, Tamil Nadu, India

Printed at
Glorious Printers, Delhi

Published by
Universities Press (India) Private Limited
3-6-747/1/A & 3-6-754/1, Himayatnagar, Hyderabad 500 029, Telangana, India

PREFACE

This book introduces readers to Theory of Computation, which is one of the fundamental pillars of Computer Science. Various standardised abstract computing models form the foundation of all programming languages. The family of programming languages has evolved in leaps and bounds, from the days of punch-cards to the modern-day special-purpose English-like languages, and this has been made possible because of the rapid evolution of various abstract models of computing.

The working of any computer program is nothing but a well-defined set of state transitions, which can be modelled mathematically. This makes it very important that students have a solid understanding of the theory of computation.

In this book, the authors have tried to present the relevant concepts in simple language, in an attempt to make the subject easy to understand. This book discusses in detail the mapping of formal grammars with the constructs of programming and natural languages to inculcate interest in students to study the subject. In fact, most books in the market have not included such discussions. The book can be used as a core textbook, or as additional reading by students pursuing undergraduate or postgraduate courses in computer science.

In addition, the authors have observed that the theory of computation forms an important part of the syllabus of various national and state-level competitive examinations. Being a high-scoring area, it is imperative that students obtain sufficient practice in solving the problems quickly. In light of this, several problems from previous competitive examinations have been included in this book, and shortcut tricks have been provided, wherever feasible.

The text is self-contained, as mathematical preliminaries are included in Chapter 1 along with several solved examples. The book then goes on to discuss the general theory of automata, properties of regular sets and regular expressions, and the basics of formal languages. Sufficient attention has also been devoted to topics such as pushdown automata and its relation to context-free languages, Turing machines and the concept of computability, including problem reducibility. The text also provides an in-depth analysis of context-free languages, covering normal forms, pumping lemma and decision algorithms.

The authors are open to receive feedback from students and the teaching community alike, to make this book even better in forthcoming editions.

Bikash Kanti Sarkar
(E-mail: bk_sarkarbit@hotmail.com)
Ambuj Kumar
(E-mail: ambuj_k@hotmail.com)

ACKNOWLEDGEMENTS

Dr Sarkar takes immense pleasure in thanking his beloved son, Sudipta, and wife, Smt Angsumita, for their inspiration and encouragement in writing the text.

Mr Ambuj Kumar would like to acknowledge his wife, Mrs Rashmi Verma, for her constant support and encouragement throughout the process of authoring and editing this book. He also hopes that his son, Adwit Kumar (who was born during the authoring of this book), develops a lifelong passion for learning new things.

The authors thank the reviewers of the text, who provided valuable comments and suggestions for improving the quality of the book, and hope that the book lives up to their high expectations. Finally, the authors would like to thank the editorial team of Universities Press for their cooperation.

Bikash Kanti Sarkar
Ambuj Kumar

To

The Almighty and My Beloved Parents

—Bikash Kanti Sarkar

To

My parents and Numerous people who came into my life and inspired me to adopt at least one good quality from them

—Ambuj Kumar

CONTENTS

1 Basic Mathematics for Theory of Computation

OBJECTIVES

At the end of this chapter, you will be able to

- Understand the mathematical tools required for studying automata theory
- Describe the mathematical concepts used to design formal versions of languages
- Explain complexity theory, which is used to deal with complexity measures
- Describe computability theory, which determines the solvability or unsolvability of problems using computational model

1.1 INTRODUCTION

In the discipline of Computer Science and Engineering, almost all the subjects included therein are directly or indirectly dependent on mathematics. **Theory of Computation** – an important subject in the field of Theoretical Computer Science[1] – deals primarily with **automata** (mathematical/abstract/conceptual models of computing devices, including real computers) and the languages recognised by automata (plural form of *automaton*). The term **computation** (execution of an algorithm) includes tasks such as text processing, language (including natural languages) compiling and performing numerical operations. Mathematical models are usually represented by **state transition graphs** (graphical representation), **state tables** (tabular representation) or **transition functions** (sets and their operations). The origin of the term *automata* and the idea of abstract model are explained later in this section.

Theory of computation discusses whether a problem (that is, a language) can be solved (recognised) by any one of the well-defined mathematical computing models, such as finite automata (FA), pushdown automata (PDA), linear bounded automata (LBA) or Turing machine (TM). In this context, we only discuss deterministic models, as all real machines are deterministic. In particular, if no deterministic mathematical model exists for solving a problem, then the problem is not solvable by the digital computer. It is important to note that FA (proposed during the 1940s–50s) is the simplest model, as less effort is required and minimal cost is incurred in solving the problem; whereas TM (studied by

[1] Theoretical Computer Science comprises topics that focus more on the abstract, logical and mathematical aspects of computing, such as the theory of computation, analysis of algorithms and semantics of programming languages. Some other fields in Computer Science are Computer Networks and Communication, Databases and Parallel Computing.

Alan Turing in the 1930s) has all the capabilities of today's computers. Certainly, our intention is to solve any problem by the simplest model possible; however, it is true that if the device is simple, its functionalities will also be simple – that is, it will be less capable of solving/recognising complex problems/languages.

We use language (known as natural language) to communicate. However, natural languages are usually not preferred for communication with real computers as such communication becomes complicated. Instead, we use programming languages (computer languages) to facilitate easy communication with digital computers. In particular, formal languages[2] are widely used to model programming languages. Formal languages are concise and simple when compared to natural languages. A **formal language** comprises symbols (letters), an alphabet (set of symbols) and a grammar (set of rules), where strings (words) are generated over the alphabet and grammar assists in generating sentences (collections of strings). So, grammar is the basis for generating language that can be accepted by the machine and for natural languages as well. Any language and its grammar can be expressed formally. It is interesting to note that *string* and *sentence* are different in natural as well as programming languages, but both are treated the same in the context of formal languages. They are known as **members** of the language.

Formally, a discussion on automata is part of automata theory, whereas language construction comes under formal language theory. However, from the above discussion, we may say that automata theory is closely related to formal language theory, where an automaton is the finite representation of a formal language, which may be an infinite set. This implies that an infinite language is represented using a finite number of states. Further, automata are often classified by the class of formal languages they can recognise. In the late 1950s, American linguist N Chomsky studied formal grammar and categorised it into four classes, known as the **Chomsky hierarchy** (discussed in Chapter 4), which are well-suited to the four types of automata, namely, FA, PDA, LBA and TM.

Thus, theory of computation comprises two major parts: (i) **automata theory**, which discusses abstract machines (the capabilities/limitations of machines) and (ii) **formal language theory**, which deals with formal grammars recognised by the machines. Apart from this, the subject also deals with the level of complexity (hard, easy, etc.) of solving problems and with how efficiently problems can be solved on models of computation, using algorithm. This is known as **complexity theory**.

Language vs Problem: In the context of formal languages and automata theory, *language* is usually treated as the *problem*. In fact, language and problem are the same when the statement of the problem relates to a **decision question** (is or is not the following true?). In other words, interpretation of strings of language by some specification/statement (for example, does a string belong to a language or not) is termed the problem. However, the treatment of the two terms is unsatisfactory when the statement of the problem is to compute or transform some input. For instance, the task of the parser in a C compiler can be thought of as a problem in the formal sense when it checks the validity of a C program. More specifically, the compiler usually has two distinct components: lexical analysis (fulfilled by FA) and syntax analysis/parsing (performed by PDA). However, the parser does more than just make a decision. It produces a parse tree, generates code and is used to solve the problem.

---

[2] Any language (natural or artificial) can be described using some formal notations (some precise definitions) which are mathematical representations (for example, first order logic, set and Kleene star). More explicitly, the strings of the language and the set of rules or grammar use formal notations. Any language with such representation is called a formal language. In the context of automata theory, a formal language deals with the form or syntax of the sentences, but not with the semantics or meaning. Examples of artificial languages are C, C++ and Java.

KEY POINTS

- Theory of Computation is not a high-level technical subject (like operating systems, database management systems or parallel computing) that is used to solve real-world problems. Instead, it provides a mathematical way to solve problems. Technical subjects, on the other hand, are devoted to practical study.
- To model the hardware of computing devices (especially real computers), the concept of automata is introduced, and the subject of automata theory discusses the mathematical modelling of computing devices (for example, digital circuits including real computers).
- To model a language or problem, a formal grammar is necessary.
- An algorithm is designed to solve a problem.
- If any mathematical computation model may be designed for solving a problem, then a standard algorithm exists to solve it. If the problem is solvable, a decision must be made about the minimum effort (or cost) required to solve it; that is, the mathematical model (automaton) suitable for solving it.
- FA and PDA are useful models for many important kinds of hardware and software. For example, the lexical analysis part of a compiler (managed by regular expressions) is designed to communicate with FA and the circuit design of computing devices is modelled based on FA, whereas the parser part of the compiler (managed mainly by context-free language) is constructed for PDA. On the other hand, TM is used for language acceptability and computation purposes. More explicitly, this device is used as the recogniser for complicated languages such as $a^nb^nc^n$, a^{2^n}, etc. Also, TM is used for computation purposes. Note that for recognising languages as well as for solving computational problems, an algorithm must exist; we usually use the algorithm and not the grammar for solving computational problems. Grammars are used primarily for generating languages to communicate with machines. For example, the C language and its compiler are used to communicate with real computers. Further, any algorithm can be implemented using the C language and communicated to the machine through the C compiler. TM is the formal version of an algorithm.

Automata – Origin: *Automata* is a Greek word that means *self-acting*. Automata or computing devices are modelled mathematically. The model hides most of the implementation details (both hardware and software) and is called an **abstract machine**. It is, in fact, a *hypothetical* model of a real computer. A typical abstract machine consists of *input, output* and the set of permitted *operations* (called transitions) which guide the machine to move from its current position to the next. It is thus a self-operating machine designed to automatically follow a predefined sequence of operations or respond to predefined instructions. Here, the input is read, symbol by symbol, until it is completely consumed. Once the input is depleted, the machine is said to have stopped.

Note: The terms *automata* and *machine* are used interchangeably in this subject. Also, designing a machine (M) for a problem (P) implies designing a procedure to solve the problem based on the capability of M.

Need for Studying Theory of Computation: The first and foremost requirement is the modelling of many important kinds of hardware and software. Hardware ranges from digital circuits to the design of devices including real computers, while software includes pattern checkers of operating systems, design of computer languages, natural language processing, formal verification, and so on (discussed in detail in subsequent chapters). Further, theory of computation discusses automata and their capabilities (that is, what type of problems can be solved and what cannot) and formal languages and their grammar to gain insight into the design of computer languages and their syntax, while also providing a brief discussion on different types of problems, such as P, NP, and so on.

Need for a Mathematical Model

- We may opt for a particular type of real machine based on the solvability of a problem by a mathematical model.
- It is significantly easier to manipulate abstract models of a computer than an actual computer.
- For the purpose of mathematical modelling of automata as well as languages, we use mathematical tools such as set, logic, graph, relation and function. In addition, the concept of complexity analysis is often used in this study.

With these factors in mind, the important mathematical concepts are discussed in this chapter.

1.2 MATHEMATICAL TOOLS FOR MODELLING AUTOMATA

In this section, the basic concepts relating to important mathematical tools such as set, logic, graph, relation and function, which are pre-requisites for studying formal languages and automata theory, are discussed.

1.2.1 Set

From the perspective of formal languages, set is used mainly to specify the strings belonging to a language; set operations are used for language operations. Further, language may be finite or infinite – this falls under the size of the set. Thus, several terminologies and operations of sets are accepted in formal languages for different purposes, as discussed below.

A **set** is a well-defined collection of distinct objects. For example, the numbers 1, 2 and 3 are distinct objects when considered separately. However, when considered collectively, they form a single set of size 3, written as {1, 2, 3}. Sets are usually denoted by capital letters, such as $A, B, C,..., N,..., S,..., Z$.

Developed at the end of the 19th century, Set Theory is one of the fundamental concepts in mathematics. It has wide-ranging applications and is used as a foundation for almost all branches of mathematics.

Describing a Set: A set can be described, or its members specified, in two ways:

- **Intentional strategy:** This strategy uses rule or semantic description. For example, A is a set whose members are the first four positive integers. This type of notation is also known as **Set Builder form**, as it uses the 'properties' of the elements of the set to represent the set. This strategy is usually used to describe sets with an infinite number of elements.
- **Extensional strategy:** Here, each member of the set is listed explicitly. An extensional definition is denoted by enclosing the list of members in curly brackets: $C = \{4, 2, 1, 3\}$, $D = \{\text{blue, white, red}\}$. However, for sets with many elements, enumeration of the members can be abbreviated. For instance, a set of the first 1000 positive integers may be specified extensionally as: {1, 2, 3,..., 1000}, where the ellipsis (...) indicates that the list continues in the obvious way. Ellipsis may also be used when sets have an infinite number of members. Thus, a set of positive even numbers can be written as {2, 4, 6, 8,...}. This type of notation is also known as **Roster form**, as it lists out explicitly all the elements of a set.

Interestingly, the notation with braces may also be used in an intentional specification of a set. For instance, the set F of the 19 smallest positive integers that are exactly four less than the nearest perfect square can be denoted as: $F = \{n^2 - 4 : n \text{ is an integer; and } 3 \leq n \leq 21\}$.

In this notation, the colon (:) means 'such that', and the description can be interpreted as '*F* is the set of all numbers of the form $n^2 - 4$, such that *n* is a whole number in the range 3 to 21 inclusive'. Sometimes, the vertical bar (|) is used instead of the colon.

Types of Sets

Countable set: A countable set is one with the same **cardinality** (number of elements) as some subset of the set of natural numbers. A countable set is either a finite set or a countably infinite set. Equivalently, a set *A* is countable if it can be enumerated in a sequence, that is, if all of its elements can be listed as a sequence: $a_1, a_2, \ldots$ Rational numbers are countable.

Uncountable set: A set is said to be uncountable if it is infinite and not countable. Real numbers are uncountable.

Countably infinite set: A set is countably infinite if its elements can be placed in a one-to-one correspondence with the set of natural numbers. In other words, we can count all the elements in the set in such a way that, even though the counting will take forever, we can get to a particular element in a finite amount of time.

Subset and superset: In set theory, a set *A* is a subset of a set *B*, or equivalently *B* is a superset of *A*, if *A* is 'contained' inside *B*; that is, all the elements of *A* are also elements of *B*. *A* and *B* may coincide. If *A* and *B* are sets and every element of *A* is also an element of *B*, then:

- *A* is a subset of (or is included in) *B*, denoted by $A \subseteq B$, or equivalently
- *B* is a superset of (or includes) *A*, denoted by $B \supseteq A$.

It is also possible that $A = B$. If *A* is a subset of *B*, but *A* is not equal to *B* (that is, there exists at least one element of *B* which is not an element of *A*), then *A* is also a **proper** (or **strict**) **subset** of *B*, written as $A \subset B$.

Null or empty (Φ) set: It is a set with no element. It is a finite set with cardinality 0.

Power set: The power set of any set *S* is the set of all subsets of *S*, including the empty set and *S* itself. It is denoted as *P*(*S*). If *S* = {x, y, z}, the subsets of *S* are: {} empty set or null set), {x}, {y}, {z}, {x, y}, {x, z}, {y, z}, {x, y, z}, and hence, the power set of *S* is {{}, {x}, {y}, {z}, {x, y}, {x, z}, {y, z}, {x, y, z}}.

Universal set (U): It is the set of all sets with respect to a set of elements. The concept of universal set comes from the term *universe*. Literally, it means that it contains all sets; that is, it is the superset of all sets. However, it is not very useful, as the objects of the sets may vary from set to set. That is why we should consider the objects/elements of the sets (for which the universal set is taken into account) as the same.

EXAMPLE 1.1

Let $A = \{0, 1\}$. Find the power set of *A* (denoted as *P*(*A*)) and the power set of *P*(*A*).

Solution Given that $A = \{0, 1\}$. Now, $P =$ Power set $P(A) = 2^A = 2^{\{0, 1\}} = \{\phi, \{0\}, \{1\}, \{0,1\}\}$. Thus, $2^P = 2^{\{\phi, \{0\}, \{1\}, \{0, 1\}\}} = \{\phi, \{\phi\}, \{0\}, \{1\}, \{0, 1\}, \{\phi, \{0\}\}, \{\phi, \{1\}\}, \{\phi, \{0, 1\}\}, \{\{0\}, \{1\}\}, \{\{1\}, \{0, 1\}\}, \{\{0\}, \{0, 1\}\}, \{\phi, \{0\}, \{1\}\}, \{\phi, \{1\}, \{0, 1\}\}, \{\phi, \{0\}, \{0, 1\}\}, \{\{0\}, \{1\}, \{0, 1\}\}, \{\phi, \{0\}, \{1\}, \{0, 1\}\}\}$ has $2^4 = 16$ elements.

Ordered pair: An ordered pair is a pair of objects such that (a, b) ≠ (b, a), where 'a' belongs to a particular set and 'b' belongs to another set. However, {a, b} is an un-ordered pair and {a, b} = {b, a}.

Cartesian product: If A and B are any two non-empty sets (may be finite/infinite), the set of all ordered pairs: (a, b) such that a $\in A$ and b $\in B$ is known as the Cartesian product of set A with set B. It is denoted by $A \times B$, where $A \times B = \{(a, b): a \in A \text{ and } b \in B\}$.

It should be noted that $A \times B \neq B \times A$. To understand this, consider an example.

Let $A = \{0, 1, 2\}$ and $B = \{3, 4\}$.

$$A \times B = \{(0, 3), (0, 4), (1, 3), (1, 4), (2, 3), (2, 4)\}$$
$$B \times A = \{(3, 0), (3, 1), (3, 2), (4, 1), (4, 1), (4, 2)\}$$

$A \times B = B \times A$ if and only if $A = B$ and it may be written as A^2. Likewise, $A \times A \times \ldots n = A^n$. The Cartesian product of a set A with itself is represented as $A \times A = \{(a, b): a \in A \text{ and } b \in A\}$.

$$\text{If } |A| = m \text{ and } |B| = n, \text{ then } |A \times B| = mn.$$

Also, $A \times \Phi = \Phi \times A = \Phi$, where A is any set and Φ is the null set. This implies that the Cartesian product of a null set with any set is a null set.

The definition of Cartesian product as specified above is also known as the **second order Cartesian product** or **Cartesian product of order 2**. Higher order Cartesian products can be defined as follows:

Cartesian product of order 3 is defined for three sets A, B and C as
$A \times B \times C = \{(a, b, c): a \in A, b \in B, c \in C\}$

Cartesian product of order 4 can be defined for four sets A, B, C and D as
$A \times B \times C \times D = \{(a, b, c, d): a \in A, b \in B, c \in C, d \in D\}$, and so on.

Some results on Cartesian product

1. $A \times (B \cup C) = (A \times B) \cup (A \times C)$
2. $A \times (B \cap C) = (A \times B) \cap (A \times C)$
3. $A \times (B - C) = (A \times B) - (A \times C)$
4. If $A \subseteq B$, then $A \times A \subseteq (A \times B) \cap (B \times A)$ and $A \times C \subseteq B \times C$ for any set C
5. If $A \subseteq B$ and $C \subseteq D$, then $A \times B \subseteq C \times D$
6. $(A \times B) \cap (C \times D) = (A \cap C) \times (B \cap D)$
7. $(A \times B) \cap (B \times A) = (A \cap B) \times (B \cap A)$
8. $A \times (B' \cup C')' = (A \times B) \cap (A \times C)$
9. $A \times (B' \cap C')' = (A \times B) \cup (A \times C)$
10. If A and B have n elements in common, then $A \times B$ and $B \times A$ have n^2 elements in common.

Applications

- To represent points in n-dimensional space; for example, $(x_1, y_1), (x_1, y_1, z_1), \ldots$
- To represent the edges of a graph; for example, $(v_1, v_2), \ldots$
- To represent relations
- To construct product automata (or machines)

Set Operations: Some basic set operations are explained below. For more details, please refer to any standard textbook on discrete mathematics.

- **Union**: The union of two sets (A and B) is a set containing all the elements that are in A or in B (or both). For example, $\{1, 2\} \cup \{2, 3\} = \{1, 2, 3\}$ $\{1, 2\} \cup \{2, 3\} = \{1, 2, 3\}$. Thus, we can write $x \in (A \cup B)$ if and only if $(x \in A)$ or $(x \in B)$. Note that $A \cup B = B \cup A$. We can define the union of three or more sets in a similar manner. If $A_1, A_2, A_3, \ldots, A_n$ are n sets, their union $A_1 \cup A_2 \cup A_3 \ldots \cup A_n$ is a set containing all the elements that are in at least one of the sets. We can write this union more compactly as $\cup_{(i = 1\ldots, n)} A_i$.

- **Intersection**: The intersection of two sets A and B, denoted by $A \cap B$, consists of all the elements that are in both A and B. For example, $\{1, 2\} \cap \{2, 3\} = \{2\}$. More generally, for sets $A_1, A_2, A_3,...,$ their intersection $\cap_i A_i$ is defined as the set consisting of the elements that are in all A_is.
- The **complement** of a set A, denoted by A^c or $\overline{A}$, is the set of all elements that are in the universal set S but not in A.
- The **difference (subtraction)** is defined as follows. The set $A - B$ consists of elements that are in A but not in B. For example, if $A = \{1, 2, 3\}$ and $B = \{3, 5\}$, then $A - B = \{1, 2\}$. Note that $A - B = A \cap B^c$.

1.2.2 Logic

Logic means mathematical reasoning (mathematical sense to understand). Here, we will focus on propositional (declarative) logic. It relates to the truth values of statements. In theory of computation, the notations and concepts of mathematical logic help to describe formal languages as well as automata in several ways. For example, connectives such as AND, OR and NOT (or their equivalent notations) are used to connect or generate new languages or automata. Likewise, the important laws of logic help fulfil the same jobs directly or indirectly. It deals with statements that are a very important part of understanding formal problems.

In this section, we briefly discuss some elementary notations and concepts of mathematical logic, which are important for formulating sensible proofs/contradictions of mathematical statements. A **statement** is an assertive sentence that can be either true or false, but not both. For example, '5 is a prime number' is a statement which is true, but '3 is a perfect square' is false.

An **open statement** is a sentence that contains variables. The true or false value depends on the values of the variables. For example, the open statement '$x^2 - 3x + 2 = 0$' is true for $x = \{1, 2\}$ and false for all other values of x. The set of all values of variable(s) for which an open statement becomes true is known as the **truth set** for that open statement. The truth or falsity of a statement is known as its **truth value.** A truth value of 'True' is usually denoted by the symbol T and the value 'False' by the symbol F. T and F are considered as opposites or complements of each other. A **simple statement** is one whose truth value is independent and does not depend on the truth value of any other statement. A **compound statement** is made up of one or more simple statements joined by connectives. For example, 'Roses are red AND the sky is blue' is a compound statement made up of two simple statements, 'Roses are red' and 'The sky is blue', connected by 'AND'. The commonly used connectives are listed in Table 1.1.

TABLE 1.1 Some common connectives and symbols used in logic

Connective	Symbol	Nature of compound statement formed
AND	$\wedge$	Conjunction
OR	$\vee$	Disjunction
NOT	$\neg$ or ~	Negation
If… then	$\Rightarrow$	Implication or Conditional
If and only if (iff)	$\Leftrightarrow$	Equivalence or Bi-conditional

The **truth table** is a table showing the relation between the truth values of a compound statement and the truth values of the simple sentences which form the compound statement. We present the truth tables (Tables 1.2 to 1.4) for all of the connectives defined above.

TABLE 1.2 Truth table for conjunction

x	y	$x \wedge y$
T	T	T
T	F	F
F	T	F
F	F	F

To summarise, a conjunction is true if and only if each of the simple statements constituting the statement is true.

TABLE 1.3 Truth table for disjunction

x	y	$x \vee y$
T	T	T
T	F	T
F	T	T
F	F	F

To summarise, a disjunction is true if at least one of the simple statements constituting it is true.

TABLE 1.4 Truth table for negation

x	$\neg x$
T	F
F	T

The truth value of a negation is the exact opposite of the truth value of the original statement. Two statements are said to be logically equivalent if they have identical truth tables. If P and Q are two logically equivalent statements, they are represented as $P \equiv Q$.

De Morgan's Laws

$\neg(x \wedge y) \equiv (\neg x) \vee (\neg y)$ *Negation of conjunction is disjunction of negation.*

The truth table is given in Table 1.5.

$\neg(x \vee y) \equiv (\neg x) \wedge (\neg y)$ *Negation of disjunction is conjunction of negation.*

The truth table is given in Table 1.6.

TABLE 1.5 Truth table for $\neg(x \wedge y)$ and $(\neg x) \vee (\neg y)$

x	y	$(\neg x)$	$(\neg y)$	$(x \wedge y)$	$\neg(x \wedge y)$	$(\neg x) \vee (\neg y)$
T	T	F	F	T	F	F
T	F	F	T	F	T	T
F	T	T	F	F	T	T
F	F	T	T	F	T	T

Observe that the truth value columns for $\neg(x \wedge y)$ and $(\neg x) \vee (\neg y)$ are identical for all possible combinations of truth values of x and y, and hence, $\neg(x \wedge y) \equiv (\neg x) \vee (\neg y)$.

TABLE 1.6 Truth table for $\neg(x \vee y)$ and $(\neg x) \wedge (\neg y)$

x	y	$(\neg x)$	$(\neg y)$	$(x \vee y)$	$\neg(x \vee y)$	$(\neg x) \wedge (\neg y)$
T	T	F	F	T	F	F
T	F	F	T	T	F	F
F	T	T	F	T	F	F
F	F	T	T	F	T	T

Observe that the truth value columns for $\neg(x \vee y)$ and $(\neg x) \wedge (\neg y)$ are identical for all possible combinations of truth values of x and y, and hence, $\neg(x \vee y) \equiv (\neg x) \wedge (\neg y)$.

The truth table for implication $x \Rightarrow y \equiv (\neg x) \vee y$ is shown in Table 1.7.

TABLE 1.7 Implication $x \Rightarrow y \equiv (\neg x) \vee y$

x	y	$(\neg x)$	$x \Rightarrow y \equiv (\neg x) \vee y$
T	T	F	T
T	F	F	F
F	T	T	T
F	F	T	T

To summarise, a true statement can only imply a true statement, whereas a false statement can imply either a true or a false statement.

Implications in proving: If we start our hypothesis with a true statement, we will always be able to prove its correctness. However, if our initial hypothesis itself is false, then the proof will be meaningless.

The truth table for equivalence $x \Leftrightarrow y \equiv (x \Rightarrow y) \wedge (y \Rightarrow x) \equiv \{(\neg x) \vee y\} \wedge \{(\neg y) \vee x\}$ is shown in Table 1.8.

TABLE 1.8 Equivalence $x \Leftrightarrow y \equiv (x \Rightarrow y) \wedge (y \Rightarrow x) \equiv \{(\neg x) \vee y\} \wedge \{(\neg y) \vee x\}$

x	y	$x \Rightarrow y$	$y \Rightarrow x$	$x \Leftrightarrow y \equiv (x \Rightarrow y) \wedge (y \Rightarrow x)$
T	T	T	T	T
T	F	F	T	F
F	T	T	F	F
F	F	T	T	T

By now, readers familiar with Boolean algebra must have realised that the concept of mathematical logic is very closely related to Boolean algebra. We list some important laws of logic below:

- **Idempotent law**: For any statement x, we have $x \wedge x \equiv x$ and $x \vee x \equiv x$.
- **Commutative laws**: For any two statements x and y, we have $x \wedge y \equiv y \wedge x$ and $x \vee y \equiv y \vee x$.
- **Associative laws**: For any three statements x, y and z, we have $(x \vee y) \vee z \equiv x \vee (y \vee z)$ and $(x \wedge y) \wedge z \equiv x \wedge (y \wedge z)$

- **Distributive laws**: For any three statements x, y and z, we have $x \vee (y \wedge z) \equiv (x \vee y) \wedge (x \vee z)$ and $x \wedge (y \vee z) \equiv (x \wedge y) \vee (x \wedge z)$
- **De Morgan's laws**: For any two statements x and y, we have $\neg(x \wedge y) \equiv (\neg x) \vee (\neg y)$ and $\neg(x \vee y) \equiv (\neg x) \wedge (\neg y)$
- **Identity laws**: For any statement x, we have $x \wedge T \equiv x$, $x \vee T \equiv T$, $x \wedge F \equiv F$ and $x \vee F \equiv x$
- **Complement laws**: For any statement x, we have $x \vee (\neg x) \equiv T$, $x \wedge (\neg x) \equiv F$, $(\neg T) \equiv F$ and $(\neg F) \equiv T$
- **Law of contrapositives**: For any two statements x and y, we have $x \Rightarrow y \equiv (\neg y) \Rightarrow (\neg x)$
- **Involution law**: For any statement x, we have $\neg(\neg x) = x$

Valid argument: An argument made up of hypotheses $H_1, H_2, H_3, \ldots, H_n$ and conclusion C is said to be a valid argument if and only if C is true whenever all of the hypotheses $H_1, H_2, H_3, \ldots, H_n$ are true. This is the basic concept of all valid proofs.

KEY POINTS

- Negation is applied on declarative statements but not on connectors such as AND, OR and NOT.
- Negation of each of the inputs with OR is NAND, that is, $(\neg x \text{ OR } \neg y) \equiv \neg(x \text{ AND } y)$.
- Negation of each of the inputs with AND is NOR, that is, $(\neg x \text{ AND } \neg y) \equiv \neg(x \text{ OR } y) \equiv (x \downarrow y)$. It is read as *neither x nor y,* that is, both are not true.

1.2.3 Graph

A graph is a mathematical concept that can be used to model many concepts derived from the real world. In automata theory, the transitions/operations (working approach) of a machine (automaton) are shown using graphs. In particular, the concepts of directed edge, self-loop and finite graph are widely used. The concept of tree is used to derive strings. Here, some basic terminology relating to graphs is discussed.

A **graph** is an ordered pair $G = (V, E)$, where V is a set of vertices (also called nodes/points/junctions) and E is a set of edges (sometimes called arcs). The edge set of a graph may be empty. An acyclic graph is a tree.

A **node** is the fundamental unit of a graph. It usually contains some information or condition. The number of nodes in a graph is termed its **order**. On the other hand, an **edge** (also known as line/arc/branch) serves as the means to providing connectivity between a pair of vertices/nodes. An edge can be represented by a pair of nodes: u and v, indicating that there is a link between u and v. The **size** of a graph is measured by the number of its edges.

Edges may be of two types: **directed** and **undirected**. An undirected edge between two vertices/nodes: u and v is usually represented as (u, v), meaning that movement is possible from u to v and vice versa, whereas $<u, v>$ represents a directed edge from u to v but in the opposite direction. So, $(u, v) = (v, u)$, that is, it is mathematically symmetric but $<u, v> \neq <v, u>$. A graph with only directed edges is called a **directed graph**. Such graphs are mainly used to represent state transition diagrams in automata theory.

An edge that starts from a vertex and moves back to itself is called a **self-edge** or **self-loop**, that is, a loop is an edge (v_i, v_f), where $v_i = v_f$. The self-loop for an undirected graph is shown in Fig. 1.1.

In Fig. 1.1, (i) and (ii) are examples of undirected loops and are equivalent to each other. Clearly, the edge in (i) meets the same vertex at two different positions; however, the edge in (ii) meets at the same position of the vertex. In both cases, the edges meet at the same vertex, so (i) and (ii) are valid for representing a self-edge/loop and are treated as equivalent. We may represent the self-loop for a directed graph in a similar manner, as shown in Fig. 1.2.

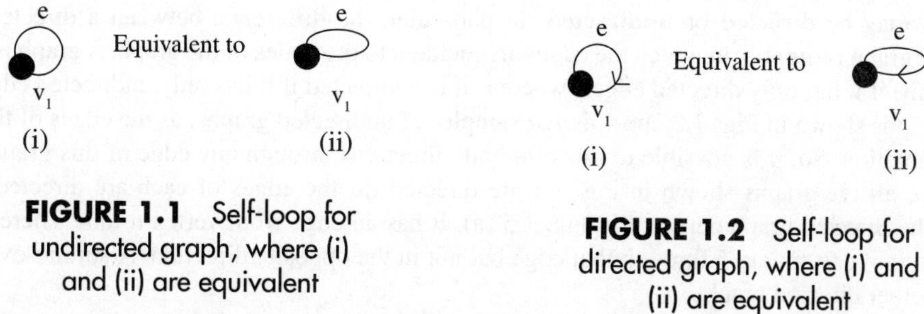

FIGURE 1.1 Self-loop for undirected graph, where (i) and (ii) are equivalent

FIGURE 1.2 Self-loop for directed graph, where (i) and (ii) are equivalent

Note: An edge is used to reach one node from another. In the case of a self-loop, the move is always to the starting node. Hence, in a self-loop, the edge may be opted for or not. If it is not opted for, the path move remains unchanged. In other words, it reports zero consideration of edge value. However, if the edge label is taken multiple times (through the self-loop), the path move is zero. This idea is accepted in designing automata.

Two vertices/nodes, x and y of graph G, are **connected** if there is an edge xy between them (edge joining); these vertices are called **adjacent** or **neighbour vertices/nodes**. The nodes x and y are termed the **end points** of the edge.

Note: Two vertices are said to be joined (with each other) if there is an edge joining them; that is, if they are adjacent vertices. Thus, two vertices that are adjacent are connected. However, two vertices that are connected need not be adjacent. For example, the vertices a and c in Fig. 1.3 are connected but not adjacent, whereas a and b are adjacent. In a graph G, a node that is not adjacent to any other node is called an **isolated node**. In Fig. 1.3, vertex d is an isolated vertex as it is not adjacent to any other vertex.

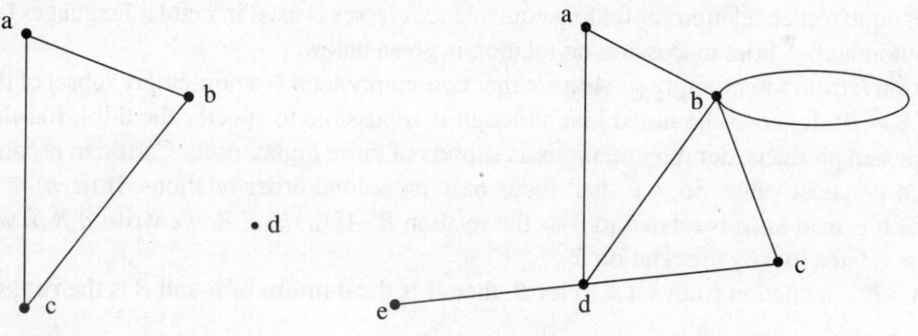

FIGURE 1.3 A graph with an isolated vertex

FIGURE 1.4 An undirected graph with no isolated vertex

A graph is **finite** if it has a finite number of vertices and a finite number of edges; otherwise it is **infinite**. If G is finite, $G(V)$ denotes the number of vertices in G and it is called the order of G. Similarly, $E(G)$ denotes the number of edges in G, and it is called the size of G.

The graph shown in Fig. 1.3 has four vertices: a, b, c and d. Here, (a, b) is a pair of vertices connecting a and b, and it represents an edge between them. Now, a and b are the **end points** of the edge (a, b). Clearly, the **neighbours** of vertex a in this graph are b and c, as there exist edges from a to b and a to c. Obviously, (a, c) and (b, c) are the other edges of this graph. Further, it is an example of finite graph and its order, $G(V)$ is 4 ($V = \{a, b, c, d\}$).

Graphs may be **directed** or **undirected.** In particular, the difference between a directed and an undirected graph is the way in which the edges are incident to the nodes of the graph. A graph is directed (or **digraph**) if it has only directed edges, whereas it is undirected if it has only undirected edges.

The graphs shown in Figs 1.3 and 1.4 are examples of undirected graphs, as the edges of the graphs have no direction. So, it is possible to move in both directions through any edge of this graph. On the other hand, all the graphs shown in Fig. 1.5 are directed (as the edges of each are directed). Let us consider the directed graph depicted in Fig. 1.5 (a). It has an edge from vertex a to b. Therefore, it is possible to move from a to b through that edge but not in the opposite direction. A similar explanation exists for other directed graphs.

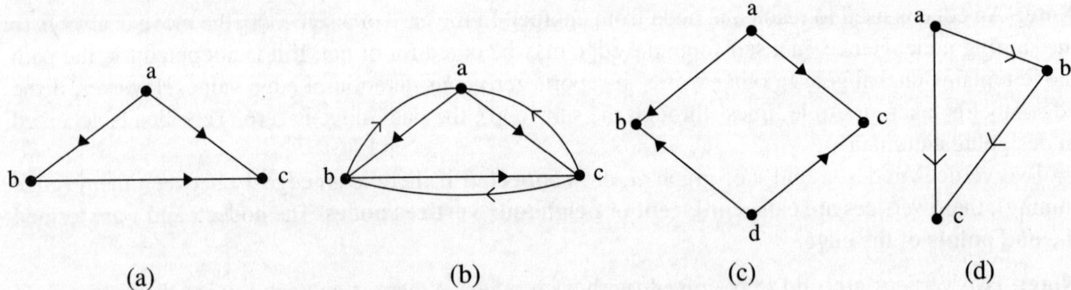

FIGURE 1.5 Some directed graphs

1.2.4 Relation

The idea of relations and their properties is used extensively in formal languages. For example, the concept of equivalence relation for finding equivalence classes is used in regular languages (equivalently in finite automata). A brief discussion on relation is given below.

A relation R from a non-empty set A to another non-empty set B is a non-empty subset of the Cartesian product $(A \times B)$. It should be noted that although it is possible to specify the third, fourth and higher order Cartesian products, defining relations as subsets of those higher order Cartesian products does not have much practical value. So, we shall focus only on second order relations. If $(a, b) \in R$, we write $a\ R\ b$, which is read as 'a is related to b by the relation R'. If $(a, b) \notin R$, we write $a \not R\ b$, which is read as 'a is not related to b by the relation R'.

If $R: A \rightarrow B$ is a relation from set A to set B, then A is the **domain** of R and B is the **range** of R.

- **Domain:** $(R: A \rightarrow B) = \{a : (a, b) \in R\}$
- **Range:** $(R: A \rightarrow B) = \{b : (a, b) \in R\}$

If $R: A \rightarrow B$ is a relation from set A to set B, its **inverse relation** R^{-1} is defined as

$$R^{-1} = \{(b, a): (a, b) \in R\}$$

A relation R from set A to itself is known as a relation on set A. A relation on set A is any subset of $A \times A$.

Types of Relations

- **Null/Void/Empty relation**: Let A be any non-empty set. Then $\phi \subseteq A \times A$, and hence, it is a relation on set A. This relation is the null or void or empty relation on set A.
- **Universal relation**: $A \times A$ is the universal relation on set A.

- **Identity relation**: $I_A = \{(a, a): a \in A \text{ for all } a \in A\}$ is the identity relation on set A. It should be noted that identity relations must contain a pairing of **every element of** A with itself.
- **Reflexive relation**: A relation R on set A is reflexive if every element of A relates to itself. R is reflexive $\Leftrightarrow (a, a) \in R$ for all $a \in A$.

  Key difference between identity and reflexive relations: I_A can contain only (a, a) pairs for all $a \in A$; however, a reflexive relation will contain all the elements of I_A and zero or more additional elements. For example, let $A = \{1, 2, 3\}$, $I_A = \{(1, 1), (2, 2), (2, 3)\}$. Let us consider a relation $R_1 = \{(1, 1), (1, 2), (2, 1), (2, 2), (3, 3)\}$. Clearly, $I_A \subseteq R_1$, and hence, R_1 is reflexive.
  Consider another relation $R_2 = \{(1, 1), (1, 2), (2, 1), (2, 2)\}$. Clearly, $I_A \not\subset R_2$, and hence, R_2 is not reflexive.

  - An identity relation is always reflexive, but every reflexive relation is not an identity relation.
  - A universal relation is always reflexive, but every reflexive relation is not a universal relation.
- **Irreflexive relation**: A relation R on a set A is irreflexive if and only if $(a, a) \notin R$ for all $a \in A$. In other words, no element is related to itself. For greater clarity, observe the following example carefully. Let $A = \{1, 2, 3\}$. Define the following relations on A: $R_1 = \{(1, 1), (2, 2), (1, 3), (3, 3)\}$, $R_2 = \{(1, 2), (2, 3)\}$, $R_3 = \{(1, 1), (2, 3)\}$. Clearly, R_1 is reflexive as every element of A is related to itself in R_1. R_2 is irreflexive because no element of A is related to itself in R_2. However, R_3 is neither reflexive nor irreflexive.
- **Symmetric relation**: A relation R on a set A is symmetric if and only if $(a, b) \in R \Rightarrow (b, a) \in R$ for all $a, b \in A$; that is, $aRb \Rightarrow bRa$ for all $a, b \in A$. In other words, R is symmetric if and only if $R = R - 1$. Identity and universal relations on any set are always symmetric.
- **Antisymmetric relation**: A relation R on a set A is antisymmetric if and only if $(a, b) \in R$ and $(b, a) \in R \Rightarrow a = b$ for all $a, b \in A$.
- **Transitive relation**: A relation R on a set A is transitive if and only if $(a, b) \in R$ and $(b, c) \in R \Rightarrow (a, c) \in R$ for all $a, b, c \in A$.
- **Equivalence relation**: A relation R on a set A is an equivalence relation if and only if it is reflexive, symmetric as well as transitive. Violation of any of these three properties will mean that the relation is not an equivalence relation.
- **Partial order relation**: A relation R on a set A is a partial order if and only if it is reflexive, antisymmetric and transitive. Violation of any of these three properties will mean that the relation is not a partial order relation.

Some examples are given below for greater clarity.

Consider the relation of **congruency** defined on a set of all possible triangles.

- Every triangle is congruent to itself, and hence, this relation is **reflexive**.
- If Δ_1 and Δ_2 are two triangles, then $\Delta_1 \cong \Delta_2 \Rightarrow \Delta_2 \cong \Delta_1$, and hence, this relation is **symmetric**.
- If Δ_1, Δ_2 and Δ_3 are three triangles, then $\Delta_1 \cong \Delta_2$ and $\Delta_2 \cong \Delta_3 \Rightarrow \Delta_1 \cong \Delta_3$, and hence, this relation is **transitive**.

Hence, the relation of congruency is an equivalence relation.

Consider the relation of greater than or equal to defined on a set of integers (Z).

- For all $x \in Z$, $x \geq x$. So, this relation is **reflexive**.
- For all $x, y \in Z$, $x \geq y$ and $y \geq x \Rightarrow x = y$. So, this relation is **antisymmetric**.
- For all $x, y, z \in Z$, $x \geq y$ and $y \geq z \Rightarrow x \geq z$. So, this relation is **transitive**.

Hence, the relation of greater than or equal to is a partial order relation.

Some interesting facts about relations are listed below:

- Universal relations are equivalence relations.
- The relation of equality is the only possible relation that is reflexive, symmetric as well as antisymmetric.
- The human relation *is brother of* is only transitive. It cannot be symmetric because if x is brother of y, then it is possible that y is sister of x.

1.2.5 Function

The concept of function is used in automata theory to describe state transitions. A concise discussion on functions is presented below to enable comprehension of the necessary concepts in the context of automata theory.

Let A and B be two non-empty sets. A relation f from A to B is said to be a **function** from A to B if and only if both of the following conditions are satisfied:

- For each $a \in A$, there exists $b \in B$ such that $(a, b) \in f$
- $(a, b) \in f$ and $(a, c) \in f \Rightarrow b = c$

In other words, a non-empty set of $(A \times B)$ is a function from A to B if each element of A appears in some ordered pair in f and no two ordered pairs in f have the same first element. A function from set A to set B is typically represented as $f: A \rightarrow B$.

A = **domain** of f and B = **range** of f.

If $f(x) = y$ for $x \in A$ and $y \in B$, then x is the **pre-image** of y, and y is the **image** of x.

Types of Functions

- **One–one function:** A function $f: A \rightarrow B$ is said to be a one–one function or an **injection** if different elements of A have different images in B. Thus, $f: A \rightarrow B$ is one–one.
  - $\Leftrightarrow a \neq b \Rightarrow f(a) \neq f(b)$ for all $a, b \in A$
  - $\Leftrightarrow f(a) = f(b) \Rightarrow a = b$ for all $a, b \in A$

  For example, the function $f(x) = (x + 2)$ defined on the set of real numbers R is one–one, because the value of $f(x)$ is unique for each x.

  Also, $f(a) = f(b) \Rightarrow (a + 2) = (b + 2) \Rightarrow a = b$, satisfying the above definition.
- **Many–one function:** A function $f: A \rightarrow B$ is said to be a many–one function, if there exist at least two elements $a, b \in A$, where $a \neq b$ but $f(a) = f(b)$. For example, the function $f(x) = x^2$ defined on the set of real numbers R is many–one because there exist multiple real numbers with the same square.

  $f(-a) = f(a) = a^2$ for all real numbers $a \in R$.

  A one–many function $f: A \rightarrow B$ *does not exist* because that will map an element $a \in A$ to at least two elements of B and will *contradict the definition of function.*
- **Onto function**: A function $f: A \rightarrow B$ is said to be an onto function or **surjection** if and only if each and every element of B has a pre-image in A. If $f: A \rightarrow B$ is an onto function, for each $b \in B$ there exists $a \in A$ such that $f(a) = b$.
- **Into function:** A function $f: A \rightarrow B$ is said to be an into function if there exists at least one element $b \in B$ for which there does not exist an element $a \in A$ such that $f(a) = b$.
- **Bijective function:** A function $f: A \rightarrow B$ is a bijective function if and only if it is one–one as well as onto. Observe that $|A| = |B|$ for a bijective function to exist from A to B.

- **Identity function:** A function $f: A \to A$ is called an identity function if and only if $f(a) = a$ for all elements $a \in A$. The identity function on a set A is usually represented as I_A.
- **Inverse of a function:** For a bijection $f: A \to B$, its inverse is defined as $f^{-1}: B \to A$, where $f(a) = b \Rightarrow f^{-1}(b) = a$, and a is called the inverse of element b. In fact, both a and b are the inverse of each other.

KEY POINTS

- A one–one function need not be onto.
- An onto function need not be one–one.
- An into function is never invertible.
- A one–one function may be into, and hence, not necessarily invertible.
- An onto function is not necessarily invertible.
- The identity function I_A is a bijection and $I_A^{-1} = I_A$.

EXAMPLE 1.2

The function $f(x) = \lfloor x \rfloor$ (floor function) defined on $R \to R$ is a classic example of a many–one function. It can also be referred to as an **infinite-to-one function**; for example, $f(4) = f(4.999) = 4$. Also note that there are infinite real numbers between 4 and 4.999. $f(x)$ is not an onto function because the RHS (right-hand side) of $f(x)$ can only have integer values, which are a subset of real numbers. So, several real numbers do not have a pre-image in R. Accordingly, f is neither invertible nor a bijection.

EXAMPLE 1.3

The function $g(x) = \frac{1}{x}$ defined on $R \to R$ is clearly one–one. However, note that there is no pre-image for zero in R because there is no real number whose reciprocal is zero. So, g is not onto and, obviously, not invertible. The interesting point to be noted is that it is because of only one element 0 that function g is not a bijection and not invertible. However, if we modify the function to be defined on $(R - \{0\}) \to (R - \{0\})$, then it is a bijection and invertible, and the inverse function will be defined as: $g^{-1}(y) = \frac{1}{y}$.

Composition of Functions: Let $f: A \to B$ and $g: C \to D$ be two functions such that $B \subseteq C$. Then, corresponding to each element $x \in A$, there exists a unique image $f(x) \in B$, whose image under g, that is, $g(f(x)) \in D$, is a unique element. This defines a function from A to D, which associates each $x \in A$ to $g(f(x)) \in D$. Thus, a formal definition can be given as follows:

If $f: A \to B$ and $g: C \to D$ are two functions such that range $(f) \subseteq$ domain (g), we can define their composition as $g \circ f: A \to D$ such that $(g \circ f)(x) = g(f(x))$ for all $x \in A$.

EXAMPLE 1.4

If $f: R \to R$ and $g: R \to R$ be two functions defined on a set of real numbers as $f(x) = (x + 1)$ and $g(x) = (5x + 1)$, then

$(g \circ f)(x) = g(f(x)) = g(x + 1) = 5(x + 1) + 1 = 5x + 6$

$(f \circ g)(x) = f(g(x)) = f(5x + 1) = 5x + 2$

Clearly, $(g \circ f) \neq (f \circ g)$. So, the composition of functions is not necessarily commutative.

We can also define $(f \circ f)$ if f is a function defined from any set A to itself.

Let us take another example, where $f: R \rightarrow R$ and $g: R \rightarrow R$ are real valued functions defined as $f(x) = x$ (identity function) and $g(x) = (x + 6)$.

Here, $(g \circ f)(x) = g(f(x)) = g(x) = x + 6$ and $(f \circ g)(x) = f(g(x)) = f(x + 6) = x + 6$.

The composition of an identity function with any function is the function itself.

Now, we list several other important results on functions:

- The composition of functions is associative, that is, $(f \circ g) \circ h = f \circ (g \circ h)$.
- The composition of two bijections is a bijection.
- If $f: A \rightarrow B$ and $g: B \rightarrow A$ are two functions such that $(g \circ f) = I_A$, then f is an injection and g is a surjection.
- If $f: A \rightarrow B$ and $g: B \rightarrow A$ are two functions such that $(f \circ g) = I_B$, then f is a surjection and g is an injection.
- Let $f: A \rightarrow B$ and $g: B \rightarrow C$ be two functions. Then
  - $(g \circ f) A \rightarrow C$ is onto $\Rightarrow g: B \rightarrow C$ is onto.
  - $(g \circ f) A \rightarrow C$ is one–one $\Rightarrow f: A \rightarrow B$ is one–one.
  - $(g \circ f) A \rightarrow C$ is onto and $g: B \rightarrow C$ is one–one $\Rightarrow f: A \rightarrow B$ is onto.
  - $(g \circ f) A \rightarrow C$ is one–one and $f: A \rightarrow B$ is onto $\Rightarrow g: B \rightarrow C$ is one–one
- The inverse of every bijection is a unique bijection.
- If $f: A \rightarrow B$ is a bijection and $g: B \rightarrow C$ is the inverse of f, then $(f \circ g) = I_B$ and $(g \circ f) = I_A$.
- If $f: A \rightarrow B$ and $g: B \rightarrow A$ are two functions such that $(f \circ g) = I_B$ and $(g \circ f) = I_A$, then f and g are bijections and $g = f^{-1}$.
- If $f: A \rightarrow B$ and $g: B \rightarrow C$ are two bijections, then $(g \circ f): A \rightarrow C$ is a bijection and $(g \circ f)^{-1} = f^{-1} \circ g^{-1}$.

1.2.6 Language

Language, as mentioned earlier, is a medium through which we communicate with each other. Human spoken language is called **natural language**. Examples include English, Hindi, Spanish, etc. However, in Computer Science, languages are used to solve problems. These are called **programming languages**. In fact, these are artificial languages as they are created to perform a task. Some examples of programming languages are Basic, Pascal, C, C++ and Java.

Whether natural or programming, any language must have a set of **symbols** and a set of **rules** to enable its manipulation. When these rules and symbols can be described mathematically, the language is termed a **formal language**. Simply speaking, the rules are more formal and precise and belong to some formal grammars (discussed later). In particular, the concept of formal language is widely employed to model languages used to communicate with the computer (computer languages). Interestingly, formal language theory primarily focuses on the syntactical aspects of languages.

Symbol: Symbols are individual objects/entities that cannot be defined; they are the atoms of the words of languages. For example, 'a', 'b', 'c',... 'z', '1', ..., '9', etc., are usually considered as symbols. These are the building blocks of formal languages and automata theory.

Alphabet: An alphabet is a finite non-empty set of symbols. Conventionally, we use notation Σ to represent the alphabet. For example, $\Sigma = \{0, 1, 2\}$. Here, Σ consists of the symbols 0, 1 and 2 only. An alphabet is thus a non-empty set of elements.

String: A language is a set of strings (words) over Σ. A string is a finite sequence of symbols. For example, $w = ab$ is a string derived from $\Sigma = \{a, b\}$. Strings are generally denoted by lowercase symbols: u, v, w, x, y, z. In practice, lowercase letters at the beginning of the English alphabet stand for symbols, whereas lowercase letters near the end of the alphabet are used for strings.

Length of a string: It is the number of positions (not the symbols) in the string and denoted by $| \, |$. If $w = ab$ is a string, $|w| = 2$. Further, $|xy| = |x|+|y|$, where x and y are strings.

Difference between number of positions and number of symbols: Let $w = ab$ and $w_1 = aa$ be two strings. Here, the number of symbols as well as the number of positions in w is 2; however, the number of symbols present in w_1 is 1 (the symbol is simply 'a'), but the total number of positions is 2; so the length of both the strings is 2 (if position is considered). On the other hand, the lengths are 2 and 1, respectively, if the number of symbols is considered to measure the length. However, as specified earlier, only the number of positions is considered.

Empty (null or zero) string: It is a string with no occurrences of symbols, and is usually denoted by the special symbol (letter) Λ or ε. Mathematically, the length of Λ ($|\Lambda|$) is 0. It is also treated as zero string, which implies zero occurrence of symbols from any alphabet. For example, a^0 is a null string as the number of occurrences of 'a' is zero. In practical use, "" is treated as the empty string in programming languages such as C, C++ and Java. However, it should not be equivalently treated as a null character, like '\0' used in C, C++ and Java. The null character is used as a string terminating character that is appended to any string (even an empty string) without adding to the length. The empty string ε has the same role as zero in natural numbers. It is used as the identity for basic operations (concatenation for strings, addition for naturals). It can be used as the terminator of tools such as regular expressions and grammar for constructing programming languages.

Substring: It is a continuous portion (p) of a string (w) such that $|p| \geq 0$. For example, let $w = abca$ be a string. Here, bc is a substring of $w = abca$, but ac is not its substring (a and c do not appear side by side in w). Likewise, abc, $abca$, ε, etc. are the substrings of w. Thus, it may be mathematically expressed as follows.

A string v is a substring of a string w if and only if there are strings x and y such that $w = xvy$. Truly, both x and y could be ε. Hence, every string is a substring of itself. Accordingly, ε is a substring of every string. Further, x and y in $w = xvy$ are, respectively, the **prefix** and **suffix** of v. In fact, the prefix of a substring (v) in a string (w) is the substring beginning from the start symbol of w up to the symbol appearing before the first symbol of v, whereas the suffix of v in w is the substring starting from the symbol appearing after the last symbol of v up to the last symbol of w.

KEY POINTS

- Every string is a substring of itself.
- Every string is the prefix as well as suffix of itself.
- A string of length n ($\neq 0$) has n possible prefixes and n possible suffixes.

Note: In this book, the symbols Λ and ε are used interchangeably to denote a null string.

Operations on strings

- **Reversal (transpose) of string**: If w represents a string, its reversal is denoted as w^R or w^T. For example, suppose $w = ab$ is a string on $\Sigma = \{a, b\}$, its reversal is ba.
- **Concatenation of strings**: If w_1 and w_2 are two strings over Σ, their concatenation is written as $w_1.w_2$ or w_1 **o** w_2, where the notations **o** and **.** stand for concatenation of strings. These are known as **juxtaposition operators**.

Suppose $w_1 = ab$ and $w_2 = aa$ are two strings over $\Sigma = \{a, b\}$. Now, one form of their concatenation is $w = w_1.w_2 = ab.aa = abaa$, whereas the other form is $w_2.w_1 = aaab$. Obviously, $w_1.w_2 \neq w_2.w_1$.

- **Concatenation of a string with an empty string**: The concatenation of any string (w) with an empty string results in w itself, that is, $\varepsilon.w = w.\varepsilon = w$. Thus, ε behaves as 1 (one) multiplication (identity) when concatenation is treated as a product operation of mathematics.
- **String intersection**: This may be thought of as a string comparison operation. Mathematically, two strings, say x and y, are equal if $x \cap y = x = y$, whereas they are unequal if $x \cap y = \Phi$. Thus, two strings are equal if they are entirely equal, otherwise they are unequal (even if some of their substrings match). For better understanding, please follow any standard string matching algorithm.

Power of alphabet: Let us take the expression 2^3. In mathematics, this gives 8, whereas it results in 222 in theory of computation. Likewise, $2^0 = 1$ in mathematics but ε in theory of computation. In the context of theory of computation, we primarily operate on symbols or strings or alphabets. Here, any symbol or string or alphabet raised to the power n means the concatenation of n-times with itself. In set theory, $\Sigma^n = \Sigma \times \Sigma \times \ldots.$ up to n times, where Σ denotes a set. Equivalently, $\Sigma^n = \Sigma\,\Sigma \ldots.$ up to n times with respect to the concatenation product[3].

Thus, if Σ is an alphabet, Σ^k ($k \geq 0$) represents a set of all possible strings of length k (known as the k-th power of the alphabet) generated over Σ. Certainly, $\Sigma^k = \Sigma\Sigma\ldots$ k-times. In other words, this expression performs concatenation (composition) of strings, one by one, k times – the length of each resulting string is k.

For better understanding, let $\Sigma = \{0, 1\}$. If $k = 0$, Σ^0 denotes a set of strings, each of length 0; that is, $\Sigma^0 = \{\varepsilon\}$ since the length of a null string is zero. If $k = 1$, $\Sigma^1 = \{0, 1\}$; that is, the resultant strings are 0 and 1, as the length of each string is 1. If $k = 2$, $\Sigma^2 = \Sigma.\Sigma = \{0, 1\}. \{0, 1\} = \{00, 01, 10, 11\}$. The resultant strings are 00, 01, 10 and 11. We may notice that each generated string is of length 2. Indeed, the results are achieved due to the concatenation product which is very similar to the Cartesian product. Likewise, we may compute Σ^3, Σ^4 and so on. Thus, if we take Σ^n over $\{0, 1\}$, the number of possible strings is 2^n (each of length n). More generally, it is $|\Sigma|^n$. Hence, if Σ has m symbols, the number of possible strings (each of length n) is m^n.

> *Difference between Σ and Σ^1*: There is a small difference between Σ and Σ^1. The former represents the alphabet ($\{0, 1\}$), whereas the latter represents a set of strings, each of length 1 (0 and 1).

KEY POINTS

- Symbols form the alphabet.
- An alphabet generates strings (words).
- A language is a collection of strings (words). Conceptually, it is a collection of sentences, each of which contains strings.
- String concatenation satisfies the associative property ($(ab)c = a(bc)$) but not the commutative property ($ab \neq ba$); where a, b and c represent strings.

[3] The concatenation product of the sets of strings A and B (written as $A.B$ or AB) is the set of strings that can be constructed by concatenating an element of A with an element of B (similar to Cartesian product, but a set of strings and not a set of pairs of elements). For example, $A = \{a, b\}$ and $B = \{c, d\}$; now, $AB = \{ac, ad, bc, bd\}$.

- For a symbol $A \in \Sigma$ and a natural number k, a^k represents the concatenation of k times of a. By convention (for symbols/strings), $a^0 = \varepsilon$ and $u^0 = \varepsilon$, where a is a symbol and u is a string.

Formal language: A language is thus a collection of sentences. A sentence, in turn, consists of words and it becomes meaningful when it follows a set of predefined rules. On the other hand, a word is formed from a selected set of symbols (called the alphabet), and follows some rule (regular expressions).

In the context of formal languages and automata theory, any language is a collection of strings, is expressed by formal grammar (set of rules expressed using mathematical representations), and is called a formal language. The current study focuses on formal languages and we will use the term language to denote formal language. Fundamentally, a language is a set (possibly infinite) of strings, all of which are generated over an alphabet Σ.

Formal languages are often called abstract languages because they do not provide detailed descriptions (including the semantic part) of the language. Examples of formal languages are given below:

- L = {w | w consists of equal number of *a*s and *b*s over {a, b}}
- L= {a^nb^n | $n \geq 0$ over {a, b}}
- L= {w | length of w is multiple of 2 over {0, 1}}
- L = {w | w is a syntactically correct C program}

Clearly, any language can be expressed using the Set Builder form as L = {w | w has property P}. It is read as '*the set of words w such that w has property P*'.

Let us consider a formal language L = {a^nb^n | $n \geq 0$ over {a, b}}. Here, ab, a^2b^2, a^3b^3, … a^kb^k are the members (strings) of L. Now, take the case of a formal version of the C programming language: L = {w | w is a syntactically correct C program}. The language has a finite alphabet, say Σ = {a–z, A–Z, 0–9, '*', '+',' ...}. Let us consider a simple C program:

void main( ) { int a, b; a=a+b; }.

In C, this is a program where each word is a string/token and *int a, b;* is a sentence. However, the program is nothing but a string in the context of formal languages. According to a formal version of C, L is a set of all valid programs (strings), L = {P_1, P_2, P_3, …}, where each P_i is a valid C program and L contains an infinite number of programs.

In C, if there is an error in any string (token) in a program, say P_i, then P_i (member of the language) is invalid. Tokens in P_i may be valid; however, one or more statements/sentences in P_i may not be correct. Then, P_i (the member) is also invalid; it does not belong to C.

In summary, a formal language consists of a set of formal rules or syntax by which the symbols can be combined into entities called sentences. Formal languages are widely used to model languages to communicate with the computer (computer languages). When a formal language is mapped to an actual computer language, each symbol of the terminal string in the formal language is treated as a token, while each string is treated as a sentence. This concept becomes clear when we study the construction of a compiler.

Closure and positive closure operations: We have seen that language is commonly denoted by the symbol L. Formally, it is a set of strings over an alphabet Σ; L is any subset of Σ^*, where Σ^* is the set of all possible strings over Σ. The superscript symbol '*' is called the **Kleene star** (or closure, free monoid or free generator); we can substitute any number (starting from 0) in place of '*'. This is a unary operator.

Thus, mathematically, $\Sigma^* = \Sigma^0 \cup \Sigma^1 \cup \Sigma^2 \cup .....,\infty$, where Σ^0 (the 0-th power of Σ) represents the selection of zero or null string (ε). In other words, * is the set of all strings of any length generated over Σ^*; it is the union of all the sets Σ^i ($i \geq 0$). So, Σ^* is the universal set (U) with respect to the alphabet Σ^*.

Now, $\Sigma\,\Sigma^* = \Sigma(\Sigma^0 \cup \Sigma^1 \cup \Sigma^2 \cup .....,\infty) = (\Sigma^1 \cup \Sigma^2 \cup \Sigma^3 \cup .....,\infty)$. It is denoted as Σ^+. Clearly, it does not include ε. Note that Σ^* denotes the closure of Σ and includes ε, whereas Σ^+ (termed positive closure) does not include ε. So, Σ^+ is identical to Σ^*, except ε. Again, $\Sigma^+ \cup \{\varepsilon\} = \Sigma^*$ and $\overline{\Sigma^*}$ (complement of Σ^*) $= \phi$. For example, let $\Sigma = \{0,1\}$. Therefore, $\Sigma^* = \{\varepsilon, 0, 1, 00, 01, 10, 11, 000, 001, 010, 011, 100, 101, 110, 111, \ldots\}$. Clearly, it is the set of strings formed by the union of all combinations of concatenation of the strings generated over $\Sigma = \{0, 1\}$.

Suppose we want to construct a language L from the set of strings (Σ^* as obtained over $\Sigma = \{0,1\}$), satisfying the property that no string contains two consecutive 0s. We would need to extract only those strings from Σ^* which consist of no two consecutive 0s (one subset from Σ^* is to be extracted to satisfy the desired property).

> **Note:** In this book, the operator: '*' is often used as the multiplication operator in algebraic expression, whereas this operator is used as closure in set/language operation (and in regular expression).

KEY POINT

- * and + are special unary operators called closure/Kleen star/star and positive closure, respectively. Both the operations are used to generate an infinite number of strings of different lengths.

Finite and infinite languages: It is important to note that we can specify some languages by listing their strings (elements) only. Such languages are called **finite languages.** Accordingly, the length of the strings is finite. However, most languages of interest are infinite, and listing all the strings is not possible; that is, an infinite number of strings of different lengths (assumed to be infinite as well) exist in the language. Naturally, we need to use the closure operation(s) for this purpose.

Operations on languages: Some basic but widely used operations on languages are discussed below:

- **Reversal of language**: If L is a language over Σ, then its reversal is $L^R = \{x^R \mid x \in L\}$. For example, let $L = \{0, 01, 001\}$. Then, $L^R = \{0, 10, 100\}$. It consists of all the strings formed by reversing the strings defined by L.
- **Complement of language**: If L is a language over Σ, then its complement is $L' = \Sigma^* - L$ over Σ. In other words, L' consists of all possible strings formed by the alphabet Σ, except the strings present in L. Let L = {strings of even number of 0s over $\Sigma = \{0, 1\}$}. This means that each valid string has an even number of 0s but it may contain an odd or even number of 1s. On the other hand, $L' = \Sigma^* - L$ gives strings which have an odd (not even) number of 0s but which may contain an odd or even number of 1s.
- **Union operation**: If L_1 and L_2 are two languages over Σ_1 and Σ_2, respectively, their union forms another new language: $L = L_1 \cup L_2$ over $\Sigma = \Sigma_1 \cup \Sigma_2$, where $\cup$ is the union operator. So, the new language consists of all the strings that belong to L_1 or L_2 or both. Thus, $L = L_1 \cup L_2$.
- **Intersection operation**: If L_1 and L_2 are two languages over Σ_1 and Σ_2, respectively, their intersection forms a new language: $L = L_1 \cap L_2$ over $\Sigma = \Sigma_1 \cup \Sigma_2$ (from the language construction point of view based on automata). It means that the new language consists of strings that satisfy the properties of both L_1 and L_2. From the set theoretic perspective, performing

union over two languages L_1 and L_2 over Σ_1 and Σ_2, we take $\Sigma = \Sigma_1 \cup \Sigma_2$ as the alphabet for the new language $L = L = L_1 \cup L_2$. On the other hand, $\Sigma = \Sigma_1 \cap \Sigma_2$ for $L = L_1 \cap L_2$; that is, the intersection of the languages gives only strings that can possibly be in the resulting set such that these are made up of characters in $\Sigma_1 \cap \Sigma_2$. However, from the FA design point of view, we use $\Sigma = \Sigma_1 \cup \Sigma_2$ for $L = L_1 \cap L_2$, since $\Sigma_1 \cap \Sigma_2 \subseteq \Sigma_1 \cup \Sigma_2$ and the intersecting method for finding FA must explicitly show the dead states that may be ignored while taking $\Sigma_1 \cap \Sigma_2$.

- **Language union and intersection:** To understand the concept of language union and intersection, let us take an example. Let L_1 be a language over $\Sigma = \{0, 1\}$ such that each string has an even number of 0s, and let L_2 be a language over $\Sigma = \{0, 1\}$ such that each string contains an even number of 1s. Now, Σ of both L_1 and L_2 is the same. So, $\Sigma = \{0, 1\}$ is also the Σ for $L = L_1 \cup L_2$. Certainly, the concept of set theory states that the new set L must contain all the strings of L_1 and L_2 and of both.
Now, consider $L = L_1 \cap L_2$. The language L_1 contains strings with an even number of 0s, but each of these strings may contain an odd number or an even number of 1s. Likewise, L_2 contains strings with an even number of 1s, but each such string may contain an odd number or an even number of 0s. However, $L = L_1 \cap L_2$ results in strings chosen from L_1 that satisfy the property for L_2, and vice versa. Thus, each string in $L = L_1 \cap L_2$ must satisfy the properties of both L_1 and L_2.

- **Concatenation**: Suppose L_1 and L_2 are the languages over Σ_1 and Σ_2, respectively, their concatenation will result in the derivation of another new language (say L). For example, $L = L_1 \text{ o } L_2$ over $\Sigma = \Sigma_1 \cup \Sigma_2$, where o is the concatenation operator. Each string in this new language consists of a string of L_1 followed by a string of L_2. It is the concatenation product of $L_1.L_2$ (very similar to the Cartesian product of L_1 and L_2). However, $L_1 \text{ o } L_2 \neq L_2 \text{ o } L_1$. In $L_1 \text{ o } L_2$, the string of L_1 appears first and then the string of L_2, whereas in $L_2 \text{ o } L_1$, the string of L_2 appears first and then the string of L_1. Suppose $L_1 = \{aa, b\}$ and $L_2 = \{c, d, e\}$, then L_1L_2 will consist of the strings $\{aac, aad, aae, bc, bd, be\}$.

Power of language: The power of a language L is defined as L^n $(n \geq 0)$, which means the n-th concatenation of L. Let us explain it mathematically. Suppose $L = \{aa, bb\}$.

- $L^0 = \{\varepsilon\}$, when $n = 0$; independent of the type of language L is.
- $L^1 = L$ (itself). In fact, the strings of L are also the strings of L^1. Here, L^1 does not imply that strings are of length 1 (like Σ^1, as discussed earlier), as the original language L may have strings of different lengths. Now, $LL = L^2 = \{aa, bb\}\{aa, bb\} = \{aaaa, aabb, bbaa, bbbb\}$. Likewise, we may compute L^3, L^4, etc.
- Let $L = \{a^nb^n \mid n \geq 0\}$. Then, L^2 (second power of L) $= L.L = \{a^ib^ia^jb^j \mid i, j \geq 0\}$. It may be noted here that the base language $L = \{a^nb^n \mid n \geq 0\}$ consists of the set of strings: $\{\varepsilon, ab, a^2b^2, a^3b^3, \ldots\}$, whereas the set of strings for $L^2 = \{a^ib^ia^jb^j \mid i,j \geq 0\}$ is $\{\varepsilon, ab, abab, a^2b^2, aba^2b^2, a^2b^2ab, a^2b^2a^2b^2, a^3b^3, \ldots\}$. Obviously, the languages (L and L^2) are different, as each satisfies a different distinct property. Each string in L satisfies the condition that an equal number of *a*s is followed by an equal number of *b*s, but no *a* appears after *b*. On the other hand, each string in L^2 consists of an equal number of *a*s followed by an equal number of *b*s, and *a* may appear after the first substring: a^ib^i but not after the second substring: a^jb^j.
Now, the closure of language L can be obtained as follows:
L^* (closure) $= L^0 \cup L^1 \cup L^2 \cup L^3 \cup \ldots\ldots\ldots\ldots\ldots\ldots = \cup_{k=0}^{\infty} L^k$

KEY POINTS

- The sets represented by the languages $L_1 = \{a^n b^n | n \geq 1\}$ and $L_2 = \{a^n b^n\}, n \geq 1$ are not equivalent. Based on the specification of the Set Builder form, $L_1 = \{a^n b^n | n \geq 1\}$ represents multiple strings of the form: $a^n b^n, n \geq 1$ in the set L_1; however, $L_2 = \{a^n b^n\}, n \geq 1$ represents only a single string of the form: $a^n b^n, n \geq 1$ in the set L_2. More explicitly, L_1 is an infinite set while L_2 is a finite set.
- String manipulation does not follow traditional mathematical operations with their usual meaning. For example, $(ab)^n \neq a^n b^n$. Here, $(ab)^n$ does not imply multiplication; rather it is concatenation – $(ab)(ab)(ab)$….. In this context, let $L_1 = \{(ab)^n | n \geq 0\}$ and $L_2 = \{a^n b^n | n \geq 0\}$ be two languages over {a, b}. Now, both the languages are not equal.

For example, for $L_1 = \{(ab)^n | n \geq 0\}$, from the structure of L_1, it is clear that each string of L_1 is a combination of (ab), whereas each string in L_2 is of the form: $a^n b^n$. For clarity about the strings of L_1, let $A = ab$, then $A^2 = AA = (ab)(ab)$. It is strictly $(ab)(ab)$ and not a^2b^2. In a similar manner, we may find A^3, A^4, and so on. On the other hand, let $w_2 \in L_2$ and $n = 2$. So, $w_2 = a^2b^2$. Here, $w_1 \neq w_2$. Likewise, $\{a^n b^n a^n b^n | n \geq 0\} \neq \{a^{2n} b^{2n} | n \geq 0\}$.

Thus, for a language L over Σ:

1. $L + L = L$
2. $L\varepsilon = \varepsilon L = L$ (ε behaves like one multiplication)
3. $L^* = L^+ \cup \{\varepsilon\}$
4. The 0-th power of any language results in a set consisting of null (ε) strings only, whereas the 0-th power of any symbol or string results in a null string only (not a set).

Empty language: An empty language (denoted by Φ) is one that does not contain any string (not even a null string). So, $\Phi = \{\ \}$ and $\Phi \neq \{\varepsilon\}$. Both are valid for any alphabet. An example of an empty language is given below.

E = {All valid English language words starting with '*a*'}
F = {All valid English language words starting with '*b*'}
Clearly, there is no valid English word that starts with '*a*' and '*b*'. Therefore, $L = E \cap F = \Phi$.

SOME BASIC OPERATIONS ON EMPTY LANGUAGES

- Concatenation of Φ and any expression (either left/right) results in Φ; that is, it behaves like zero multiplication. For example, $\Phi L = L\Phi = \Phi$. Hence, $\Phi^2 = \Phi.\Phi = \Phi$, $\Phi^3 = \Phi.\Phi.\ \Phi = \Phi$, and so on. Also, $(p, \Phi) = \Phi$ in the context of Cartesian product.
- $\Phi + L = L + \Phi = L$.
- $\Phi^* = \Phi^0 \cup \Phi^1 \cup \Phi^2 \cup \ldots\ldots = \{\varepsilon\} \cup \Phi \cup \Phi \cup \ldots = \{\Phi\}$. Here, Φ^0 represents the selection of the null string from Φ, and it is $\{\varepsilon\}$ since $L^0 = \{\varepsilon\}$ by convention. However, $\Phi^+ = \Phi^1 \cup \Phi^2 \cup \ldots = \Phi = \{\ \}$.

Note: Closure of any language L (except Φ) is infinite; however, closure of $\Phi = \{\Phi\}$ is finite. Also, closure of $L = \{\Phi\}$ is finite.

Empty string vs empty language: An empty string (ε) should not be confused with an empty language ($\varnothing$). $\varnothing$ is a formal language (a set of strings) that contains no strings (not even the empty string). More specifically, $\varnothing$, $\{\varepsilon\}$ and ε are three different elements, where

- $\varnothing$ is a (unique) set with no elements
- $\{\varepsilon\}$ is a set with one element (the null string)
- ε is a string of length 0.

EXAMPLE 1.5

Consider sets of binary strings $A = \{0, 00, 000\}$ and $B = \{11\}$. Find the language denoted by each of the following: (a) A^0 (b) A^1 (c) $A \cup A^2$ (d) AB^2 (e) $(AB)^2$ (f) A^*

Solution (a) $A^0 = \{\varepsilon\}$. (b) Given that $A = \{0, 00, 000\}$. Now, $A^1 = A$. So, $A^1 = \{0, 00, 000\}$. (c) Given that $A = \{0, 00, 000\}$. Now, $A^2 = AA = \{0, 00, 000\}. \{0, 00, 000\} = \{00, 000, 0000, 00000\}$. Therefore, $A \cup A^2 = \{0, 00, 000, 0000, 00000, 000000\}$. (d) $B = \{11\}$. Now, $B^2 = BB = \{11\}\{11\} = \{1111\}$. Thus, $AB^2 = \{0, 00, 000\}\{1111\} = \{01111, 001111, 0001111\}$, since $A = \{0, 00, 000\}$. (e) Using $A = \{0, 00, 000\}$ and $B = \{11\}$, we get $(AB)^2 = \{011011, 0110011, 01100011, 0011011, 00110011, 001100011, 00011011, 000110011, 0001100011\}$. (f) $A^* = A^0 \cup A^1 \cup A_2 \; ..... = \{\varepsilon, 0, 00, 000, 0000, 00000, 000000, \ldots\}$.

EXAMPLE 1.6

Let $L_1 = \{001, 10, 111\}$ and $L_2 = \{\varepsilon, 10, 110\}$ be two languages over $\{0, 1\}$. Find $L_1 \cup L_2$ and $L_1 \cap L_2$.

Solution For the given languages L_1 and L_2, we get $L_1 \cup L_2 = \{\varepsilon, 001, 10, 110, 111\}$ and $L_1 \cap L_2 = \{10\}$.

EXAMPLE 1.7

Let $L = \{0, 11\}$ over $\{0, 1)$. Find L^*.

Solution Given that $L = \{0, 11\}$. Thus, $L^0 = \{\varepsilon\}$, $L^1 = = L = \{0, 11\}$, $L^2 = LL = \{0, 11\} \{0, 11\} = \{00, 011, 110, 1111\}$, and so on. Hence, $L^* = L^0 \cup L^1 \cup L^2 \cup .....$

Note: The number of elements in L^i $(i \geq 0)$ is n^i, where n is the number of elements in L.

EXAMPLE 1.8

Give a language for which $L^n = L$ $(n \geq 0)$.

Solution The language is $L = \{\varepsilon\}$. Now, $L^2 = \{\varepsilon\} \{\varepsilon\} = \{\varepsilon\} = L$, $L^3 = \{\varepsilon\} \{\varepsilon\} \{\varepsilon\} = \{\varepsilon\} = L$, and so on. Thus $L^n = L$.

Note: The closure of any language, except empty ($\varnothing$) language and $L = \{\varepsilon\}$, is infinite.

EXAMPLE 1.9

Let Σ be a finite non-empty alphabet and 2^{Σ^*} be the power set of Σ^*. Then prove that 2^{Σ^*} is uncountable but Σ^* is countable.

Solution Here, 2^{Σ^*} is the power set of Σ^*. Now, Σ^* is a countably infinite set but the power set of a countably infinite set is uncountable. Hence, 2^{Σ^*} is uncountable but Σ^* is countable.

1.2.7 Pigeonhole Principle

Also known as the Dirichlet drawer principle or box principle, it states that if there are more pigeons than pigeonholes, then there must be at least one pigeonhole with at least two pigeons in it. In terms of set theory, this principle can be expressed as follows:

Let X and Y be any two finite sets such that $|X| < |Y|$. Then, a function $f: X \rightarrow Y$ cannot be one–one; that is, there must exist at least two elements, x_1 and $x_2 \in X$, such that $f(x_1) = f(x_2)$.

Generalised pigeonhole principle: This states that if n number of pigeonholes are occupied by $(kn + 1)$ or more pigeons, where k is a positive integer, then at least one pigeonhole is occupied by $(k + 1)$ or more pigeons.

Some examples are given below for clarity.

EXAMPLE 1.10

How many persons need to be present in a room to ensure that there are at least two persons born in the same month?

Solution Treat $n = 12$ possible calendar months as 12 pigeonholes. Insert $k = 1$ and $n = 12$ in the generalised pigeonhole principle. Clearly, we need $(kn + 1) = 13$ persons.

So, at least 13 (or more) persons need to be present in the room.

EXAMPLE 1.11

There are 50 baskets of apples. Each basket contains no more than 24 apples. Show that there are at least three baskets containing the same number of apples.

Solution Each basket can have either 1, 2, 3 or … 24 apples. So, we have $n = 24$ pigeonholes. 50 baskets are to be fitted in 24 pigeonholes. Apply the generalised pigeonhole principle: take $k = 2$; if we have $kn + 1 = 49$ baskets, then there are at least three baskets with the same number of apples. In this case, we have 50 (1 more than the required number of baskets), and hence, we will definitely have at least three baskets with an identical number of apples.

EXAMPLE 1.12

Show that among any four integers, one can find two numbers so that their difference is divisible by three.

Solution When an integer is divided by three, there are only three possible values of remainders: 0, 1 and 2. This maps to three pigeonholes, while a set of four integers maps to four pigeons. So, if we choose any four integers, at least two will always give the same remainder when divided by three. Let us represent those numbers as $(3m + r)$ and $(3k + r)$.

Difference $= (3m + r) - (3k + r) = 3(m - k) =$ multiple of 3. This proves the given statement.

EXAMPLE 1.13

Suppose every point in a real plane is coloured red or blue. Then, for any distance $d > 0$, there are two points exactly at distance d from one another that are of the same colour.

Solution Let us consider any equilateral triangle whose side length is d. Place this triangle anywhere in the plane (Fig. 1.6).

By the pigeonhole principle, two of the vertices must have the same colour, as there are three vertices. These vertices are at distance d from each other, as required.

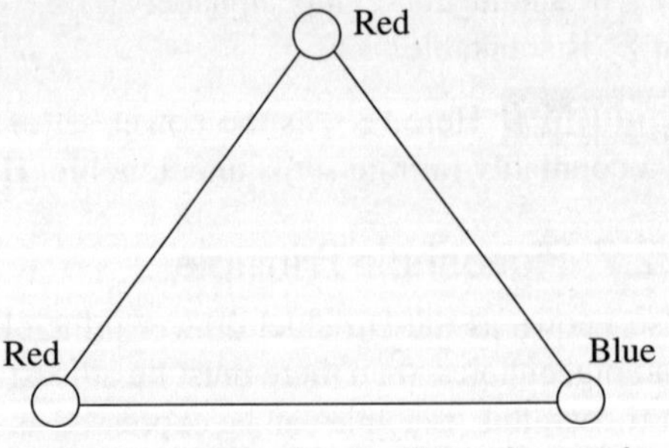

FIGURE 1.6 An equilateral triangle with three vertices

EXAMPLE 1.14

A row of houses is randomly assigned distinct numbers between 1 and 50 (inclusive). How many houses must be present to ensure that there are five houses numbered consecutively?

Solution Split the numbers into 10 pigeonholes: 1–5, 5–10, 11–15, 16–20...
So, there must be at least $10 \times 4 + 1 = 41$ 'pigeons' or 41 houses.

1.2.8 Mathematical Induction

This is an extremely powerful technique for proving the correctness of several mathematical statements. We first define some basic terms.

- **Statement:** Any sentence or description that can be judged to be true or false. Examples include (i) There are 7 days in a week. (ii) $(x + 1)$ is a factor of $(x^2 - 3x + 2)$. (iii) $A \cap B = B \cap A$. Clearly, (i) and (iii) are true but (ii) is false.
- **Mathematical statement:** A statement involving mathematical relations; for example, $P(n)$ is a mathematical statement; $n(n + 1)(n + 2)$ is divisible by 6. P(1) is true and so is P(2). In general, the above statement is true for all natural numbers n.
  - **The First Principle of Mathematical Induction**
    Let $P(n)$ be a statement involving the natural number n such that (i) P(1) is true; that is, $P(n)$ is true for $n = 1$ and (ii) $P(k + 1)$ is true whenever $P(k)$ is true; that is, $P(k)$ is true $\Rightarrow P(k + 1)$ is true for any $k \geq 1$.Then, $P(n)$ is true for all natural numbers n.
  - **The Second Principle of Mathematical Induction**
    This is a more generalised version of the first principle. Let $P(n)$ be a statement involving the natural number n such that (*i*) P(1) is true; that is, $P(n)$ is true for $n = 1$ and (ii) $P(m + 1)$ is true whenever $P(k)$ is true for all k, where $1 \leq k \leq m$. Then, $P(n)$ is true for all natural numbers n.

Some proofs using mathematical induction are given below.

EXAMPLE 1.15

$n(n + 1)(n + 2)$ is a multiple of 6 for all natural numbers n.

Proof Let $P(n)$ be the mathematical statement : $n(n + 1)(n + 2)$ is a multiple of 6.

Insert $n = 1$ in the statement to obtain $1 \times 2 \times 3 = 6$, which is clearly a multiple of 6. So, P(1) is true.

Now, let us assume that $P(k)$ is true; $k(k + 1)(k + 2)$ is a multiple of 6 for some natural number $k \geq 1$. Check the validity of $P(k + 1)$.

$(k + 1)(k + 2)(k + 3) = k(k + 1)(k + 2) + 3(k + 1)(k + 2) = (\text{multiple of } 6) + 3(k + 1)(k + 2)$ [by our initial assumption]

Now, observe the number $(k + 1)(k + 2)$:

If k is even, then $(k + 1)(k + 2) = (\text{even} + 1)(\text{even} + 2) = \text{odd} \times \text{even} = \text{even number}$
If k is odd, then $(k + 1)(k + 2) = (\text{odd} + 1)(\text{odd} + 2) = \text{even} \times \text{odd} = \text{even number}$
So, $(k + 1)(k + 2)$ is an even number, and hence, $3(k + 1)(k + 2)$ is a multiple of 6. Thus, $P(k)$ is true $\Rightarrow P(k + 1)$ is true.

Hence, by the **first** principle of mathematical induction, $P(n)$ is true for all natural numbers n, and this proves the given statement.

EXAMPLE 1.16

$10^{2n-1} + 1$ is divisible by 11 for all natural numbers n.

Proof Let P(n) be the mathematical statement : $10^{2n-1} + 1$ is divisible by 11.
First, insert $n = 1$ in the statement to obtain $10 + 1 = 11$, which is a multiple of 11. So, P(1) is true.

Now, let us assume that P(k) is true; $10^{2k-1} + 1$ is divisible by 11 for some natural number $k \geq 1$.
Check the validity of P($k + 1$).

$10^{2(k+1)-1} + 1 = (100 \times 10^{2k-1}) + 1 = 99 \times 10^{2k-1} + 10^{2k-1} + 1$

$= 99 \times 10^{2k-1}$ + (multiple of 99) = (multiple of 99)

Now, P(k) is true $\Rightarrow$ P($k + 1$) is true.

Hence, by the first principle of mathematical induction, P(n) is true for all natural numbers n, and this proves the given statement.

EXAMPLE 1.17

A string generating function F_n ($n \geq 1$) is defined as follows:

$F_1 = 0$; $F_2 = 1$; $F_n = F_{n-1}F_{n-2}$ for $n \geq 3$; that is, $F_3 = 10$; $F_4 = 101$ and so on.

Prove that no F_i contains either 00 or 111 as a substring. **[UGC NET 2010]**

Proof This problem has two parts, and we will use the first and second principles of mathematical induction to prove both these parts separately.

Part 1: We first prove that F_i does not contain 00 as a substring.

Define P(n) as the statement: F_n does not contain 00 as a substring.

Clearly, $F_1 = 0$; $F_2 = 1$; $F_3 = F_2F_1 = 10$; $F_4 = F_3F_2 = 101$. So, P(1), P(2), P(3), P(4) are true.

If we are able to prove that every F(i) starts with 1 for every $i \geq 2$, then we can easily apply the second principle of mathematical induction here.

Define another mathematical statement Q(n) : The string F(n) starts with 1 for all $n \geq 2$.
Clearly Q(2) is true and obviously Q(3) and Q(4) are also true.

Now, assume that Q(k) is true for some $k \geq 2$.

Consider the string F($k + 1$) = F(k)F(k – 1)

Since we assumed Q(k) to be true, F(k) starts with 1, implying that F(k + 1) also starts with 1. So, Q(k) is true $\Rightarrow$ Q(k + 1) is true.

Hence, by the first principle of mathematical induction, Q(n) is true for all $n \geq 2$.

Now, returning to the original problem, let us apply the next step of the second principle of mathematical induction. Assume that P(k) is true for all k, where $1 \leq k \leq m$, for some integer $m \geq 4$.

Consider the string F(m + 1) = F(m) F(m – 1). By our induction hypothesis, neither F(m) nor F(m – 1) contains 00 as a substring. One possibility remains: what if F(m) ends with 0 and F(m – 1) begins with 0?

However, F(m – 1) begins with 1 whenever $m \geq 3$, as already proved above. [For values of $k < 3$, we have already seen the correctness.]

Thus, all possibilities of F(m + 1) containing 00 are eliminated. So, the mathematical statement P(m + 1) is true whenever P(k) is true for $1 \leq k \leq$ m. Hence, by the second principle of mathematical induction, F(n) cannot contain 00 as a substring.

Part 2: We now attempt to prove that F_i does not contain 111 as a substring. Define P(n) as the statement: *F_n does not contain 111 as a substring.* Clearly, $F_1 = 0$; $F_2 = 1$; $F_3 = F_2F_1 = 10$;

$F_4 = F_3F_2 = 101$; $F_5 = F_4F_3 = 10110$.

Observe from Part 1 of this problem that every F_i starts with 1 for $i \geq 2$. If we can prove that no F_i ends with 11 for $i \geq 2$, then we have accomplished our goal. Observe that F_3 ends with 10, F_4 ends with 01 and F_5 ends with 10.

The next level: $F_6 = F_5F_4$, and hence, F_6 ends with 01. $F_7 = F_6F_5$, and hence, F_7 ends with 10. $F_8 = F_7F_6$, and hence, F_8 ends with 01.

Define Q(n) as the mathematical statement: *Q(n) ends with 10 when n is odd and Q(n) ends with 01 when n is even for $n \geq 3$.*

The base statements Q(3), Q(4), Q(5), Q(6), Q(7) and Q(8) are true, as discussed above.

Assume the statement Q(k) to be true for some integer value $k \geq 3$.

Consider the string F(k + 1) = F(k) F(k – 1). If k is even, then both (k + 1) and (k – 1) are odd. If k is odd, then both (k + 1) and (k – 1) are even. So, for even values of (k + 1), the string F(k + 1) ends with the same set of digits as F(k – 1); that is; it ends with 01.

For odd values of (k + 1), the string F(k + 1) ends with the same set of digits as F(k – 1), that is, it ends with 10. So, Q(k) is true $\Rightarrow$ Q(k + 1) is true.

Hence, by the first principle of mathematical induction, Q(n) is true for all $n \geq 3$.

So, every F_i ends with 01 or 10. Now, returning to the original problem, let us set up the induction hypothesis : P(k) is true for all values of $1 \leq k \leq m$ for some integer $m \geq 4$.

Consider the string F(m + 1) = F(m) F(m – 1). By our induction hypothesis, neither F(m) nor F(m – 1) contains 111 as a substring. So, the only possibility for F(m +1) to contain 111 as a substring is by combining the last digit(s) of F(m) and the first digit(s) of F(m – 1).

However, we have proved that F(m – 1) will always start with 1 (when $m \geq 3$) and F(m) will always end with 10 or 01 (when $m \geq 3$). So, we cannot have a substring of 111 by combining the last digit(s) of F(m) and the first digit(s) of F(m – 1). Thus, the mathematical statement P(m + 1) is true whenever P(k) is true for $1 \leq k \leq m$.

Hence, by the second principle of mathematical induction, F(n) cannot contain 111 as a substring.

Counterexamples: The concept of counterexamples is an extremely powerful technique for disproving an invalid statement. Quite often, readers will encounter problems of the type "prove or disprove" while studying any field of mathematics and computing.

Often, it is much easier to find a counterexample rather to find a technically correct proof for a statement. There is no general/universal way to learn this technique. It can be mastered only by practice and sharp observation. We present a few solved problems below to illustrate the technique of using counterexamples.

EXAMPLE 1.18

If p is a prime number, then (p + 7) is also prime.

Solution This can be disproved using several counterexamples. Take $p = 2$ (the smallest prime number). However, (p + 7) = 9, which is not prime. Here, we decided to look at smallest possible value of p.

Another obvious counterexample is choosing $p = 7$. Clearly, 7 is a prime number, but (7 + 7) = 7 × 2, which clearly contradicts the definition of prime numbers.

Readers are encouraged to find as many counterexamples as they wish. In fact, there are an infinite number of counterexamples to disprove the given statement.

EXAMPLE 1.19

We wish to construct a valid triangle whose sides have length = a positive integer. Such a triangle can have any even positive integer value as its perimeter.

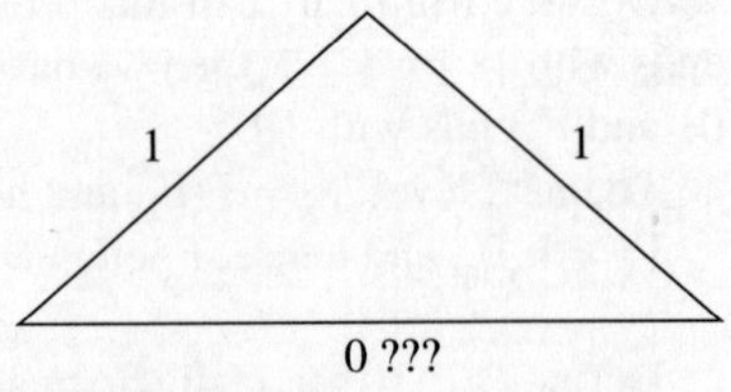

FIGURE 1.7 A triangle with three vertices

Solution Let the lengths of the three sides of the triangle be x, y, z (Fig. 1.7). Consider drawing a triangle with perimeter = 2 (an even positive integer). We need to find three positive integers x, y and z that satisfy the condition $x + y + z = 2$.

This is impossible because the smallest positive integer value is 1, and hence, at least one of x, y, z must be zero or negative, which contradicts the given condition. So, we have found a counterexample to disprove the given statement.

EXAMPLE 1.20

If A, B and C are sets satisfying the condition $A \cap B = B \cap C = C \cap A = \phi$, then $A \cap B \cap C \neq \phi$.

Solution It is very easy to find counterexamples to disprove this statement. Let $A = \{1, 2\}$, $B = \{3, 4\}$ and $C = \{5, 6\}$. Clearly, $A \cap B = B \cap C = C \cap A = \phi$, but $A \cap B \cap C = \phi$, which contradicts the given statement.

EXAMPLE 1.21

If $x + y > z$, then $x^2 + y^2 > z^2$ for any non-zero real numbers x, y, z.

Solution If we think of Pythagorean triplets, we can easily disprove this statement. We know that (a, b, c) is a Pythagorean triplet if it satisfies the condition $a^2 + b^2 = c^2$, that is, if a, b and c are lengths of the sides of a right-angled triangle with c as the longest side. Every Pythagorean triplet is a counterexample to the given statement. Consider $x = 3$, $y = 4$ and $z = 5$. Clearly, $3 + 4 > 5$, but $3^2 + 4^2 = 5^2$, and hence, the given statement is disproved.

EXAMPLE 1.22

The function $f(n) = (n^2 - n + 41)$ always generates a prime integer value for all integer values of n.

Solution Disproving this statement requires *sharp observation*. The beauty of this function is that it generates prime numbers for all integers from $n = -39$ to $n = 40$, and hence, may lead to confusion that it is a valid formula. However, insert $n = 41$ and $f(41) = 41^2 = 1681$, which is not prime.

Alternatively, try any number of the form $41k$, where k is an integer. $f(41k) = (41k)^2 - 41k + 41 = 41 \times (41k^2 - k + 1)$, which is not a prime integer. So, it is just a matter of observing closely to find a counterexample for this statement.

EXAMPLE 1.23

Given two matrices that satisfy the property $AB = BA$ and $A^2 + B^2 = I$. Then, both A and B are diagonal matrices.

Solution We can think of various combinations of square matrices of size 2×2, and try out matrices made up of 0s and 1s only.

$$\text{Let } A = I_2 = \begin{bmatrix} 1 & 0 \\ 0 & 1 \end{bmatrix} \text{ and } B = \begin{bmatrix} 0 & 1 \\ 0 & 0 \end{bmatrix}$$

$$AB = \begin{bmatrix} 1 & 0 \\ 0 & 1 \end{bmatrix} * \begin{bmatrix} 0 & 1 \\ 0 & 0 \end{bmatrix} = \begin{bmatrix} 0 & 1 \\ 0 & 0 \end{bmatrix} \text{ and } BA = \begin{bmatrix} 0 & 1 \\ 0 & 0 \end{bmatrix} * \begin{bmatrix} 1 & 0 \\ 0 & 1 \end{bmatrix} = \begin{bmatrix} 0 & 1 \\ 0 & 0 \end{bmatrix}$$

Also, observe that $A^2 = I_2$ and $B^2 = \begin{bmatrix} 0 & 1 \\ 0 & 0 \end{bmatrix} * \begin{bmatrix} 0 & 1 \\ 0 & 0 \end{bmatrix} = \begin{bmatrix} 0 & 0 \\ 0 & 0 \end{bmatrix}$

Clearly, $A^2 + B^2 = I$. However, observe that B is not a diagonal matrix, and hence, we have found a counterexample.

EXAMPLE 1.24

Given a string S, a prefix P (of S), and a suffix X (of S). If P and X are palindromes and |P| + |X| = |S|, then S is a palindrome string.

Solution An easy trick is to construct a prefix and suffix of unequal lengths.
Consider P = *abcba,* X = *ded* and S = *abcbaded.* Both P and X are palindromes.
|P| = 5, |X| = 3, |S| = 8. Clearly, |P| + |X| = |S|, but S is not a palindrome string.

EXAMPLE 1.25

For an integer n, the numbers $(n^{17} + 9)$ and $(n + 1)^{17} + 9$ are relatively prime to each other.

Solution This is a problem of historical interest, commonly known as the **Seventeen conjecture.** This statement has a counterexample for the gigantically large value of n.

$n = 8424432925592889329288197322308900672459420460792433$ is the first counterexample for this statement. This is an instance of a very hard-to-find counterexample.

> **Tip:** Several hard-to-prove/disprove statements involving prime numbers form a huge set of open/unsolved problems.

1.3 MATHEMATICAL CONCEPTS FOR COMPLEXITY ANALYSIS

From the perspective of language theory, complexity analysis requires us to deal with classes of problems (formal languages) such as P, NP, NP-C and NP-Hard (explained in Chapter 7). For this purpose, the essential points of complexity analysis are discussed below.

1.3.1 Function

This concept was explained earlier in the chapter. If x and y are two variables such that for some value of $x(x \in A)$ there exists a definite, finite, real value of $y(y \in B)$, then it is said to be a real, single-valued function of x, where A and B are two sets of real values. The set A is called the domain and the set B is called the range (co-domain) of the function. It is mathematically denoted as:

$$y = f(x), x(x \in A), y(y \in B)$$

Functions involve constants and variables that may be connected by algebraic operators, such as +, −, × and ÷, and/or non-algebraic operators, such as trigonometric, logarithm, inverse, exponential, and so on. A function may be polynomial or non-polynomial.

Polynomial: Given a non-negative integer k, a polynomial with a variable n of degree k is denoted as $P(n) = \sum_{i=0}^{k} a_i n^i$, where each a_i is a constant (also termed coefficient) of the polynomial and $a_k \neq 0$.

A polynomial is asymptotically **positive** (non-negative) if and only if the highest degree term $a_k > 0$ appears in that polynomial. For example, in $P(n)$, the degree is k, since $a_k \neq 0$; however, it is asymptotically positive if $a_k > 0$.

Mathematically, the degree of a function can be obtained as follows:

$$\lim_{n\to\infty} \frac{nf'(n)}{f(n)} = d \tag{1.1}$$

A simple example of a polynomial with one variable is given below:

$P(n) = 5/2 + 3n + 4n^2 + 5n^3$

Here, the degree of $P(n) = 3$. This can be easily verified by Equation (1.1).

Examples of non-polynomial but non-exponential functions

- $f_1(n) = 5n^{-2} + 1$ .... degree of $f_1(n) = -2$ (as per Equation (1.1))
- $f_2(n) = 5x^{-1/2} + 1$ .... degree of $f_2(n) = 1/2$ (as per Equation (1.1))
- $f_3(n) = log^k n + 1$ $(k \geq 0)$ .... degree of $f_3(n) = 0$ (as per Equation (1.1))

Function $f_3(n)$ is a logarithmic function as it is a function of variable n containing $logn$. More specifically, it is a **poly-logarithmic function**.

Examples of non-polynomial but exponential functions

- A function of the form $f(n) = k^n (k > 1)$ is genuinely exponential.
- Some functions are termed super-exponentials; for example, $f(n) = n^n$, $f(n) = n!$.

> **Note:** Some functions which *grow faster* than polynomial functions but not quite exponentially, such as $O(n^{log\,n})$, are also (incorrectly) called exponential in the context of complexity theory. For example, n^{100}.

Equation: When a function $f(n)$ equals zero, it is known as an equation ($f(n) = 0$).

Growth: This explains how the value of a function changes as the input size of the problem increases.

Growth Function: This function is used to estimate the number of steps an algorithm uses as its input grows. So, the total number of steps performed by the algorithm may be an expression. Examples include c (constant), n (linear), $log_2 n$ (logarithm), n^2 (quadratic) and n^3 (cubic).

Ideally, the lower order terms in the expression are ignored to choose the growth function, since the highest degree term alone is sufficient to express the rate of growth of a complete function. For instance, $f(n) = 2n^2 + 3\sqrt{n} + 7$. In this example, the highest degree term with a positive constant is $2n^2$. Therefore, the simpler function here is n^2. In other words, we may ignore the lower degree terms as well as the constant coefficients, as these are dominated by the higher degree terms.

Some important inequalities: Let a and b be two real variables. Then,

- $|a+b| \leq |a| + |b|$. This is known as **triangle inequality**. It may be used to find a pair of values: (c, n_0) for expressing the 'big O' notation.
- $|a - b| \leq |a|+|b|$
- Let $f(x)$ and $g(x)$ be two real function and suppose $f(x) \geq g(x)$

  $\Rightarrow f(x) > g(x)$ or $f(x) = g(x)$

  Hence, $g(x) < f(x)$ or $g(x) = f(x) \Rightarrow g(x) \leq f(x)$

  In other ways, $f(x) \geq g(x)$

  $-f(x) \leq -g(x) \Rightarrow g(x) \leq f(x)$

Note: Let *a*, *b* and *c* be three variables. Now, $a \leq b \leq c$ implies that *b* cannot be less than *a*, and *b* cannot be greater than *c*. This concept will help us to apply and understand the concept of asymptotic notation while measuring the complexity of an algorithm.

1.3.2 Asymptotic Notation

The complexity function of any algorithm (involving any number of variables) is finally expressed in terms of asymptotic notation. More specifically, the order of growth of the running time of an algorithm is expressed in terms of asymptotic notation. It also allows us to compare the relative performance of alternative algorithms. Asymptotic notations are mathematical notations. Examples include O (big O), θ (theta) and Ω (omega).

An **asymptote** is a line or curve represented by a function $g(n)$ that approaches a given curve (say, $f(n)$) arbitrarily closely, and the distance between them approaches zero as they tend to infinity. More specifically, they do not meet at any finite distance. So, we cannot state that the function $f(n)$ is equal to the function $g(n)$, as these functions do not result in the same value for every input. However, we can say that they grow equivalently as n increases (that is, their asymptotic growth rates are equal), and their rates of growth are usually bounded by a constant factor based on the concept of asymptotic notation: big O (O), big omega (Ω) and theta (Θ).These notations are explained below. Indeed, measuring the complexity of an algorithm by using asymptotic notation only provides an inexact or approximate amount of time for an algorithm. However, this is the most useful practical measure of the performance of an algorithm.

Big Oh (O) Notation: This gives the least upper bound function for measuring the worst case complexity of an algorithm. It is expressed as $f(n) \leq c \cdot g(n)$ for positive constants c and n_0 and for all $n \geq n_0$, where n_0 is the minimum possible value; that is, the value of $f(n)$ always lies on or below $c \cdot g(n)$ for all $n \geq n_0$. Thus, $f(n)$ is bounded above by a constant multiple of $g(n)$. This means that $f(n)$ grows asymptotically no larger than $g(n)$. More explicitly, the complexity function $f(n)$ can never be greater than the specified $g(n)$. The inequality $f(n) \leq c \cdot g(n)$ can be expressed by asymptotic big O notation as, $f(n) = O(g(n))$. It gives the least upper bound function. This upper bound need not be polynomially bounded above.

Note: When $f(n)$ is expressed as $f(n) = O(g(n))$, it means that $f(n) \leq c \cdot g(n)$; that is, $g(n)$ is the lowest upper bound function for $f(n)$. There may be functions greater than $g(n)$ that are also greater than $f(n)$, but $g(n)$ is the smallest one for which $f(n)$ is smaller.

The = sign means $\subseteq$ (set inclusion; subset). So, it is a one-way equality in this connection. That is why we may write $f(n) = O(g(n))$ but not $O(g(n)) = f(n)$.

In any case, $f(n)$ may never take values more than approximately those of the $g(n)$ operation (Fig. 1.8).

Big Omega (Ω) Notation: This gives the greatest lower bound function for measuring the best case complexity of an algorithm. It is expressed as $c \cdot g(n) \leq f(n)$ for positive constants c and n_0 and for all $n \geq n_0$, where n_0 is the minimum possible value; that is, the value of $f(n)$ always lies on or above $c \cdot g(n)$ for all $n \geq n_0$. Thus, $f(n)$ is bounded below by a constant multiple of $g(n)$. So $f(n)$ grows asymptotically no smaller than $g(n)$; that is, $f(n)$ is asymptotically greater than or equal to $g(n)$. It is expressed in asymptotic big omega notation as $f(n) = \Omega(g(n))$. It gives the greatest lower bound function.

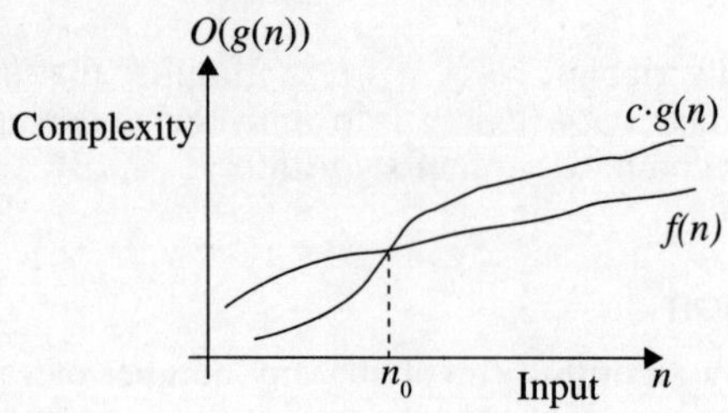

FIGURE 1.8 A sketch for big Oh notation

Note that the growth function $g(n)$ is only a lower bound on $f(n)$; that is, it should be as large as the function of n for which $f(n) = \Omega(g(n))$. In any care, $f(n)$ may never take values less than approximately those of the $g(n)$ operation (Fig. 1.9).

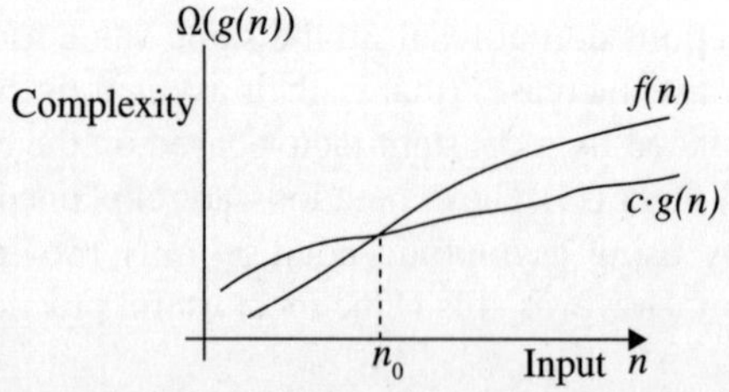

FIGURE 1.9 A sketch for big omega notation

> **Note:** When $f(n)$ is expressed as $f(n) = \Omega(g(n))$, $c \cdot g(n) \leq f(n)$, that is, $g(n)$ is the greatest lower bound function for $f(n)$. Again, there may be functions less than $g(n)$ that are also less than $f(n)$, but $g(n)$ is the greatest one for which $f(n)$ is greater.

Theta (Θ) Notation: It is expressed as $c_1 \cdot g(n) \leq f(n) \leq c_2 \cdot g(n)$ for positive constants c_1 and c_2 and n_0 for all $n \geq n_0$, where n_0 is the minimum possible value; that is $f(n)$ always lies between $c_1 \cdot g(n)$ and $c_2 \cdot g(n)$. Thus, $f(n)$ grows asymptotically no smaller and no larger than $g(n)$. It is expressed by asymptotic theta (Θ) notation as $f(n) = \Theta(g(n))$.

Concisely, $O(g(n))$ can be read as 'order at most $f(n)$'; Ω $(g(n))$ as 'order at least $g(n)$' and $\Theta(g(n))$ as 'order of exactly $g(n)$' (Fig. 1.10).

So, we can write mathematically Θ $(g(n)) = O(g(n)) \cap \Omega(g(n))$, if $f(n)$ can be expressed as $\Omega(g(n))$ ($\cap$ is an intersection operation).

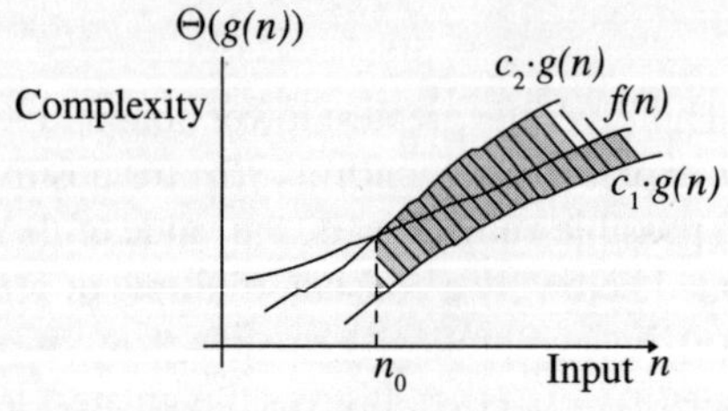

FIGURE 1.10 A sketch for theta notation

KEY POINTS

- The asymptotic representation of the running time of an algorithm is an inexact measurement expressed in terms of growth function and asymptotic notation.
- The term 'polynomially bounded' is a different definition of the word 'polynomial' from the one we had previously defined – $P(n)=\sum_{i=0}^{k} a_i n^i$, where each a_i is a constant (also termed coefficient) of the polynomial and $a^k \neq 0$. In the case of polynomially bounded, the word 'polynomial' is used to group together functions that are bounded from above by polynomials. So, *log n* and *n*log*n*, which are not polynomials in the original sense, are polynomials by the alternative definition, as they are bounded from above by, for example, *n* and n^2, respectively. Here is a simple list of examples in this respect.
Polynomially bounded above: Let us see how the functions *logn*, *n*logn* and $n^{5.7}$ are polynomially bounded above.

$$log\ n = O(n),\ n\ log\ n = O(n^2),\ n^{5.7} = O(n^6);\ \text{when } k < 0,\ n^k = O(n^0).$$

Note: A function *f*(*n*) is polynomially bounded if $f(n) = O(n^k)$ for some constant *k*. The exponent *k* need not be positive. A running time of $O(1/n)$ is still a polynomial time. Remember also that big O notation provides only an upper bound. Therefore, a running time of $O(n^k)$, where *k* is negative, is also O(1) ($O(n^0)$). In complexity analysis, it is assumed that functions are asymptotically non-negative. If we take an asymptotically negative function, then that function can be bounded above as $O(n^0)$.

Polynomial time algorithm: A polynomial time algorithm is one whose order-of-magnitude time performance (*f*(*n*)) is bounded from above by a polynomial function of *n*, where *n* is the size of its input. It must have an upper bound of n^k (*k* is zero or a positive integer). Thus, *f*(*n*) is polynomially bounded if $f(n) = O(n^k)$ for some constant *k*. Lower bounds do not matter.

Suppose an algorithm runs in $O(2^n)$ time. This means that the complexity of the algorithm cannot be bounded above at any cost by any polynomial. So, it is definitely not polynomial time. Some examples of polynomial time problems (with known polynomial time algorithms) are searching an unordered list, searching an ordered list, sorting a list, multiplying integers (even though there is a gap) and finding a minimum spanning tree in a graph (even though there is a gap).

Exponential time algorithm: This is an algorithm whose order-of-magnitude time performance is not bounded from above by a polynomial function of *n*. Examples of exponential functions include $O(2^n)$, $O(n!)$, $O(n^n)$. It is also called non-polynomial time algorithm. An example of a non-polynomial time problem (with no known polynomial time algorithms) is Towers of Hanoi – we can prove that any algorithm that solves this problem must have a worst-case running time that is at least $2^n - 1$ and listing all permutations (all possible orderings) of *n* numbers.

Tractable problem: A problem that is solvable by a polynomial time algorithm, that is, the upper bound is polynomial.

Intractable problem: A problem that cannot be solved by a polynomial time algorithm. The lower bound is exponential, that is, not bounded from above.

1.4 ADDITIONAL SOLVED PROBLEMS

PROBLEM 1

If the alphabet is {a}, what is the complement of a* and a^+?

Solution The complement of a* is { }, whereas the complement of a^+ is {ε}.

PROBLEM 2

Prove by mathematical induction that every integer of the form 6^n, $n > 0$ has 6 as the digit in the unit place.

Solution Let P(n) be the mathematical statement : 6^n has 6 as the digit in the unit place.

Insert $n = 1$ to obtain 6; thus, 6 is the digit in the unit place. So, P(1) is true. Let us assume that P(k) is true for some $k > 0$. This means that 6^k has 6 as the digit in the unit place. In other words, we can say $6^k = 10m + 6$, where m is a positive integer.

Let us check the correctness of P($k + 1$).

$6^{k+1} = 6 \times 6^k = 6\{10m + 6\} = 60m + 36 = 60m + 30 + 6 = 10(6m + 3) + 6$

Now, the remainder when the above number is divided by 10 is 6. Hence, the digit in the unit place of 6^{k+1} is also 6, if it is assumed that the digit in the unit place of 6^k is 6.

So, P(1) is true and P($k + 1$) is true whenever P(k) is true for $k > 0$. Hence, by the first principle of mathematical induction, the given statement is true.

PROBLEM 3

Let A be a finite set of size n. The number of elements in the power set of $A \times A$ is:

(a) 2^{2^n} (b) 2^{n^2} (c) $(2^n)^2$ (d) $(2^2)^n$ (e) None of these **[GATE 1993]**

Solution The number of elements in $A \times A = n^2$. We know that for any set of size k, the number of elements in its power set is given by 2^k. So, using $k = n^2$ in the above formula, we get 2^{n^2}. So the correct option is (b).

PROBLEM 4

The number of functions from an m element set to an n element set is:

(a) $m + n$ (b) m^n (c) n^m (d) mn **[GATE 1998]**

Solution Let A be the set with m elements and B be the set with n elements. We need to count the number of functions from A to B. Each element of A needs to map to exactly one element of B.

For each element of A, there are n number of options for mapping it. So, the total number of options $= n \times n \times n \times .... \times n$ [m times] $= n^m$. So the correct option is (c).

PROBLEM 5

If P, Q and R are subsets of the universal set ∪, then $(P \cap Q \cap R) \cup (P' \cap Q \cap R) \cup Q' \cup R'$ is:

(a) $Q' \cup R'$ (b) $P \cup Q' \cup R'$ (c) $P' \cup Q' \cup R'$ (d) Universal set **[GATE 2008]**

Solution Use notations of Boolean algebra for a superfast solution. The given expression $= (P \cap Q \cap R) \cup (P' \cap Q \cap R) \cup Q' \cup R'$

$= PQR + P'QR + Q' + R'$ [converting to Boolean algebra]

$= (P + P')QR + Q' + R' = QR + Q' + R'$

$=(Q + Q')(R + Q') + R' = R + Q' + R'$

$= 1 + Q' = 1 =$ Universal set

So the correct option is (d).

PROBLEM 6

The statement (~p) → (~q) is logically equivalent to which of the statements given below?

I. p → q II. q → p III. (~q) ∨ p IV. (~p) ∨ q

(a) I only (b) I and IV only (c) II only (d) II and III only **[GATE 2017]**

Solution The given expression = (~p) → (~q) = p ∨ (~q). This matches with III. Based on the options given, the answer should be (d). To verify it, check the expression for II. q → p = (~q) ∨ p = p ∨ (~q). So, the given expression turns out to be equivalent to both II and III. So, the correct answer is (d).

PROBLEM 7

Let T be a binary relation defined as $T = \{(a, b) \in R \text{ if } a - b \leq 3\}$. Determine whether T is reflexive, symmetric, antisymmetric or transitive. **[JNU PhD Fellowship Exam 2016]**

Solution *Reflexivity:* For every real number x, $(x - x) = 0 \leq 3$. So, $(x, x) \in T$ for all $x \in R$. T is reflexive.

Symmetry: For any two real numbers x and y, $(x - y) \leq 3$ does not mean $(y - x) \leq 3$. For example, if $x = 0$ and $y = 4$. Clearly, $x - y = 0 - 4 = -4 \leq 3$. But, $y - x = 4 - 0 = 4 > 3$. T is not symmetric.

Antisymmetry: Counterexamples exist for this case as well. Take $x = 2$, $y = 3$. Clearly, $x - y = 2 - 3 = -1 \leq 3$ and so $(2, 3) \in T$. Also, $y - x = 3 - 2 = 1 \leq 3$ and so $(3, 2) \in T$. But, $2 \neq 3$. So, T is not antisymmetric.

Transitivity: Counterexamples exist for this case as well. Take $x = 10$, $y = 8$, $z = 6$. Clearly, $x - y = 10 - 8 = 2 \leq 3$ and so $(10, 8) \in T$. Also, $y - z = 8 - 6 = 2 \leq 3$ and so $(8, 6) \in T$. But, $x - z = 10 - 6 = 4 > 3$ and so $(10, 6) \notin T$. So, $(x, y) \in T$ and $(y, z) \in T$ does not necessarily imply $(x, z) \in T$. So, T is not transitive.

PROBLEM 8

$S = \{(1,2), (2,1)\}$ is a binary relation on set $A = \{1, 2, 3\}$. Is it irreflexive? Add the minimum number of ordered pairs to S to make it an equivalence relation. Give the modified S. **[GATE 2002]**

Solution $A = \{1, 2, 3\}$, $S = \{(1, 2), (2, 1)\}$. Observe that $(1, 1) \notin S$, $(2, 2) \notin S$ and also $(3, 3) \notin S$. To summarise, $\{(x, x) \notin S$ for all $x \in A\}$, and hence, S is irreflexive. To make S reflexive, we need to satisfy the conditions of reflexivity, symmetry and transitivity.

Reflexivity: As S is purely irreflexive, we need to include (1, 1), (2, 2) and (3, 3). This makes S reflexive by including three more ordered pairs.

Symmetry: Modified $S = \{(1, 1), (1, 2), (2, 1), (2,2), (3,3)\}$. Clearly $(1, 2) \in S \Leftrightarrow (2, 1) \in S$. Also, $(1, 1) \in S \Leftrightarrow (1, 1) \in S$. $(2, 2) \in S \Leftrightarrow (2, 2) \in S$. $(3, 3) \in S \Leftrightarrow (3, 3) \in S$. So, symmetry is satisfied. No more ordered pairs need to be included to make S symmetric.

Transitivity: $(1, 1) \in S$ and $(1, 2) \in S \Rightarrow (1, 2) \in S$.
$(1, 2) \in S$ and $(2, 1) \in S \Rightarrow (1, 1) \in S$.
$(1, 2) \in S$ and $(2, 2) \in S \Rightarrow (1, 2) \in S$.
$(2, 1) \in S$ and $(1, 1) \in S \Rightarrow (2, 1) \in S$.

$(2, 1) \in S$ and $(1, 2) \in S \Rightarrow (2, 2) \in S$.
$(2, 2) \in S$ and $(2, 1) \in S \Rightarrow (2, 1) \in S$.

Transitivity is also satisfied. No more ordered pairs need to be included to make *S* transitive.

Now, *S* is reflexive, symmetric as well as transitive. Thus, three more ordered pairs need to be added to make *S* an equivalence relation.

Modified $S = \{(1, 1), (1, 2), (2, 1), (2, 2), (3, 3)\}$

PROBLEM 9

Let *F* be the set of one-to-one functions from the set {1,2,….., n} to the set {1,2,…..*m*} where $m \geq n \geq 1$.

(a) How many functions are members of *F*?

(b) How many functions *f* in *F* satisfy the property $f(i) = 1$ for some i, $1 \leq i \leq n$?

(c) How many functions *f* in *F* satisfy the property $f(i) < f(j)$ for some $1 \leq i \leq j \leq n$? **[GATE 1997]**

Solution (a) Let $A = \{1,2,\ldots.., n\}$ and $B = \{1,2,\ldots.., m\}$, where $m \geq n \geq 1$.

We have to consider only one–one functions $f: A \to B$. So, once we have chosen an element from set *B* to match with an element of set *A*, the same element cannot be chosen again.

$f(1)$ can be any one of *m* elements. Once $f(1)$ is chosen, $f(2)$ has $(m - 1)$ options.

Similarly, $f(3)$ has $(m - 2)$ options. $f(4)$ has $(m - 3)$ options.

…

$f(k)$ has $(m - k + 1)$ options.

…

$f(n)$ has $(m - n +1)$ options.

Number of members of $F = m*(m - 1)*(m - 2)*(m - 3)*\ldots *(m - k + 1)*\ldots*(m - n + 1) = \dfrac{m!}{(m-n)!} = mP_n$.

(b) To count the number of functions where $f(i) = 1$ for some i, $1 \leq i \leq n$. Fix $f(1) = 1$ and then count the number of possible functions. Any one of $f(i)$ can take the value 1, so the count obtained in the previous step needs to be multiplied by *n*. Fix $f(1) = 1$. Now, we need to count the number of one–one functions from set $A_1 = \{2, 3, 4, 5, \ldots\ldots., n\}$ to $B_1 = \{2, 3, 4, 5, \ldots\ldots\ldots.., m\}$.

$|A_1| = (n - 1)$ and $|B_1| = (m - 1)$ and $m \geq n \geq 1$.

Number of such functions $= \dfrac{(m-1)!}{\{(m-1)-(n-1)\}} = \dfrac{(m-1)!}{(m-n)!}$. [Repeat the logic of (a), replacing *m* with $(m - 1)$ and replacing *n* with $(n - 1)$]

Total number of functions = where $f(i) = 1$ for some i, $1 \leq i \leq n = \dfrac{n(m-1)!}{(m-n)!}$.

(c) We shall simply count the number of functions for which $f(i) > f(j)$ for all $1 \leq i \leq j \leq n$ and then subtract this count from the number obtained in (a). Any sequence which is not strictly always decreasing has at least one pair of increasing elements in it. If we pick any set of *n* distinct elements of set *B*, there is only one way to arrange them in strictly decreasing order.

The total number of ways of picking up subsets of *n* elements from a set of *m* elements =

$$mC_n = \frac{m!}{n!(m-n)!}.$$

Required number of functions $= \dfrac{m!}{(m-n)!} - \dfrac{m!}{n!(m-n)!} = \dfrac{m!(n!-1)}{n!(m-n)!}$.

PROBLEM 10

Let p and q be propositions. Using only the truth table, decide whether p ⇔ q does not imply p → ~q is true or false. **[GATE 1994]**

Solution We need to check for the truth value of the expression ~{(p ⇔ q) → (p → ~q)}. If all truth values turn out to be true, then the given statement is true, else false.

Simply construct the truth table, as follows. The key is to use small expressions, so as to eliminate chances of committing mistakes. We represent True by the symbol T, and False by the symbol F.

p	q	(p ⇔ q)	~q	(p → ~q)	(p ⇔ q) → (p → ~q)	~{(p ⇔ q) → (p → ~q)}
F	F	T	T	T	T	F
F	T	F	F	T	T	F
T	F	F	T	T	T	F
T	T	T	F	F	F	T

The expression ~{(p ⇔ q) → (p → ~q)} is False for three cases and True for one case.

So, ~{(p ⇔ q) → (p → ~q)} is neither a tautology nor a contradiction. Hence, the given statement is neither true nor false, it is multi-valued.

PROBLEM 11

A relation R is defined on the set of integers as $x\,Ry$ iff $(x+y)$ is even. Which of the following statements is true?

(a) R is not an equivalence relation
(b) R is an equivalence relation having one equivalence class
(c) R is an equivalence relation having two equivalence classes
(d) R is an equivalence relation having three equivalence classes **[GATE 2000]**

Solution *Reflexivity*: For all integers x, $(x + x) = 2x$ is even, so R is reflexive.

Symmetry: $(x + y)$ is even ⇒ $(y + x)$ is even. So, R is symmetric.
Transitive: $(x + y)$ is even and $(y + z)$ is even ⇒ $(x + y + y + z)$ is even ⇒ $(x + z) + 2y$ is even ⇒ $(x + z)$ is even. So, R is transitive.

Clearly, R is an equivalence relation. For a set of natural numbers, the sum of even numbers always give an even number, the sum of odd numbers always gives an even number and the sum of any even and any odd number always gives an odd number. So, R must have two equivalence classes: one for even and one for odd.

$\{\ldots,-4,-2,0,2,4,\ldots\},\{\ldots,-3,-1,1,3,\ldots,\}$. The correct option is (c).

PROBLEM 12

Consider the following statements:
S1: There exist infinite sets A, B and C such that $A \cap (B \cup C)$ is finite.
S2: There exist two irrational numbers x and y such that $(x + y)$ is rational.
Which of the following is true about S1 and S2?

(a) Only S1 is correct
(b) Only S2 is correct
(c) Both S1 and S2 are correct
(d) Neither S1 nor S2 is correct **[GATE 2001]**

Solution Observe that both S1 and S2 pertain to the existence of particular values which satisfy the given statements, and not about the general correctness of both the statements for all cases.

Let A = {set of all prime numbers}, B = {set of all natural numbers which are multiples of 2} and C = {set of all natural numbers which are multiples of 3}. Clearly, all three sets are infinite.

$B \cup C$ = {set of all numbers which are multiples of 2 or 3}

$A \cap (B \cup C) = \{2\}$, as 2 is the only prime number which is a multiple of 2 or 3.

So, there exist values satisfying S1. Options (b) and (d) are eliminated.

Now, check the correctness of S2. Let $x = 3 + \sqrt{2}$ and $y = 3 - \sqrt{2}$. Clearly, both x and y are irrational numbers. But $(x + y) = 4$, which is rational. So, there exist values satisfying S2 as well. So the correct option is (a).

PROBLEM 13

Let $f: A \to B$ be a function, and let E and F be subsets of A. Consider the following statements about images.

S1:$f(E \cup F) = f(E) \cup f(F)$ S2:$f(E \cap F) = f(E) \cap f(F)$

Which of the following is true about S1 and S2?

(a) Only S1 is correct (b) Only S2 is correct

(c) Both S1 and S2 are correct (d) Neither S1 nor S2 is correct **[GATE 2001]**

Solution $E \subseteq A$ and $F \subseteq A$.

Let $y \in f(E \cup F) \Rightarrow y \in f(E)$ or $y \in f(F) \Rightarrow y \in f(E) \cup f(F) \Rightarrow f(E \cup F) \subseteq f(E) \cup f(F)$.

$y \in f(E) \cup f(F) \Rightarrow y \in f(E)$ or $y \in f(F) \Rightarrow y \in f(E \cup F) \Rightarrow f(E) \cup f(F) \subseteq f(E \cup F)$.

So, $f(E \cup F) = f(E) \cup f(F)$. S1 is correct. Options (b) and (d) are eliminated.

For S2, a counterexample exists. Let $A = B = \{1, 2, 3\}$. Let $f1(1) = 2, f1(2) = 3, f1(3) = 1$.

Let $f2$ be a constant function defined as $f2(x) = 2$. Let $E = \{1, 2\}$ and $F = \{3\}$.

Clearly, $E \cap F = \varphi$, and hence, $f(E \cap F) = \varphi$. $f(E) = f(\{1, 2\}) = \{2, 3\}$.

$f(F) = f(\{3\}) = \{2\}$, and hence, $f(E) \cap f(F) = \{2\} \neq f(E \cap F)$.

S2 is false. So the correct option is (a).

PROBLEM 14

How many onto (or surjective) functions are there from an n-element ($n \geq 2$) set to a two-element set?

(a) 2^n (b) $2^n - 1$ (c) $2^n - 2$ (d) $2(2^n - 2)$ **[GATE 2012]**

Solution Let us represent the sets as $A = \{a_1, a_2, a_3, \ldots a_n\}$ and $B = \{0, 1\}$. Each element of set A can map to either 0 or 1. The total number of all possible functions from A to $B = 2 \times 2 \times 2 \times \ldots 2$ [n times] $= 2^n$.

Now, there are only two non-onto functions possible from A to B. *[Observe this argument carefully]*. They are the constants functions $f(x) = 1$ and $g(x) = 0$.

In the case of f, the element '0' present in B has no pre-image in A, and hence, it is not onto. In the case of g, the element '1' present in B has no pre-image in A, and hence, it is not onto. All other possible functions will ensure that at least one element of A is mapped to 0, and at least one element of A is mapped to 1. So, the number of onto functions $= 2^n - 2$. The correct option is (c).

PROBLEM 15

Let X and Y be finite sets and $f: X \to Y$ be a function. Which one of the following statements is TRUE?

(a) For any subsets A and B of X, $|f(A \cup B)| = |f(A)| + |f(B)|$

(b) For any subsets A and B of X, $f(A \cap B) = f(A) \cap f(B)$

(c) For any subsets A and B of X, $|f(A \cap B)| = \min\{|f(A)|, |f(B)|\}$

(d) For any subsets S and T of Y, $f^{-1}(S \cap T) = f^{-1}(S) \cap f^{-1}(T)$ **[GATE 2014]**

Solution Let $X = Y = \{1, 2, 3, 4, 5\}$, $A = \{1, 2\}$ and $B = \{2, 3\}$. $A \cup B = \{1, 2, 3\}$.

Let $f(1) = 3, f(2) = 3, f(3) = 4, f(4) = 1, f(5) = 2$.

$f(A \cup B) = f(\{1, 2, 3\}) = \{3, 4\}, f(A) = f(\{1, 2\}) = \{3\}, f(B) = f(\{2, 3\}) = \{3, 4\}$

Clearly, $|f(A \cup B)| = 2, |f(A)| = 1, |f(B)| = 2$, and hence, $|f(A \cup B)| \neq |f(A)| + |f(B)|$. So, option (a) is eliminated.

Consider another example. Let $X = Y = \{1, 2, 3, 4, 5\}$, $A = \{1, 2\}$, $B = \{3, 4\}$. Let $f(1) = 1, f(2) = 3, f(3) = 1, f(4) = 3, f(5) = 5$.

$f(A \cap B) = f(\varphi) = \varphi, f(A) = f(\{1, 2\}) = \{1, 3\}, f(B) = f(\{3, 4\}) = \{1, 3\}$

$f(A) \cap f(B) = \{1, 3\} \neq \varphi$. So, option (b) is eliminated.

Using the same example as above, $|f(A \cap B)| = 0$, $|f(A)| = 2$, $|f(B)| = 2$, and hence, $|f(A \cap B)| \neq \min\{|f(A)|, |f(B)|\}$. So, option (c) is eliminated.

We are left with only option (d).

$f^{-1}(S \cap T) = \{x \mid x \in X \text{ and } f(x) \in S \text{ and } f(x) \in T\} = \{x \mid x \in X \text{ and } f(x) \in S\} \cap \{x \mid x \in X \text{ and } f(x) \in T = f^{-1}(S) \cap f^{-1}(T)$. The correct option is (d).

PROBLEM 16

Find the number of onto functions (surjective functions) from set $X = \{1, 2, 3, 4\}$ to set $Y = \{a, b, c\}$.

[GATE 2015]

Solution Figure 1.11 shows one possible onto function from X to Y.

As it is a function, each element of set X maps to exactly one element in set Y.

As it is an onto function, each element of Y must have at least one pre-image in set X.

X has four elements and Y has three elements; therefore, two elements of X will map to the same element in Y, and the remaining two elements of X will map to the remaining two elements of Y.

Based on these observations, we can perform the following analysis.

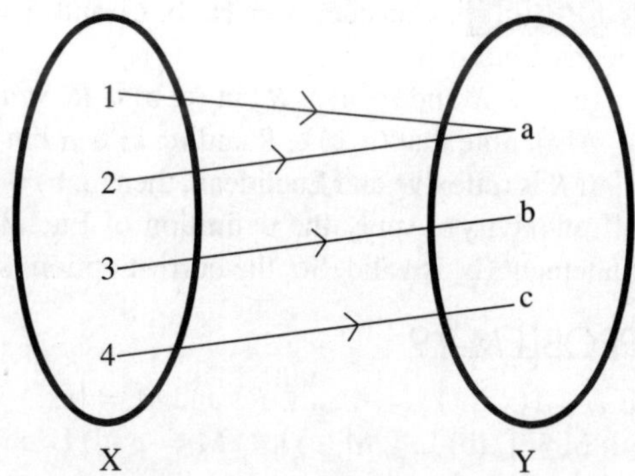

FIGURE 1.11 An onto function from X to Y

Number of ways of choosing two elements (out of four) from set X which will map to the same element of $Y = {}^4C_2$. Number of ways of choosing the common element (out of three) from set Y which will map to the two elements chosen in the previous step $= {}^3C_1$

Now, we are left with two elements in X and two elements in Y. Number of ways of mapping them = 2. Hence, the total number of onto functions $= {}^4C_2 \times {}^3C_1 \times 2 = 6 \times 3 \times 2 = 36$.

PROBLEM 17

Consider the following expressions: (i) False (ii) Q (iii) True (iv) P V Q (v) ~Q V P

The number of expressions given above that are logically implied by $P \wedge (P \rightarrow Q)$ is ________.

[GATE 2016]

Solution $P \wedge (P \rightarrow Q) = P \wedge (\sim P \vee Q) = (P \wedge \sim P) \vee (P \wedge Q) = (P \vee Q)$.

Observe carefully: This problem is asking us to find out which expression implied by $P \wedge (P \rightarrow Q)$ is a tautology.

{P ∧ (P → Q)} → false = (P ∨ Q) → false = ~(P ∨ Q) ∨ false = ~(P ∨ Q) = not necessarily a tautology [P = Q = true, makes it false]

{P ∧ (P → Q)} → Q = (P ∨ Q) → Q = ~(P ∨ Q) ∨ Q = (~P ∧ ~Q) ∨ Q = (~P ∨ Q) ∧ (~Q ∨ Q) = (~P ∨ Q) = not necessarily a tautology [P = true, Q = false, makes it false]

{P ∧ (P → Q)} → true = (P ∨ Q) → true = ~(P ∨ Q) ∨ true = true [tautology]

{P ∧ (P → Q)} → (P ∨ Q) = (P ∨ Q) → (P ∨ Q) = ~(P ∨ Q) ∨ (P ∨ Q) = true [tautology]

{P ∧ (P → Q)} → (~Q ∨ P) = (P ∨ Q) → (~Q ∨ P) = ~(P ∨ Q) ∨ (P ∨ ~Q) = (~P ∧ ~Q) ∨ (P ∨ ~Q) = {(~P ∨ P) ∧ (~Q ∨ P)} ∨ ~Q

= (~Q ∨ P) ∨ ~Q = (~Q ∨ P) = not necessarily a tautology [P = false, Q = true, makes it false]

So, two expressions are logically implied by P ∧ (P → Q).

PROBLEM 18

A binary relation $R \subseteq (S \times S)$ is said to be Euclidean if for every a, b, c ∈ S, (a, b) ∈ R and (a, c) ∈ R implies (b, c) ∈ R. Which of the following statements is valid?

(a) If R is Euclidean, (b, a) ∈ R and (c, a) ∈ R, then (b, c) ∈ R, for every a, b, c ∈ S.
(b) If R is reflexive and Euclidean, (a, b) ∈ R implies (b, a) ∈ R, for every a, b ∈ S.
(c) If R is Euclidean, (a, b) ∈ R and (b, c) ∈ R, then (a, c) ∈ R, for every a, b, c ∈ S.
(d) None of the above **[CMI PhD Entrance 2015]**

Solution Consider S = {a, b, c} and R = {(a, b), (a, c), (b, a), (b, b), (b, c), (c, a), (c, c)}. R is clearly Euclidean.

(c, a) ∈ R and (b, a) ∈ R but (c, b) ∉ R. *Statement (a) is invalid.*

Also, note that (a, c) ∈ R and (c, a) ∈ R but (a, a) ∉ R. *Statement (c) is also invalid.*

If R is reflexive and Euclidean, then (a, b) ∈ R obviously means that (a, a) ∈ R and (b, b) ∈ R (because of reflexivity). Apply the definition of Euclidean relation over (a, b) and (a, a) to obtain (b, a) ∈ R. Statement (b) is valid. So, the correct option is (b).

PROBLEM 19

If L = {(x, y) | y = e^x, x ∈ R} and M = {(x, y) | y^2 = x, x ∈ R}, which of the following is correct?
(a) M ⊆ L (b) L ⊆ M (c) L ∩ M = φ (d) L ∪ M = L **[SAU Delhi M.Phil/PhD Entrance 2013]**

Solution One of the sets L and M may be a subset of the other if and only if there exists some (x, y) satisfying both conditions $y = e^x$ and $y^2 = x$. Eliminate x from these two equations to obtain

For any real value of y, $y^2 \geq 0$, and hence $e^{y^2} \geq 1$.

Case 1: $y \leq 0$. Clearly $e^{y^2} \geq 1$ and $y \leq 0$. So, $e^{y^2} > y^2$.

Case 2: $0 < y < 1$. For this case $e^{y^2} > 1$, So, $e^{y^2} > y$.

Case 3: $y \geq 1$.

It is known from exponential series that $e^k > k$ for all real numbers k. So, $e^{y^2} > y^2$.

Also, when $y \geq 1$, $y^2 \geq y$. So, $e^{y^2} > y$.

So, we conclude that for all real values of y, $e^{y^2} > y$. There is no value of (x, y) which occurs in both L and M. So, L ∩ M = φ. The correct option is (c).

PROBLEM 20

Let R be a binary relation over a set S. R is an equivalence relation if it is reflexive, transitive and symmetric. A relation is called a partial order if it is reflexive, transitive and antisymmetric. (Notation: Let aRb denote an order pair $(a, b) \in R$). The relation R is said to be well ordered if R is a partial order and there does not exist an infinite descending chain (with respect to R) within S. An infinite sequence x_1, x_2, ... of elements of S is called an infinite descending chain if for all i we have $x_{i+1} R x_i$ and $x_i \neq x_{i+1}$. Take $S = N \times N$ and let the binary relation R over S be such that $(i_1, j_1)\ R\ (i_2, j_2)$ if and only if either $(i_1 < i_2)$ or $((i_1 = i_2) \wedge (j_1 \leq j_2))$. Which of the following statements is true of R?

(a) R is an equivalence relation but not well ordered
(b) R is a partial order but not well ordered
(c) R is a partial order and well ordered.
(d) R is an equivalence relation and well ordered
(e) R is neither a partial order nor an equivalence relation **[TIFR PhD Entrance 2012]**

Solution *Reflexivity:* Clearly $(i, j)\ R\ (i, j)$ as it satisfies the condition $(i = i)$ and $(j \leq j)$. So, R is reflexive.

Symmetry: Let $(i, j)\ R\ (m, n)$. This means either $(i < m)$ or $(i = m$ and $j \leq n)$. This can imply $(m, n)\ R\ (i, j)$ if and only if $i = m$ and $j = n$, which is not always necessarily true. So, R is not symmetric. Therefore, R is clearly not an equivalence relation. Options (a) and (d) are eliminated.

Now, let us focus on the properties of partial order.

Antisymmetry: $(i, j)\ R\ (m, n)$ and $(m, n)\ R\ (i, j)$ is possible if and only if $(i < m$ AND $m < i)$ or $\{(i = m)$ AND $(j \leq n)$ AND $(n \leq j\}$. All these conditions together imply that $i = m$ and $j = n$. So, R is antisymmetric.

Transitivity: $(i, j)\ R\ (m, n)$ and $(m, n)\ R\ (p, q)$

Case 1: $(i < m)$ and $(m < p)$ clearly implies $i < p$, and hence, $(i, j)\ R\ (p, q)$.

Case 2: $(i = m$ and $j \leq n)$ and $(m < p)$ clearly implies $i < p$, and hence, $(i, j)\ R\ (p, q)$.

Case 3: $(i < m)$ and $(m = p$ and $n \leq q)$ clearly implies that $i < p$, and hence, $(i, j)\ R\ (p, q)$.

Case 4: $(i = m$ and $j \leq n)$ and $(m = p$ and $n \leq q)$ clearly implies that $i = p$ and $j \leq q$, and hence, $(i, j)\ R\ (p, q)$.

So, R is transitive and partially ordered.

Now, we need to choose between options (b) and (c). There are only infinite ascending chains as $(x, y)\ R\ (p, q)$ implies that either $p > x$ or $q > y$ whenever $(x, y) \neq (p, q)$. So, no infinite descending chains exist, and hence, the relation is well ordered. The correct option is (c).

PROBLEM 21

Consider a family F of subsets of $\{1, 2, 3,...n\}$ such that for any two distinct sets A and B in F, we have $A \subset B$ or $B \subset A$ or $A \cap B = \phi$. Which of the following statements is true? (*Hint*: What does the Venn diagram of this family look like?) **[TIFR PhD Entrance 2016]**

(a) $|F| \leq 2n$ and there exists a family F such that $|F| = 2n$
(b) $|F| \leq n^2$ and there exists a family F such that $|F| = n^2$
(c) $|F| \leq 2n^2$ and there exists a family F such that $|F| = 2n^2$
(d) $|F| \leq 2^{n-1}$ and there exists a family F such that $|F| = 2^{n-1}$
(e) None of the above

Solution Draw the Venn diagram for this family F (Fig. 1.12).

The smallest grey circles represent subsets containing exactly one element each. The next level of subsets has exactly two elements each. The next level has exactly four elements each. The next level of subsets has exactly eight elements each. Continuing in this fashion, we end up with the entire set.

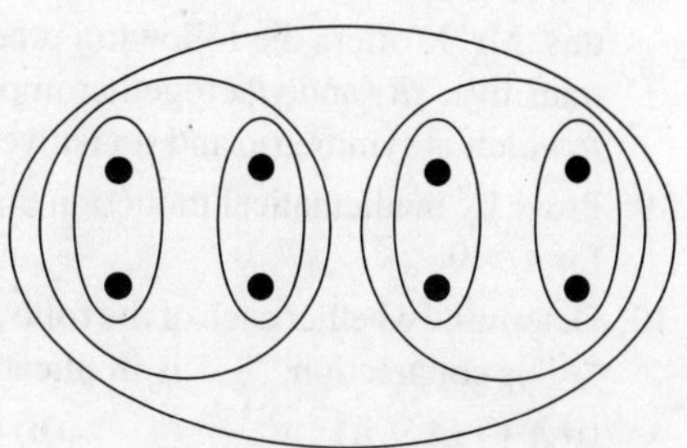

FIGURE 1.12 Venn diagram

It is clear that for any two subsets created in this manner, either they are disjoint or one is a strict subset of the other. Let $2^k \le n < 2^{k+1}$ for some whole number k. For n elements, the number of subsets with zero elements (the empty set φ) = 1.

Number of subsets with 1 element = $n/2^0$.
Number of strictly disjoint subsets with two elements = $n/2^1$.
Number of strictly disjoint subsets with 2^2 elements = $n/2^2$.
Number of strictly disjoint subsets with 2^3 elements = $I/2^3$.
...
Number of strictly disjoint subsets with 2^k elements = $n/2^k$.
Total number of possible members of set

$$F = 1 + \sum_{i=0}^{k}\left(\frac{n}{2^i}\right) = 1 + n\left\{1 + \frac{1}{2} + \frac{1}{2^2} + \frac{1}{2^3} + \dots + \frac{1}{2^k}\right\} 1 + \left[\frac{n\left\{1 - \frac{1}{2(k+1)}\right\}}{1 - \frac{1}{2}}\right]$$

$$= 1 + 2n\left\{1 - \frac{1}{2^{(k+1)}}\right\}.$$

When n is a proper power of 2, that is, $n = 2^k$, the value of the above expression = 1 + $2n\left\{1 - \frac{1}{2n}\right\}$ = $2n$.

When $2^k < n < 2^{k+1}$, the value of the above expression is strictly less than $2n$. So, $|F| \le 2n$ and there exists a family F such that $|F| = 2n$. The correct option is (a).

EXERCISES

1. Define length of string. Find the length of the strings $u = abcag$ and $w = gh$.
2. Perform all possible operations on the strings given in Q1.
3. Define prefix and suffix of a string.
4. Define concatenation of languages L_1 and L_2. Also, define Kleene star.
5. Give an algorithm to decide whether $|L| \ge 100$.
6. Show that proposition C is a logical consequence of the formula $A \wedge (A \rightarrow (B \vee C)) \wedge (B \rightarrow \sim A)$ using truth tables.
7. The Fibonacci sequence $\{f_1, f_2, f_3, \dots, f_n\}$ is defined by the following recurrence: $f_{n+2} = f_{n+1} + f_n$, $f_2 = f_1 = 1$. Prove by induction that every third element of this sequence is even.
8. Mr. X claims the following: If a relation R is both symmetric and transitive, then R is reflexive. For this, Mr. X offers the following proof: "From xRy, using symmetry we get yRx. Now, because R is transitive, xRy and yRx together imply xRx. Therefore, R is reflexive". Give an example of a relation R which is symmetric and transitive but not reflexive. **[GATE 1999]**
9. Prove by mathematical induction that every integer of the form 5^n has 5 as the digit in the unit place for $n > 0$.
10. Determine whether each of the following is a tautology, a contradiction, or neither ("∨" is disjunction, "∧" is conjunction, "→" is implication, "¬" is negation, and "↔" is bi-conditional (if and only if)).

    (i) $A \leftrightarrow (A \vee A)$ (ii) $(A \vee B) \leftrightarrow B$ (iii) $A \wedge (\neg (A \leftrightarrow B))$

11. Prove by induction that the expression for the number of diagonals in a polygon of n sides is $n(n-3)/2$.
12. Let R be a symmetric and transitive relation on a set A. Then
    (a) R is reflexive, and hence, an equivalence relation
    (b) R is reflexive, and hence, partially ordered
    (c) R is reflexive, and hence, not an equivalence relation
    (d) None of the above
    Give a proper explanation for the correct answer.
    [GATE 1995, Pondicherry University PhD Entrance 2013]
13. Let A and B be sets and let A^c and B^c denote their complements. The set $(A - B) \cup (B - A) \cup (A \cap B)$ is equal to
    (a) $A \cup B$ (b) $A^c \cup B^c$ (c) $A \cap B$ (d) $A^c \cap B^c$
    Give a proper explanation for the correct answer. **[GATE 1996]**
14. Suppose X and Y are sets and $|X|$ and $|Y|$ are their respective cardinalities. It is given that there are exactly 97 functions from X to Y. From this, one can conclude that
    (a) $|X| = 1, |Y| = 97$ (b) $|X| = 97, |Y| = 1$ (c) $|X| = |Y| = 97$ (d) None of these
    Give a proper explanation for the correct answer. **[GATE 1996]**
15. Which of the following is false?
    (a) $((x \rightarrow y) \wedge x) \rightarrow y$ (b) $((\sim x \rightarrow y) \wedge (\sim x \wedge \sim y)) \rightarrow x$
    (c) $x \rightarrow (x \vee F)$ (d) $((x \vee y) \Leftrightarrow (\sim x \rightarrow \sim y))$
    Give a proper explanation for the correct answer. **[GATE 1996]**
16. Let R_1 and R_2 be two equivalence relations on a set. Consider the following assertions:
    (i) $R_1 \cup R_2$ is an equivalence relation (ii) $R_1 \cap R_2$ is an equivalence relation
    Which of the following is correct?
    (a) Both assertions are true
    (b) Assertion (i) is true but assertion (ii) is not true
    (c) Assertion (ii) is true but assertion (i) is not true
    (d) Neither (i) nor (ii) is true
    Give a proper explanation for the correct answer. **[GATE 1998, 2005, Karnataka SET 2014]**
17. Consider the following relations:
    $R1$ (a,b) iff (a+b) is even over the set of integers
    $R2$ (a,b) iff (a+b) is odd over the set of integers
    $R3$ (a,b) iff a.b > 0 over the set of non-zero rational numbers
    $R4$ (a,b) iff $|a - b| \leq 2$ over the set of natural numbers
    Which of the following statements is correct?
    (a) $R1$ and $R2$ are equivalence relations, $R3$ and $R4$ are not
    (b) $R1$ and $R3$ are equivalence relations, $R2$ and $R4$ are not
    (c) $R1$ and $R4$ are equivalence relations, $R2$ and $R3$ are not
    (d) $R1$, $R2$, $R3$ and $R4$ are all equivalence relations
    Give a proper explanation for the correct answer. **[GATE 2001]**

18. The propositional statement $(P \rightarrow (Q \vee R)) \rightarrow ((P \wedge Q) \rightarrow R$ is
    (a) Satisfiable but not valid (b) Valid (c) A contradiction (d) None of the above
    Give a proper explanation for the correct answer. **[GATE 2004]**
19. Let $f: B \rightarrow C$ and $g: A \rightarrow B$ be two functions, and let $h = f \text{ o } g$. Given that h is an onto function, which of the following is true?
    (a) f and g should both be onto functions (b) f should be onto but g need not be onto
    (c) g should be onto but f need not be onto (d) Neither f nor g needs to be onto
    Give a proper explanation for the correct answer. **[GATE 2005]**
20. Let R be the relation on the set of positive integers such that aRb if and only if a and b are distinct and have a common divisor other than 1. Which one of the following statements about R is true?
    (a) R is symmetric and reflexive but not transitive
    (b) R is reflexive but not symmetric and not transitive
    (c) R is transitive but not reflexive and not symmetric
    (d) R is symmetric but not reflexive and not transitive
    Give a proper explanation for the correct answer. **[GATE 2015]**
21. Let X and Y denote the sets containing 2 and 20 distinct objects, respectively, and F denote the set of all possible functions defined from X to Y. Let f be randomly chosen from F. Find the probability of f being one-to-one. **[GATE 2015]**
22. A function $f: N^+ \rightarrow N^+$, defined on the set of positive integers N^+, satisfies the following properties:

$$f(n) = \begin{cases} f\left(\dfrac{n}{2}\right), \text{if } n \text{ is even} \\ f(n+5), \text{if } n \text{ is odd} \end{cases}$$

    Let $R = \{i : \exists j : f(j) = i\}$ be the set of distinct values that f takes. Find the maximum possible size of R. **[GATE 2016]**
23. Which of the following is true for functions $X \xrightarrow{f} Y \xrightarrow{g} Z$:
    (a) If f is injective and g is injective, the composition gof must be surjective
    (b) If f is surjective and g is injective, the composition gof cannot be injective
    (c) If f is injective and g is surjective, the composition gof must be injective
    (d) If f is surjective and g is surjective, the composition gof cannot be surjective
    Give a proper explanation for the correct answer. **[Gujarat SET 2014]**
24. Let R denote the set of real numbers. Let $f: R \times R \rightarrow R \times R$ be a bijective function defined by $f(x, y) = (x + y, x - y)$. The inverse function of f is given by
    (a) $f^{-1}(x,y) = \left[\dfrac{1}{x+y}, \dfrac{1}{x-y}\right]$ (b) $f^{-1}(x, y) = (x - y, x + y)$
    (c) $f^{-1}(x,y) = \left[\dfrac{x+y}{2}, \dfrac{x-y}{2}\right]$ (d) $f^{-1}(x, y) = [2(x - y), 2(x + y)]$
    Give a proper explanation for the correct answer. **[Karnataka SET 2014]**
25. If A and B are subsets of a universal set S, A is a subset of B and $A \neq B$, then

(a) $B' \cap A = \varphi$ (b) $A \cap B = B$ (c) $B \cap A' = \varphi$ (d) $B \subset (A \cap B)$

Give a proper explanation for the correct answer. **[Maharashtra SET 2011]**

26. Let A be the set of all natural numbers that are perfect squares, and B be the set of all prime numbers. Then $A \cap B =$

(a) The set of all composite numbers (b) {1} (c){0} (d) φ.

Give a proper explanation for the correct answer. **[IIT Chennai PhD Entrance 2014]**

27. Prove that in any sequence of 105 integers, there will always be a sub-sequence of consecutive elements in the sequence whose sum is divisible by 99. **[ISI Kolkata JRF 2015]**

28. Let S be a subset of {10, 11, 12 . . . , 98, 99} containing 10 elements. Show that there will always exist two disjoint subsets A and B of S such that the sum of the elements of A is the same as that of B. Give an example to show that this property does not hold if S contains 7 elements. **[ISI Kolkata JRF 2011]**

29. Out of a group of 21 persons, 9 eat vegetables, 10 eat fish and 7 eat eggs; 5 eat all three. How many persons eat at least two out the three dishes? **[GATE 1993]**

30. Consider the sequence $s_1 = s_2 = 1$. $s_i = 1 + \min\{s_{i-1}, s_{i-2}\}$. Prove by induction on n that $s_n = \left\lceil \frac{n}{2} \right\rceil$. **[GATE 2000]**

31. Prove using mathematical induction that the digit in the unit place of every integer of the form 5^n, $n > 0$ is 5.

32. Prove by mathematical induction that the digit in the unit place of every integer of the form 4^n, $n > 0$ is either 4 or 6.

33. Consider the function $h: N \times N \to N$ so that h (a, b) = $(2a + 1)2^b - 1$ where $N = \{0, 1, 2, 3,\dots\}$ is the set of whole numbers.

(a) Prove that the function h is an injection (one–one).

(b) Prove that it is also a surjection (onto). **[GATE 2001]**

34. Let $U = \{1, 2, ..., n\}$. Prove that $|\{(S,T) : S, T \subseteq U, S \cap T = j\}| = 3^n$. **[IIT Hyderabad PhD Entrance 2013]**

35. Show that for any integer $p > 3$, the integers p, $(p + 2)$ and $(p + 4)$ cannot be simultaneously prime. **[ISI Kolkata JRF 2014]**

36. Consider the set $A = \{1, 2, 3, 4\}$. A subset $S \subseteq A \times A$ is said to be symmetric if (x, y) $\in S$ whenever (y, x) $\in S$. For example, {(1, 1), (2, 3), (3, 2)} is a symmetric subset of $A \times A$, whereas {(1, 1), (2, 3), (3, 1)} is not. **[ISI Kolkata JRF 2017]**

(a) Determine how many subsets of $A \times A$ are symmetric.

(b) What is the probability that a randomly chosen subset of $A \times A$ is symmetric?

37. Which of the following is an equivalence relation on the set of all functions from Z to Z?

(a) {(f, g) | f (x) – g (x) = 1, x ∈ Z} (b) {(f, g) | f (0) = g (0) or f (1) = g (1)}

(c) {(f, g) | f (0) = g (1) and f (1) = g (0)} (d) {(f, g) | f (x) – g (x) = k for some k ∈ Z}

[UGC NET 2018]

38. A survey shows that 63% of Indians like Punjabi food, whereas 76% like South Indian food. If x% of Indians like both Punjabi and South Indian food, then

(a) $23 \le x \le 63$ (b) $23 \le x \le 69$
(c) $39 \le x \le 69$ (d) $39 \le x \le 63$ **[SAU Delhi M.Phil/PhD Entrance 2017]**

39. Which of the following relations on $\{1, 2, 3\} \rightarrow \{5, 6, 7, 8\}$ in an injective function?
(a) $1 \rightarrow 6, 2 \rightarrow 7, 3 \rightarrow 5$ (b) $1 \rightarrow 7, 2 \rightarrow 7, 3 \rightarrow 5$
(c) $1 \rightarrow 8, 2 \rightarrow 5, 3 \rightarrow 8$ (d) $1 \rightarrow 6, 2 \rightarrow 7, 3 \rightarrow 5, 1 \rightarrow 8$ **[Gujarat SET 2017]**

40. Let *N* be the set of natural numbers. Consider the following sets:
P: Set of rational numbers (positive and negative) *Q*: Set of functions from $\{0, 1\}$ to *N*
R: Set of functions from *N* to $\{0, 1\}$ *S*: Set of finite subsets of *N*
Which of the above sets are countable? **[GATE 2018]**

41. What is the number of different reflexive relations on a set with *n* elements?
[HCU PhD Entrance 2016]

42. For a set *A*, define P(*A*) to be the set of all subsets of *A*. For example, if $A = \{1, 2\}$, then $P(A) = \{\varnothing, \{1, 2\}, \{1\}, \{2\}\}$. Let $f: A \rightarrow P(A)$ be a function and *A* is not empty. Which of the following must be true?
(a) *f* cannot be one-to-one (injective) (b) *f* cannot be onto (surjective)
(c) *f* is both one-to-one and onto (bijective) (d) No such *f* is possible
(e) If such a function *f* exists, then *A* is infinite **[TIFR PhD Entrance 2017]**

43. Let *X* be a set of size *n*. How many pairs of sets (*A*, *B*) are there to satisfy the condition $A \subseteq B \subseteq X$?
[TIFR PhD Entrance 2010]

44. A relation *R* is defined on the set of positive integers as *xRy* if $2x + y \le 5$. The relation *R* is
I. Reflexive II. Symmetric III. Transitive
(a) I only (b) II only (c) III only (d) I and II only
[Punjab Tech University PhD Entrance 2010]

45. Consider the set *A* containing *n* elements. What is the total number of injective functions from *A* onto itself? **[JNU M.Phil/PhD Entrance 2012]**

46. Let *X* and *Y* be two sets having *n* elements each. What is the total number of bijective functions from *X* to *Y*? **[Karnataka SET 2015]**

47. In a set of people, the relation *x is not older than y* satisfies which property?
(a) Transitivity (b) Reflexivity
(c) Antisymmetry (d) Symmetry **[Gujarat SET 2016]**

48. Consider the set of all functions $f: \{0,1,\ldots,2014\} \rightarrow \{0,1,\ldots,2014\}$ such that $f(f(i)) = i$, for $0 \le i \le 2014$. Consider the following statements.
P: For each such function it must be the case that for every i, $f(i) = i$,
Q: For each such function it must be the case that for some i, $f(i) = i$,
R: Each such function must be onto.
Which one of the following is correct?
(a) *P*, *Q* and *R* are true (b) Only *Q* and *R* are true
(c) Only *P* and *Q* are true (d) Only *R* is true **[GATE 2014]**

49. Let S denote the set of all functions f: $\{0, 1\}^4 \rightarrow \{0, 1\}$. Denote by N the number of functions from S to the set $\{0,1\}$. The value of $\log_2 \log_2 N$ is ______________. **[GATE 2014]**
50. Let S be a set of n elements. The number of ordered pairs in the largest and the smallest equivalence relations on S are:
    (a) n and n (b) n^2 and n (c) n^2 and 0 (d) n and 1
    [GATE 2007, Rajasthan CS Lecturer Recruitment 2014]
51. Let E, F and G be finite sets. Let $X = (E \cap F) - (F \cap G)$ and $Y = (E - (E \cap G)) - (E - F)$. Which one of the following is true?
    (a) $X \subset Y$ (b) $X \supset Y$
    (c) $X = Y$ (d) $X - Y \neq \varphi$ and $Y - X \neq \varphi$ **[GATE 2006]**

2 Finite Automata

OBJECTIVES

At the end of this chapter, you will be able to

- Explain the central concepts of finite automata, acceptors and transducers (Moore machine and Mealy machine)
- Describe deterministic and non-deterministic finite automata and their equivalence
- Reduce the number of states in finite automata
- List the applications of finite automata

2.1 INTRODUCTION

Automata are mathematical models of real computers or their components. These models perform computations on input, moving through a series of *states* or *configurations* (explained in detail later). In the case of string acceptance (language recognition), the computation of an automata reaches an accepting state for an input string, beginning from the initial state. At each stage of the computation, a transition function/relation determines the next configuration on the basis of the current (or current as well as the earlier) configuration.

Several types of automata, namely, finite state machine (FSM) or finite automata (FA), pushdown automata (PDA), linear bounded automata (LBA) and Turing machine (TM) are studied in the theory of automata. It is interesting to note that each of the specified types of automata has a finite number of states (memory) but the first type is conventionally named the finite state machine as its memory is limited to a finite number of possible states; that is, no other memory (or a very small volume of tape memory) is theoretically considered.

From an application point of view, each specific type of automaton has its own advantage in solving problems and plays an important role in applications such as designing digital circuits, text processing, language (including natural language) processing (compiling), designing operating systems (state model of the operating system), computing numerical operations and artificial intelligence. TM is the most general but also the most complex among these for computations, whereas FA is the simplest mathematical model.

The concept of finite automata was introduced by DA Huffman in 1954 and by Moore in 1956. This model contains transition functions and a finite number of states – it is the abstraction of a computer with finite resources. The main purpose of this model is to check whether a string is contained within some language.

It may be noted here that FSM is not really memory-less; it has memory capacity (in the form of tape) that is fixed in size and cannot be expanded like in PDA, LBA and TM. The circuits of real computers, such as combinational (without memory) and sequential (with a very small amount of memory), are constructed based on the concept of FSM. For better understanding of digital circuits, a block diagram is shown in Fig. 2.1.

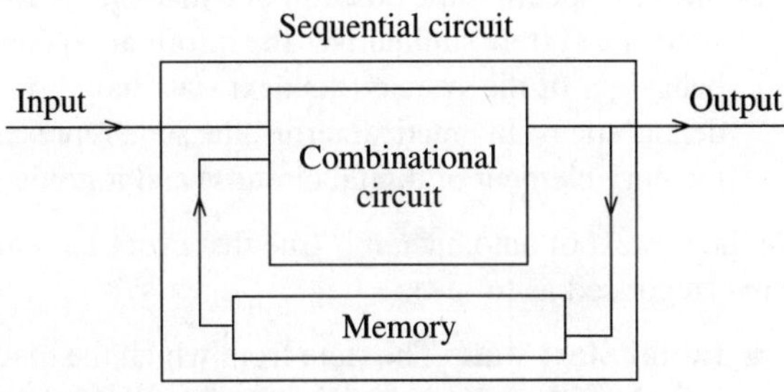

FIGURE 2.1 A simple block diagram of a digital circuit

Examples of combinational circuits include encoders, pattern detectors (for example, matching binary instructions for addition, multiplication, etc.) and circuits for performing complement operations (1's complement, 2's complement), whereas binary serial adders (with a small amount of memory to store the carry bit), parity bit generators and sequence detectors fall under sequential circuits. Some circuits like the pattern detector are acceptors (recognisers)[1] with final states and these are usually called FA. On the other hand, some are finite machines with output (for example, encoders, circuits for complement operations, parity bit generators, binary adders) and known as transducers (or converters). Thus, FSM works well for logic circuit design.

Certain well-known Unix tools (Lex, grep, awk, etc.) and pattern-matching tools of editors, where the number of states remains small, are designed based on FA. Further, the behaviour of FSM is also observed in many real-world devices; for example, traffic signal machines (as transducers), video games, network protocol analysers (as transducers), vending machines (as transducers) where the machine sends items such as snacks, alcohol, cigarettes and lottery tickets automatically to customers after they insert a credit card. In fact, all of the above-mentioned tools and devices can be formally described by state machines that store the status of something at a given time and operate on the input to change the status and/or cause an action or output to take place for any given change. So, state machines are also useful to developers and for system maintenance.

2.2 BASIC CHARACTERISTICS OF AUTOMATA

A computer is basically a state machine, and each machine instruction is the input that changes one or more states and may cause other actions to take place. The computer's data register stores the state. At any moment in time, a computer system can be seen as a very complex set of states. When this state machine is mathematically modelled, it is called automata. In other words, an automaton is an abstract model of a digital computer, and it is assumed to operate in a discrete time frame. Every automaton can be described by the following basic characteristics:

- **Input:** In a discrete instant, an automaton takes input from a set of input symbols (Σ). For any type of automaton, Σ is the finite non-empty set of input symbols.
- **Input tape:** The space in which the input symbols are stored.
- **Temporary storage:** Automata may have a temporary storage device, consisting of an unlimited number of cells, each capable of holding a single symbol from an alphabet (not necessarily the same as the input alphabet). The automaton can read and change the contents of the storage cells.

[1] An acceptor receives a string (w) as input and determines whether w belongs to a language (L). So, it is a language recognition device. Automata are called language acceptors/recognisers.

- **State:** It specifies the position of a machine at any instance during certain computations. In fact, the state of a system summarises the information concerning *past input* that is required to determine the behaviour of the system (the next state based on subsequent input). Letters such as p and q usually denote states. In practical terms, the state represents the storage element (flip-flop circuit – the basic memory element of digital circuits) and it holds some binary value.

In the context of automata, it is true that every automaton remains in a state at any moment; these states are categorised as follows:

- **Initial/Start state:** The state from which the machine begins its computations. A practical example is the boot program, which is stored in ROM; boot-up starts from here. So, the boot program itself is an initial state, whereas the operating system itself is also a state. We normally represent such states by the symbol q_0.
- **Final state:** It represents acceptance of an input, when the machine moves to such a state (the desired computation is performed successfully). More specifically, an input string is accepted if the machine remains in the final state on reading the last symbol of the input. We usually denote the final state by the symbol q_f. It is important to note that a string: $w \in L$ (language) if the corresponding machine (not the language L) halts in an accepting state, say q_{accept}.
- **Non-final state:** When the machine is in this state, the processed string is not accepted (it is rejected). The symbol q_i $(1 \leq i \leq n)$ is generally used to represent the non-final state.
- **Dead/Trap state:** It is a rejecting state that is essentially a dead end. Once the machine enters a dead state, it cannot reach any accepting state. This state is used to indicate a '*crash*', arising from some abnormal situation (error) in which the machine cannot carry out its mission as expected. Even if sufficient input is found later, the machine cannot move out this state; that is, no further input can help the machine escape from this state. So, the string may not be accepted. The utility of such a state is to make the user aware of the strings from which the machine may never recover. The symbol D is used in this chapter to represent the dead/trap state.

HALTING STATE

For all of the above-mentioned states (including the dead state), the machine halts/stops. So these can be treated as halting states. However, the word *halt* (halt state) is used specifically in the context of TM to signal the end of computation.

- **State relation** (Transition function or next state function): The next state of an automaton at any instant of time is generally determined by the *present state* and the *current input*. The input is usually read from the tape. In particular, state relation controls the next move for an input (the state the automaton will move to). The size of the input parameters varys from automaton to automaton. In the same way, the output of the function also varies. The symbol δ is used to denote a transition function.

Note: State movements in automata due to input means movement from one register to another.

- **Output relation:** Output is related only to the state (in the case of acceptors) or to some symbol called the output symbol (in the case of converters such as transducers).

Configuration: This term is used to refer to a particular state of the control unit, input file and temporary storage. The transition of an automaton from one configuration to the next is called a move. In fact, the move relation is associated with the configuration of the machine (instantaneous description; discussed later) and there is a small difference between move relation and state relation.

2.3 BASIC CATEGORIES OF AUTOMATA

Modern computers consist of three main components: the central processing unit, memory (theoretically unbounded in size) and input–output devices[2]. Theoretically, TM is treated as an abstract model of a real computer with all three components. However, there exist other kinds of automata with less power than that of TM. For example, FSM, PDA and LBA. Their transition functions vary and this makes them complex.

The CPU can be modelled by FSM, as it executes a finite number of possible inputs (strings), whereas a real computer is supposed to execute an infinite number of inputs; this is not possible in FSM as it has finite memory size. The most important criterion in designing an automaton is to minimise the number of states, that is, to design a *minimum state automaton*.

State Transition Graph: The actions of automata can be graphically depicted as consisting of states (nodes/vertices) and directed arcs labelled with inputs (indicating transitions). Such a graphical representation is called a state transition graph/diagram and it often provides a better way to understand the logic. The graph is a directed graph. In general, a transition graph is preferred to describe automata. However, the computation of automaton can also be described by tabular representation, called the state table. In practice, the transition table is constructed from the transition diagram. The state table is used in FSM to show the behaviour of a sequential circuit, consisting of the present state, next state (based on present input) and output. In this book, transition diagrams are frequently used to denote automata and vice versa.

2.4 FINITE STATE MACHINE AND ITS TYPES

The concept of FSM falls under the behaviour of reactive systems. A reactive system is one that changes its actions, output and status based on some information (input). To model a reactive system with FSM, first we must identify the states and operations to be performed. Almost all electronic devices today are based on FSM. Further, it is the restricted model of an actual computer. Some of the essential circuits of real computers, such as combinational and sequential circuits, are constructed based on the concept of FSM.

FSM can be categorised into the following groups:

- FSM as acceptor, usually called finite automata (FA)
- FSM with output, usually called transducer or converter

Acceptor and Transducer: Acceptors check the acceptance or rejection of a string. For this, the presence of a final state in automata is essential. FA as acceptor requires one or more final states to accept string(s). On the other hand, transducers are used to produce some new output for the given input – they are mainly used as coding devices; for example, Moore and Mealy machines. As they function as simple converters, the number of final states in such automata is zero.

Thus, we can conclude that the number of final states in FSM may be zero or more. If the number of final states in FSM is zero, it is known as a transducer, whereas if the number of final states is one or more, it is called FA. Note that TM can also be considered as a transducer.

[2] Input–output devices are the means by which information is communicated to and from the outside world. Examples include terminals, card readers and line printers. Technically, the heads associated with such devices can move backwards and forward. In particular, tapes of TM can be used for this purpose, and are hence regarded as input–output devices. On the other hand, tapes of FA and PDA are simply input devices and not input–output devices.

2.4.1 Informal Picture

Consider a very simplified version of the login process in a computer. This process may be viewed using an informal FA model to understand the behaviour of FA. The process is stated briefly below.

Description of the login problem: Initially, the computer waits for a username to be typed. This can be considered as a state of the system, called the *start state*. When the username is entered, the computer checks whether it is valid or not. If the username does not match exactly, the computer remains in the start state. If, however, the username matches exactly (valid), the computer moves to some other state and waits for the password. If the password entered is valid, it moves to the *final state* (the user has successfully logged in to the system); otherwise, it moves back to the start state. Ignoring detailed state transitions (symbol by symbol), the process is shown graphically in Fig. 2.2. This shows that the role of FA with the final state is to check the acceptance of strings.

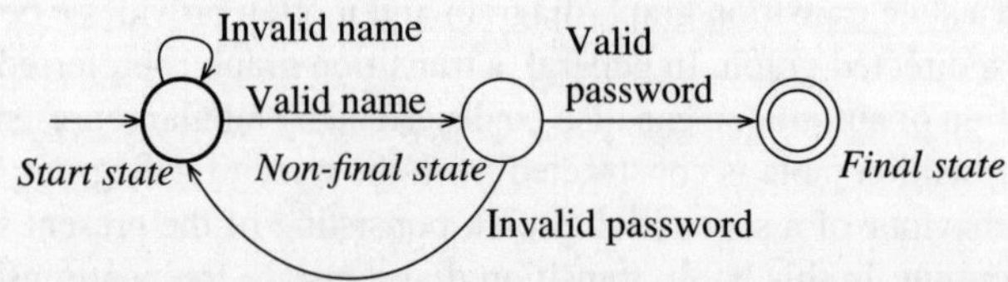

FIGURE 2.2 A finite automaton (transition diagram) representing the login process

2.4.2 High-level Description

As mentioned earlier, FA can be likened to a real computer with a central processor of fixed capacity. Such a machine has no auxiliary memory to remember historical data; that is, it stores only the current state but not the path through which it travelled to reach that state. The machine receives an input string and stores it in the INPUT tape by means of a movable reading head. We assume that the size of the tape is very large but fixed (both the sides are bounded by end markers), and the input in the tape is not erasable. The movement of the head here is unidirectional; it can move only in the forward direction.

An FA consists of three components: an input tape, a reading head and a finite control box (FCB) (Fig. 2.3).

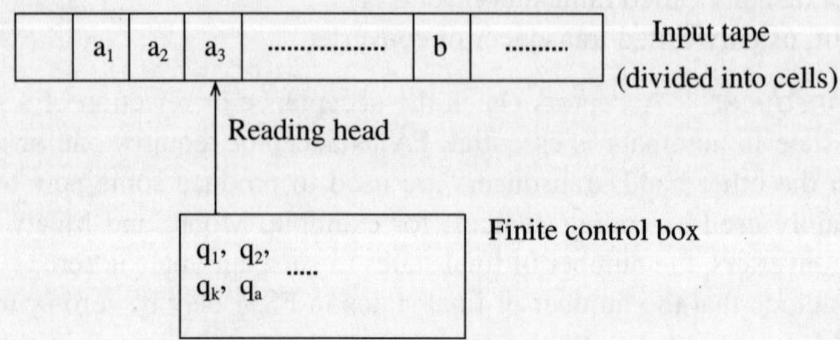

FIGURE 2.3 A high-level model of FA

Here, a_1, a_2,..., b are the input symbols on the tape, and q_1, q_2, ... are the states of the FA. The FCB, with the help of the movable reading head, can sense the symbol that is written at a position on the tape. It also changes the state if required. In fact, the FCB has a finite number of states (memory). In practical terms, the FCB may be viewed as the CPU (controller) of a real computer, where states are registers that are limited in number.

FA can be of two types: deterministic finite automata (DFA) and non-deterministic finite automata (NDFA).

Note: With reference to a real computer, the computational power of FSM may be viewed as equivalent to the computational power of a CPU (with in-built chip/memory). The CPU has a memory of limited size, bounded at the two ends of the tape. In other words, it is equivalent to the FCB plus limited tape size of the theoretical model. Thus, it is assumed that each input in FA is bounded by the two end points of the tape, which behave as markers.

2.4.3 Deterministic Finite Automaton (DFA)

A DFA moves to a unique next state on reading an input. It thus accepts a word only if there exists one path from q_0 (start state) to a state in F (set of accept/final states) labelled with the input word. Fundamentally, the empty string is not allowed in DFA for a state move. A formal definition is given below.

Formal Definition: A DFA is represented by a five-tuple[3] notation, $A = (Q, \Sigma, \delta, q_o, F)$, where

(i) Q is a finite non-empty set of states
(ii) Σ is a finite non-empty set of symbols, called the alphabet of the language, that the automaton

Note: If the size of the alphabet becomes very large (closer to infinite), the transition rules as well as the required number of states may be large, which may require a lot of memory.

(iii) δ is the transition function, defined as, $\delta: Q \times \Sigma \rightarrow Q$. This transition function states that Λ move[4] (or ε move) is not allowed in DFA, where the symbols Λ (and ε or epsilon) denote a null/empty string. In this book, both the symbols are used interchangeably.
For example, a DFA reads an input symbol (say, $a \in \Sigma$) in a state (say, $q_i \in Q$) and moves to another state (say, $q_j \in Q$). Mathematically, it is expressed as $\delta: q_i \times a \rightarrow q_j$, which can be viewed as $\delta(q_i, a) = q_j$ for all q_i and $q_j \in Q$, where q_i is the present state and q_j is the next state. Note that the next state may be the present state itself.
(iv) q_0 is the start state or the state in which the automaton remains when no input has been entered ($q_0 \in Q$). It is not a set.
(v) F is a subset of states of Q ($F \subseteq Q$), called the final or accept state. Any final state is denoted by q_f, and it belongs to Q. If the initial state is the final state, an empty string is also accepted.

The states that are not final are called non-final or intermediate states. Thus, all the states except q_0 and q_f in an FA are considered as intermediate states.

Acceptance and Rejection of Strings: A string that is processed over a path from the start (initial) state to the final state of an FA is an accepted string; otherwise it is rejected (it remains in the non-final state (halt state) when processing of that string is completed). This is true for PDA and LBA as well. However, a non-accepting string in TM may not be halted (known as the halting problem and discussed in Chapter 6).

---

[3] A tuple is a framework that represents a part/component. Mathematically, a tuple is a finite ordered list (sequence) of elements.

[4] Λ move (ε move) denotes a transition from one state to another without consuming any input – an empty string. Importantly, Λ move (with respect to any automaton) will be assumed to be explicit. It may be noted that Λ is neither an alphabet symbol nor a string of blanks.

Let us now consider an example. Suppose q_0 and q_f are, respectively, the start and final states of an FA. Assume that q_1 and q_2 are two non-final states of the same machine. Now, a string w = '*abb*' described through the path $q_0 - a - q_1 - b - q_2 - b - q_f$ is accepted by this FA, as the FA moves to the final state q_f for '*abb*' and simultaneously halts there. However, the strings '*a*' and '*ab*' are simply rejected by the same FA as, on reading such strings, the automaton does not reach the final state. But, the FA halts for both the strings.

Extended Transition Function: δ^* is sometimes denoted as the extended transition function that shows the processing of an input string (w) moving to a state (final, non-final or the same) through a number of intermediate states. Here, $|w| \geq 0$.

With respect to DFA, δ^* is expressed as: $\delta^* (q, w) = \delta^* (q, xa) = \delta (\delta^*(q, x)\ a) = p$, where $w = xa$ is the processing string, q is the start state and p is the last state. Note that the base case of δ^* for DFA is $\delta^* (q, \varepsilon) = q$. Certainly, the DFA results in at most one next state in each step. On the other hand, δ^* of NFA may result in multiple states in each step.

Hence, using the extended function, a string $x \in \Sigma^*$ (string over Σ) is accepted by FA (say, M) if $\delta^* (q_0, x) \in$ F; that is, if the machine moves to the final state for x. On the other hand, a string is not accepted (it is rejected) if it remains in any non-final state after being processed.

Configuration and Transition Function: Configuration shows the present state of the machine and the unread part of the string to be processed, whereas a transition is a rule or function that allows the machine to perform an operation. Thus, configuration applies a transition.

Instantaneous Description (ID): This shows how an automaton processes a string to accept or reject it. More specifically, the description shows how an automaton goes from configuration to configuration to process an input, using the move relation $\vdash$, where the RHS to $\vdash$ states the output configuration (to be performed next). The configuration is obtained by applying a specific transition rule, which must be mentioned. For IDs, we apply the transition relations of automata. With respect to FA, ID is described by an extended transition function (denoted as δ^* or $\hat{\delta}$).

> **Note:** Suppose the strings $w_1 = ab$ and $w_2 = aba$ are accepted by a machine M. M starts at the initial state q_0 to process w_1 and moves to its final state q_f to accept it. Likewise, for $w_2 = aba$, M does not start at q_f but at q_0, and proceeds further to process the string: *aba*.

EXAMPLE 2.1

Prove that $\delta(q, xy) \equiv \delta(\delta(q, x), y)$

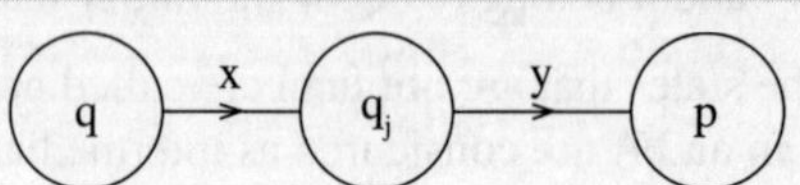

FIGURE 2.4 $\delta(q, xy) \equiv \delta(\delta(q, x), y) = p$

Solution A basic property of the transition function with respect to FA is $\delta(q, aw) = \delta (\delta(q, a), w))$, where q is a state, a is the single input and w is a string. By applying this property, we can express $\delta(q, xy)$ as $\delta(q, xy) \equiv \delta (\delta(q, x), y)$. Let us assume that from state q, the machine moves to state q_j (through several intermediate states) on reading the substring x of w. Next, from state q_j, the machine moves to state p, on reading the rest of w (that is, y). Thus, $\delta(q, xy) \equiv \delta (\delta(q, x), y) = p$ (Fig. 2.4).

Based on this idea, we may easily prove that if $\delta(q, x) = \delta(q, y)$, then $\delta(q, xz) = \delta(q, yz)$.

2.4.4 Non-deterministic Finite Automaton (NDFA or NFA)

Before defining NDFA, let us first explain the term *non-determinism*. This term actually implies 'guess' or 'random choice'. If the choice is proper, then any computation for which the choice has been made can be solved successfully. Otherwise, it will be problematic – the valid string may be rejected or the result obtained may be incorrect. Thus, an ambiguous situation occurs. However, in the context of a non-deterministic machine (a fully theoretical idea-based machine), it is assumed that the choice (or guess) to solve a problem is always correct.

Interestingly, the idea of NDFA (extension of DFA) was first proposed by Rabin and Scott in 1959. An NDFA allows a choice of zero or more next states for each input symbol; it shows more outgoing arcs with the same input symbol. On reading an input symbol, the machine may move to a number of next states; there is no unique next state. In addition, the empty string (Λ) is explicitly allowed as input in NDFA for a state move, as it is a non-deterministic characteristic. If it is allowed in NDFA, it is named ε NDFA (or ε NFA).

HOW DOES ε MOVE CAUSE NON-DETERMINISM?

ε move states that the machine may move to some other state without consuming any input from the present state. However, the machine may stay in the present state as well, without consuming any input. Thus, ε move behaves as a guess for deciding whether to move or to stay. We can express the causes of non-determinism as follows.

ε move is an insignificant concept in automata theory. Note that ε may be considered at any position of a given string, since concatenation of the null string at any position does not change the string at all. So, we may implicitly consider it at any state of the FA, and it does not have any effect on the diagram. For example, $\delta(q, \varepsilon) = q$. However, it does not assert an ε move, unless it is explicitly specified to move from one state to another. Thus, the presence of implicit and explicit ε moves causes non-deterministic behaviour; this is why it is not allowed in DFA.

Now, based on the inclusion and exclusion of ε move in a state, NDFA may be categorised as NDFA with ε move and NDFA without ε move (discussed later in this chapter).

String Acceptability: We know that each state in NDFA can have multiple transitions for an input symbol. However, an NDFA accepts a word if there exists at least one path from q_0 (start state) to a state in F (set of accept/final states) labelled with the input word; the paths enable the machine to parse all the strings that it recognises.

Formal Definition: An NDFA is represented by a five-tuple, $A = (Q, \Sigma, \delta, q_0, F)$, where

(i) Q is a non-empty finite set of states.

(ii) Σ is a non-empty finite set of symbols, that we call the alphabet of the language, accepted by the automaton.

(iii) The most generalised form of δ (transition function) is defined as $\delta: Q \times (\Sigma \cup \{\varepsilon\}) \rightarrow 2^Q$. Certainly, $\varepsilon \notin \Sigma$. The transition function makes sure that ε move is allowed but it may be made specific for NDFA with or without ε transition (on the basis of the presence or absence of the ε transition).
Physical implication of the transition function: The function states that, on reading an input from a state (say, $q_i \in Q$), an NDFA (M) may move to one of 2^Q subsets of Q (where 2^Q is the power set of Q). In the worst case, one of the subsets contains all the states of Q. The function also states that the receiving input may be ε (a real input symbol that does not belong to Σ) in a state. For better understanding, we may use the transition function of DFA, $\delta: Q \times \Sigma \rightarrow Q$ (for example, $\delta(q_i, a) = q_j$, where q_i and q_j belong to Q). Likewise, $\delta: Q \times (\Sigma \cup \{\varepsilon\}) \rightarrow 2^Q$ (for example, $\delta(\{q_i, a) = \{q_j, q_k, \ldots, q_r\}$, where the sets $\{q_i\}$ and $\{q_j, q_k, \ldots, q_r\}$ belong to the power set of Q: P(Q)).

(iv) q_0 is the start state, that is, the state in which the automaton stays when no input has been processed ($q_0 \in Q$). It is not a set.
(v) $F \subseteq Q$ (subset of Q) is the set of final states.

Analysis of the Transitions of DFA and NDFA: In DFA, a transition is as follows: δ: Q × Σ → Q. This states that the inputs are, respectively, a state from Q and an input symbol from Σ, and the outcome is only a state of Q. So, each transition can be expressed as (*p*, *a*), and it results in only one outcome (state). In other words, the outcome is at most one. This implies that the transition may not be specified at all for some input(s); the machine may move to a dead/trapped state for these inputs. Each such transition is often treated as an empty (or Φ) transition.

A transition function (δ) in NDFA may be as follows: δ: Q × (Σ ∪ {ε}) → 2^Q. This states that the inputs are, respectively, a state from Q and an input symbol from Σ or ε, but the outcome is a subset of states (taken from 2^Q: a power set denoted as P(Q)). So, each transition (*p*, *a*) may have multiple state outcomes.

Suppose the set Q of an NDFA consists of three states (q_0, q_1 and q_2). Clearly, the number of subsets is $8 = 2^Q$, where Q in 2^Q represents the number of states present in Q (here, it is 3). The 8 subsets are, respectively, Φ (empty state[5]), $\{q_0\}$, $\{q_1\}$, $\{q_2\}$, $\{q_0, q_1\}$,$\{q_0, q_2\}$,$\{q_1, q_2\}$ and $\{q_0, q_1, q_2\}$. The subsets collectively represent the power set of Q (P(Q)).

Thus, on reading an input, an NDFA may move to one or more possible states, represented by a subset of P(Q).

THEOREM 2.1

For every NDFA, there exists an equivalent DFA.

Proof The NDFA does not obey restrictions in moves; each NDFA is an ambiguous FA. On the other hand, each DFA is an unambiguous FA. Further, based on the formal definition of NDFA, a DFA can be viewed as a subset of NDFA in the context of state moves, and all subsets have equivalent power. In other words, every DFA can be seen as a special kind of NDFA in which there is exactly one next state for each input state and input symbol. Hence, for every NDFA, there exists an equivalent DFA.

The theorem is illustrated later in this chapter using NDFA to DFA construction algorithms, along with examples.

Corollary Every NDFA is not a DFA but the converse is true.

An NDFA can show moves of multiple paths from state(s), but it is true that it can simultaneously show a single path as well, as the NDFA is the superset. Now, if the NDFA always shows moves only of the correct paths, then it is treated as a DFA. Hence, every NDFA is not a DFA but the converse is true. Further, every NDFA can be converted to an equivalent DFA. So, both have the same power.

Advantages and Disadvantages

Advantages: To highlight the advantages of NDFA, we should first be aware of the disadvantages of DFA. One major disadvantage is that designing DFA for certain problems becomes very complex.

---

5 The empty state (Φ) is a mathematical concept. It is the state of a list with zero items in it. The state is usually used in non-deterministic automata. However, one could argue that the state labelled ∅ (representing the empty state; mathematically, it is the empty set of states: { } or no configuration for NDA) in the power set construction may equivalently represent the dead state (D) for a DFA. In this book, the notations Φ and D are used interchangeably to denote the dead state for DFA.

Further, the number of states is usually higher in DFA, as it is based on deterministic behaviour. So, we often prefer to design NDFA to solve problems. Some of the advantages of NDFA are listed below.

- Designing an NDFA is much easier than designing an DFA, as the automaton can move to multiple states for a given input from any state. Further, ε move (in the case of ε NFA) is allowed from one state to another; this is not possible in DFA. Hence, the number of states in NDFA is naturally lower than that in DFA, which results in smaller space (memory) requirements as well. In short, it is compact compared to DFA.
- NDFA reduces the complexity of the mathematical work to establish many important properties, such as 'union of two regular languages is regular', 'concatenation of two regular languages is also regular', etc.

Disadvantages: Although NFAs often need fewer states than the corresponding DFAs and they are easy to construct, designing real machines from an NDFA is very difficult, as direct mapping from the NDFA to the real machines (which are completely deterministic in nature) is not a simple task, unless we use the concept of backtracking. More specifically, NFAs are not well-suited to language recognition because non-determinism can cause the machine to make incorrect choices. In general, when an NFA is used for language recognition, it is necessary to backtrack and explore all possible choices. This is a time-consuming activity; in fact, an exponential time problem.

Thus, to model any real machine from an NDFA, the first requirement is to convert the NDFA to DFA because "For every NDFA, there is an equivalent DFA (subset of NDFA)". Examples of NDFA and the standard conversion methods from NDFA to DFA are discussed in Section 2.9.1.

Note: Languages that are recognised by FA are called regular languages. In this context, there exists a strong equivalence: "For every regular language, there is a finite state automaton, and vice versa" (discussed in Chapter 3).

2.5 GRAPHICAL SHAPES USED TO DRAW FINITE AUTOMATA

The graphical shapes used for drawing the state diagram of FA, along with a short description of each, are shown in Fig. 2.5.

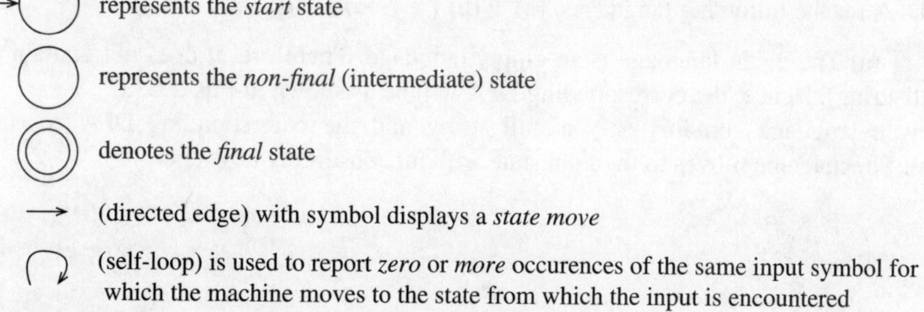

FIGURE 2.5 Graphical shapes for drawing the state transition diagram

Note: The loop implies that the edge label may or may not be used; that is, zero or more times. Zero is the case when the edge label is not used (the edge is not considered at all). On the other hand, the edge label may be considered multiple times. However, in both cases, the machine does not move from the state in which the self-loop is considered. Importantly, the shapes presented in Fig. 2.5 are used to design automata other than FA as well.

Important points for designing a state transition diagram

- While designing the state diagram for an FA (especially for DFA), all the symbols of Σ must be considered explicitly as inputs in each state of the automaton.
- FA has a finite number of states. So, if it is necessary to accept an infinite number of strings (the language has an infinite number of strings), the FA must have a cycle (loop). This is true for other types of automata as well.
- A self-loop denotes zero or more occurrences of symbols. In the case of zero occurrence, we may assume the case to be $\delta(q_i, \varepsilon) = q_i$, that is, an implicit transition.
- In DFA, multiple transitions for the same input from a state are not allowed. However, any undefined transition (other than implicit ε transition) implies that the automaton moves to the trapped/dead state (D) for such input(s).
- In NDFA, multiple transitions for the same input from a state are allowed. ε transition (other than the implicit ε transition) is also allowed. Any undefined transition implies that the machine moves to the empty state (Φ) for such input(s). As noted earlier, the empty state (Φ) for NDFA may result in a dead state (D) for DFA after conversion.
- Technically, a DFA has a transition for every input symbol in its alphabet from each of its states. On the other hand, we need not show all such transitions including the dead state for NDFA. So, NDFA may have several transitions from a state or no transition at all.
- Interestingly, the dead state may or may not arise for all problems solvable by FA. It depends solely on the nature of the problem.

2.6 DESIGNING SOME BASIC FINITE AUTOMATA

Designing a DFA to solve a problem (P) means solving P mathematically (if possible) by applying the capability of FA.

EXAMPLE 2.2

Draw the DFA for the following languages: (a) Φ (b) { ε }, (c) Σ* (d) Σ^+.

Solution (a) The given language is an empty language. Therefore, it does not contain any string (even a null string). Hence, the corresponding DFA will be as shown in Fig. 2.6 (a).

(b) The given language contains only a null string and the corresponding DFA is as shown in Fig. 2.6 (b). The machine moves to the final state without consuming any input.

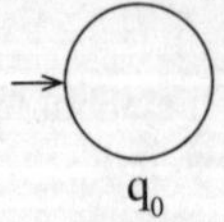

FIGURE 2.6 (a) Transition diagram (DFA) for null language

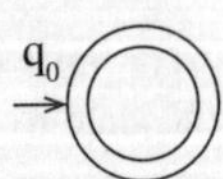

FIGURE 2.6 (b) Transition diagram (DFA) for null string

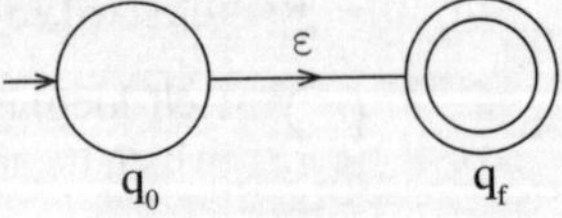

FIGURE 2.6 (c) An equivalent ε NFA (for Fig. 2.6 (b)) for null string

Note: For the start state of the machine, it is assumed that no input (null string) is processed, and we need not show any explicit ε move for it in the DFA. Thus, if the initial state is the final state, it will be assumed that the null string is also accepted, and again we need not show an ε move for it in the DFA. The shape of the final state (in the initial state) implicitly indicates the ε move for accepting a null string. This assumption is applicable to other types of automata too.

Figure 2.6 (c) is an NDFA because the machine may jump to q_f from q_0 without moving the read head; that is, $\delta(q_0, \varepsilon) = q_f$. More specifically, the machine goes to state q_f from q_0 without consuming any input (on the basis of guess). Hence, the transition diagram is mathematically non-deterministic or NDFA.

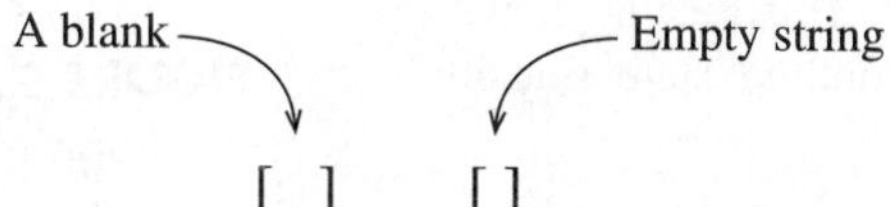

FIGURE 2.6 (d) Pictorial representation of blank and empty strings

The difference between a blank and an empty string is shown in Fig. 2.6 (d). Technically, an empty string does not contain anything (even a blank), whereas a blank represents a symbol.

(c) The given language (L) is Σ* over alphabet Σ. So, any string including ε is a valid string of L, and the equivalent DFA is shown in Fig. 2.6 (e). Here, the initial state is the final state as the null string is also valid.

This DFA equivalently generates the language $L = \{w \mid w \in \Sigma^* \text{ and } |w| = k, k \geq 0\}$.

(d) The given language (L) is Σ^+ over Σ. So, any string excluding ε is a valid string of L, and the equivalent DFA is shown in Fig. 2.6 (f). Here, the initial state is not the final state as the null string does not belong to L.

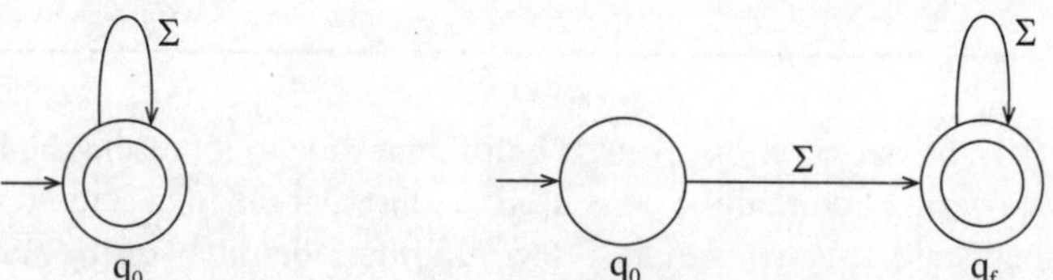

FIGURE 2.6 (e) DFA for Σ* **FIGURE 2.6 (f)** A DFA for Σ^+

Tip: First, draw the simple path(s) (the path(s) of the FA for the shortest possible string(s)) and then extend the paths where possible.

EXAMPLE 2.3

Draw the state transition graphs and transition tables for a DFA that recognises sets of strings over Σ = {0,1} such that (a) each string ends with substring 00 and (b) possesses an even number of 1s.

Solution (a) *Problem analysis*: The problem states that every accepted string ends with 00. This suggests that the length of the valid string is ≥ 2. In addition, any string ending with '1' or a single '0' is rejected.

Designing the procedure of the proposed automaton: Here, we are interested in two consecutive 0s. For this purpose, a minimum of three states (q_0, q_1, q_f) is essential to accept the smallest string: 00, and the path of acceptance can be described as $q_0 - 0 - q_1 - 0 - q_f$ (directed from q_0 to q_f), where q_0, q_1 and q_f are assumed, respectively, as the start state, non-final state and final state. This means that for any valid string, the proposed automaton (say, M) starts from q_0, travels the path $q_0 - 0 - q_1 - 0 - q_f$ and reaches the final state. Further, M must remain at q_f for input '0' as such input at q_f would imply that the string should end with 00.

In particular, whenever '1' is read by M in any one of the states q_0, q_1 and q_f, it will return to q_0, expecting to travel the path $q_0 - 0 - q_1 - 0 - q_f$ (to obtain a valid string), since it may still expect to obtain 00 from the rest of the string. Thus, there is no chance of the presence of a dead state in M. This idea is shown in Fig. 2.7 (a) and the corresponding state transition table is presented in Table 2.1.

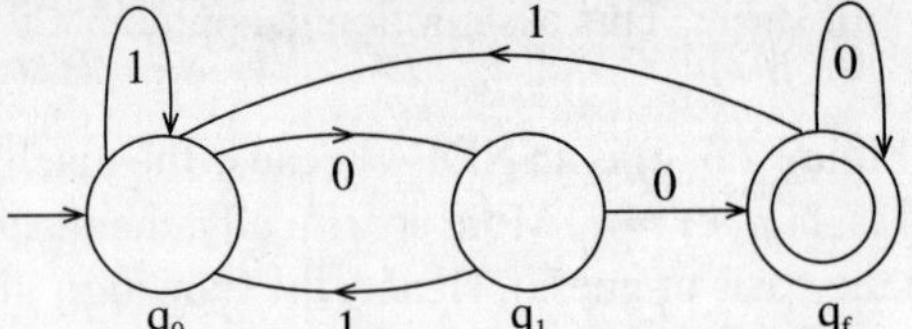

FIGURE 2.7 (a) A DFA (transition diagram) for Example 2.3 (a)

Note: A complete DFA diagram shows all the possible paths that can be processed by the machine. However, to process any string, the machine starts from the initial state. Further, repetition of states in a machine occurs due to the presence of repetitions of patterns, which is managed by the reuse of registers in real digital devices.

TABLE 2.1 Transition table for the DFA of Example 2.3

State	Input (0)/Next state	Input(1)/Next state
$\rightarrow q_0$	q_1	q_0
q_1	q_f	q_0
$*q_f$	q_f	q_0

Note: In Table 2.1, the operator '→' is used before the state to represent the initial state, whereas the symbol '*' appearing before the state is used to represent the final state. If a state q is the initial as well as the final state, it is denoted as $\rightarrow(*q)$. Further, identities of the states are shown in the first column, and the second column implies the state output for the given input from the state used in the first column. This practice is followed throughout the book.

Thus, the designed DFA is $M = (\{q_0, q_1, q_f\}, \{0, 1\}, \delta, q_0, \{q_f\})$, where δ is shown in the diagram as well as in the table. The transitions/rules are as follows:

$\delta(q_0,0) = q_1$
$\delta(q_0,1) = q_0$
$\delta(q_1,0) = q_f$
$\delta(q_1,1) = q_0$
$\delta(q_f,0) = q_f$
$\delta(q_f,1) = q_0$

Instantaneous description

Acceptance of string: Let $w = 0100$.

Initial configuration: $(q_0, 0100) \vdash (q_1, 100)$ by Transition rule 1

$\vdash (q_0, 00)$ by Transition rule 2
$\vdash (q_1, 0)$ by Transition rule 1
$\vdash (q_f)$ by Transition rule 3

Thus, $(q_0, 0100) \overset{*}{\vdash} (q_f)$, where $\overset{*}{\vdash}$ denotes zero or more move relations.

Note that instead of writing $(q_0, 0100)$, we can write it as $\delta(q_0, 0100)$, and the above relation may be expressed as $\delta^*(q_0, 0100) = q_f$

Rejection of string: Let $w = 01001$.

Initial configuration: $(q_0, 0100) \vdash (q_1, 100)$ by Transition rule 1
$\vdash (q_0, 00)$ by Transition rule 2
$\vdash (q_1, 0)$ by Transition rule 1
$\vdash (q_f, 1)$ by Transition rule 3
$\vdash (q_0)$ by Transition rule 6

$(q_0, 01001) \overset{*}{\vdash} (q_0)$, where $\overset{*}{\vdash}$ denotes zero or more move relations.

One may notice that q_0 is a non-final state, and so the string 01001 is rejected.

(b) *Problem analysis:* The problem states that the number of 0s in the valid string is not important. It may be zero, odd or even. However, the number of 1s must be even. This emphasizes that the proposed automaton (say, M) begins at the start state and reaches the final state, traversing an even number of 1s.

Design procedure: The valid strings consist of an even number of 1s but the number of 0s may be odd or even. Hence, strings with the following characteristics should be recognised by the proposed automaton:

(i) Strings with no 1s, like ε, 0, 00, … are valid, as each of these possesses an even number of 1s. To describe such strings, a single state (say, q_0), which may be considered as both the initial and the final state, is sufficient.

(ii) Strings consisting of both 1s and 0s, but having an even number of 1s are also valid. In order to accept such strings, two states are necessary: one is the initial state as well as final state (q_0) (already considered and added), and the other must be a non-final state (say, q_1).

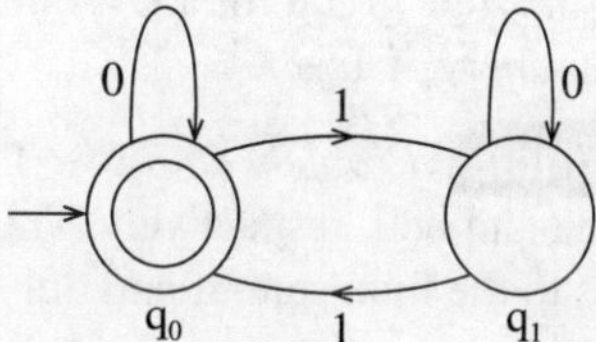

FIGURE 2.7 (b) A DFA for Example 2.3

In brief, to process a string, M passes from q_0 to q_1 or from q_1 to q_0 when '1' is read, but remains in its current state when '0' is read (as the number of 0s is not considered). The corresponding state transition graph and state table are shown in Fig. 2.7 (b) and Table 2.2, respectively. No dead state arises here.

Table 2.2 Transition table for the DFA of Example 2.3

State	Input (0)/Next state	Input(1)/Next state
$\rightarrow(*q_0)$	q_0	q_1
q_1	q_1	q_0

Thus, $M = (\{q_0, q_1\}, \{0, 1\}, \delta, q_0, \{q_0\})$, where δ is specified in the diagram as well as in the table.

FA AND THE INPUT END MARKER IN ITS TAPE

In FA, it is assumed that the end of the string is not marked explicitly, so a tentative decision is taken after each input symbol. Further, the size of the tape in FA is fixed. Hence, it may be assumed that the string is ended by the tape and therefore the head cannot proceed when it reaches the end of the tape. However, a right-end marker (such as '\$') may be used in the tape to indicate the end of input for real implementation.

ε-INPUT IS NOT USED IN FA TO END THE INPUT

As in FA, so it may be for DFA or NFA. Let us take DFA. If Λ (ε move) is used in DFA to end the processing of input, it conceptually assures the idea of NFA. The reason is that an implicit transition $\delta(q_i, \varepsilon) = q_i$ $(i = 1, 2,...)$ may be considered at each state in a DFA. Further, if an ε move such as $\delta(q_i, \varepsilon) = q_f$ is explicitly made to move from q_i to q_f, then two paths exist for the ε input at q_i. This causes non-determinism. As a result, a valid string may not be treated as valid. For example, let us add a new state, say $q'f$ (as the final state) and consider q_f as a non-final state, instead of the final state, as shown in Fig. 2.7 (a), such that $q_f - \varepsilon - q'f$. The problem is that there exists an implicit transition: $\delta(q_f, \varepsilon) = q_f$ but $\delta(q_f, \varepsilon) = q'f$ is now added. So, two ε moves are now present at q_f. However, the machine may choose any of them. If it chooses the correct one ($q_f - \varepsilon - q'f$), it shows the valid string as valid. On the other hand, if it chooses the other one, that is, $\delta(q_f, \varepsilon) = q_f$, then it reports a valid string as invalid.

That is why it is difficult to control the non-determinism aspect of DFA if ε move is used to end the processing of a string. However, it may be controlled by PDA, as the stack symbol is used in the transition of PDA.

EXAMPLE 2.4

Draw the state graph for a DFA that recognises a set of strings containing an even numbers of 0s and an odd numbers of 1s.

Solution *Problem analysis:* The problem states that the number of 0s and 1s in a valid string should be even and odd, respectively. This implies that the proposed automaton starts from its initial state and moves to the final state if and only if it traverses an even number of 0s but an odd number of 1s.

Designing procedure: For designing such an FA, inclusion of the necessary states are explained below step-wise.

Step 1: Let us first add the start state (say, q_0) to the proposed FA (say, M). This q_0 may not be the final state as the number of 1s must be odd in a valid string. If q_0 becomes the final state, then the null string is accepted. Hence, it is essential to add some new state(s) in the FA to process valid strings.

Step 2: If input '1' is read from q_0, it should reach its final state (say, q_1), as the number of 1s and 0s traversed will be odd and even, respectively. However, M moves back to q_0 from q_1 on receiving input '1', as such a string does not contain an odd number of 1s. Thus, two states in total have been added thus far, and the latest inclusion q_1 is necessary for the final state.

Step 3: Let us assume that M is at q_0 and the input is '0'. So the next state may not be q_0 or q_1. If it is so, the specified conditions for a valid string are violated. Hence, another distinct non-final state (say, q_2) must be added to move q_2 from q_0 for input '0'. Thus, the total number of states added so far is three for the proposed automaton M.

Step 4: Now, if '1' is read at q_2, M should not return to q_0 or q_1 to maintain the acceptance property of the string. Hence, one more non-final state (say, q_3) must be added to the automaton.

Therefore, a minimum of four states are necessary to construct an automaton to satisfy the given properties. The state diagram of the proposed automaton to accept the valid strings is shown in Fig. 2.8.

Thus, M = ({q_0,q_1, q_2, q_3}, {0, 1}, δ, q_0, {q_1}), where δ is pictorially specified in the diagram.

Also, in the proposed automaton, q_0 is the start state, q_1 is the final state; q_2 and q_3 are non-final states. Conceptually, no dead state may occur in this problem as the required constraints for the string can be satisfied from the current string.

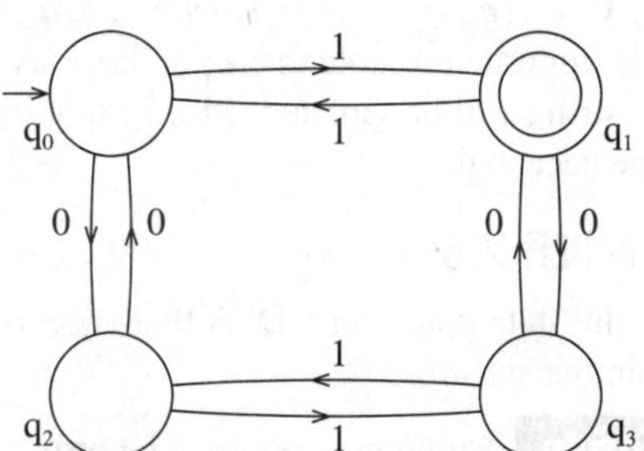

FIGURE 2.8 State diagram for accepting strings with an even number of 0s and odd number of 1s

Note: The graph can be redesigned to solve problems such as even number of 0s and 1s, even number of 1s and odd number of 0s, etc. Let us take the case of even number of 0s and 1s. Here, the initial state is obviously the final state as well because a string with size zero has an even (0) number of 0s and an even (0) number of 1s.

USE OF NEGATION (OR COMPLEMENT) LOGIC FOR DRAWING THE DFA

One may often perform negation logic (¬) to draw a DFA from an existing one. Let us take an example of negation of logical AND: ¬ (even (0) AND even (1)). It is equivalent to odd (0) OR odd (1) by De Morgan's laws as negation of even is odd. Now, if one designs a DFA for (even (0) AND even (1)), the DFA for negation of (even (0) AND even (1)) can easily be drawn by taking the complement. This is discussed in detail later in this chapter.

EXAMPLE 2.5

Draw a DFA to accept strings beginning and ending with the same symbol over $\Sigma = \{a,b\}$.

Solution Null strings are not valid here as such strings would not begin and end with the same symbol over $\Sigma = \{a, b\}$. In fact, the valid strings are single (*a*), single (*b*), *a* and then zero or more *a*s followed by one or more *b*s followed by *a*s; *b* and then zero or more *b*s followed one or more *a*s followed by *b*s. Acceptance of all such strings is shown in Fig. 2.9.

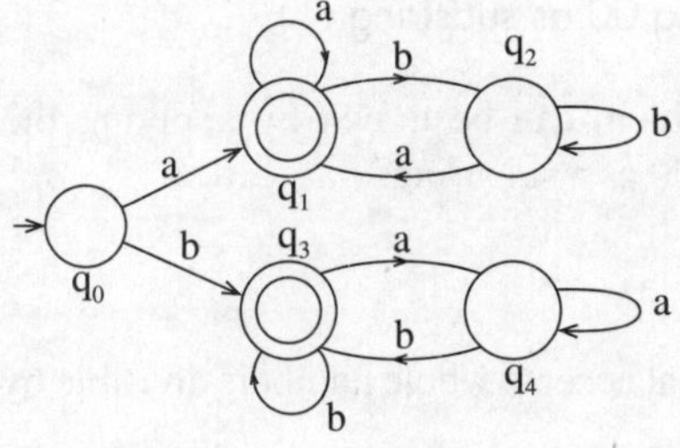

FIGURE 2.9 State diagram for strings beginning and ending with the same symbol

Thus, M = $(\{q_0, q_1, q_2, q_3, q_4\}, \{a, b\}, \delta, q_0, \{q_1, q_3\})$, where δ is implicitly shown in the diagram.

It is important to note that q_1 and q_3 may not be merged. If they are merged, the acceptance criterion of the string will be violated. More explicitly, strings beginning and ending with different symbols will also be accepted.

EXAMPLE 2.6

Draw the state graph for a DFA that accepts all strings over $\Sigma = \{0,1\}$ which end in 1 but which do not contain the substring 00.

Solution *Problem analysis:* The problem states that the accepted strings must satisfy the following constraints: (*i*) strings should end in a '1' and (*ii*) strings should not contain 00 as substring. These two constraints ensure that the proposed automaton (say, M) starts in the initial state to process valid strings and reaches the final state if '1' is received at the end, but no two consecutive 0s are found over the path of traversal.

Design procedure: Let us first add the start state (say, q_0) to the proposed automaton. As the accepted string ends in a '1', the start state q_0 may not be the final state. If so, the null string is also accepted; that is, the acceptance property of the string is violated in such a case (Fig. 2.9 (b)).

Hence, some more states are essential for processing the valid strings. If M reads '1' at q_0, it should reach the final state. Thus, the most recently added state is to be considered as the final state (q_f). Certainly, M remains at q_f if '1' is received here. Further, for input '0' from states q_0 and q_f, respectively, M must not move to q_0. If it does move to q_0, the invalid string consisting of substring 00 may also be accepted. So, one more non-final state (say, q_1) must be added to M to resolve this situation.

Now, if '1' is encountered at q_1, it reaches the final state because it satisfies the validity condition of the string. Thus, a total of three states (excluding the dead state) is required for the proposed automaton.

The complete state diagram is shown in Fig. 2.9 (a).

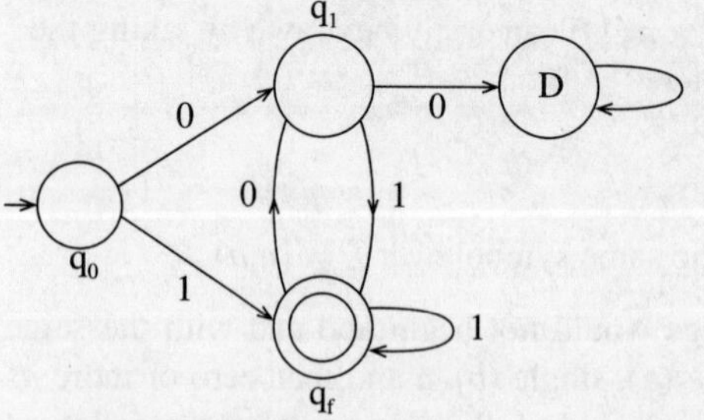

FIGURE 2.9 (a) A DFA for accepting strings ending with 1 but not containing 00 as substring

FIGURE 2.9 (b) An incorrect DFA for the given problem

> **Note:** The DFA for this problem can be drawn by applying the concept of negation for the constraint: strings containing 00 as a substring.

EXAMPLE 2.7

Draw the state graph for a DFA that accepts whole numbers divisible by 5 over $\Sigma = \{0,1,2,3,4,5,6,7,8,9\}$.

Solution *Problem analysis:* Any whole number can be viewed as a string of one or more digits taken from Σ. Additionally, a number that ends with '0' or '5' is divisible by 5. This suggests that the proposed automaton starts from its initial state and reaches its final state if it traverses a path ending with either '0' or '5'.

Design procedure: In order to design an automaton (say, M) to solve the given problem, first take a state q_0 as the start state. The null string does not represent any number; so q_0 is not considered the final state. Hence, the proposed automaton needs at least one more state at this stage.

In order to process any valid string (number), M starts from q_0 and stays there for different digits (inputs), except for '0' and '5'. On receiving '0' or '5' at q_0, it will reach the final state, say q_f as M needs to traverse a path ending with '0' or '5' at q_0 to accept the string. Hence, out of these two states, one is the start state and the other must be the final state. Obviously, M remains at q_f for both inputs '0' and '5' but moves back to the non-final state (q_0) for the remaining inputs. The complete state graph is shown in Fig. 2.10 (a) .

1,2,3,4,6,7,8,9
0,5
1,2,3,4,6,7,8,9
0,5
q_0
q_f

FIGURE 2.10 (a) A DFA for accepting numbers divisible by 5

Thus, M = $(\{q_0, q_f\}, \{0, 1, 2, 3, 4, 5, 6, 7, 8, 9\}, \delta, q_0, \{q_f\})$, where δ is given in the diagram.

However, if we consider natural numbers (strictly positive integers), then two diagrams are possible: one that considers the leading 0s and the other that ignores both leading 0s and the number 0 itself (Fig. 2.10 (b)).

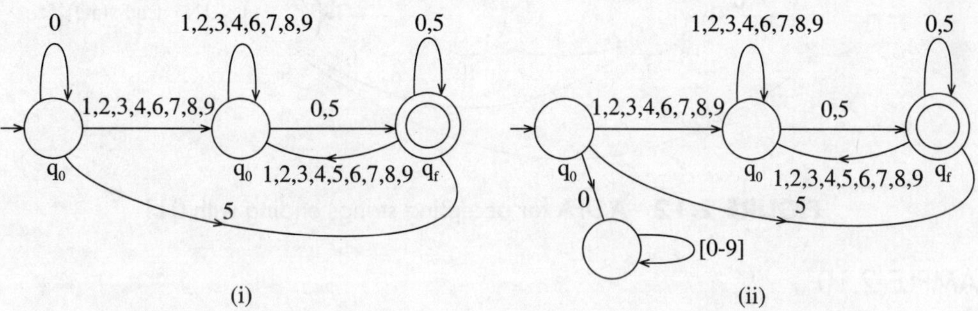

FIGURE 2.10 (b) A DFA for accepting positive numbers divisible by 5

EXAMPLE 2.8

Design a DFA that accepts strings over $\Sigma = \{0,1\}$, containing 011 as a substring.

Solution *Problem analysis:* The condition to be satisfied here for any valid string is the presence of at least one substring: 011. Undoubtedly, 011 is the smallest-length valid string. To satisfy this condition, the minimum number of states in the proposed automaton (say, M) should be four, as the length of such a substring is 3.

Design procedure: To accept a string, M starts from its initial state (say, q_0), and traverses a path described by substring: 011 to reach the final state (say, q_f). Further, for any input (either '0' or '1') received at q_f, M moves back to q_f as the prefix of the present string already consists of 011 as substring. However, if M receives '0' at q_1 and q_2, respectively, it moves back to q_1, expecting to obtain a valid string. On the other hand, M stays at q_0 for '1'. The state graph of the automaton is shown in Fig. 2.11.

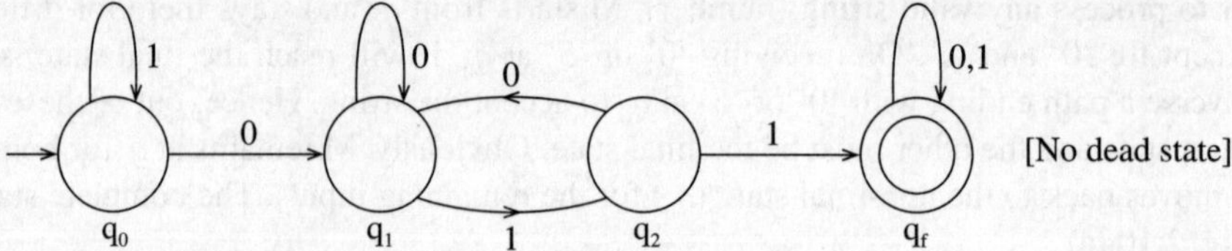

FIGURE 2.11 A DFA (transition diagram) for accepting strings with 011 as substring

Tip: First draw the simple path and then extend it.

EXAMPLE 2.9

Design a DFA that recognises strings over $\Sigma = \{0, 1\}$, ending with 011.

Solution *Problem analysis and design procedure:* The condition to be satisfied here is to accept strings ending with substring 011. First, draw the simple paths for the simple strings and then extend them. The corresponding automaton can be redesigned from Fig. 2.11, as shown in Fig. 2.12.

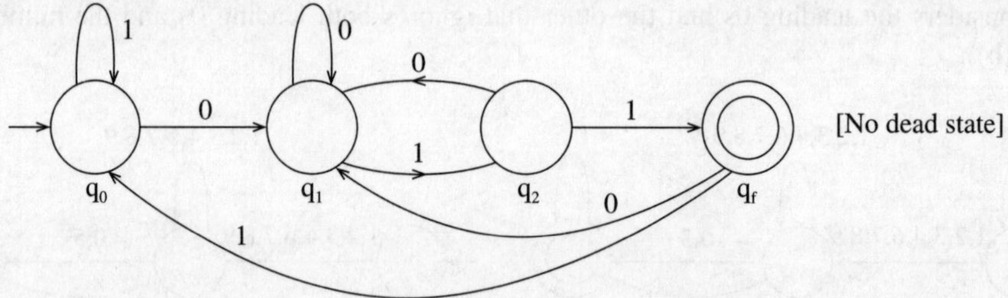

FIGURE 2.12 A DFA for accepting strings ending with 011

EXAMPLE 2.10

Design a DFA that accepts strings over $\Sigma = \{a, b\}$, not containing three consecutive *b*s (as a substring).

Solution *Problem analysis:* The following hints can be drawn from the problem:

- String with no '*b*' is accepted.
- String with single '*b*' at any position is accepted.
- String with two consecutive *b*s is accepted.
- String with three consecutive *b*s is not accepted and correspondingly the automaton moves to the dead state (D) for such a string.

Design procedure: From these observations, it can be concluded that the proposed automaton (say, M) needs three states (excluding the dead state), and all these states must be the final state. The initial state is the start state as well as the final state. The state graph corresponding to the machine is given in Fig. 2.13.

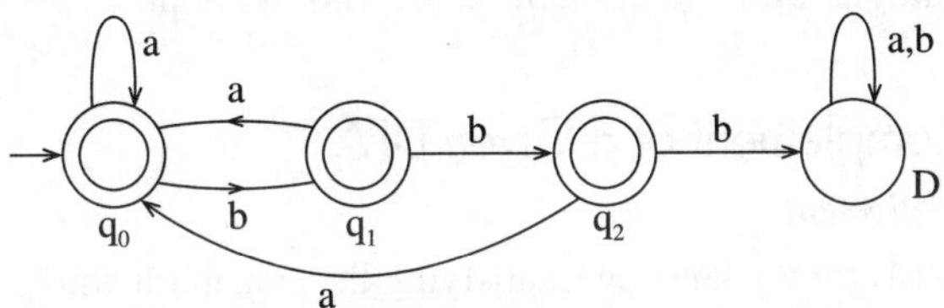

FIGURE 2.13 A DFA for accepting strings not containing three consecutive *b*s

It is interesting to note that the DFA can be designed by applying the idea of negation as well.

Note: In this automaton, it is shown that for input '*a*' at q_1 and q_2, the automaton moves to q_0 but not to q_1 and q_2, respectively, as it is possible to move to the dead state after three consecutive *b*s. Now, if it moves to q_1 and q_2, respectively, the designed DFA may direct strings without three consecutive *b*s to move to the dead state (which should not be allowed). Such a faulty machine would move to the dead state for a string like $w = ababb$.

EXAMPLE 2.11

Design a DFA to accept a string $\{w \in \Sigma^* \mid w \text{ has } n \text{ occurrences of 0s, where } n \bmod 5 \text{ is } 3\}$.

Solution Here, we concentrate on the number of 0s but there is no restriction on the number of other symbols in Σ. According to the given property, strings with a number of 0s, such as, 3, 8, 13, 18, 23, and so on, are valid. A DFA based on the given property is shown in Fig. 2.14.

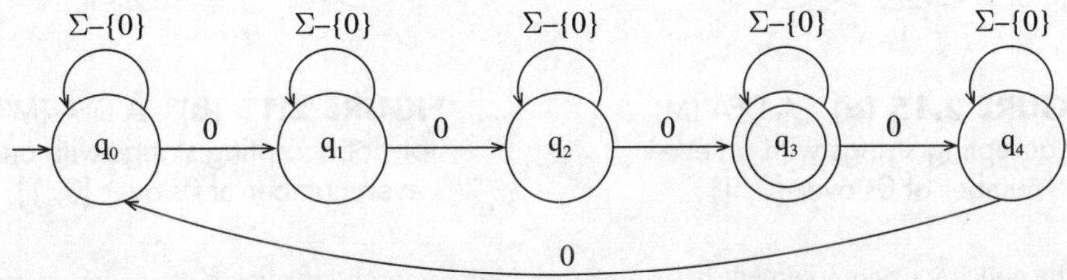

FIGURE 2.14 A DFA for accepting strings with occurrences of 0s, *n* mod 5 = 3

2.7 BASIC OPERATIONS ON DFA

We can build new DFAs from existing ones, performing operations such as complement, union, intersection, subtraction and concatenation. DFAs are closed under these operations. In other words, some new regular languages are formed from existing regular languages recognised by DFAs. To explain the idea behind 'closed under operation', the term *closed* with respect to set is first defined.

A set is closed under an operation if application of that operation on any two elements of the set produces another element of the same set. The concept of closure operation with respect to DFA means that new DFAs are generated from existing DFAs.

Note: The DFA is equivalent to NDFA. So, closure operations may also be applied on NDFAs, though it may be difficult. Therefore, to make the arguments as simple as possible, we assume DFAs.

2.7.1 Finding the Complement of a Given DFA

Procedure to find the complement

Step 1: Construct a DFA (M) for the language, satisfying the constraint and ignoring the given negation criteria.

Step 2: Perform complement of the designed DFA (M) to convert to M′ in order to satisfy the desired constraint.

Key points of the procedure: To find the complement of a DFA:

- Only the states are changed; the non-final states become the final states and vice versa. The dead state (if any) becomes the final state in the complemented DFA. The initial state remains the same for both the FA.
- Edges are unchanged. Thus, Σ remains unchanged.
- Mathematically, let M = (Q, Σ, δ, q_0, F) be a DFA. Then, its complement $\overline{M}$ = (Q, Σ, δ, q_0, Q-F).

For example, DFA (M) accepts an even number of 0s over {0, 1}. The DFA corresponding to this is shown in Fig. 2.15 (a). Based on the procedure discussed above, the complement of this is shown in Fig. 2.15 (b), which validates strings not accepting an even number of 0s.

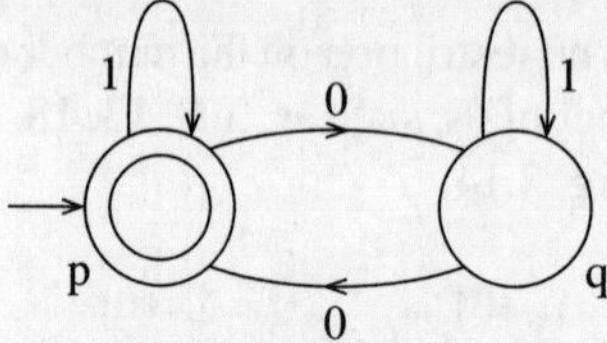

FIGURE 2.15 (a) A DFA (M) for accepting strings with an even number of 0s over {0, 1}

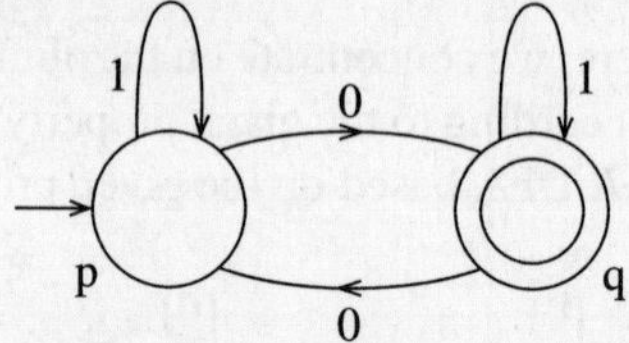

FIGURE 2.15 (b) A DFA (M′) for not accepting strings with an even number of 0s over {0, 1}

Note: The concept of complement helps to solve complex problems using the negation property. For instance, suppose we need to design a DFA that recognises strings such that each valid string does not contain three consecutive *b*s (as a substring) over {a, b}. To solve it, first draw a DFA that recognises strings with three consecutive *b*s as substring. The diagram is shown in Fig. 2.15 (c).

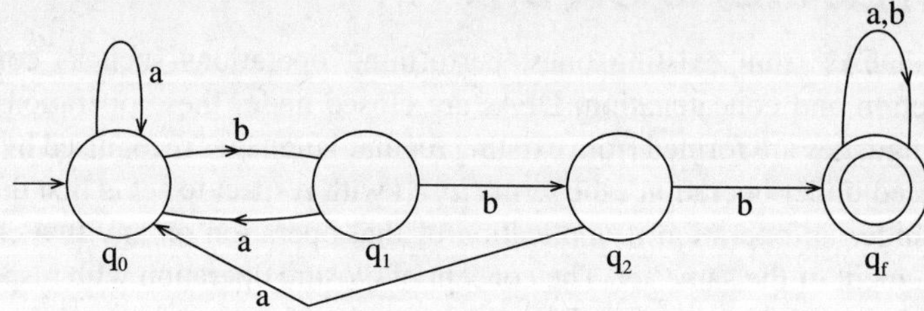

FIGURE 2.15 (c) A DFA (M) accepting three consecutive *b*s over {a, b}

According to the given property, as soon as the machine receives three consecutive *b*s, it moves to the final state and remains there for inputs *a* and *b*. Now, based on the concept of complement, the complemented machine appears as shown in Fig. 2.15 (d).

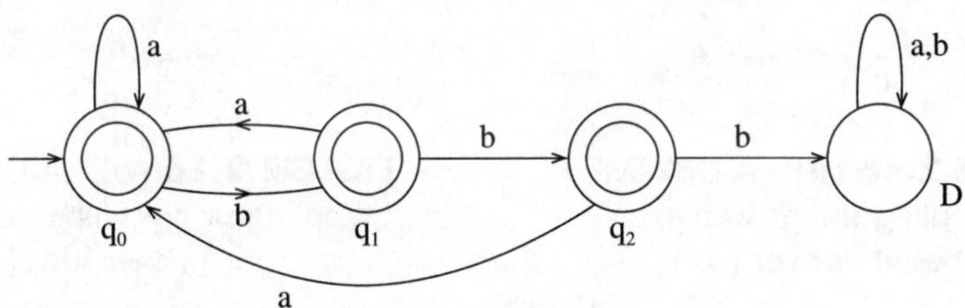

FIGURE 2.15 (d) A DFA (M′) for not accepting three consecutive *b*s over {a, b}

One may note that the final state (q_f) of Fig. 2.15 (c) is treated as the dead state (D) for the complement of M, as there is no path to move to some other state from this state on reading any input.

2.7.2 Product Automata

To obtain a product automaton, we need at least two automata. A product automaton of two automata $M_1 = (Q, \Sigma, \delta, q_0, F)$ and $M_2 = (Q', \Sigma, \delta', q'_0, F')$ is an automaton $M = (Q_M, \Sigma, \delta_M, (q_0, q'_0), F_M)$, where $Q_M = Q \times Q'$, δ_M is the transition function expressed as $\delta_M : Q_M \times \Sigma \rightarrow Q_M$, assuming that both are working on the same alphabet Σ.

More specifically, $\delta_M((q, q'), c) = (\delta(q, c), \delta(q', c))$, where $c \in \Sigma$ and $(q, q') \in Q_M$ such that $q \in Q$ and $q' \in Q'$. Further, $F_M \subseteq Q_M$ and F_M varies from operation to operation as follows:

- F_M for $L(M_1) \cap L(M_2)$ is $(F \times F')$; for the intersection of two DFAs
- F_M for $L(M_1) \cup L(M_2)$ is $(Q \times F') \cup (F \times Q')$; for the union of two DFAs

In particular, product automata deals with the Cartesian product of sets of states.

Intersection of two DFAs: The intersection of two FAs results in product automaton.

Procedure

Input: $M = (Q_M, \Sigma_1, \delta_M, q_M, F_M)$ and $L = (Q_L, \Sigma_2, \delta_L, q_L, F_L)$ are two DFAs.

Output: R (resulting automaton) = $M \cap L$

Method: The following steps are necessary to construct $R = M \cap L$.

Step 1: Find $\Sigma = \Sigma_1 \cup \Sigma_2$, where Σ is the new set of alphabets of R.

Step 2: Compute $Q_M \times Q_L$ that represents the set performed by the Cartesian product on Q_M and Q_L, such that $(q_M, q_L) \in Q_M \times Q_L$ is the new start state of R.

Step 3: Construct $F_M \times F_L$, the set of final states performed by the Cartesian product on F_M and F_L.

Step 4: The new transition function δ for R is defined as follows:

$$\delta((p, q), a) = (\delta_M(p, a), \delta_L(q, a)), \text{ where } p \in Q_M \text{ and } q \in Q_L$$

One may note that some of the states obtained from $Q_M \times Q_L$ may not be reachable from the start state and may therefore be removed from the resulting machine.

The above procedure is explained below in Figs 2.16 (a) and 2.16 (b).

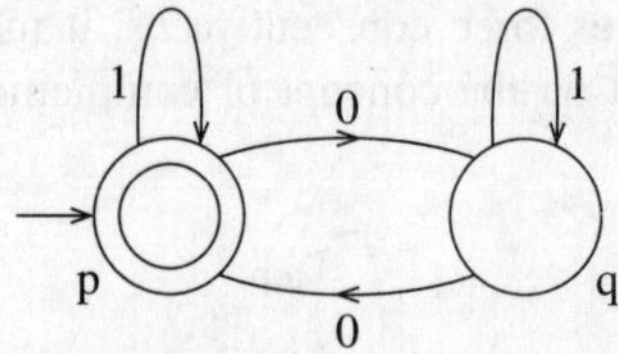

FIGURE 2.16 (a) A DFA (M) for accepting strings with even number of 0s over {0, 1}

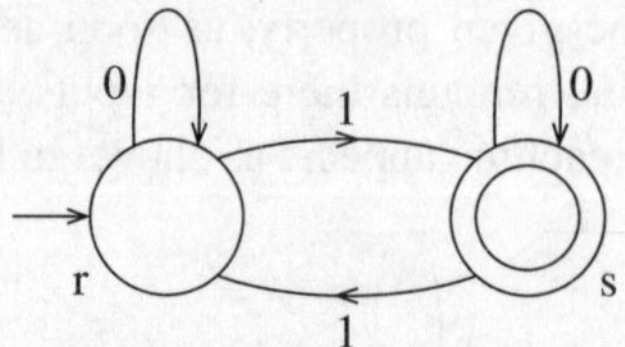

FIGURE 2.16 (b) A DFA (L) for accepting strings with odd number of 1s over {0, 1}

Here, $Q_M = \{p, q\}$, start state: $q_M = p$, set of final states: $F_M = \{p\}$.
$Q_L = \{r, s\}$, start state: $q_L = r$, set of final states: $F_L = \{s\}$.

Finding the resultant automaton (R = M ∩ L): Now, $Q_M \times Q_L = \{(p, r), (p, s), (q, r), (q, s)\}$ and $F_M \times F_L = \{(p, s)\}$.

The start state of R is $(q_M, q_L) = (p, r)$, whereas (p, s) is the only final state. Next, let us apply the transition function $(\delta\ (p, q), a) = (\delta_M\ (p, a),\ \delta_L\ (q, a))$ for each input (a) of $\{0, 1\}$ to obtain the following:

(i) $\delta((p, r),1) = (\delta_M(p,1),\ \delta_L(r,1)) = (p, s)$
(ii) $\delta((p, r),0) = (\delta_M(p,0),\ \delta_L(r,0)) = (q, r)$
(iii) $\delta((p, s),1) = (\delta_M(p,1),\ \delta_L(s,1)) = (p, r)$
(iv) $\delta((p, s),0) = (\delta_M(p,0),\ \delta_L(s,0)) = (q, s)$
(v) $\delta((q, r),1) = (\delta_M(q,1),\ \delta_L(r,1)) = (q, s)$
(vi) $\delta((q, r),0) = (\delta_M(q,0),\ \delta_L(r,0)) = (p, r)$
(vii) $\delta((q, s),1) = (\delta_M(q,1),\ \delta_L(s,1)) = (q, r)$
(viii) $\delta((q, s),0) = (\delta_M(q,0),\ \delta_L(s,0)) = (p, s)$

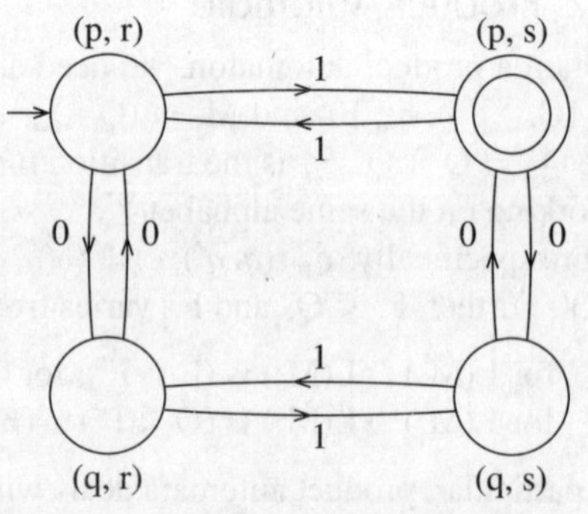

FIGURE 2.16 (c) A DFA (R) for accepting strings with odd number of 1s AND even number of 0s

Thus, the resultant DFA combining both the above DFAs is shown in Fig. 2.16 (c). It is, indeed, the intersection of two DFAs: M and L.

Interestingly, R = M ∩ L = L ∩ M.

STRATEGY FOR CONSTRUCTING A SET OF FINAL STATES (F) FOR THE PRODUCT MACHINE

- *Intersection operation*: For $L_1 \cap L_2$, $F = \{(p, q) \mid p \in F_1 \text{ AND } q \in F_2\}$, where F_1 is the set of final states for L_1 and F_2 is the set of final states for L_2. The new machine M accepts $L_1 \cap L_2$.
- *Union operation*: For $L_1 \cup L_2$, $F = \{(p, q) \mid p \in F_1 \text{ OR } q \in F_2\}$, where F_1 is the set of final states for L_1 and F_2 is the set of final states for L_2. The new machine M accepts $L_1 \cup L_2$.
- *Difference operation*: For $L_1 - L_2$, $F = \{(p, q) \mid p \in F_1 \text{ AND } q \notin F_2\}$, where F_1 is the set of final states for L_1 and F_2 is the set of final states for L_2. The new machine M accepts $L_1 - L_2$.

Note: The strategy discussed above simplifies the design of many complex DFAs, satisfying multiple conditions joined by operations, such as AND, OR, minus, etc. For example, the design of DFA satisfying conditions, such as number of *as* divisible by *m* AND number of *bs* divisible by *n*, joined by OR operation, and so on, can be performed easily. However, the number of final states becomes more for OR (union) than for AND operations. For instance, each state except (q, r) will be the final state of the DFA for the above two DFAs (as shown in Figs 2.16 (a) and (b)), satisfying the condition of even number of 0s OR odd number of 1s.

Tip: Let $\Sigma = \{a, b\}$. Assume that an FA accepts valid strings over Σ, each of which begins with '*a*' (rule 1), whereas each valid string for another FA ends with '*b*' (rule 2). Now, if we need an intersection of these two rules, each resultant string must satisfy both the rules – each must begin with '*a*' but end with '*b*'. We may handle other cases in a similar manner.

Note: More operations (properties) on DFAs (equivalently over regular languages) are discussed in Chapter 3. It should be noted that the design of DFAs can be simplified using regular expressions (also discussed in Chapter 3).

2.8 BASIC TIPS FOR DESIGNING FINITE AUTOMATA

On the basis of the DFAs we have discussed thus far used for solving problems, we may highlight some basic tips to design FAs (especially DFAs):

1. First, understand the given problem clearly; that is, grasp the necessary characteristics to be designed in the automaton.
2. Next, draw a simple diagram considering the base cases of the given properties, and then extend the diagram for each input symbol of Σ from every state to consider all possibilities. Try to visualise the problem in terms of mathematical induction to generate larger valid strings – such a line of thought may come in handy for solving many problems.
3. In the case of negation (not containing a symbol or substring), one may first draw the DFA ignoring the negation. The complement of the drawn figure will be the resulting automaton.

Some tricky and interesting FAs are discussed below:

(i) Let us attempt to design a DFA that accepts strings with '*ab*' or '*bba*' as substring. In this example, the two specified substrings are not subsets of each other. So, while designing the state diagram for a simple case of this problem, two separate paths need to be considered: one for '*ab*' and another for '*bba*'. However, later on, both paths may meet in a certain state.

(ii) Let us take another example to design a DFA that recognises strings with '*ab*' or '*aba*' as substring. In this example, the larger substring '*aba*' includes the smaller substring '*ab*'. So, draw a DFA for the smaller substring, and then extend it after proper analysis. You may also prefer to visualise the problem as follows: design a DFA for the larger string and then introduce the appropriate final state(s) to handle the smaller substring.

(iii) Making a slight modification to (ii), if the problem states that every string accepted by a DFA can be viewed as a combination of any number of substrings '*ab*' or '*aba*' (assuming zero combination also), then it is necessary to design the DFA focussing on both the substrings (Fig. 2.17 (a)). However, if we design an automaton (Fig. 2.17 (b)) based on only one substring, then some valid strings (for example, '*abab*') are ignored.

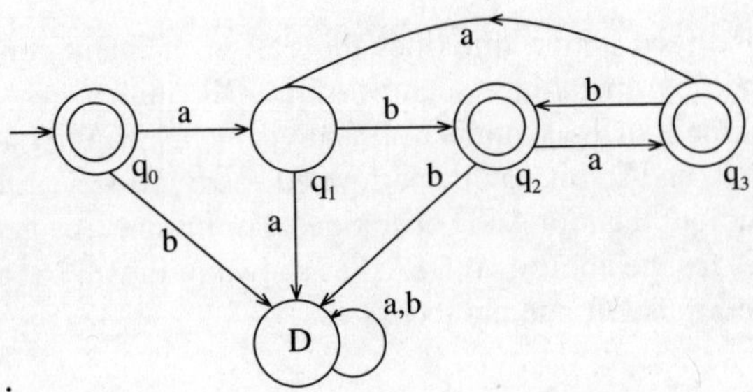

FIGURE 2.17 (a) A correct DFA for accepting strings with a combination of substrings 'ab' and 'aba'

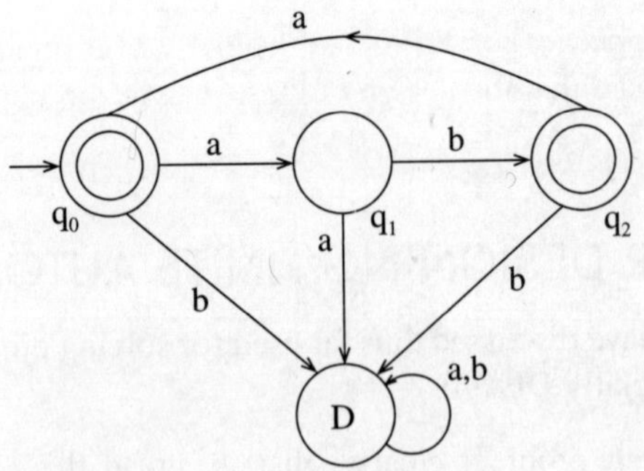

FIGURE 2.17 (b) An incorrect DFA for accepting strings with a combination of substrings 'ab' and 'aba'

Note that if the machine moves to state q_0 from q_2 on reading '*a*', it moves to the dead state for input '*b*' from q_0 and the string: *abab* is invalidated. However, strings like *abab* are valid. So, Fig. 2.17 (b) is incorrect.

(iv) For a DFA to recognise strings over Σ such that a particular symbol of Σ appears at least in '*r*' places, out of the last *m* places ($m > r$), the number of possible arrangements of places for that particular symbol in the accepted strings may be specified as $(n-1)!({}^mC_r + {}^mC_{r+1} + \ldots + {}^mC_m)$, where *n* is the total number of distinct symbols in Σ, ${}^mC_r$ results in the number of possible strings with particular symbols in *r* places out of the last *m* places (it is the problem of choosing *r* items out of *n* items, where all the items are treated as identical); ${}^mC_{r+1}$ is the number of possible strings with particular symbols in $(r+1)$ places, out of *m* places; and so on. Finally, ${}^mC_m$ represents the number of possible strings with particular symbols in *m* places, out of *m* places. Further, we need to arrange the rest of the symbols $(n-1)$, except the particular symbol, among themselves. This can done in $(n-1)!$ ways.

For better understanding, here is an example: design a DFA to recognise strings over Σ = {0, 1}, which contains at least two zeros in its last four places. Clearly, in this example, $n = 2$, $r = 2$, $m = 4$. The possible combinations are $(2-1)!\,({}^4C_2 + {}^4C_3 + {}^4C_4)$. Note that designing the DFA will be easier if we consider the tree structure in this regard. In fact, the possible ending strings are 0011, 0101, 0110, 1100, 1001 and 1010, each with exactly two zeros [${}^4C_2 = 6$ strings]. Each of the strings 0001, 0010, 0100 and 1000 has exactly three zeros [${}^4C_3 = 4$ strings]. The string 0000 has exactly four zeros [${}^4C_4 = 1$ string]. Similarly, we may think about the case of *at most* logic.

2.9 NDFA AND ITS CATEGORIES

We have seen that the essential restriction we make when defining deterministic definite automata is that there should be exactly one possible transition for each input symbol in each state. We drop this restriction to obtain a larger class of automata, called non-deterministic finite automata (NDFA). An NDFA may have several options about where to move next (during computation), or it may have no option at all and be stuck. Therefore, designing NDFAs for problems is much easier than designing DFAs. However, the idea of determinism of DFA is important when implementing a program (or physical machine) to recognise a regular language.

As we have mentioned earlier, there are two types of NDFA: NDFA with Λ move (or ε move) and NDFA without Λ move. These types are now discussed in detail. It should be noted that the notations Λ move and ε move can be used interchangeably.

2.9.1 NDFA Without ε Move

In this type of NDFA, the transition function δ is defined as δ: Q × Σ → 2^Q, where Q denotes the set of states and Σ is the alphabet. There must be at least one path in NDFA from the start state to the final state to accept valid strings of the given language for which the NDFA is drawn.

Let us now illustrate the construction of an NDFA taking a language that accepts strings ending in 01 over {0, 1}. One possible NDFA for solving this problem is given in Fig. 2.18.

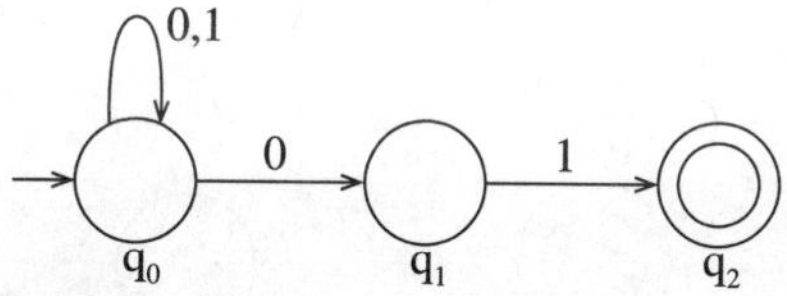

0,1
0
1
q_0
q_1
q_2

FIGURE 2.18 An NDFA to accept strings ending in 01

We can see that there are two paths for the machine (M) when input 0 is read at state q_0: it may move to state q_0 or it may move to q_1. So, it does both.

Let us take a string (w) = 001. The string can be parsed by the machine as

Path 1: $q_0 - 0 - q_0 - 0 - q_1 - 1 - q_2$ (accepted)

or

Path 2: $q_0 - 0 - q_1 - 0$ .... (not accepted)

So, the concept of probability occurs here. Anyway, it is assumed that a non-deterministic machine always choses the correct path (based on guess) at each step, and hence, the first one is taken into account to validate the string. However, we cannot ensure it.

Note: If a self-loop labelled 0 and 1 is taken at q_1 (instead of q_0), the diagram shows that strings like 011 (strings not ending in 01) may be validated. However, it is not valid according to the given criterion, as the path will parse only those strings that are recognised by the machine.

Problems with NDFAs: As shown in Fig. 2.18, to validate string w = 001, one of the two paths may be selected. Accordingly, the valid string may not be accepted if the right path is not chosen, as there is no guarantee that the correct path will be selected. However, if the machine attempts to parse the string further by adopting a backtracking approach (if not accepted in earlier attempts), it becomes a time-consuming job. Hence, any NDFA is problematic.

EXAMPLE 2.12

(a) Draw an NDFA for the language L over {a, b} such that each valid string begins with '*a*'.

Solution To better understand the difference between DFA and NDFA, let us first draw a DFA for the given problem (Fig. 2.19 (a)). Here, the dead state (D) is also shown.

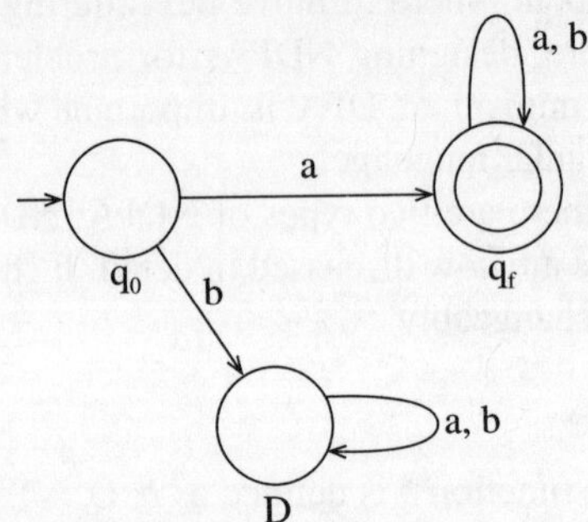

FIGURE 2.19 (a) A DFA to accept strings beginning with 'a'

Now, if the dead state is not shown explicitly, it may be assumed as an NDFA for the given language (Fig. 2.19 (b)).

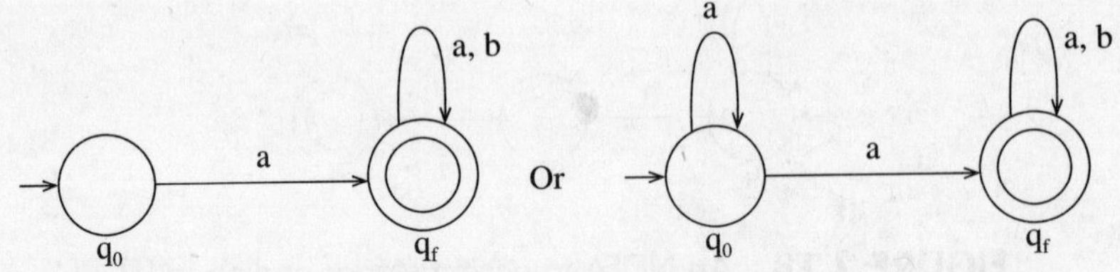

FIGURE 2.19 (b) An NDFA to accept strings beginning with 'a'

We may note that the state diagram, as shown in Fig. 2.19 (b), looks like a DFA. However, it is not a DFA because the transition from state q_0 for input '*b*' is not shown explicitly. In fact, for input '*b*' at q_0, the machine moves to the dead state with respect to the DFA. For DFA, it is necessary to show all the transitions explicitly, whereas this is not required for NDFA. However, we often ignore the move to dead state for DFA too. Therefore, such a criterion does not strongly differentiate between DFA and NDFA. Thus, the main criterion should be the presence of multiple transitions or the same input from the same state.

> **Note:** If a self-loop labelled '*a*' and '*b*' is taken at q_0 (instead of q_f), Fig. 2.19 (b) will show that strings beginning with '*b*' are also allowed, but this is not correct according to the given criterion. Further, if input '*b*' is ignored at q_f, then information about '*b*' in the automata will not be sufficient; that is, it may be assumed that the alphabet contains only {a}. Again, if '*a*' is not considered at q_f, some valid strings beginning with '*a*' may not be recognised by the converted DFA. So, while designing the NDFA, provide the minimum information (inputs) to describe the language recognised by the NDFA.

(b) Draw an NDFA for the language L over {a, b} such that each valid string ends with '*a*'.

Solution As the string ends with '*a*', one possible NDFA for the given language is shown in Fig. 2.20.

Note that the machine moves to two different states (q_0 and q_f) from state q_0 for input '*a*'. Moreover, it is assumed that the machine moves to the final state for the last '*a*' in the string. For other *a*s (not the last one) and *b*s, it moves to q_0. So the last one is determined correctly by the machine in a non-deterministic manner.

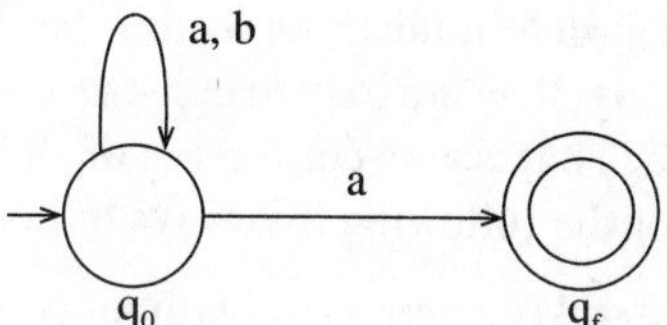

FIGURE 2.20 An NDFA to accept strings ending with 'a'

(c) Draw an NDFA for the language L over {a, b} such that each valid string begins with '*a*' and ends with '*b*'.

Solution As the string begins with '*a*' and ends with '*b*', one possible NDFA for the given language is shown in Fig. 2.21.

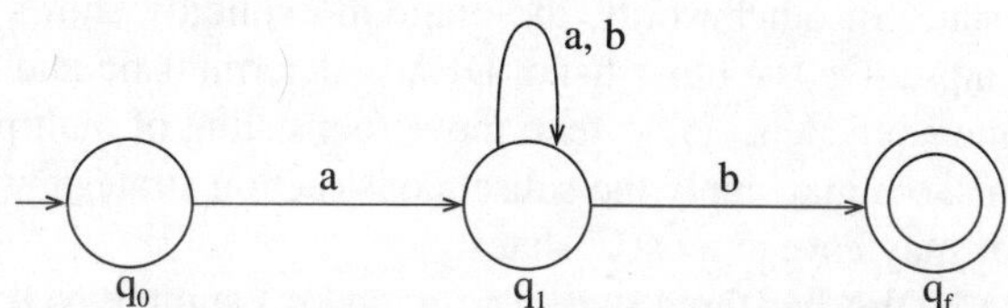

FIGURE 2.21 An NDFA to accept strings beginning with 'a' and ending with 'b'

Clearly, all the transitions at q_0 and q_f are not shown explicitly. Further, two moves are shown for '*b*' at q_1.

Note: If the self-loop labelled '*a*' and '*b*' is taken at q_0 (instead of q_1), the diagram shows that strings beginning with '*b*' are also allowed.

(d) Draw the state graph for an NDFA that accepts strings over $\Sigma = \{a, b\}$ if each string can be viewed as a concatenation (or combination) of any number of substrings: *ab* or *aba*.

Solution Any number of substrings implies zero or more substrings. Hence, the null (empty) string will also be accepted in the automaton. One possible NDFA is shown in Fig. 2.22.

The state graph shows that if the machine receives '*b*' at q_1 (a non-final state), it may move to q_0 (which is the start as well as the final state) or q_2 (non-final state) to satisfy the characteristics of the given problem. Thus, $M = (\{q_0, q_1, q_2\}, \{a, b\}, \delta, q_0, \{q_0\})$, where δ is as is shown in Fig. 2.22.

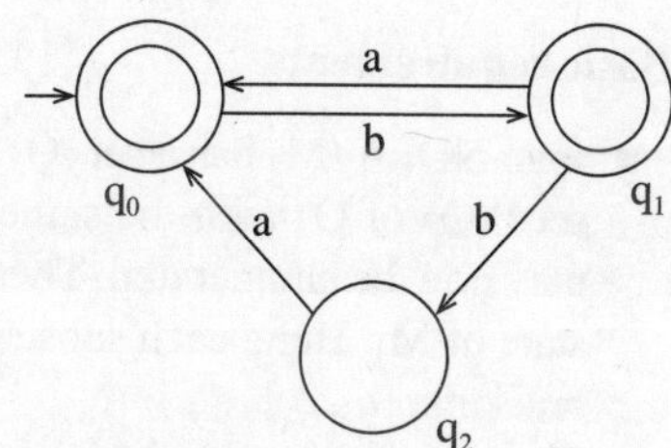

FIGURE 2.22 An NDFA to accept string combinations of *ab* or *aba*

Note: Designing NDFA for certain problems may be complex. However, designing NDFA based on regular expressions may make the task easier (discussed in Chapter 3).

KEY POINTS

- In DFA, there exists *at most* one path from the start state to the final state to accept an input.
- In NDFA, there exists *at least* one path from the start state to the final state to accept an input.

Subset Construction: A Conversion Strategy from NDFA (Without ε Move) to DFA: We have already seen that the transition function in DFA returns *at most* one state for a state and an input, whereas an NDFA may have a number of output states from a state on reading the same input; that is, it is non-deterministic in nature. So, NDFAs often need *fewer* states than the corresponding DFAs. Accordingly, they are easy to construct. But any practical implementation usually requires deterministic behaviour. That is why conversion of an NFA to its equivalent DFA is often preferred. However, we may use one of the following two ways to implement an NDFA:

- Backtracking approach (without converting NDFA to DFA)
- Subset construction, a deterministic approach (converting NDFA to DFA)

Procedure to construct DFA from NDFA (without ε move): We have seen that the transition function for NDFA is δ: Q × Σ → 2^Q, where Q denotes the set of states in NDFA. The transition function reports that NDFA may move to one of 2^Q possible subsets of Q from a state for an input. Note that several moves in NDFA from a state (a subset consisting of states) for an input are not shown in the form of a single state. In other words, the diagram explicitly shows multiple paths from the same state for the same input. On the other hand, DFA is deterministic and multiple state moves are not allowed. Therefore, the equivalent single state move (consisting of multiple states) must be shown during conversion. For this, we may apply the subset construction strategy, where the equivalent DFA for an NDFA with n states may have at most 2^n states.

For clarity, suppose an NDFA has three states (q_0, q_1 and q_2) and the NDFA moves to states q_1 and q_2 for an input 'a' from state q_0. Now, state moves in NDFA are shown separately for the same input. However, this is not allowed in DFA. In the context of DFA, the states q_1 and q_2 must be *merged* into a single state, and the move must be shown accordingly. The same technique will be followed for other inputs and states as well. This technique may ultimately cover all possible subsets of $\{q_0, q_1, q_2\}$ in the *worst case*. As a result, the number of states in DFA increases as compared to NDFA.

Thus, the idea behind the subset construction strategy is to define a DFA whose states are the subsets of the set of all states of the given NFA. Each final state of the DFA is a subset that contains at least one final state of the NFA. The conversion strategy uses the concept of subset because transition of NDFA results in multiple states (for each state and input) which are, indeed, subsets of a power set. Before outlining the steps used to construct the state table of DFA from the state table of NDFA, the preliminaries for the conversion are given below:

Basic requirements

- If an NDFA (M) has a set (Q) of n states, its equivalent DFA (M_1) may have up to 2^n states (a power set P(Q) of Q) with the same initial state (q_0) as in NDFA. All possible subsets (2^n subsets) of Q may now be enumerated. These are the possible states of M_1, that is, each subset of Q represents one state of M_1. Here, each subset (that is, single state) may be represented by enclosing them in square brackets, $[q_0, q_1]$.
- The alphabet set Σ of M is also the alphabet set for M_1.
- The set of final states (F′) of M_1 is constructed as follows: F′ consists of the subsets of Q, each of which includes at least one final state of M.
- The transition function, δ' of M_1 is defined as:

$\delta'([q_i, q_j, \ldots], a) = \delta(q_i, a) \cup \delta(q_j, a) \cup \ldots$, where δ is the transition function for M, $\{q_i, q_j, \ldots\}$ is a subset of Q and $a \in \Sigma$.

To show all possible transitions/operations of M_1 in tabular form (with columns for each distinct input), the following steps may be adopted.

Note: Instead of enumerating all possible subsets, one may prefer the *lazy construction* strategy (to save time), where the subset includes only those that are required in the solutions.

DFA state table construction steps

Step 1: Consider the initial state as the present state, and obtain the next states for the possible inputs of Σ (placed in different columns) by applying δ' (as specified above).

Step 2: Choose a state (as the present state) from the next state column if it is not already processed as the present state. If all the states in the next state column are new, then *arbitrarily* select a state as the present state and apply δ' for that state.

Step 3: Construction of the table stops when no new state appears in the subsequent columns. Otherwise, go to *Step 2*.

> **Note:** A DFA (M_1) may have 2^n (or exponential) number of states in the worst case for an NDFA (M) with n states. This may cause difficulties when we try to implement the DFA and result in a very long construction time. However, it is evident that for most real languages/problems, both DFA and NFA have approximately the same number of states and exponential behaviour is rarely seen. So, the number of states (m) in the converted DFA from an NDFA with n states may be bounded as: $n \le m \le 2^n$. Now, if the worst case situation arises (2^n states), we may prefer on-the-fly subset construction, which makes use of simulation – most text editing programs use this concept. We may take one specific worst case formal example in which an NFA with $(n + 1)$ states has no equivalent DFA with fewer than 2^n states. The specific problem deals with the language of 0s and 1s such that the n-th symbol from the end is 1. The problem is discussed later in this chapter (see Example 2.13 (c)).

EXAMPLE 2.13

(a) An NDFA is given in Fig. 2.23 (a). Convert it to an equivalent DFA.

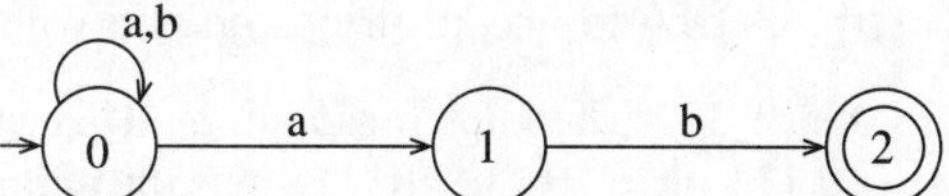

FIGURE 2.23 (a) An NDFA to accept strings ending with *ab* over {a, b}

Solution In this example, the NDFA (M) has states Q = {0, 1, 2}. The state transition table for the given NDFA is presented in Table 2.3 (a).

TABLE 2.3 (a) State table corresponding to Fig. 2.23 (a)

State	Input (a)/Next state	Input (b)/Next state
→ 0	0, 1	0
1	Φ	2
*2	Φ	Φ

Let us first find the power set: P(Q) of Q. The elements of P(Q) are, respectively, Φ, {0}, {1}, {2}, {0, 1},{0,2}, {1,2}, {1,2, 3}. The equivalent DFA (M_1) may possess up to 2^3 states that are, respectively, D (dead state: resulting from Φ[6] state), [0], [1], [2], [0, 1], [0, 2], [1, 2], [0, 1, 2]. Here, each state of DFA

[6] In the context of NDFA, Φ state indicates that there is no transition at all. However, this does not necessarily represent the dead state for NDFA. On the other hand, during construction of the DFA table from NDFA, if the next state is obtained as Φ, this resulting Φ is treated as the dead state for DFA.

is enclosed by '[ ]' to represent a single state. For example, [0, 2] denotes a single state. However, it is sometimes expressed as [02] or {0, 2}.

The state [0] is the initial state of M_1 as it is the initial state of M. Further, $\Sigma = \{a, b\}$ of the given NDFA is also the Σ of the equivalent DFA. Let us now construct the set of final states as $F' = \{[2], [0, 2], [1, 2], [0, 1, 2]\}$, since F′ of DFA consists of subsets, each with the final state 2 of NDFA. Finally, based on the three steps (specified in the conversion strategy), the state table is shown in Table 2.3 (b) for the equivalent DFA of the given NDFA.

TABLE 2.3 (b) State table of the converted DFA

State	Input (a)/Next state	Input(b)/Next state
→[0]	[0, 1]	[0]
[0,1]	[0, 1]	[0, 2]
*[0,2]	[0,1]	[0]

The equivalent DFA for Table 2.3 (b) is shown in Fig. 2.23 (b).

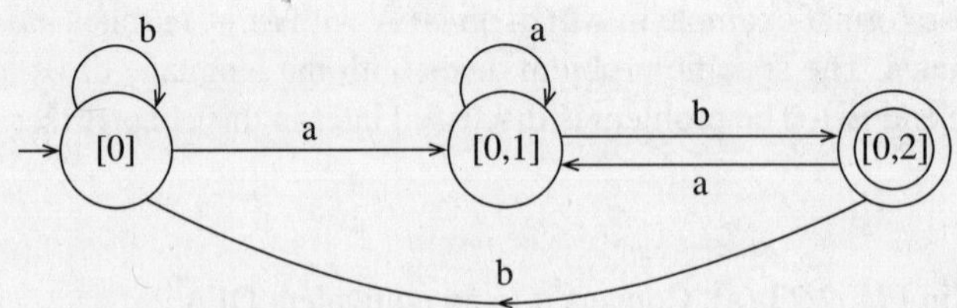

FIGURE 2.23 (b) A DFA to accept strings ending with *ab* over {a, b}

Let us take the transition from state [0, 1] of the DFA for input '*a*'. This can be expressed as $\delta'([0, 1], a) = \delta(0, a) \cup \delta(1, a) = \{0,1\} \cup \Phi = \{0,1\} = [0,1]$, since the union of any set (S_1) with Φ (null) state is S_1 itself. Here, δ' is the transition relation of the DFA, whereas δ is the transition function of the given NDFA. Obviously, the converted DFA does not move to state Φ in this transition as $\{0, 1\} \cup \Phi = \{0, 1\}$ is treated as a single state. The state Φ with respect to NDFA is treated as D (dead state) with respect to DFA.

(b) Find a DFA equivalent to the NDFA M, as shown in Fig. 2.19 (b).

Solution The transition table corresponding to the first transition graph presented in Fig. 2.19 (b) is given in Table 2.4 (a).

TABLE 2.4 (a) A state table correponding to Fig. 2.19 (b)

State	Input (a)/Next state	Input (b)/Next state
$\rightarrow q_0$	q_f	ϕ
$*q_f$	q_f	q_f

The converted DFA table is given in Table 2.4 (b).

TABLE 2.4 (b) DFA state table for Table 2.4 (a)

State	Input (a)/Next state	Input (b)/Next state
$\rightarrow[q_0]$	$[q_f]$	$\delta'([q_0], b) = \delta([q_0], b) = \Phi = D$
$*[q_f]$	$[q_f]$	$[q_f]$

(c) Design an NDFA and a DFA to accept strings over {0, 1} with the third bit from the right as 1.

Solution (*i*) Let us first draw an NDFA for the given problem (Fig. 2.23 (c)).

FIGURE 2.23 (c) An NDFA to accept strings over {0, 1} with the third bit from the right end as 1

(*ii*) We now attempt to draw the DFA for the same problem. For this, we may consider the following possible simple but valid strings: (a) 100 (b) 101 (c) 110 (d) 111. For these strings, we may demand four final states to construct the DFA, named q_{100}, q_{101}, q_{110} and q_{111}. Thus, assuming q_0 as the start state and satisfying the four simple strings, a simple DFA is depicted in Fig. 2.23 (d).

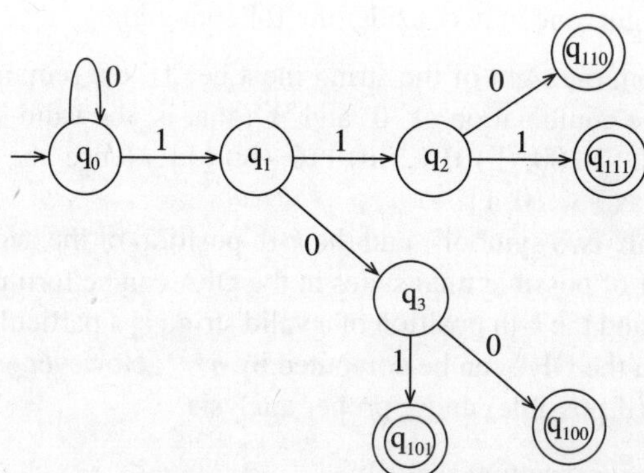

FIGURE 2.23 (d) An incomplete DFA to accept simple strings over {0, 1} with the third bit from the right as 1

Finally, the extended form of the above simple DFA is depicted in Fig. 2.23 (e), considering all possible cases.

In Fig. 2.23 (e), if input '1' is encountered in the final state q_{101}, the machine moves to q_2, whereas it moves to state q_1 from q_{100} and q_{110} for input '1' to accept the next strings. For input '1' at q_{111}, the machine moves to state q_{111} itself. Further, from states q_{100} and q_{101}, on reading input '0', the DFA moves to states q_0 and q_3, respectively, for the purpose of recognising the next valid strings. However, from state q_{111}, on reading input 0, the DFA moves to q_{110} to accept other possible valid strings.

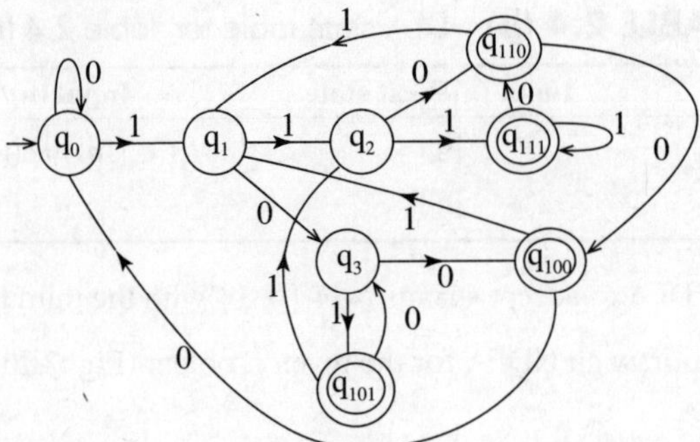

FIGURE 2.23 (e) A complete DFA to accept strings over {0, 1} with the third bit from the right as 1

Note: If we convert the given NDFA, as shown in Fig. 2.23 (c), by applying the subset construction approach, an identical DFA (as shown in Fig. 2.23 (e)) will be obtained. The interesting point is that the NDFA has $(n + 1)$ $(3 + 1 = 4)$ states but its equivalent DFA has 2^n states $(2^3 = 8)$. In other words, it is no fewer than $2^3 = 8$. This is a *worst case* example for which DFA has no fewer than 2^n states.

Tips for the *k*-th bit from the right: To design a DFA to accept strings over {0, 1} containing a '1' at the third position from the right, one may consider the following hints:

- As the third position from the right of the string must be '1', the remaining two symbols after the third can be any possible combination of '0' and '1'; that is, the valid string may end with any of the following substrings: (i) 100, (ii) 101, (iii) 110, (iv) 111. Hence, the number of final states may be considered as 4 (2^2), as $\Sigma = \{0, 1\}$.
- Thus, if Σ consists of only two symbols, and the k-th position of the accepted string is a particular symbol of Σ, the number of possible final states in the DFA can be formulated as: 2^{k-1}. Similarly, if Σ consists of n symbols and the k-th position of a valid string is a particular symbol of Σ, the number of possible final states in the DFA can be computed by n^{k-1}. However, some of the final states may be reduced by merging (if possible) under proper analysis.

Disadvantages of the subset construction strategy

- The strategy requires 2^n states in the worst case, where n is the number of states in the NDFA. Therefore, space (and cost) complexity is exponential.
- This technique needs $\Sigma \times 2^n$ transitions in total to convert NDFA to DFA, where Σ is the alphabet of NFA and n is the number of states. Thus, the time complexity to fill the transition table is $O(\Sigma \times 2^n)$, which is exponential time.

2.9.2 Epsilon NFA (ε NFA)

This is a special kind of NFA, where ε move is allowed. So it is an extended model of NFA. Thus, an NFA with one or more ε transitions is called ε NFA. The transition function δ for ε NFA is defined as δ: $Q \times \Sigma \cup \{\varepsilon\} \rightarrow 2^Q$, where Q denotes the set of states and Σ is the alphabet.

ε NFA is useful in many cases such as for constructing FAs from regular expressions, combining a number of FAs, and so on. Interestingly, ε transitions are convenient but do not increase the power of FAs. Further, as they show non-deterministic behaviour, we need to convert to an equivalent DFA.

Conversion Strategy from NDFA (With ε Move) to DFA: For the purpose of conversion (to eliminate ε transitions), we need knowledge of ε CLOSURE.

ε CLOSURE(s) (denoted as ECLOSURE(s)) = set of NFA states reachable from NFA state s on ε transitions. This gives the following equivalents.

(i) *Equivalence based on transitive property*: Figure 2.24 (a) is equivalent to Fig. 2.24 (b), eliminating the ε transition. Thus, if a state s can reach state r through ε move and there is a transition from r to q on an input $a \in \Sigma$ $(a \neq \varepsilon)$, then consider the transition from s to q through input a.

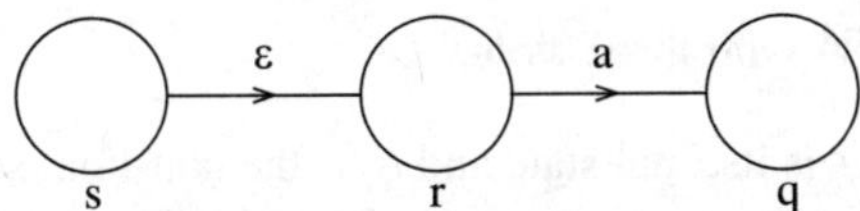

FIGURE 2.24 (a) A part of NDFA with ε transition

FIGURE 2.24 (b) Equivalent DFA for Fig. 2.24 (a), eliminating the ε transition

(ii) *Equivalence for final state*: For the transition graph (Fig. 2.24 (c)), we may consider the state s as the final state, since q is a final state and the machine moves to q from the non-final state s on reading ε. We make state s an accepting state if s can reach some accepting state q through the ε move.

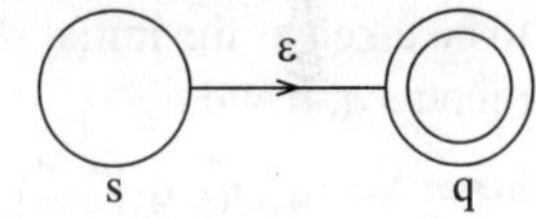

FIGURE 2.24 (c) A part of NDFA with ε transition

The procedure for converting ε NFA to equivalent DFA is presented below.

Procedure: Let $M_E = (Q_E, \Sigma, \delta_E, q_{0E}, F_E)$ be an ε NFA, where Q_E is a set of states, Σ is a set of symbols, δ_E is the transition from $Q_E \times \Sigma\, (\cup \{\varepsilon\})$ to 2^{Q_E}, q_{0E} is the start state and F_E is the set of final states. The equivalent DFA $M_D = (Q_D, \Sigma, \delta_D, q_{0D}, F_D)$ can be obtained by applying the following steps.

/* *Finding the start state of* M_D */

Step 1: If q_{0E} is the start state of ε NFA, then ECLOSURE (q_{0E}) is the start state of DFA:

That is, q_{0D} = ECLOSURE (q_{0E}). // *It is a set of states.*

/* *Finding transitions of* M_D *(new states) beginning from start state* q_{0D} */

Step 2: Find the new states of the DFA (transitions of DFA) by applying the transition of ε NFA. Let $\{q_i, q_j, \ldots, q_k\}$ be a state in DFA. The transition of the DFA from this state on input 'a', $\delta_D(\{q_i, q_j, \ldots q_k\}, a)$ is computed as follows:

$$\delta_D(\{q_i, q_j, \ldots, q_k\}, a) = \text{ECLOSURE}(\delta_E\{q_i, q_j, \ldots, q_k\}, a).$$

This step continues till a new state (that is, new subset of Q_E) is found (by processing all the inputs in each state).

/* *Finding the set of final states of* M_D */

Step 3: Construct the set of final states of DFA as follows: $\{q_i, q_j, \ldots, q_k\}$ is a final state of DFA if it contains at least one final state of ε NFA, and it is included in F_D.

These steps are explained with an example.

EXAMPLE 2.14

Convert the ε NFA shown in Fig. 2.25 (a) to its equivalent DFA.

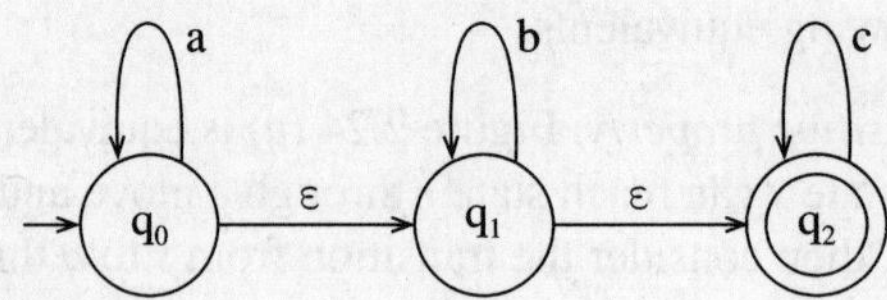

FIGURE 2.25 (a) An ε NFA with three states

Solution Here, q_0 is the start state of the given NFA. q_2 is its final state and q_1 is the non-final state. Also, $\Sigma = \{a, b, c\}$. Now, q_1 moves to q_2 (final state) through ε move, so q_1 is also treated as the final state. Further, q_0 moves to q_1 through ε move, so q_0 is again the final state. Thus, all three states, q_0, q_1 and q_2, are the final states.

Let us apply *Step 1* of the procedure to obtain ECLOSURE $(q_0) = \{q_0, q_1, q_2\}$ = A (new state). This is to be taken as the initial state of the DFA. Next, apply *Step 2* to state A = $\{q_0, q_1, q_2\}$ to compute δ_D for inputs *a*, *b* and *c*.

State = A= {q₀, q₁, q₂}

- Input = '*a*'

$\delta_D(\{q_0, q_1, q_2\}, a) = \text{ECLOSURE}(\delta_E\{q_0, q_1, q_2\}, a)$

$= \text{ECLOSURE}(\delta_E(q_0, a) \cup \delta_E(q_1, a) \cup \delta_E(q_2, a)) = \text{ECLOSURE}(q_0 \cup \Phi \cup \Phi)$

$= \text{ECLOSURE}(q_0) = \{q_0, q_1, q_2\}$ = A (already generated)

- Input = '*b*'

$\delta_D(\{q_0, q_1, q_2\}, b) = \text{ECLOSURE}(\delta_E\{q_0, q_1, q_2\}, b)$

$= \text{ECLOSURE}(q_1) = \{q_1, q_2\}$ = B (a new state of DFA)

- Input = '*c*'

$\delta_D(\{q_0, q_1, q_2\}, c) = \text{ECLOSURE}(\delta_E\{q_0, q_1, q_2\}, c)$

$= \text{ECLOSURE}(q_2) = \{q_2\}$ = C (a new state of DFA)

Here, two distinct sets of states are obtained: $\{q_1, q_2\}$ (B) and $\{q_2\}$ (C). So, we apply *Step 2* separately on these two sets, and obtain the following output.

State = B = {q₁, q₂}

- Input = '*a*', '*b*', '*c*'

$\delta_D(\{q_1, q_2\}, a) = \phi$, $\delta_D(\{q_1, q_2\}, b) = \{q_1, q_2\}$= B, $\delta_D(\{q_1, q_2\}, c) = \{q_2\}$ = C

State = C = { q₂}

- Input = '*a*', '*b*', '*c*'

$\delta_D(\{q_2\}, a) = \phi$, $\delta_D(\{q_2\}, b) = \phi$, $\delta_D(\{q_2\}, c) = \{q_2\}$ = C

Note that no new state is generated at this stage.

Here, all three states (A, B and C) are final states (as per *Step 3* of the procedure). Further, A is the initial as well as the final state. The state operations are presented in Table 2.5 (state table). Also, the equivalent DFA is shown in Fig. 2.25 (b). Here, Φ is the dead state.

TABLE 2.5 Transition table for the given operations

δ	a	b	c
→(*A)	A	B	C
*B	ϕ	B	C
*C	ϕ	ϕ	C

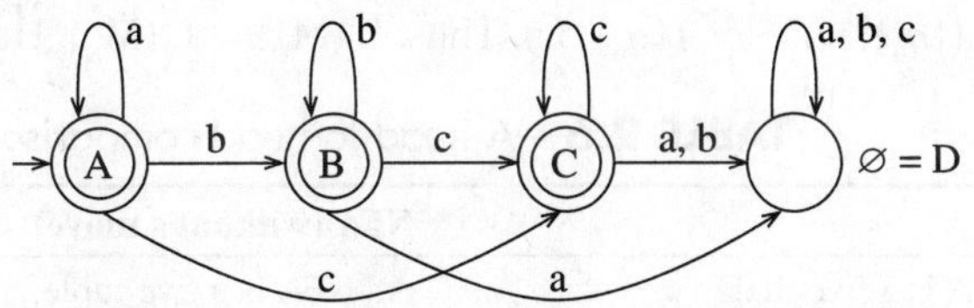

FIGURE 2.25 (b) A DFA equivalent to ε NFA shown in Fig. 2.25 (a)

From the figures, it is clear that the accepted language is $L = \{a^k b^m c^n \mid k, m, n \geq 0\}$ over $\{a, b, c\}$; that is, no 'a' may appear after 'b' and 'c'; and no 'b' may appear after 'c'.

Differences in equivalent DFA with ε NFA and without ε NFA: In the case of NFA without ε move, the start state of the constructed DFA contains only the start state of NFA; whereas the start state (q_D) of the constructed DFA from ε NFA contains ECLOSURE (q_0), where q_0 is the start state of ε NFA.

THEOREM 2.2

If there exists an NFA, $M_N = (Q_N, \Sigma, \delta_N, q_0, F_N)$ that accepts the language L (M_N), there exists an equivalent DFA, $M_D = (Q_D, \Sigma, \delta_D, q_0, F_D)$ such that L (M_D) = L (M_N).

Proof Prove that $\delta^*_D(q_0, w) = \delta^*_N(q_0, w)$.

We know that if Q_N represents the states of NFA, then the power set of Q_N, which contains the set of subsets of Q_N, comprises the states of DFA, denoted by Q_D. The DFA interprets each set as a single state in DFA.

Basis Consider a string $w = \varepsilon$, where $|w| = 0$. Now $\delta^*_D(q_0, \varepsilon) = \{q_0\}$, by δ^*_D of DFA, whereas $\delta^*_N(q_0, \varepsilon) = \{q_0\}$ by δ^*_N of NFA. Hence, $\delta^*_D(q_0, \varepsilon) = \delta^*_N(q_0, \varepsilon)$, when $w = \varepsilon$.

Induction hypotheses Let us now assume that $\delta^*_D(q_0, \varepsilon) = \delta^*_N(q_0, \varepsilon)$ for some w, where $|w| = n$. Now, we need to prove that the statement $\delta^*_D(q_0, \varepsilon) = \delta^*_N(q_0, \varepsilon)$ is true for some w, where $|w| = n + 1$.

Inductive proof Let $w = xa$, where a is the last symbol of w and x is the remaining string of w. So, $|x| = n$ and $|xa| = n + 1$

By using the δ^*_N for NFA, we get $\delta^*_N(q_0, w) = \delta^*_N(q_0, xa) = \delta_N(\delta^*_N(q_0, x), a)$ (1)

Now, x is the string to be processed and after consuming string x, let the states of the machine be $\{p_i, p_j, \ldots, p_k\}$, that is, $\delta^*_N(q_0, x) = \{p_i, p_j, \ldots, p_k\}$.

Substituting this in Equation (1), we get $\delta^*_N(q_0, w) = \delta^*_N(q_0, xa) = \delta_N(\delta^*_N(q_0, x), a) = \delta_N(\{p_i, p_j, \ldots, p_k\}, a)$

$= \delta_N(p_i, a) \cup \delta_N(p_j, a) \cup \ldots (p_k, a)$ (2)

Now, for DFA, we know that $\delta^*_D(q_0, w) = \delta^*_D(q_0, xa) = \delta_D\ \delta^*_D((q_0, x), a)$ (3)

Equation (3) states that x is the string to be processed, and after consuming x, let the states of the machine be $\{p_i, p_j, \ldots, p_k\}$, that is, $\delta^*_D(q_0, x) = \{p_i, p_j, \ldots, p_k\}$.

Substituting this in Equation (3), we get

$$\delta^*_D(q_0, w) = (\delta^*_D\{p_i, p_j, \ldots, p_k\}, a) = \delta_N(p_i, a) \cup \delta_N(p_j, a) \cup \ldots\ \delta_N(p_k, a). \quad (4)$$

Now, by comparing Equations (2) and (4), we get

$\delta^*_D(\{q_0\}, w) = \delta^*_N(\{q_0\}, w)$. Thus, L (M_D) = L (M_N). Hence, the theorem is proved.

TABLE 2.6 A head-to-head comparison of DFA, NFA and ε NFA

DFA	NFA (without ε move)	ε NFA
A DFA is a five-tuple $M = (Q, \Sigma, \delta, q_0, F)$ where ■ Q is a set of finite states ■ Σ is the set of input alphabets ■ δ: $Q \times \Sigma$ to Q ■ q_0 is the start state ■ $F \subseteq Q$ is the set of final states	An NFA is a five-tuple $M = (Q, \Sigma, \delta, q_0, F)$ where ■ Q is a set of finite states ■ Σ is the set of input alphabets ■ δ: $Q \times \Sigma$ to 2^Q ■ q_0 is the start state ■ $F \subseteq Q$ is the set of final states	An ε NFA is a five-tuple M = = $(Q, \Sigma, \delta, q_0, F)$ where ■ Q is a set of finite states ■ Σ is the set of input alphabets ■ δ: $Q \times (\Sigma \cup \{\varepsilon\})$ to 2^Q ■ q_0 is the start state ■ $F \subseteq Q$ is the set of final states
■ There can be zero or one transition from a state on an input symbol	■ There can be zero, one or more transitions from a state on an input symbol	■ There can be zero, one or more transitions from a state with or without giving any input
■ The number of transitions is higher	■ The number of transitions is lower	■ Relatively more transitions when compared with NFA
■ Difficult to construct	■ Easy to construct	■ Easy to construct using regular expressions

2.10 MINIMISATION OF FINITE AUTOMATA

As noted earlier, circuits of real computers are constructed based on the concept of FSM. On the other hand, FSMs are mathematically represented by a number of states, some of which may be redundant. However, states are costlier to implement. So, our main aim is to eliminate the redundant states to reduce the cost. In other words, we need to use less hardware (flip-flops). Consequently, such minimisation results in less space and easier processing of input in less time. Also, it is interesting to note that a language produced by a DFA is always unique; however, the converse is not true; that is, there may exist multiple DFAs for a language, where some of the DFAs are redundant. So, it is necessary to find the minimised DFA or to minimise the number of states of a given DFA. *Thus, there exists a unique minimal DFA for every DFA.*

For the purpose of minimising the FA (especially DFA), two standard methods are used: equivalence theorem and table-filling algorithm. These are discussed below. However, before minimisation, we need to identify the unreachable states (if any) and remove them, and merge the equivalent states.

2.10.1 DFA Minimisation Using Equivalence Theorem

Before applying this theorem, let us look at the necessary terminology.

Unreachable states: The states in an automata are unreachable if it is not possible to reach these states from the initial state. For example, from Fig. 2.26, it is obvious that we cannot visit state 2; such a state is said to be unreachable. So, we can simply remove such states from the automaton without changing its behaviour (see Fig. 2.26 on the right).

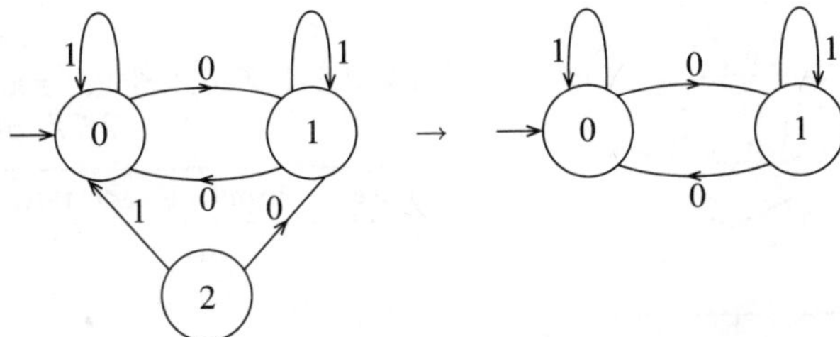

FIGURE 2.26 DFA with unreachable state and its removal

However, removing unreachable states from FA is not enough. We should merge some/all non-distinguishable states into a single state; that is, merge the equivalent states.

Equivalent (or non-distinguishable or indistinguishable) states: Two states p and q are said to be equivalent if $\delta(p, w) = q_f \in F$ (set of final states) and $\delta(q, w) = q'_f \in F$, where w is a string over alphabet Σ. Obviously, q_f need not be equal to q'_f. However, both p and q are equivalent if $\delta(p, w) \notin F$; then $\delta(q, w) \notin F$, but both the resulting states will be in the same group (part). We may use the same concept for the set of final states (if it is partitioned). Thus, two states p and q are equivalent if $\delta(p, w)$ and $\delta(q, w)$ are the final states or if both are non-final states. As it is difficult to test: $\delta(p, w)$ and $\delta(q, w)$ for all strings: w (as there may be an infinite number of strings), we will use the concept of k-equivalent ($k \geq 0$); that is, 0-equivalent, 1-equivalent, 2-equivalent, and so on, in recursive manner. For 0-equivalent, $|w| = 0$; | for 1-equivalent, $|w| = 1$, and so on. More explicitly, no input (0 input) is considered for 0-equivalent; one input symbol is taken for 1-equivalent. In this way, two input symbols are considered for 2-equivalent. Interestingly, a 2-equivalent string can be found from 1-equivalent by adding one input symbol with the string (of length 1) to the 1-equivalent. Likewise, a 3-equivalent string can be obtained from a 2-equivalent string. Hence, k-equivalent strings can be found from $(k - 1)$-equivalent strings. Thus, it is a *recursive approach*.

Note: Two states p and q are equivalent if $\delta(p, w)$ and $\delta(q, w)$ are the final states or if both are non-final states. In any case, they will be in the same group (set). If not, then non-deterministic behaviour occurs.

Algorithm for DFA minimisation using equivalence theorem

Step 1: All the states Q of the DFA (M) are divided into two parts: final states and non-final states, and the partition is denoted by Π_0. This is known as the 0th equivalent. The 0-equivalent is divided into two parts: a set of non-final states and a set of final states, as these two sets satisfy the concept of equivalent states for, respectively, a set of final states and a set of non-final states for input ε. Take a counter k and initialise it to 0.

Step 2: Increment k by 1 and find the k-th equivalent. Here, equivalence checking of the states in each partition of Π_{k-1} for each input is carried out.

Step 3: If $\Pi_k \neq \Pi_{k-1}$, repeat *Step 2*; else go to *Step 4*.

Step 4: Consider each partition of Π_k as a single state in the reduced DFA. The partition in Π_k that contains the initial state of the original DFA is taken as the initial state of the reduced DFA, whereas the partitions which contain at least one final state of the original DFA are treated as the final states of the reduced automata. Interestingly, there may be a number of final states in the reduced machine.

For example, consider a DFA, as shown in Fig. 2.27 (a). The equivalent state table is shown in Table 2.7. Now, we will find its equivalent minimal DFA.

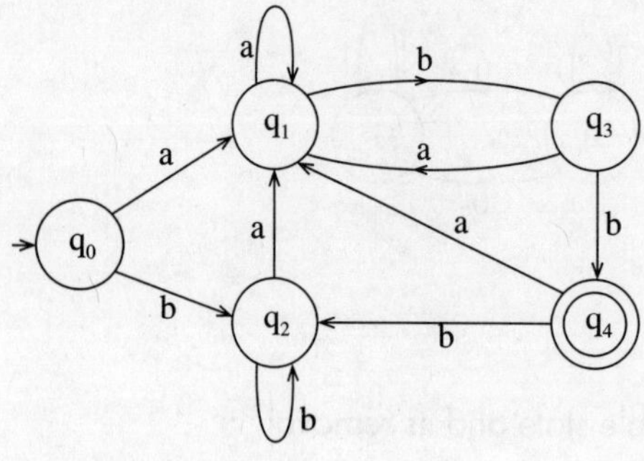

FIGURE 2.27 (a) A DFA over {a, b}

TABLE 2.7 State table for DFA shown in Fig. 2.27 (a)

State	Input(a)/Next state	Input(b)/Next state
$\rightarrow q_0$	q_1	q_2
q_1	q_1	q_3
q_2	q_1	q_2
q_3	q_1	q_4
$*q_4$	q_1	q_2

Π_0 (0-equivalent) = set of non-final states and set of final states $=\{\{q_0, q_1, q_2, q_3\},\{q_4\}\}$. For the Π_1 (1-equivalent), we need to find equivalent states in each group (partition) of Π_0 (0-equivalent) for single input. Here, there are two partitions (groups): $\{q_0, q_1, q_2, q_3\}$ and $\{q_4\}$.

Consider partition $\{q_0, q_1, q_2, q_3\}$. Each state in $\{q_0, q_1\}$ gives q_1 on input '*a*' and $q_1 \in \{q_0, q_1, q_2, q_3\}$. Likewise, for input '*b*', q_0 gives q_2 and q_1 gives q_3, but both q_2 and q_3 are in the group $\{q_0, q_1, q_2, q_3\}$. So, they are 1-equivalent.

To check the 1-equivalence of q_2, we may consider either $\{q_0, q_2\}$ or $\{q_1, q_2\}$, as $\{q_0, q_1\}$ is 1-equivalent. From the table, we get $\{q_0, q_2\}$ is 1-equivalent. So, $\{q_0, q_1, q_2\}$ will be in the same group. Now, let us check the 1-equivalence for $\{q_3\}$. For this, consider $\{q_0, q_3\}$. For input '*a*', each $\{q_0, q_3\}$ goes to the q_1 that is in the same group, whereas q_0 goes to $q_2 \notin F$ for input 'b' and q_3 goes to $q_4 \in F$. So, q_3 is not 1-equivalent, that is, it will not be in group $\{q_0, q_1, q_2, q_3\}$. Thus, $\Pi_1 = \{\{q_0, q_1, q_2\}, \{q_3\}, \{q_4\}\}$.

To find Π_2 (2-equivalent), we need equivalent states in each group (partition) of Π_1 (1-equivalent) for a single input. First consider the partition $\{q_0, q_1, q_2\}$. Let us take $\{q_0, q_1\}$. For input '*a*', q_0 and q_1 give to $q_1 \in \{q_0, q_1, q_2\}$, but for input '*b*', q_0 gives to q_2 and q_1 gives to q_3, and q_2 and q_3 belong to separate groups. So, we separate them. However, $\{q_0, q_2\}$ is 2-equivalent. On the other hand, $\{q_3\}$ and $\{q_4\}$ cannot be split further, as each of them is single. So, $\Pi_2 = \{\{q_0, q_2\}, \{q_1\}, \{q_3\}, \{q_4\}\}$.

In the same way, we obtain $\Pi_3 = \{\{q_0, q_2\}, \{q_1\}, \{q_3\}, \{q_4\}\}$. So, $\Pi_2 = \Pi_3$ and we now stop splitting. Thus, the reduced DFA has four states: $[q_0, q_2]$, $[q_1]$, $[q_3]$, $[q_4]$, where $[q_0, q_2]$ is the initial state and $[q_4]$ is the final state (Fig. 2.27 (b)).

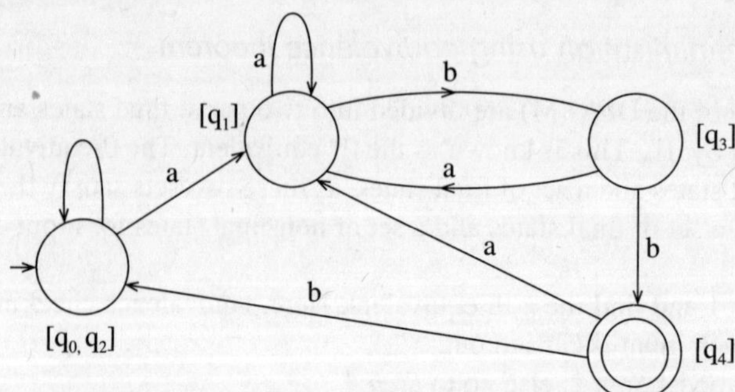

FIGURE 2.27 (b) Reduced DFA of Fig. 2.27 (a)

Note that the reduced machine moves to state $[q_0, q_2]$ on reading input '*b*' at $[q_0, q_2]$, although the original machine moves to state q_2 from q_0 for this input. This is because states q_0 and q_2 are not separate; that is, they have been merged into one state, $[q_0, q_2]$. This is true for state q_4 and input '*b*' as well.

2.10.2 Myhill–Nerode Theorem/Table-filling DFA Minimisation Algorithm

For minimisation of DFA, we can also use the Myhill–Nerode theorem (named after its inventors), which provides a theoretical background that enables us to minimise a DFA. This theorem states that a language L is regular if and only if R_L (a relation on strings by the rule that $x\,R_L\,y$) has a finite number of *equivalence classes*, and moreover, that the number of states in the smallest DFA recognising L is equal to the number of equivalence classes in R_L. In particular, this implies that there is a unique minimal DFA with a minimum number of states. The DFA minimisation algorithm based on this theorem is known as the *table-filling algorithm*, discussed below.

The algorithm first removes the unreachable states, if any. Next, it merges the equivalent states. The current approach is slightly different from the original theory.

Given: An FA (M) with initial state q_0, final state F and transition function δ.

Step 1: Remove the unreachable states.

Step 2: Mark the distinguishable pairs of states.

To achieve this, we first mark all pairs (p, q), where $p \in$ F and $q \notin$ F, as distinguishable. Then we proceed as follows:

repeat
for all non-marked pairs: (p, q) *do*
for each letter *a do*
if the pair $\delta\,(p, a)$, $\delta(q, a)$ is marked *then* mark (p, q)
until no new pairs are marked

Step 3: Construct the reduced automaton, M′.

We first determine the equivalence classes of the indistinguishability relation. For each state q, the equivalence class consists of all the states p for which the pair (p, q) is not marked in *Step 2*.

The states of M′ are the equivalence classes. The initial state q'_0 (for M′) is the equivalence class that contains q_0. The final states F′ (for M′) are the equivalence classes that consist of the final states of M. The transition function δ' (for M′) is defined as follows: to determine $\delta'(X, a)$, for some equivalence class X, pick any $q \in X$, and set $\delta'(X, a) = Y$, where Y is the equivalence class that contains $\delta(q, a)$.

Time complexity analysis of the algorithm: For the DFA M, let the number of states of M be n and the size of the input alphabet Σ be k. The time required for initialisation of the table is $O(n^2)$ and the rest of the algorithm needs $O(kn^2)$. Hence, the total required time is $O(n^2) + O(kn^2) = O(kn^2)$.

EXAMPLE 2.15

Apply the table-filling algorithm to the automaton given in Fig. 2.28.

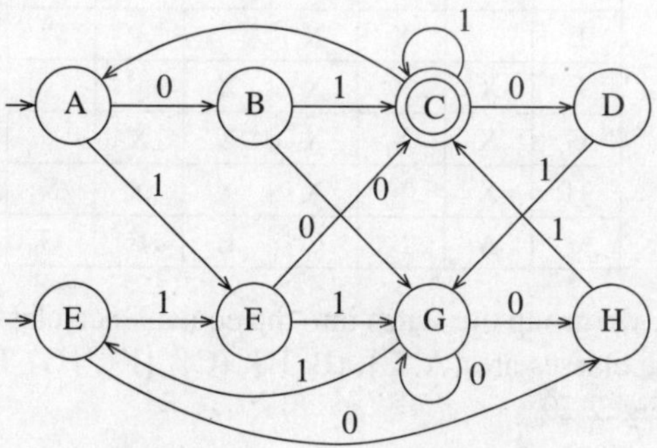

FIGURE 2.28 A simple DFA

Solution In Step 1, we have one unreachable state, D. We remove this state and proceed to Step 2. Here, the pairs of final and non-final states are marked and we obtain Table 2.8.

TABLE 2.8 A table corresponding to Fig. 2.28

B						
C	X	X				
E			X			
F			X			
G			X			
H			X			
	A	B	C	E	F	G

In the first iteration, we examine all unmarked pairs. For example, for pair (A, B), we obtain δ(A, 1) = F and δ (B, 1) = C; the pair (C, F) is marked, so we mark (A, B) too. After performing this for all the pairs, we obtain Table 2.9.

TABLE 2.9 A table after the first iteration

B	X					
C	X	X				
E		X	X			
F	X	X	X	X		
G		X	X		X	
H	X		X	X	X	X
	A	B	C	E	F	G

In the next iteration, we examine the remaining pairs. For example, we will mark pair (A,G) because δ (A, 0) = B and δ (G, 0) = G, and the pair (B, G) is marked. When we have finished the task, we obtain Table 2.10.

TABLE 2.10 A minimised table after the second iteration

B	X					
C	X	X				
E		X	X			
F	X	X	X	X		
G	X	X	X	X	X	
H	X		X	X	X	X
	A	B	C	E	F	G

Finally, on applying Step 3, we group the states into the equivalence classes. Since (A, E) is equivalent and (B, H) is equivalent, the classes are {A, E}, (B, H}, (C}, {F}, {G} The minimal automaton M′ of Fig. 2.28 is now shown in Fig. 2.29.

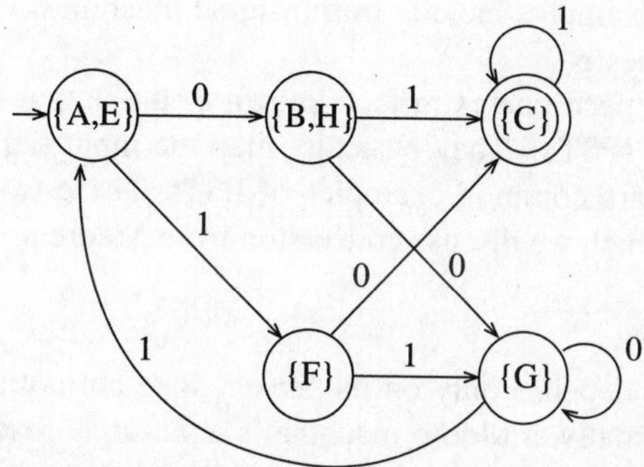

FIGURE 2.29 Minimal DFA of the DFA shown in Fig. 2.28

Note: Minimal DFA is always unique. However, minimal NFA is not necessarily unique. Two NFAs are given below in Fig. 2.30 (a) and (b). Both are different but each is minimal.

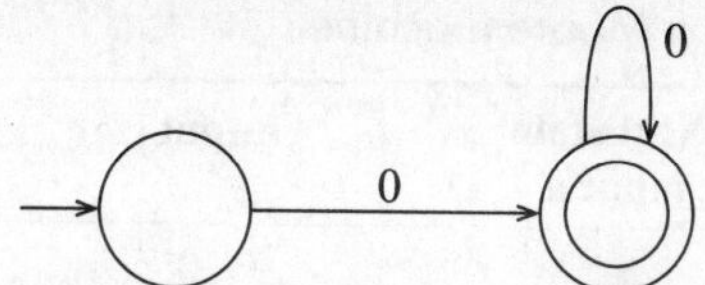

FIGURE 2.30 (a) An NFA to accept strings with one or more 0s

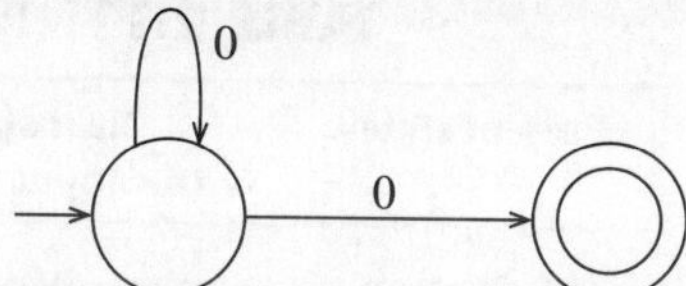

FIGURE 2.30 (b) Another NFA to accept strings with one or more 0s

Summary of DFA and NDFA

- All the transitions in DFA are deterministic and each transition leads to exactly one state. On the other hand, the transitions in NDFA may be non-deterministic; that is, a transition may lead to a subset of states.
- For each transition in DFA, all possible symbols should be defined. However, it is not necessary in NDFA.
- Designing a DFA is more difficult, as the number of states is more than in NDFA.
- Practical implementation on the basis of DFA is feasible, which is not possible for NDFA unless it is converted to DFA.
- For every NDFA, there is an equivalent DFA (in power).
- For any language, there may be many DFAs and NDFAs, but the smallest DFA (with respect to the number of states) is always preferred.

2.11 FINITE STATE MACHINE WITH OUTPUT

Thus far, we have seen how FSMs are used to accept strings as well as languages. In short, they are used as language acceptor called FA, which is used in many applications, such as modelling sequential and combinational circuits of computers, constructing the lexical analyser of the compilation process and designing cellular automata, where pattern matching is the essential goal. Here, we will discuss the applications of FSM with output (instead of accepting or rejecting strings), known as coding devices, and how these are fulfilled. Examples include traffic signal machines, video games, network protocol analysers and vending machines.

The machines technically need one extra tape known as the output tape. Such machines are often called finite state transducers (FSTs). They basically map the input sequence to the output sequence. Moore and Mealy machines are common examples of FSTs. These two machines are first described below with examples, after which we discuss conversion from Moore to Mealy and the reverse.

2.11.1 Moore Machine

An FSM in which the output depends only on its present state but not on the present input symbol is called a Moore machine. Formally, a Moore machine is a six-tuple machine M = (Q, Σ, Δ, δ, λ, q_0), where Q is a finite non-empty set of states; Σ is the input alphabet; Δ is the output alphabet (output symbols respective to input symbols) and Δ need not necessarily be Σ; δ is the transition function defined as $\delta : Q \times \Sigma \rightarrow Q$; λ is the output function defined as $\lambda(q_i) = x$, where $q_i \in Q$, $x \in \Delta$ and q_0 is the start state of M and $q_0 \in Q$.

Thus, it is indeed a type of output producer. An example of input conversion along with the necessary operations of such a machine is presented in Table 2.11.

TABLE 2.11 Transition table for a Moore machine

Present state	Next state (for input a = 0)	Next state (for input a = 1)	Output (Δ)
$>q_0$	q_3	q_1	0
q_1	q_1	q_2	1
q_2	q_2	q_3	0
q_3	q_3	q_0	0

Here, $\Sigma = \Delta = \{0, 1\}$. Let us consider an input string $w = 011$. Now, based on Table 2.11, the input string 011 is to be converted into 0001. Actually, the resultant string is obtained by applying the specific output function $\lambda(q_i) = x$. The way by which output value x changes corresponding to states q_i is presented below with a brief explanation.

Initially, the state is q_0. So, $\lambda(q_0) = 0$. Next, the machine moves to q_3 on reading '0' at q_0. Now, $\lambda(q_3) = 0$. Accordingly, the output is 0. Then, on receiving '1' at q_3, the machine moves to q_0, with the output '0' ($\lambda(q_0) = 0$). Finally, for the current input '1' at q_0, the machine moves to q_1, with '1' as output.

Comment: In the case of a Moore machine, we should not instruct the machine to receive null as input in any state. Otherwise, the machine could fall into an infinite loop. For instance, in the example discussed above, if the proposed Moore machine reads null input at q_1, it falls into an infinite loop.

2.11.2 Mealy Machine

An FSM in which the output depends on its present state as well as the present input symbol is called a Mealy machine. This is six-tuple M = (Q, Σ, Δ, δ, λ, q_0), where Q is a finite non-empty set of states; Σ is the input alphabet; Δ is the output alphabet (output symbols respective to input symbols) and need not necessarily be Σ; δ is the transition function defined as $\Delta : Q \times \Sigma \rightarrow Q$; λ is the output function defined as $\lambda(q_i, a) = x$, where $q_i \in Q$, $a \in \Sigma$ and $x \in \Delta$ and q_0 is the start state of the machine, $q_0 \in Q$.

In a Mealy machine, for every state and every input, there will be an output. An example with input conversion is shown in Table 2.12 .

TABLE 2.12 Transition table for Mealy machine

Present state	Next state (for input a = 0)	Output (Δ)	Next state (for input a = 1)	Output (Δ)
$>q_0$	q_3	1	q_1	0
q_1	q_1	0	q_2	1
q_2	q_2	0	q_3	0
q_3	q_3	1	q_0	0

Let us consider an input string $w = 011$. Consulting Table 2.12, the input string 011 is to be converted to 100:

- For input 0 in 011, $\delta(q_0, 0) = q_3$ and $\lambda(q_0, 0) = 1$, since the machine starts processing at q_0. This means that the machine moves to q_3 on reading '0' at q_0 and produces output '1'.
- For the next input 1, $\delta(q_3, 1) = q_0$ and $\lambda(q_3, 1) = 0$. This means that on receiving '1' at q_3, the machine moves to q_0 and simultaneously gives the output '0'.
- Finally, on reading input '1' at q_0, the machine moves to q_1, resulting in output '0'.

2.11.3 Transformation of Moore Machine to Mealy Machine

The technique to convert a Moore machine to a Mealy machine is explained here with an example. The operation of a Moore machine for a certain kind of input conversion is described in Table 2.13.

TABLE 2.13 A transition table for a Moore machine

Present state	Next state (for input a = 0)	Next state (for input a = 1)	Output (Δ)
$>q_0$	q_3	q_1	0
q_1	q_1	q_2	1
q_2	q_2	q_3	0
q_3	q_3	q_0	0

Transformation strategy: In order to transform a Moore machine into an *equivalent* Mealy machine, we need to apply the transformation function: $\lambda(\delta(q, a)) = x$ for each state in the present state column of the given Moore table, where, $q \in Q$, $a \in \Sigma$, $x \in \Delta$.

Step 1: Take the starting state q_0 of the Moore machine as the starting state of the Mealy machine.

Step 2: Apply the $\lambda(\delta(q, a)) = x$ function to generate the next states for possible inputs of Σ.

Step 3: Put an unprocessed state in the present state column from subsequent columns of the current row or earlier rows and proceed to Step 2. If multiple unprocessed states are present in subsequent columns, arbitrarily select any one. If no unprocessed state is present in the subsequent columns, go to *Step 4*.

Step 4: Stop construction of the table.

Let us illustrate this with an example. First, take the first row of the Moore machine table to obtain the first row of the Mealy machine table. Clearly, q_0 is in the *present state column* of the first row in the Moore table, and the next state from q_0 for inputs 0 and 1 are, respectively, q_3 and q_1. So, $\lambda(\delta(q_0, 0)) = 0$, since $\delta(q_0, 0) = q_3$ and $\lambda(q_3) = 0$; likewise, $\lambda(\delta(q_0, 1)) = 1$, since $\delta(q_0, 1) = q_1$ and $\lambda(q_1) = 1$. Note that the outputs 0 and 1 are obtained from the last column (output) of the Moore table.

In this manner, the remaining rows of the Mealy table are generated and shown in Table 2.14.

TABLE 2.14 Converted Mealy machine table

Present state	Next state (for input a = 0)	Output (Δ)	Next state (for input a = 1)	Output (Δ)
$>q_0$	q_3	0	q_1	1
q_1	q_1	1	q_2	0
q_2	q_2	0	q_3	0
q_3	q_3	0	q_0	0

2.11.4 Transformation of Mealy Machine to Moore Machine

The transformation of a Mealy machine to a Moore machine is explained with the help of an example, given in Table 2.15.

TABLE 2.15 A Mealy machine table

Present state	Next state (for input a = 0)	Output (Δ)	Next state (for input a = 1)	Output (Δ)
$>q_1$	q_3	0	q_2	0
q_2	q_1	1	q_4	0
q_3	q_2	1	q_1	1
q_4	q_4	1	q_3	0

Conversion strategy: To convert a Mealy machine to a Moore machine, we need to apply the output function $\lambda(q_i, c) = x$, where $q_i \in Q$, $c \in \Sigma$ and $x \in \Delta$.

Initially, take the Mealy machine state table as the Moore table for the purpose of conversion. Next, apply the following basic steps to convert the Mealy machine to a Moore machine:

Step 1: Extract the states from the present state table, in which each state is associated with different outputs.

Step 2: If no such state is found in the table, then go to *Step 4*; else split each of those states as q_{ia}, q_{ib}, where state q_i produces two output values, say $a \in \Delta$ and $b \in \Delta$, for two different inputs of Σ.

Step 3: Construct the new state table applying the output function $\lambda(q_i, c) = x$. Take the constructed table as the present state table and go to *Step 1*.

Step 4: Stop construction of the table.

Step 5: Make a single output column (instead of two or more output columns) to obtain a Moore table.

For example, in Table 2.15, states q_2 and q_4 are identified as states with different outputs. Now, split the states on the basis of their different output associations. State q_2 is associated with two different output values, 0 and 1. So, two new states can be formed from q_2 and taken as: q_{20} and q_{21} (the naming convention here is taken according to the output values). In the same way, new states from q_4 can be considered: q_{40} and q_{41}. After splitting the identified states, the transformed table is shown in Table 2.16.

TABLE 2.16 The converted table

Present state	Next state (for input a = 0)	Output (Δ)	Next state (for input a = 1)	Output (Δ)
$>q_1$	q_3	0	q_{20}	0
q_{20}	q_1	1	q_{40}	0
q_{21}	q_1	1	q_{40}	0
q_3	q_{21}	1	q_1	1
q_{40}	q_{41}	1	q_3	0
q_{41}	q_{41}	1	q_3	0

Naturally, states like q_2 and q_4 are not present in the new table (as these have already been split). Two separate rows for q_2 (one for q_{20} and the other for q_{21}) are shown in Table 2.16. Likewise, two separate states, q_{40} and q_{41}, have been taken for q_4. This strategy is repeated until all the *next* states in the last processed table are found with identical outputs. For this particular example, Table 2.16 has no state associated with different output values. Therefore, splitting of states is stopped, and construction of the corresponding Moore table is carried out, as shown in Table 2.17.

TABLE 2.17 The converted Moore table

Present state	Next state (for input a = 0)	Next state (for input a = 1)	Output (Δ)
$>q_1$	q_3	q_1	1
q_{20}	q_1	q_{40}	0
q_{21}	q_1	q_{40}	1
q_3	q_{21}	q_1	0
q_{40}	q_{41}	q_3	0
q_{41}	q_{41}	q_3	1

Now, to report the output associated with each state in Table 2.17, please refer to Table 2.16. For example, the output associated with q_1 in Table 2.16 is always 1; so, insert '1' in the output column of Table 2.17. Similarly, consider the other states.

Comparison of Moore machine and Mealy machine

- The length of output in the Mealy machine is the same as its input, as output depends on the present state and present input. On the other hand, the length of output in the Moore machine is one greater than that of its input, as output depends only on present state.

- The Mealy machine usually has fewer states than the Moore machine. This results in faster reaction to inputs. More specifically, they do not have to wait for the clock; that is, the output depends on the present state as well as the present input. So, they need not wait for the next state. In particular, if the number of states in the Mealy machine is n and the number of outputs is m, then the corresponding Moore machine has no more than $mn + 1$ states.
- Asynchronous outputs and asynchronous feedback are dangerous.
- Moore machines are safer to use for the following reasons:
  - Output changes at the clock edge (always one cycle later).
  - It requires more logic to decode state into outputs.
- The Moore model is very easy to code, whereas the Mealy program is complex.

KEY POINT

- If a less powerful hardware configuration is targetted, then the Mealy machine is preferred.

FA (Acceptor) vs FSM with Output (Transducer)

- In a transducer, there is one output for each input symbol (each edge consists of an input/output), whereas no output is associated with each symbol in an acceptor. More explicitly, when we need to convert an input string to an output string, we prefer to use the transducer. On the other hand, if we need to check the acceptance of a string, we prefer the acceptor (pattern checker). Thus, the acceptor is simpler than the transducer.
- An acceptor can be represented as a transducer by applying the output for each input. However, converting the transducer to an acceptor is not of practical interest.
- Two tapes are used in such transducers: one is the Input tape and the other is the Output tape. Accordingly, two heads are used: one is the Read head to read input from the Input tape and the other is the Write head to write output on the Output tape. In fact, no R/W head (like in TM) is used to perform both read and write operations simultaneously on the same tape. Further, both the heads cannot move backwards.

EXAMPLE 2.16

(a) Design a Mealy machine to convert a binary number to its 1's complement form.

Solution The given problem is to convert any binary number into its 1's complement form. This can be accomplished if the machine shows output '1' for input '0' and vice versa.

To design such a machine, one state is sufficient, and it is the start state (say, q_0). In fact, if input '0' is encountered at q_0, its corresponding output will be '1' (which is represented here as 0/1). On the other hand , if '1' is received at q_0, the corresponding output will be '0' (which is represented as 1/0). The input/output is shown graphically in Fig. 2.31 (a).

0/1
1/0
q_0

FIGURE 2.31 (a) Mealy machine for 1's complement of binary number

Here, the content stored earlier at q_0 is not used to compute the complement. Thus, if input is 1101, the output is 0010.

Input: 1101

State and input: q_0–1– q_0 –1– q_0–0 – q_0 – 1 – q_0

Output: 0 0 1 0

Note: In the state diagram of the Mealy machine, an edge (arrow) shows input as well as output, in the format: *input / output*; whereas the state diagram of the Moore machine shows output in the format: *state / output*.

(b) Design a Moore machine to convert a binary number to its 1's complement form.

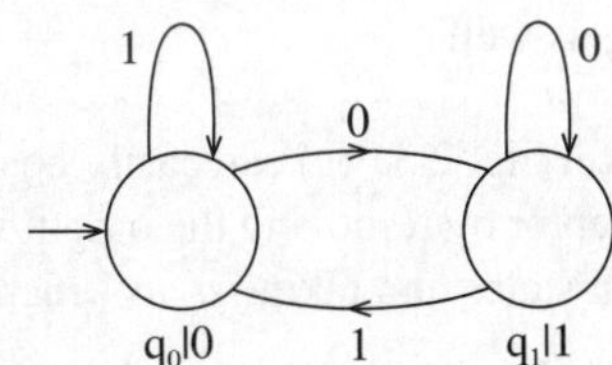

FIGURE 2.31 (b) Moore machine for 1's complement of binary number

Solution We know that output by the Moore machine depends only on state and one Moore machine for the given problem is shown in Fig. 2.31 (b). Here, for input 0, the machine moves to state q_1 from state q_0 and the resulting output is 1, that is, input 0 is stored in q_1. Likewise, input 1 and null are stored at q_0 to give output 0.

Thus, if the input to the Moore machine is 1101, its 1's complement will be 00010. Importantly, the size of the input here is 4, whereas the size of the output is $(4 + 1) = 5$.

Now, referring to the schemes that produce output followed by machines, as shown in Figs 2.31 (a) and (b), we obtain 0010 and 00010, respectively, as output for the input $w = 1101$. Thus, if the length of the input string is n, the length of the output string produced by the Mealy machine will be n, while it would be $(n + 1)$ for the Moore machine (as the output depends only on state).

FSM WITH OUTPUT AS AN ACCEPTOR

FSM with output is a special kind of FA. Acceptance of string can also be checked by applying this type of FSM; that is, they may behave like a normal FA. For example, when a string ends with 00, the output may be terminated by a special output symbol such as 'x' (especially for the last 0 of '00') to indicate that it is a valid string; otherwise, some other symbol (except 'x') may be used for other inputs. Similarly, examining the even or odd number of 1s in a string may be performed and it is shown below.

However, a normal FA cannot perform string conversion except acceptance of string, as the provision of an output for each input is ignored here.

(c) Construct a Mealy machine for finding 2's complement of a given binary number. Assume the string is read from LSB to MSB and the end carry is discarded.

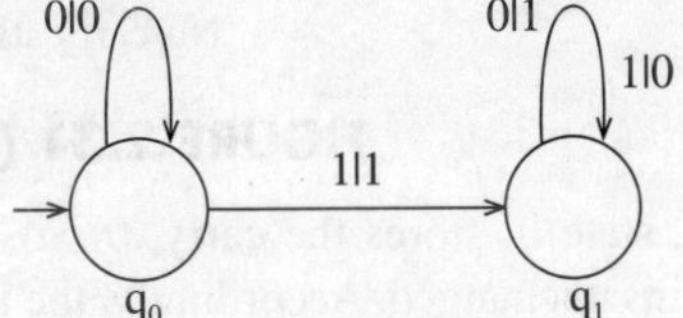

FIGURE 2.31 (c) Mealy machine for 2's complement of binary numbers

Solution Here, $\Sigma = \{0, 1\}$, $\Delta = \{0, 1\}$.

To construct the Mealy machine to find 2's complement of a binary number (N), we may use the following basic steps:

Step 1: Find the first unprocessed 1 of N from the LSB. If it is found, then keep it unchanged and traverse the rest of N (towards MSB) by converting 0 to 1 and 1 to 0.

Step 2: For the binary substring (starting from LSB towards MSB) consisting of 0s, keep it unchanged.

The logic based on the above-mentioned steps is shown in Fig. 2.31 (c).

(d) Design a Mealy machine to verify an even number of 1s (outputting even if the number of 1s is even, else odd).

Solution The state transition diagram for the given problem is shown in Fig. 2.31 (d).

Here, the null string is ignored. If one wants to consider it, then a new start state must to be added to the machine so that it connects to the present start state through ε move. The new start state will the result in even output.

An equivalent Moore machine is shown in Fig. 2.31 (e), which is capable of operating the null string as well.

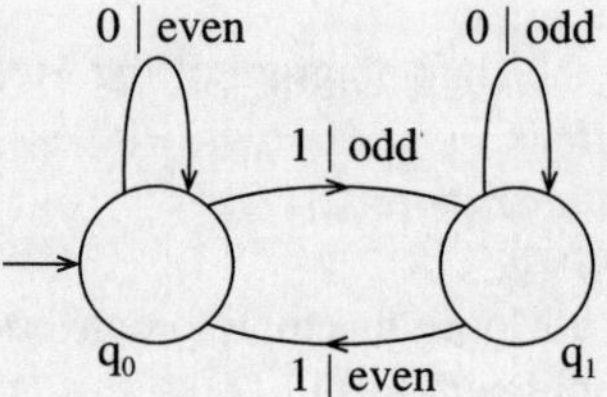

FIGURE 2.31 (d) Mealy machine for checking even number of 1s or 0s

> **Note:** Fig. 2.31 (e) can easily be represented as an acceptor by removing the output values associated with states and taking q_0 as accepting state.

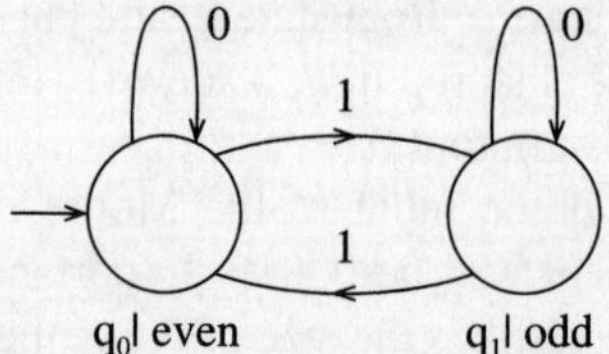

FIGURE 2.31 (e) Moore machine for checking even number of 1s or 0s

(e) Design a Mealy machine for a serial binary adder.

Solution Here, the possible combinations of input strings are 00, 01, 10 and 11, whereas the outputs are simply 0 or 1. Obviously, two input symbols (one from each terminal tape) for each combination are simultaneously received and the output corresponding to the combination (as shown in Fig. 2.31 (f)) is produced and stored in an output tape. We may assume that the input symbols (one from each tape) are read from the right or the inputs in each tape are written in reverse order. As it is an abstract model, we do not focus on how the string is actually read or written in reverse order.

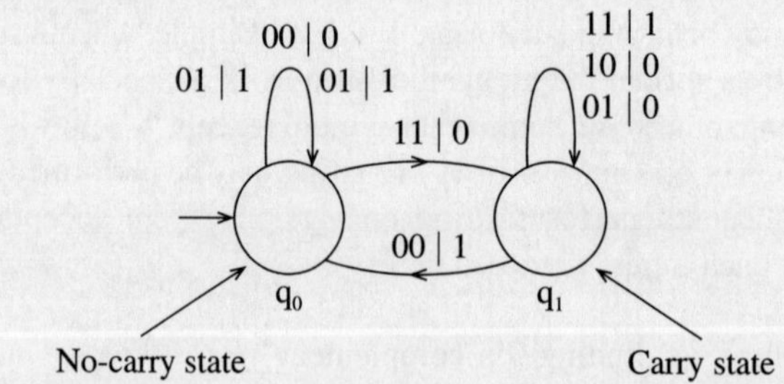

FIGURE 2.31 (f) Mealy machine for serial binary adder

Here, state q_0 stores the carry, 0. So, for any of the combinations 00, 01, 10 with carry 0, the carry remains unchanged. Accordingly, the machine remains at q_0 for these inputs. However, as soon as carry 1 is produced for 11, the machine moves to state q_1 and the state stores the carry, 1. At q_1, for any of the combinations 10, 01, 11 with carry 1, the carry remains unchanged and the machine remains at q_1 for these input strings.

> **Note:** During addition, 0 is appended just before the MSB of each number; for example, 14 = 1110, but it will be 01110. Of greater interest, the state diagram for a binary subtractor is shown in Fig. 2.31 (g).

(f) Design a Mealy machine for computing Exclusive-OR.

Solution In the case of Exclusive-OR for binary inputs, one output for two binary input symbols is obtained. The inputs (*x*, *y*) and the corresponding outputs (O) are as follows:

- $x = 0, y = 0, O = 0$
- $x = 0, y = 1, O = 1$
- $x = 1, y = 0, O = 1$
- $x = 1, y = 1, O = 0$

A simple state diagram of the Mealy machine is shown in Fig. 2.31 (h).

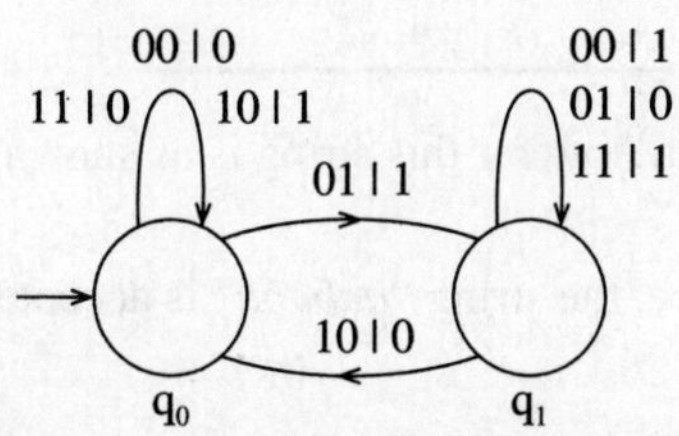

FIGURE 2.31 (g) Mealy machine for serial binary subtractor

FIGURE 2.31 (h) Mealy machine for computing Exclusive-OR

(g) Design a Moore machine for traffic signal.

Solution The state diagram of a Moore machine corresponding to the traffic signal problem is shown in Fig. 2.31 (i).

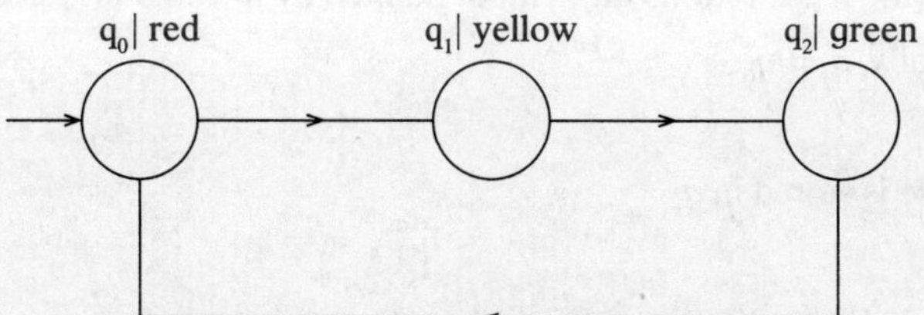

FIGURE 2.31 (i) Moore machine for traffic signal

2.11.5 FA-based String Matching Algorithm

Finite automata provides a very natural algorithm for string matching. We can understand it easily using an example. Consider the DFA shown in Fig. 2.32 defined over an alphabet $\Sigma = \{a, b\}$.

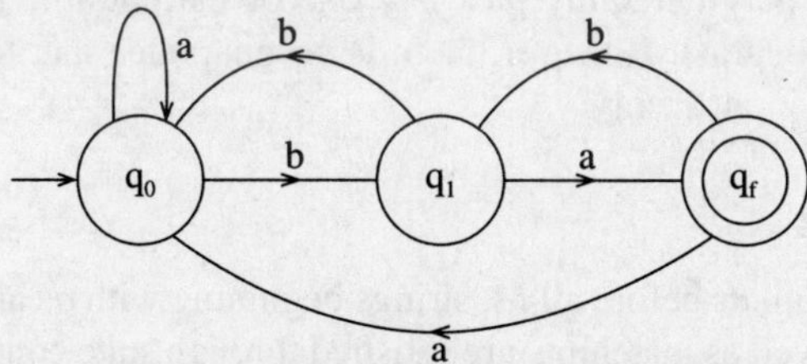

FIGURE 2.32 An FA for matching strings

Time complexity: The time to construct a DFA for solving string matching problem is $O(n|\Sigma|)$, where n is the length of the shortest string to be matched and $|\Sigma|$ is the number of symbols present in Σ. The state transition table for this DFA is given in Table 2.18.

TABLE 2.18 Transition table for Fig. 2.32

Present state	Input symbol = a	Input symbol = b
$\rightarrow q_0$	q_0	q_1
q_1	q_f	q_0
$* q_f$	q_0	q_1

Consider the processing of string "*aababa*".The state transition flow for this string is as shown below:

$q_0 q_0 q_0 q_1 q_f q_1 q_f$

Processing of string stops at state q_f (a final state), and hence, the string "*aababa*" is accepted by the DFA.

Algorithm for matching strings by constructed DFA

Parameters

ip: input pointer //initially, *ip* = 0; it accesses the content of a location pointed to by *ip* in string *w;* Q: set of states; F: set of final states, a subset of Q; q_0: (initial state of M) $\in$ Q; *w*: string over alphabet Σ

while (*w* is not entirely consumed) *do*

begin

$a = w(ip)$ // $a \in \Sigma$ and $w(ip)$ returns the symbol pointed by *ip* and it is stored at *a*.

$\delta(q_i, a) = q_j$ // initially, $q_i = q_0$.

$ip = ip + 1$

$q_i = q_j$ // output state is stored in q_i

end

If ($q_i \in$ F), then *w* is valid // when the string is completely processed.

else invalid.

IMPLEMENTING STRING MATCHING ALGORITHM USING NDFA

In the case of NDFA, we can use the same model as discussed above; however, there is a problem in the case of NDFA, namely, movement of more than one state from a given state for a specific input symbol. To tackle this, we may randomly pick any one of the possible transitions, assuming that random selection is more accurate. However, there is no guarantee that this algorithm will always accurately accept any valid input string.

Power and limitations of FA

- Ordering (appearance of all *a*s before all *b*s, strings beginning with *a*s and ending with *b*s, etc.) and finite measure value, as well as matching, are satisfied through state control of FA (or state change).
- Further, the measure value (expressible as arithmetic progression or linear measures like a^n but not like a^{n^2} or a^{2^n}, which are non-linear expressions of *n*) without any comparison can be resolved by FA for an infinite language (using a loop formed by a finite number of states).

To overcome the limitations of FA, measure constraint as well as comparison for infinite language can be resolved by more powerful automata like PDA or higher. However, in the case of PDA (where stack is used), the counting criteria for infinite languages must be linear. So, an infinite language like a^{n^2} or a^{2^n} may not be recognised by PDA.

Claim for equivalent power of NDFA and DFA: For proving equivalence of NDFA and DFA, we may take the help of the definite method for converting any NDFA to an equivalent DFA.

2.12 ADDITIONAL SOLVED PROBLEMS

PROBLEM 1

Design a DFA to accept language L = $\{w \mid w$ contains alternating sequences of 0s and 1s$\}$ over $\{0, 1\}$.

Solution A DFA for the given problem is shown in Fig. 2.33. According to the given property, it may be assumed that if two consecutive identical symbols are encountered, the machine will move to the dead state.

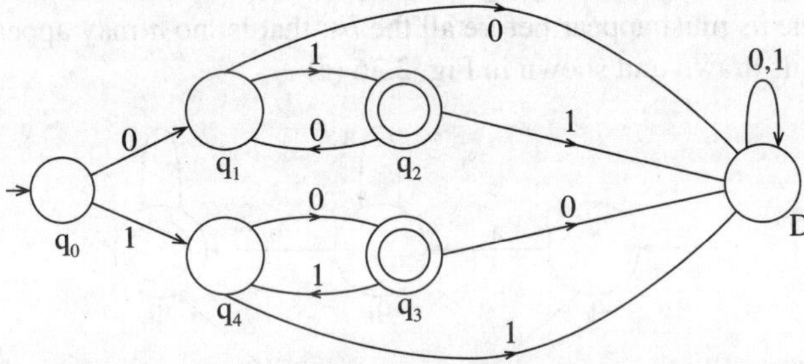

FIGURE 2.33 A DFA to accept strings consisting of alternating sequences of 0s and 1s

PROBLEM 2

Design a DFA, L = $\{a^n b \mid n \geq 0\}$ over $\{a, b\}$.

Solution According to the given language, the number of *as* in the valid string may be zero or more, but the number of *bs* is only one. Further, all *as* must appear before *b*; that is, no *a* may appear after a *b*. With these points in mind, the corresponding DFA is shown in Fig. 2.34.

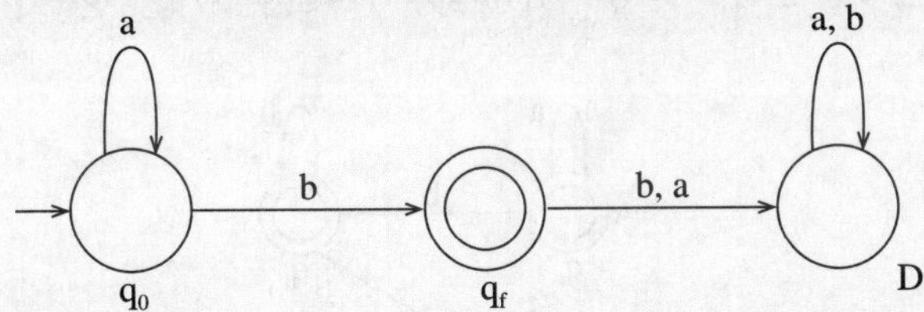

FIGURE 2.34 DFA accepting L = $\{a^n b \mid n \geq 0\}$ over $\{a, b\}$

PROBLEM 3

Design the DFA for L = $\{ w \mid w \in \{a, b, c\}$ and w contains the pattern *abac*$\}$.

Solution From the given language, it is true that a string is valid if it contains the pattern *abac*. Based on this, a DFA is drawn and shown in Fig. 2.35.

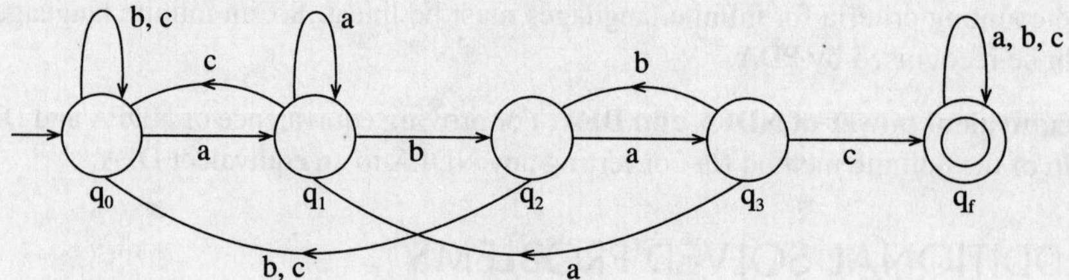

FIGURE 2.35 DFA accepting pattern *abac*

PROBLEM 4

Design the DFA for $L = \{a^n b^m \mid n, m > 0\}$ over $\{a, b\}$.

Solution According to the given language, the number of *a*s and *b*s in the valid string may be one or more. Further, all the *a*s must appear before all the *b*s; that is, no *a* may appear after a *b*. With these points in mind, a DFA is drawn and shown in Fig. 2.36 (a)

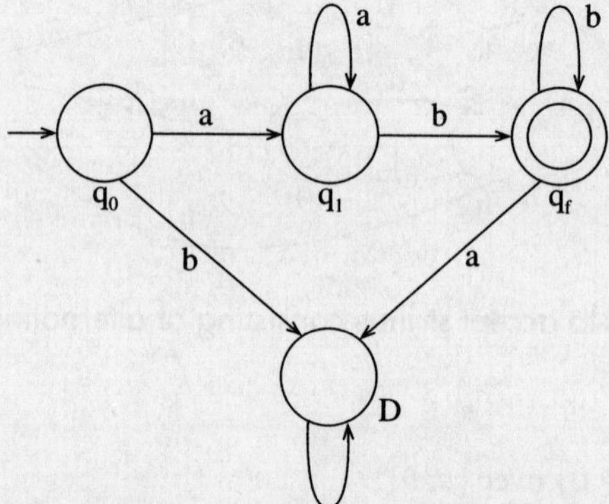

FIGURE 2.36 (a) DFA accepting $L = \{a^n b^m \mid n, m > 0\}$ over $\{a, b\}$

Note: For $L = \{a^n b^m \mid n, m \geq 0\}$ over $\{a, b\}$, the state diagram is as shown in Fig. 2.36 (b).

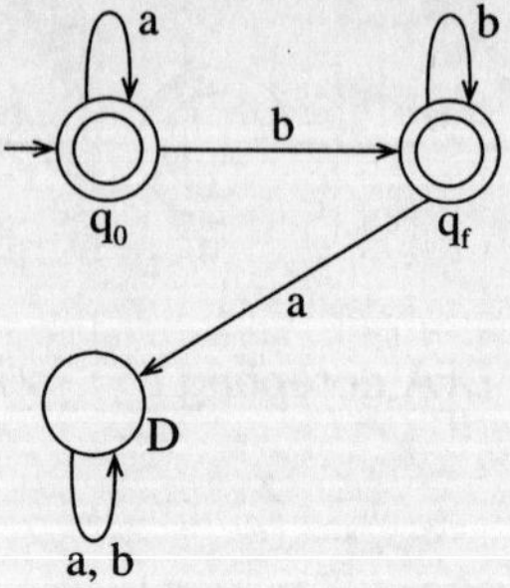

FIGURE 2.36 (b) DFA accepting $L = \{a^n b^m \mid n, m \geq 0\}$ over $\{a, b\}$

PROBLEM 5

Design a DFA that recognises a set of strings on $\sum = \{a, b\}$ with prefix *ab*.

Solution The proposed DFA recognises only strings that have the prefix *ab*. So, if any string starting with '*b*' is not valid, that is, the DFA must move to the dead state for such a string. Likewise, if '*a*' is encountered just after the first '*a*', the machine goes to the dead state. Keeping these points in mind, a DFA is drawn and shown in Fig. 2.37.

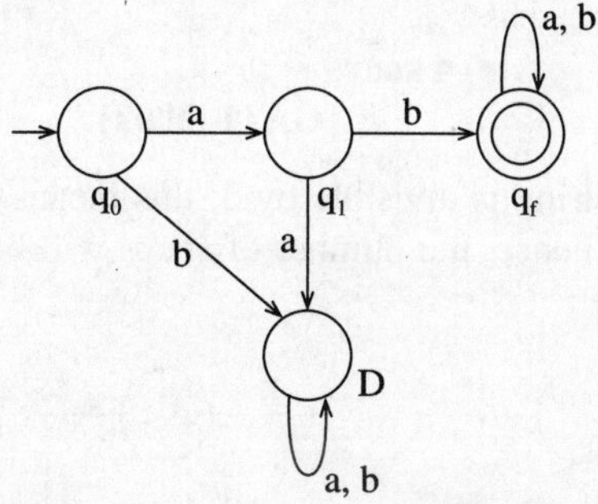

FIGURE 2.37 DFA recognising strings with prefix *ab*

PROBLEM 6

Design a DFA over {0, 1} to accept strings such that each block of five symbols has the last two symbols as 0.

Solution Here, we must focus on each block of size 5 such that the last two symbols of the block are 0s. However, the other three symbols in each block may be 0s or 1s. Further, if 1s are encountered for the fourth or fifth symbols in the block, the machine must go to the dead state. Following this idea, a DFA is drawn and shown in Fig. 2.38.

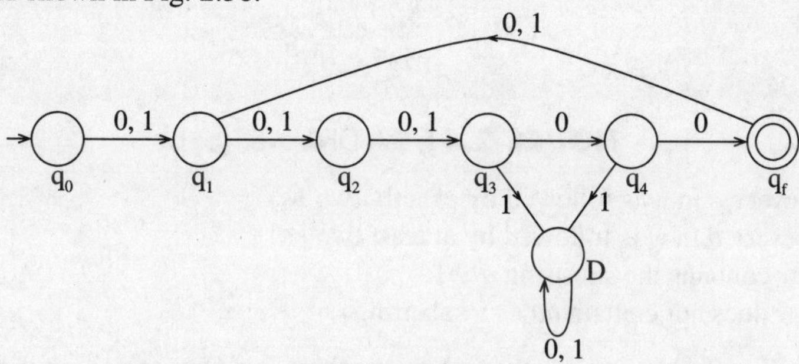

FIGURE 2.38 DFA to accept strings such that each block of 5 symbols has the last symbols as 0s

PROBLEM 7

Design a DFA to accept signed integers.

Solution Any integer starting with '+' or ' – ' will be treated as a signed integer. However, we generally ignore the '+' sign before any positive integer. Here, it is assumed that '+' or '–' will appear explicitly before each signed integer, and a DFA based on this constraint is shown in Fig. 2.39. Hoever, during real implementation of this problem, we may assume that '+' will not appear explicitly before any positive integer.

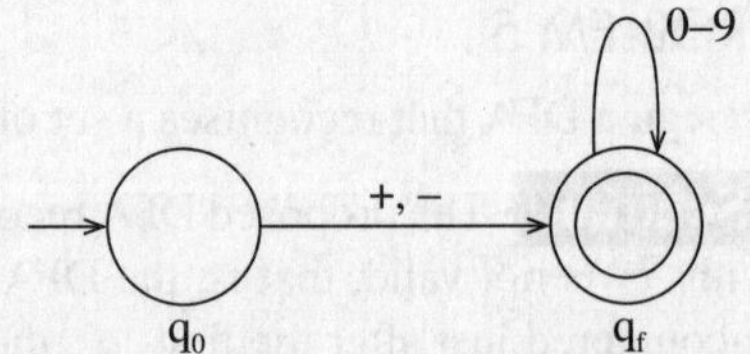

FIGURE 2.39 DFA recognising signed integers

PROBLEM 8

The smallest finite automaton that accepts the language {x | length of x is divisible by 3} over $\sum$ has

(a) 2 states (b) 3 states (c) 4 states
(d) 5 states **[GATE 2002]**

Solution The length of the string is divisible by 3; this means that the number of input symbol-based transitions must be 3, and hence, the number of states will be at least 4. The DFA is shown in Fig. 2.40. The correct option is (c).

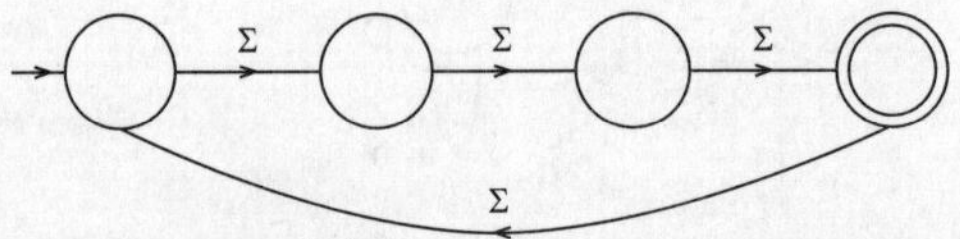

FIGURE 2.40 Smallest FA

PROBLEM 9

Which language is recognised by the following finite state machine?

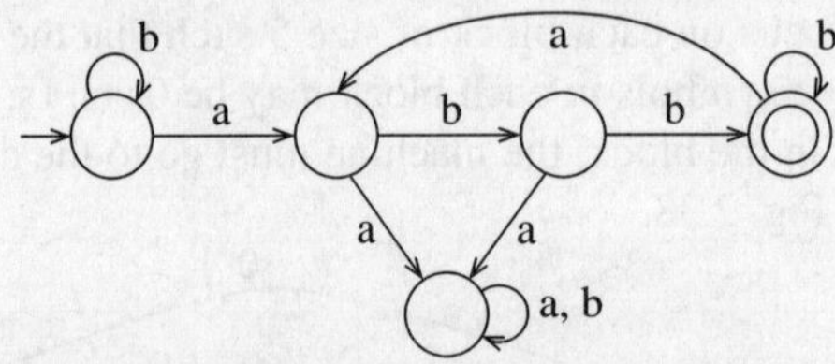

FIGURE 2.41 A DFA over {a, b}

(a) $\{w \in \{a, b\}^* \mid$ every a in w is followed by exactly two $bs\}$
(b) $\{w \in \{a, b\}^* \mid$ every a in w is followed by at least two $bs\}$
(c) $\{w \in \{a, b\}^* \mid w$ contains the substring $abb\}$
(d) $\{w \in \{a, b\}^* \mid w$ does not contain aa as a substring$\}$ **[GATE 2005]**

Solution From Fig. 2.41, it is clear that whenever there is a state transition through symbol '*a*', it must be followed by exactly two *b*s for the transition to stay in a valid (non-halt) state. Hence, option (a) is ruled out because the string *abbbbb* is accepted by this automaton. Option (b) is more accurate, and is, in fact, equivalent to Option (c) as well.

Also, observe that whenever an '*a*' is followed by another '*a*', the automaton goes into a dead state. So, Option (b) is also correct. Hence, there are multiple correct options here: (b), (c) and (d).

PROBLEM 10

Given the language L = {*ab*, *aa*, *baa*}*, which of the following strings are in L*?

1) abaabaaabaa 2) aaaabaaaa 3) baaaaabaaaab 4) baaaaabaa
(a) 1, 2 and 3 (b) 2, 3 and 4 (c) 1, 2 and 4 (d) 1, 3 and 4 **[GATE 2012]**

Solution Try to break the strings into contiguous pieces and check if each piece is valid or not.
abaabaaabaa – see the breakup as highlighted – accepted string.
aaaabaaaa – see the breakup as highlighted – accepted string.
baaaaabaaaab – see the breakup as highlighted – but the last b is not valid. So, 3 is not accepted.
baaaaabaa – see the breakup as highlighted – accepted string.
So, only string 3 is not accepted. The correct option is (c).

PROBLEM 11

Construct a DFA that accepts a language over {0,1} such that valid strings contain neither the substring 00 nor the substring 11. **[GATE 2000]**

Solution Given that the valid strings in the language contain neither the substring 00 nor the substring 11. We know that neither P nor Q ≡ ¬P AND ¬Q. Hence, we may first draw DFA for strings accepting the substrings 00 and 11. Next, the complement of the drawn DFA is to be taken to solve the given problem.

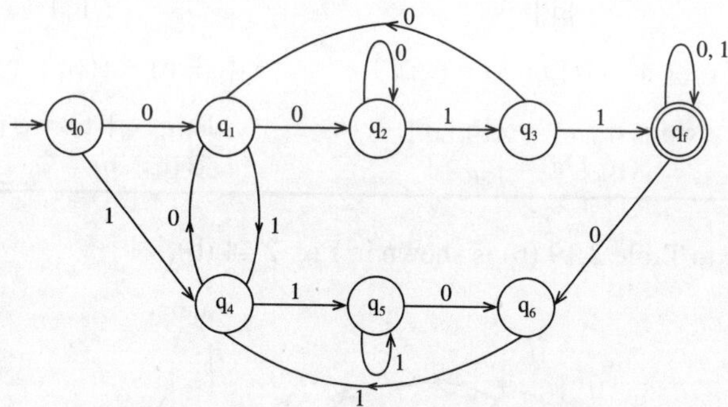

FIGURE 2.42 DFA accepting strings containing substrings 00 and 11

Now, take the complement of Fig. 2.42 to satisfy the give properties, that is, to make the final states as the non-final states and the non-final states as the final states.

PROBLEM 12

Design a DFA to recognise binary number whose decimal equivalent is divisible by 2.

Solution The decimal equivalent of a binary number is divisible by 2 if the binary number ends with 0. A DFA is shown in Fig. 2.43.

FIGURE 2.43 DFA to recognise binary numbers divisible by 2

PROBLEM 13

Design an NDFA to accept strings not containing two consecutive *b*s over {*a*, *b*}. Then, convert it into an equivalent DFA.

Solution An NDFA for accepting strings not containing two consecutive *b*s is shown in Fig. 2.44 (a).

TABLE 2.19 (a) State table for NDFA shown in Fig. 2.44 (a)

State	Input(a)/Next state	Input(b)/Next state
$\rightarrow q_0$	q_0	q_f
q_f	q_0, q_f	Φ

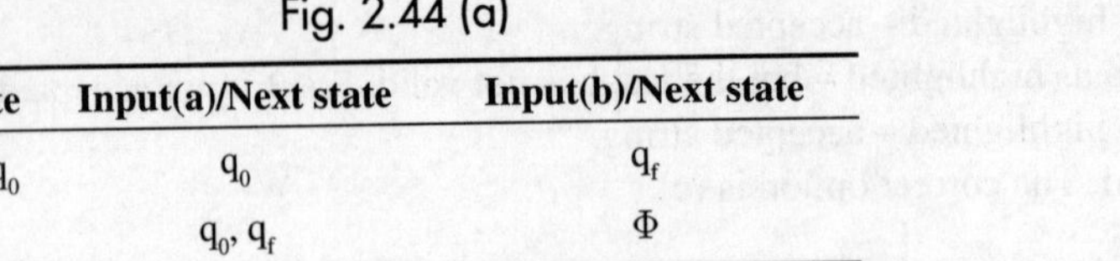

FIGURE 2.44 (a) An NDFA for accepting strings not containing two consecutive *b*s

Applying the subset construction strategy to convert NDFA to DFA, we obtain the state table shown in Table 2.19 (b) of the equivalent DFA for the NDFA shown in Fig. 2.44 (a).

TABLE 2.19 (b) State table of DFA for the NDFA shown in Fig. 2.44 (a)

State	Input(a)/Next state	Input(b)/Next state
$\rightarrow[q_0]$	$[q_0]$	$[q_f]$
$*[q_f]$	$\delta'([q_f], a) = \delta(\{q_f\}, a) = [q_f, q_0]$	$\delta'([q_f], b) = \delta(\{q_f\}, b) = \Phi = D$
$[q_f, q_0]$	$\delta'([q_f, q_0], a) = \delta(\{q_f\}, a) \cup \delta(\{q_0\}, a) = [q_f, q_0]$	$\delta'([q_f, q_0], b) = \delta(\{q_f\}, b) \cup \delta(\{q_0\}, b) = \Phi \cup q_f = [q_f]$

The DFA corresponding to Table 2.19 (b) is shown in Fig. 2.44 (b).

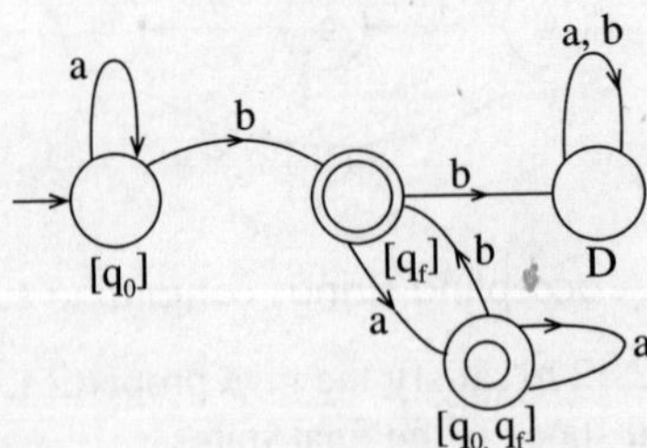

FIGURE 2.44 (b) A DFA for accepting strings not containing two consecutive *b*s

PROBLEM 14

Design a four-state NFA to recognise strings ending with *abb* over $\{a, b\}$.

Solution The NFA is very intuitive and is shown in Fig. 2.45.

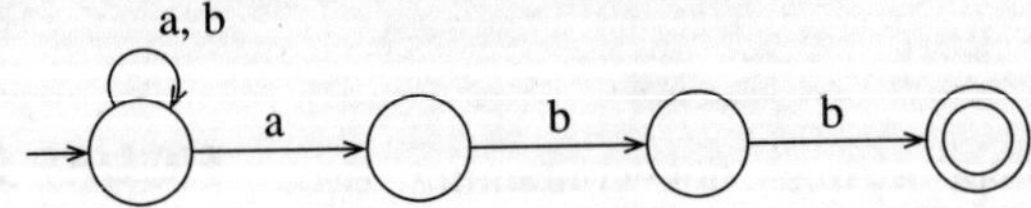

FIGURE 2.45 An NFA to recognise strings ending with *abb* over $\{a, b\}$.

PROBLEM 15

(a) Design an ε NFA for L = $\{0^k \mid k$ is divisible by 2 or 3$\}$. Then convert it into an equivalent DFA.

Solution An ε NFA for the given language is shown in Fig. 2.46 (a).

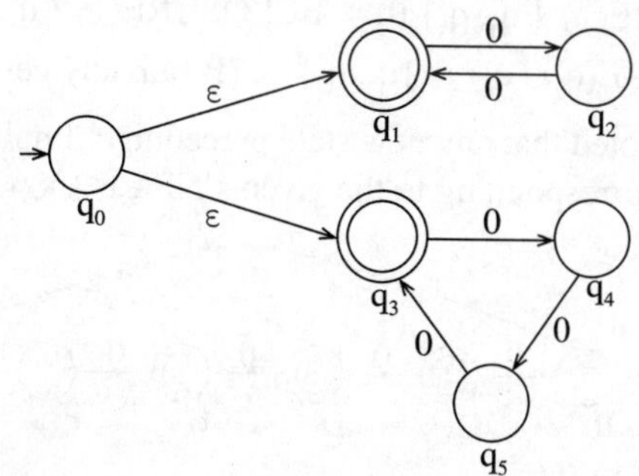

FIGURE 2.46 (a) An ε NFA for L = $\{0^k \mid k$ is divisible by 2 or 3$\}$

To convert the given ε NFA to DFA, we must follow the ECLOSURE strategy discussed earlier.

Now, based on the strategy, we get

ECLOSURE(q_0) = $\{q_0, q_1, q_3\}$ = *A (state) // *as the initial state of DFA*

Now, for state A = $\{q_0, q_1, q_3\}$, compute δ_D for input 0

δ_D ($\{q_0, q_1, q_3\}$,0) = ECLOSURE(δ_E $\{q_0, q_1, q_3\}$,0)

= ECLOSURE(δ_E (q_0, 0) ∪ δ_E (q_1,0) ∪ δ_E (q_3, 0)) = ECLOSURE(Φ ∪ q_2 ∪ q_4)

= $\{q_2, q_4\}$ = B (state)

For state B = $\{q_2, q_4\}$ and input 0

δ_D ($\{q_2, q_4\}$,0) = ECLOSURE(δ_E $\{q_2, q_4\}$,0)

= ECLOSURE(δ_E (q_2, 0) ∪ δ_E (q_4,0) = ECLOSURE(q_1∪ q_5)

= $\{q_1, q_5\}$ = *C (state)

For state C = $\{q_1, q_5\}$ and input 0

δ_D ($\{q_1, q_5\}$,0) = ECLOSURE(δ_E $\{q_1, q_5\}$,0)

= ECLOSURE(δ_E (q_1, 0) ∪ δ_E (q_5,0) = ECLOSURE(q_2 ∪ q_3)

= $\{q_2, q_3\}$ = *D (state)

For state D = $\{q_2, q_3\}$ and input 0

δ_D ($\{q_2, q_3\}$,0) = E-CLOSURE(δ_E $\{q_2, q_3\}$,0)

= ECLOSURE(δ_E (q_2, 0) ∪ δ_E (q_3,0) = ECLOSURE(q_1∪ q_4)

= $\{q_1, q_4\}$ = *E (state)

For state E = $\{q_1, q_4\}$ and input 0

δ_D ($\{q_1, q_4\}$,0) = ECLOSURE(δ_E $\{q_1, q_4\}$,0)

= ECLOSURE(δ_E (q_1, 0) ∪ δ_E (q_4,0) = ECLOSURE(q_1∪ q_4)

= $\{q_2, q_5\}$ = F (state)

For state F = $\{q_2, q_5\}$ and input 0

δ_D ($\{q_2, q_5\}$,0) = ECLOSURE(δ_E $\{q_2, q_5\}$,0)

$= \text{ECLOSURE}(\delta_E(q_2, 0) \cup \delta_E(q_5, 0) = \text{ECLOSURE}(q_1 \cup q_3)$

$= \{q_1, q_3\} = {}^*G$ (state)

For state F = $\{q_1, q_3\}$ and input 0

$\delta_D(\{q_1, q_3\}, 0) = \text{ECLOSURE}(\delta_E\{q_1, q_3\}, 0) = \text{ECLOSURE}(\delta_E(q_1, 0) \cup \delta_E(q_3, 0))$

$= \text{ECLOSURE}(q_2 \cup q_4) = \{q_2, q_4\} = {}^*B$ (already generated)

So, we stop here. It may be noted that any new state preceding * implies the final state. Hence, the equivalent DFA corresponding to the given ε NFA is shown in Fig. 2.46 (b).

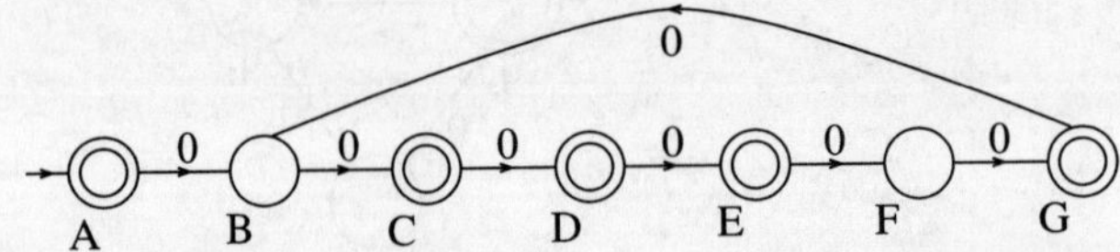

FIGURE 2.46 (b) A DFA for L = $\{0^k \mid k \text{ is divisible by 2 or 3}\}$

However, discarding redundant states, the reduced DFA is shown in Fig. 2.46 (c).

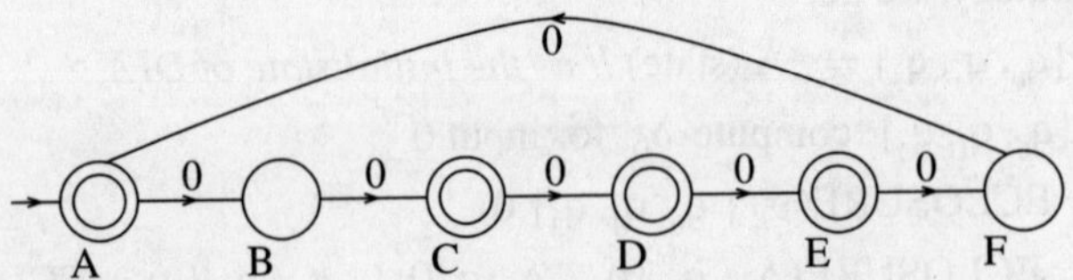

FIGURE 2.46 (c) A DFA for L = $\{0^k \mid k \text{ is divisible by 2 or 3}\}$

(b) Design an NFA to accept strings with an even number of 0s over {0, 1}.

Solution

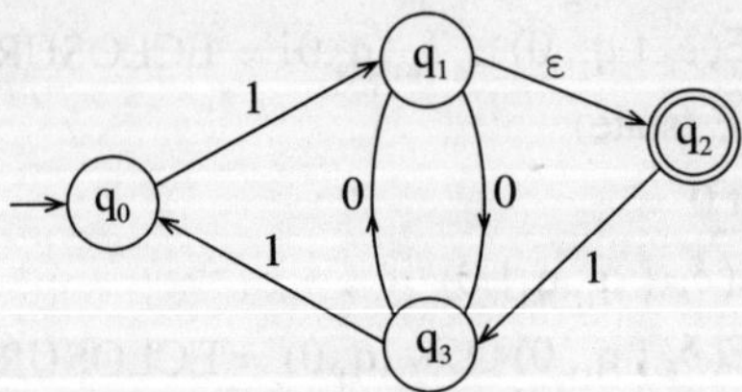

FIGURE 2.46 (d) An NFA to accept strings with an even number of 0s over {0, 1}

(c) Design an NFA to accept an even number of 0s and an odd number of 1s over {0, 1}.

Solution

FIGURE 2.46 (e) An NFA accepting strings with an even number of 0s and an odd number of 1s over {0, 1}

PROBLEM 16

Design an NDFA for the language L = $\{0^m1^n0^p0 \mid m, n, p \geq 0$ over $\{0, 1\}\}$ using only three states.

Solution An NDFA for the given problem is shown in Fig. 2.47.

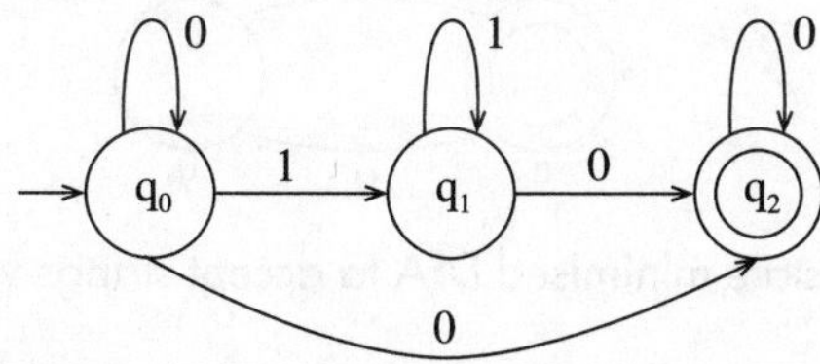

FIGURE 2.47 An NDFA to accept L = $\{0^m1^n0^p0 \mid m, n, p \geq 0$ over $\{0, 1\}\}$

PROBLEM 17

(a) Design an NDFA to accept an even number of 0s over {0, 1} (ignoring null string as valid string). Find the set of reachable states for the input string 11.

(b) Convert the NDFA into an equivalent DFA.

Solution One possible NDFA for accepting the valid strings is shown in Fig. 2.48 (a).

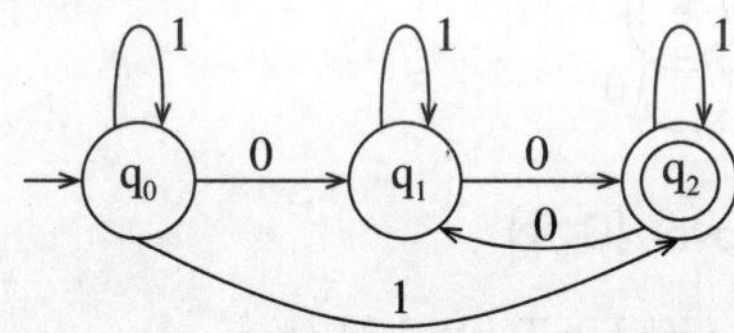

FIGURE 2.48 (a) NDFA to accept strings with an even number of 0s

TABLE 2.20 (a) State table for NDFA shown in Fig. 2.48 (a)

State	Input(0)/Next state	Input(1)/Next state
→(q_0)	q_1	q_0, q_2
q_1	q_2	q_1
*q_2	q_1	q_2

The set of reachable states for the input string 11 is $\delta(q_0,11) = \delta(\delta(q_0,1),1) = \delta(\{q_0,q_2\},1)$

$$= \delta(q_0,1) \cup \delta(q_2,0) = \{q_0, q_2\}$$

By consulting the transitions of the NDFA (displayed in Table 2.20 (a)), an equivalent DFA table is created as shown in Table 2.20 (b). The equivalent DFA is shown in Fig. 2.48 (b).

TABLE 2.20 (b) State table for DFA

State	Input(0)/Next state	Input(1)/Next state
→[q_0]	[q_1]	[q_0, q_2]
[q_1]	[q_2]	[q_1]
*[q_2]	[q_1]	[q_2]
*[q_0, q_2]	[q_1]	[q_0, q_2]

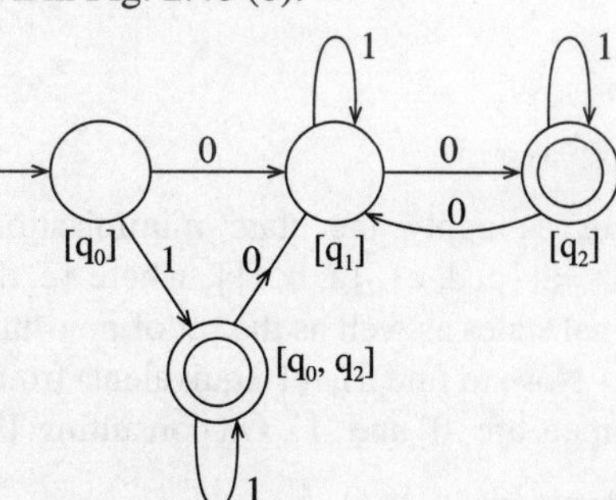

FIGURE 2.48 (b) A DFA to accept strings with an even number of 0s

One may note that the DFA shown in Fig. 2.48 (b) is not a minimised one. It can be minimised using any state minimisation approach, and the resulting DFA is shown in Fig. 2.48 (c).

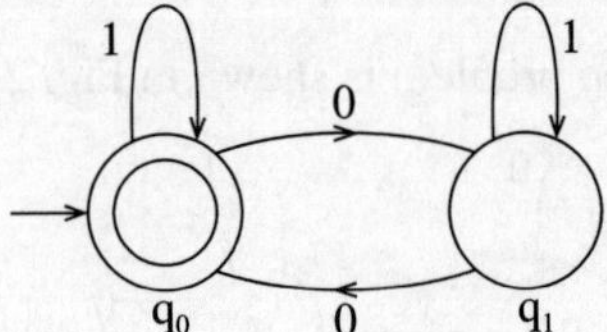

FIGURE 2.48 (c) A state minimised DFA to accept strings with an even number of 0s

PROBLEM 18

Find the reduced DFA of the one shown in Fig. 2.49 (a).

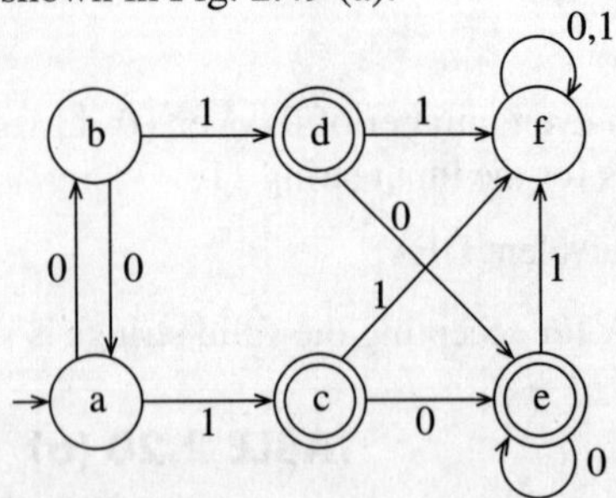

FIGURE 2.49 (a) A DFA over {0, 1}

Solution The state table equivalent to Fig. 2.49 (a) is presented in Table 2.21 (a).

TABLE 2.21 (a) Equivalent state table for Fig. 2.49 (a)

q	δ(q,0)	δ(q,1)
a	b	c
b	a	d
c	e	f
d	e	f
e	e	f
f	f	f

Let us apply the state minimisation algorithm to the DFA above. Here, the 0-th equivalent, $\Pi_0 = \{\{c, d, e\}, \{a, b, f\}\}$, where {c, d, e} is the set of final states. In particular, Π_0 contains the set of final states as well as the set of non-final states.

Now, to find Π_1 (1-equivalent) from Π_0, we need to consider a single input for the states in Π_0. The inputs are '0' and '1'. On consulting Table 2.21 (a), we obtain the following:

Partition: {a, b, f}

- Input = '0' a – 0 – b $\notin$ F, b – 0 – a $\notin$ F, f – 0 – f $\notin$ F
- Input = '1': a – 1 – c $\in$ F, b – 1 – d $\in$ F, f – 1 – f $\notin$ F

Thus, {a, b, f} may not be in the same group, as the machine goes to the non-final state from *f* on reading input '1', whereas the machine moves to the final state from the other two states for input '1'. Hence, the new groups:{a, b}, {f} are constructed from the single group:{a, b, f}.

Partition: {c, d, e}: This group need not be split, as each of these moves to the final state for input '0' and the non-final state for input '1'. Therefore, $\Pi_1 = \{\{c, d, e\}, \{a, b\}, \{f\}\}$.

Let us now attempt to find the 2-equivalent: Π_2. In this regard, we need to consider strings, each of length 2, which are 00, 01, 10, 11 and 10. We may apply a single input for these states (groups of states) in Π_1, which gives the following:

Partition: {c, d, e}

- Input = 0 : c – 0 – e ∈ {c, d, e}, d – 0 – e ∈ {c, d, e}, e – 0 – e ∈ {c, d, e}
- Input = 1 : c – 1 – f ∈ { f }, d –1 – f ∈ {f}, e –1 – f ∈ {f}

So, this partition need not be split (as each of these moves to the final state for input '0'and the non-final state for input '1').

- Partition: {f} may not be split, as it has a single state.
- Partition:{a, b} may not be split, since for input = '0', a – 0 – b ∈ {a, b}, b – 0 – a ∈ {a, b}; and for input = '1'; a – 1 – c ∈ {c, d, e}, b – 1 – d ∈ {c, d, e}

However, we may carry out direct verification over the strings: 00, 01, 10, 11 for the groups, {a, b}, {f}, {c, d, f}. For example, for group {a, b}: a – 00 – a ∉ F, b – 00 – b ∉ F; a – 01 – d ∈ F, b – 01 – c ∈ F and so on.

Anyway, Π_2 = {(c, d, e), (a, b),(f)}. Hence, $\Pi_1 = \Pi_2$, so stop partitioning.

Now, the equivalent states (in the form of partitions) are named A = {a, b}, B = {c, d, e} and C = {f}. Here, the initial state is A = {a, b}, as the initial state '*a*' of the original DFA is present. Further, B = {c, d, e} is the final state of the reduced DFA, as the final states of the original DFA are present in B. Finally, C = {f} is the non-final (intermediate) state of the reduced machine.

Thus, there are three states in the reduced DFA, shown in Fig. 2.49 (b). Also, its equivalent table is given in Table 2.21 (b).

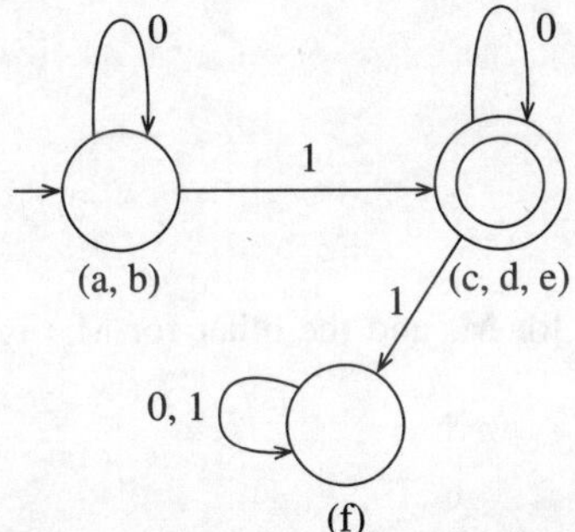

FIGURE 2.49 (b) A minimal DFA of the DFA shown in Fig. 2.49 (a)

TABLE 2.21(b) The state table for Fig. 2.49 (b)

Q	δ(q,0)	δ(q,1)
(a, b)	(a, b)	(c, d, e)
(c, d, e)	(c, d, e)	(f)
(f)	(f)	(f)

PROBLEM 19

Design a DFA to accept strings beginning with '*a*' and ending with '*b*' over {*a*, *b*} by applying the logic of product automata.

Solution As we need to apply the logic of product automata, we must first draw two DFAs: one to accept strings beginning with '*a*' and the other to accept strings ending with '*b*'. These are shown in Figs 2.50 (a) and (b), respectively.

FIGURE 2.50 Two distinct DFAs over {a, b}: **(a)** M_1: accepting strings beginning with 'a'; **(b)** M_2: accepting strings beginning with 'b'

Here, it is not possible for machine M_1 to move forward if the string begins with '*b*', as it first checks the starting symbol, irrespective of what comes next. So, M_1 moves to the dead state in this case. On the other hand, M_2 may move forward even if '*a*' is encountered after '*b*', as M_2 does not see what may come later on. So, no dead case may arise for M_2. The dead state D is denoted as Φ while computing the transition.

Let us now apply the product strategy to these two machines.

Obviously, $M_1' = (\{p, r, \Phi\}, \{a, b\}, \delta, p, \{q\})$ and $M_2' = (\{r, s\}, \{a, b\}, \delta', r, \{s\})$.

Now, $M = M_1 \cap M_2 = (Q_M, \{a, b\}, \delta_M, (p, r), F_M)$, where $Q_M = \{p, r, \Phi\} \times \{r, s\}$, $F_M = \{q\} \times \{s\} = \{q, s\}$ and is : $(\delta_M (p, p'), a) = (\delta (p, a) \cup \delta(p', a))$, where $p \in \{p, q\}$ and $p' \in \{r, s\}$.

Now, $Q_M = \{ p, q, \Phi\} \times \{r, s\} = \{ (p, r), (p, s), (q, r), (q, s), (\Phi, r), (\Phi,s)\}$.

Also,

- $(\delta_M (p, r), a) = (\delta (p, a) \cup \delta' (r, a)) = (q, r)$
- $(\delta_M (p, r), b) = (\delta (p, b) \cup \delta' (r, b)) = (\Phi, s)=\Phi$
- $(\delta_M (p, s), a) = (\delta (p, a) \cup \delta' (s, a)) = (q, r)$
- $(\delta_M (p, s), b) = (\delta\, \Phi, s) = \Phi$
- $(\delta_M (q, r), a) = (\delta (q, a) \cup \delta' (r, a)) = (q, r)$
- $(\delta_M (q, r), b) = (\delta (q, b) \cup \delta' (r, b)) = (q, s)$
- $(\delta_M (q, s), a) = (\delta (q, a) \cup \delta' (s, a)) = (q, r)$
- $(\delta_M (q, s), b) = (\delta (q, b) \cup \delta' (s, b)) = (q, s)$

The equivalent DFA satisfying both the conditions (one for M_1 and the other for M_2) is shown in Fig. 2.50 (c).

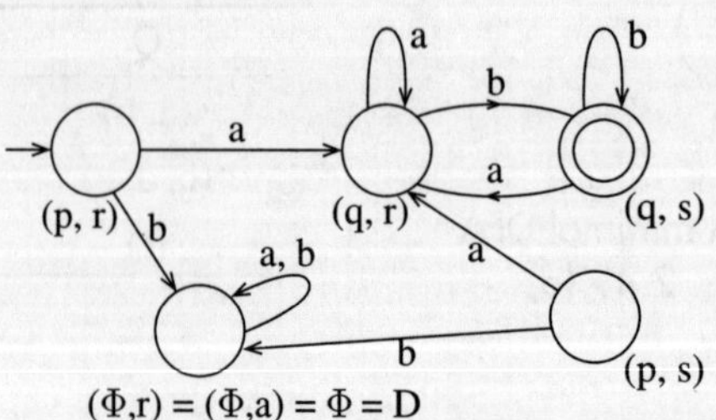

FIGURE 2.50 (c) A product FA of two automata shown in Figs 2.50 (a) and (b)

One may note that the state (p, s) shown in the resultant DFA is unreachable from the start state (p, r). So, it can be discarded from the figure. Further, if we perform $M' = M_2 \cap M_1$ the final DFA is as is shown in Fig. 2.50 (d), but it also satisfies both the criteria.

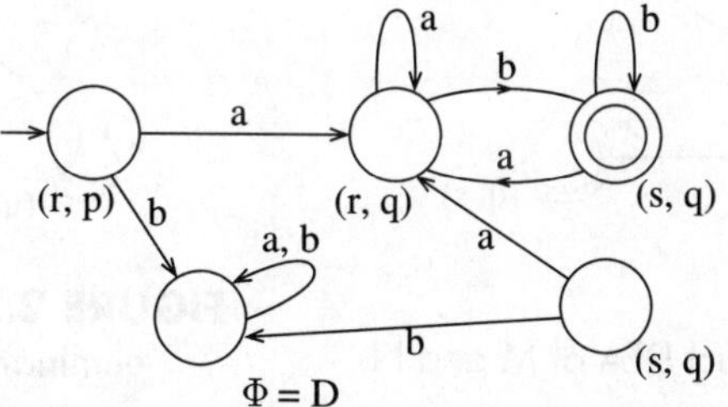

FIGURE 2.50 (d) A DFA (M) for $M_2 \cap M_1$

State (p, s) in Fig. 2.50 (c) and state (s, q) in Fig. 2.50 (d) are unreachable states, so we remove them from the respective figures.

PROBLEM 20

Consider the following DFAs, M and N, given in Fig. 2.51 (a). Describe the languages accepted by the machines. Find the number of states in a minimal DFA that accepts the language L(M) ∩ L(N).

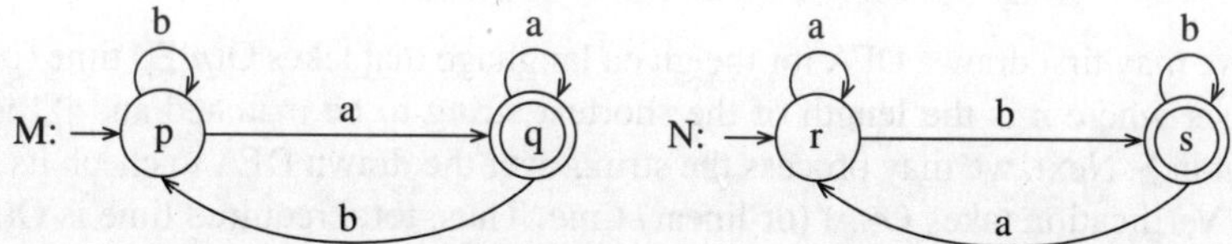

FIGURE 2.51 (a) DFAs M and N

Solution From the diagrams, it is clear that machine M accepts strings ending with '*a*', whereas machine N accepts strings ending with '*b*' over $\{a, b\}$.

M has two states, p and q, where p is the start state and q is the accepting state. N has two states, r and s, where r is the start state and s is the accepting state. Now, performing cross-product on the sets of states, $\{p, q\} \times (r, s\}$, we get $\{(p, r), (p, s), (q, r), (q, s)\}$ as the set of states of the product machine, where (p, r) is the initial state and (q, s) is the final state. Next, on the basis of the transition relations, we find the product machine (DFA) shown in Fig. 2.51 (b).

Clearly, (q, s) (although it is an accepting state in the product machine) is the unreachable state. So, we may discard it from the diagram and the result is shown in Fig. 2.51 (c). So, the product machine does not accept any string (even null string). So, L = L(M) ∩ L(N) = Φ.

To find the minimal DFA of the product DFA, let us denote (p, r) as A, (p, s) as B and (q, r) as C. On the basis of these notations, the state diagram corresponding to the DFA is shown in Table 2.22.

TABLE 2.22 State table for Fig. 2.51 (c)

State	Input(a)/Next state	Input(a)/Next state
→A	C	B
B	C	B
C	C	B

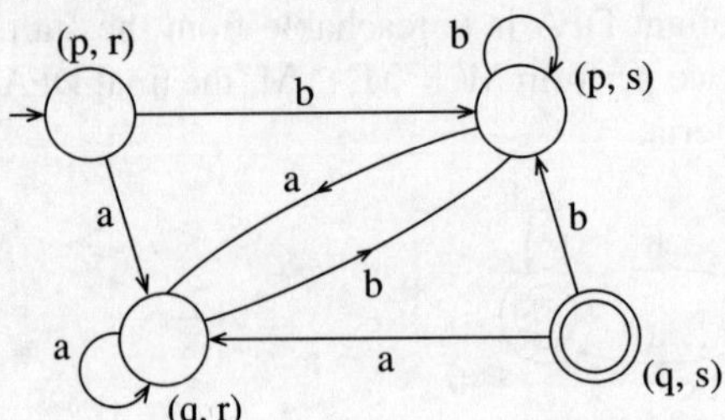

FIGURE 2.51 (b) Product DFA of M and N

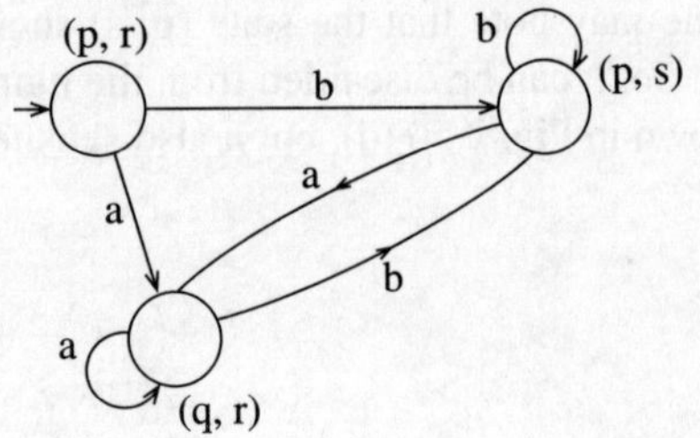

FIGURE 2.51 (c) Product DFA after eliminating unreachable state

On applying equivalence theorem to minimise the DFA, we get $\Pi_0 = \{\{A, B, C\}, \{\ \}\}$. Consulting the state table, we obtain $\Pi_1 = \{\{A, B, C\}, \{\ \}\}$. Thus, $\Pi_0 = \Pi_1$ and stop to find the next equivalent. So, the group: {A, B, C} is denoted as E. Certainly, no final state is present here; that is, the product machine recognises an empty language.

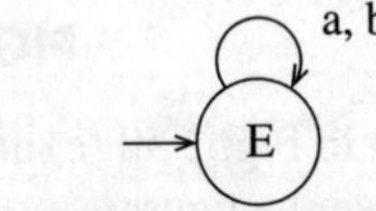

FIGURE 2.51 (d) Minimal DFA of Fig. 2.51 (c)

PROBLEM 21

Present your idea for testing membership in a regular language.

Solution We may first draw a DFA for the given language that takes $O(n|\Sigma|)$ time (polynomial time) in the worst case, where n is the length of the shortest string to be matched and $|\Sigma|$ is the number of symbols present in Σ. Next, we may process the string over the drawn DFA to check its acceptability by the automaton. Verification takes $O(m)$ (or linear) time. Thus, total required time is $O(n|\Sigma|) + O(m) = O(n|\Sigma|)$ (polynomial time).

PROBLEM 22

Design DFAs for the following problems (each with a small variation) over $\{a, b\}$.

(i) String whose last two symbols are *b*. (ii) String with substring *bb* (string with two consecutive *b*s) (iii) String in which each *b* appears as a pair (*bb*). (iv) String which does not contain two consecutive *b*s.

Solution (i) The state diagram for accepting strings with the last two symbols as '*b*' is shown in Fig. 2.52 (a). From the diagram, it is clear that all possible strings consisting of substring *bb* at the end are valid.

(ii) The state diagram to accept strings with substring *bb* is shown in Fig. 2.52 (b). The diagram shows that as soon as the first substring *bb* is found in a string, that string is accepted, and any subsequent combinations of '*a*' and '*b*' will be accepted.

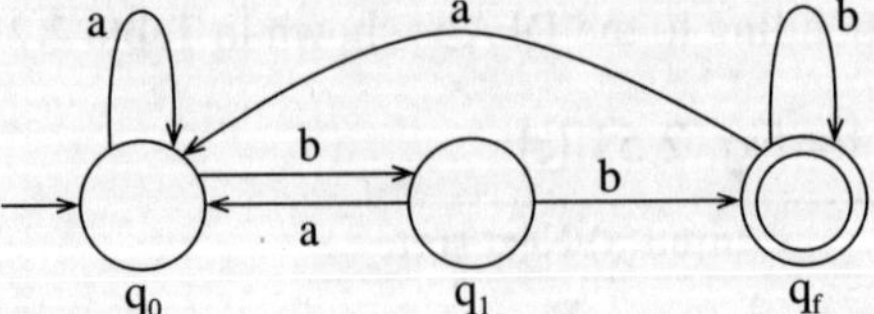

FIGURE 2.52 (a) DFA to recognise strings whose last two symbols are *b*

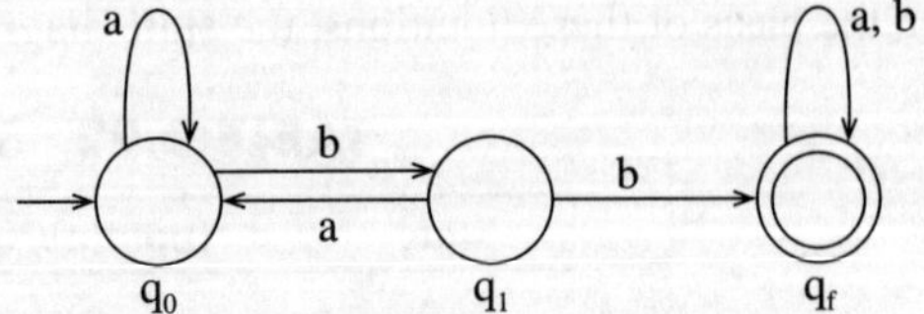

FIGURE 2.52 (b) DFA to accept strings with substring *bb*

(iii) The state diagram (shown in Fig. 2.52 (c)) suggests that a string is accepted if and only if '*b*' appears in the string as a pair. For example, *bb*, *bbbb*, *bbabb*, etc. are valid, whereas *babb*, *bbbabb*, etc. are not valid. Therefore, if '*a*' is inputted at state q_1, there is no chance to obtain a pair. The FA then goes to the dead state.

(iv) The state diagram shown in Fig. 2.52 (d) ensures that the DFA does not accept any string consisting of two consecutive *b*s.

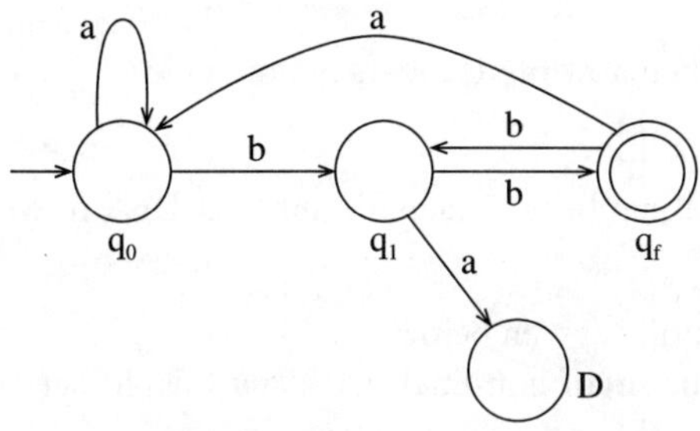

FIGURE 2.52 (c) DFA to accepting strings with *b*s as pair

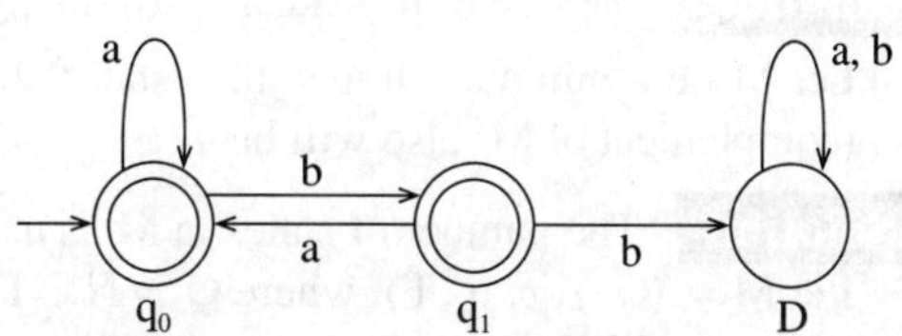

FIGURE 2.52 (d) DFA that does not contain two consecutive *b*s in strings

Note: To solve such a problem (designing the DFA to satisfy the negation property) in an easier way, we may first design the DFA to satisfy the positive property and then take its complement.

PROBLEM 23

Draw a DFA for the language L over $\{a, b\}$ such that each valid string begins with '*a*' and ends with '*b*'. Next, find its complement. Also, describe the language recognised by the complemented DFA.

Solution A DFA to recognise the given language is shown in Fig. 2.53 (a).

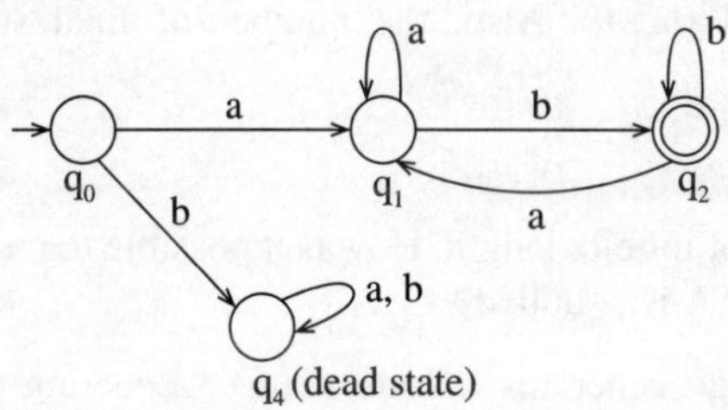

FIGURE 2.53 (a) A DFA (M) to recognise strings beginning with 'a' and ending with 'b'

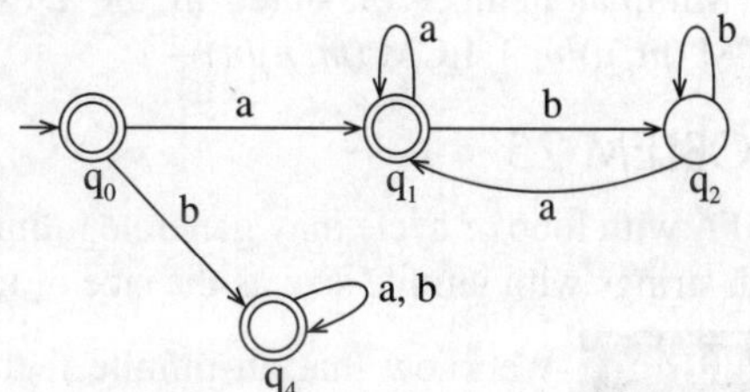

FIGURE 2.53 (b) The complemented DFA of Fig. 2.53 (a)

Its complement is shown in Fig. 2.53 (b). Clearly, all the states (including the dead state: q_4), except q_2, will be the final states in the complemented automata. We may describe the given language and its complement as follows.

Let us take two conditions P and Q such that P denotes "string begins with *a*" and Q denotes "string ends with *b*". Now (*P* AND *Q*) is true for Fig. 2.53 (a), whereas $\neg$ (*P* AND *Q*) = $\neg P$ OR $\neg Q$ is true for Fig. 2.53 (b). In other words, if any one of $\neg P$ (string not beginning with '*a*') and $\neg Q$ (string not ending with '*b*') or both are true, then the string is valid. Thus,

- ε string is also valid, since it does not begin with '*a*' and end with '*b*'.
- Strings that reach q_1 are valid, since $\neg P$ is not true but $\neg Q$ is true.
- Strings that reach q_4 are valid, since $\neg P$ is true.
- However, strings that reach q_2 are not valid, as both $\neg P$ and $\neg Q$ are not true.

Therefore, the complemented DFA looks as is shown in Fig. 2.53 (b).

PROBLEM 24

(a) If we describe a DFA with '*n*' states and an alphabet Σ, then how big is the transition table?

Solution The size of the 2D array (transition table) is $n \times |\Sigma|$.

(b) Let M be a minimal DFA with *n* states. Justify that the number of states in minimal DFA of M′ (complement of M) also will be *n*.

Solution The number of states in M′ is *n*. The justification is given below.

Let $M = (Q, \Sigma, \delta, q_o, F)$, where $Q = N \cup F$; here, N is the set of non-final states and F is the set of final states of M. Suppose $|N| = n_1$ and $|F| = n_2$. $|Q| = |N| + |F| = n_1 + n_2 = n$ (say), where | | is used to find the size of the set.

Now, $M' = (Q', \Sigma, \delta, q_o, F'))$, where $Q' = N' \cup F'$; N′ is the set of non-final states and F′ is the set of final states of M′. According to the idea of complement, $N' = Q - F$ and $F' = Q - N$, where Q, N and F denote, respectively, the set of states, the set of non-final states and the set of final states of M.

Thus, $|Q'| = |N'| + |F'| = |Q - F| + |Q - N| = (n - n_2) + (n - n_1) = (2n - (n_1 + n_2)) = 2n - n = n$

Therefore, the number of states in M′ also will be *n*. More explicitly, the number of states in M′ remains unchanged; only the non-final states of M will become the final states of M′ with the same start state.

(c) Let Σ be the alphabet. Find the minimum number of states in the DFA to recognise strings of length divisible by (i) *m* and *n* (ii) *m* or *n*, where *m* and *n* are positive integers. Also, find the number of final states in each case.

Solution (i) Minimal number of states in the DFA is: LCM (m, n). Also, the number of final states is 1. (ii) Minimal number of states in the DFA is: LCM (m, n). Also, the number of final states is (LCM $(m, n)/m$ + LCM $(m, n)/n$) – 1.

PROBLEM 25

An FA with loop or cycle may generate infinite strings of infinite length. How is it possible to recognise such strings with infinite size as the size of memory in FA is bounded?

Solution We know that an infinite regular language generates infinite strings, repeating a finite number of patterns. The size of memory in any FA is limited. So, it is not possible to recognise valid strings if their length exceeds the size of memory, although, modern computers may be able to accept more, as their in-built memory increases. However, there still exists the concept of bound. We may theoretically claim that strings of infinite size for infinite languages may be made accepted by FA if it allows a finite number of repeating patterns.

PROBLEM 26

Is language $L = \{a^n b^n \mid n \leq 100^{10}\}$ accepted by FA?

Solution Yes, we may theoretically claim that the given L is accepted by FA, as *n* is bounded above and so L is finite. Matching of *a*s and *b*s is possible through state and finite memory. Actually, some content is to be saved for each transition, *for example,* a → b, for this transition, *a* and *b* are to be saved. Likewise, the other cases are also true. However, $L = \{a^n b^n \mid n \geq 0\}$ is infinite as *n* is unbounded. Hence, $L = \{a^n b^n \mid n \geq 0\}$ may not be accepted by FA, as memory is limited. In particular, the usual capability (memory and the transition relation) of FA does not enable us to design any FA for $L = \{a^n b^n \mid n \geq 0\}$.

PROBLEM 27

What is the probability that the NFA shown in Fig. 2.54 successfully accepts the string *aba*? Assume that the NFA randomly resolves the ambiguity at each step.

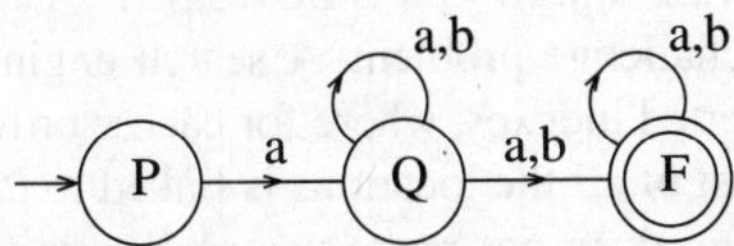

FIGURE 2.54 An NFA over {*a, b*}

Solution We list all possible paths that can be taken

P→a→Q→b→Q→a→F [accepted]

P→a→Q→b→F→a→F [accepted]

P→a→Q→b→Q→a→Q [Stuck]

No other paths are possible. Two paths accept the string, and one path causes it to be stuck.

Required probability = 2/(2 + 1) = 2/3

PROBLEM 28

Design a DFA for the language L = {$a^n b^m$ | $(n + m) > 1$ and $(n + m)$ is divisible by 2}.

Solution Here, Σ = {a, b}. From the given language, it is clear that no 'a' will appear after 'b'. Now, analysing the language properties, one DFA is drawn, as shown in Fig. 2.55.

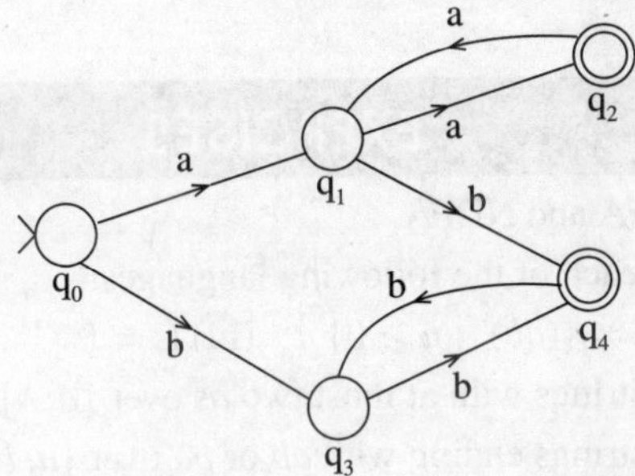

FIGURE 2.55 A DFA for L = {$a^n b^m$ | $(n + m) > 1$ and $(n + m)$ is divisible by 2}

2.13 RESEARCH DIRECTIONS

There are several interesting research problems in the domain of automata. One very important application of automata model is in software design and testing.

Any software can be modelled as a state transition diagram, where each state corresponds to an independent standalone module of the software. Unit testing is mainly concerned with testing of standalone modules. Again, each standalone module can be thought of as an automaton in itself. So, a software can be visualised as an automaton, whose each state is also an automaton itself! Thus, this leads to automaton of automata.

One huge potential of this approach could be to implement a general purpose software that can read the entire source code of another software and generate a finite automata diagram from it. If this can be accomplished, it will then become feasible to eliminate several bugs in the code just by analysing the generated automaton. In general, any software crashes because it encounters a state that it cannot handle. This is nothing but a dead state or a halt state.

Application on the web and other online text processing: On the web and for other online text processing, searching is an open challenge problem. A search engine (an example of text searching) usually uses technology called inverted indexes, where for each word appearing on the web (consisting of around 10^8 different words), a list of all the locations is linked to the places where this word occurs. However, inverted index technology does not make use of FA, and CRAWLERs take more time to search text, but the application of an FA-based technique may be useful in this respect.

Robotics and artificial intelligence: Imagine that you want to design a robot who can prepare a cup of tea for you. The entire process of the robot understanding your order for tea, finding the kitchen in your house, going into the kitchen, and preparing tea can be visualised as a finite automaton! The next time you watch a science fiction movie featuring robots, observe carefully. Continue to think and visualise – there are a huge number of situations that can be modelled as a finite automaton.

Some project-based problems

1. Design a model for ATM machine.
2. Model a simple hand gesture–based recognition machine.
3. Model a system for cryptography, implementing encryption and decryption.
4. Design a model for representing images.
5. Design a model for communication protocol.

EXERCISES

1. Prove the equivalence of DFA and NDFA.
2. Design a DFA to recognise each of the following languages:
   (i) L= $\{a^n b \mid n \geq 0\}$, (ii) L = $\{(ab)^n \mid n \geq 0\}$ (iii) L = $\{w \mid w \in \Sigma^*$ and $|w| = 3k, k \geq 0\}$
3. Design a DFA that accepts strings with at most two *as* over $\{a, b\}$.
4. Design a DFA that accepts strings ending with *ab* or *ba* over $\{a, b\}$.
5. Design a DFA to accept the language L = $\{w \mid |w| \bmod 5 \neq 0\}$ over $\{a, b\}$.
6. Design a DFA that accepts strings having at least one *a* over $\{a, b\}$.
7. Design a DFA that accepts strings having exactly one *a* over $\{a, b\}$.
8. Design a DFA that accepts strings having four *as* over $\{a, b\}$.
9. Design a DFA and NDFA to accept strings over $\{0,1\}$ that end in 0 but do not have 11 as substring.
10. Design an NDFA to accept the language L = $\{a^* \cup b^*\}\ a^*$.
11. Design a DFA to accept positive integers divisible by 3.

Hint:

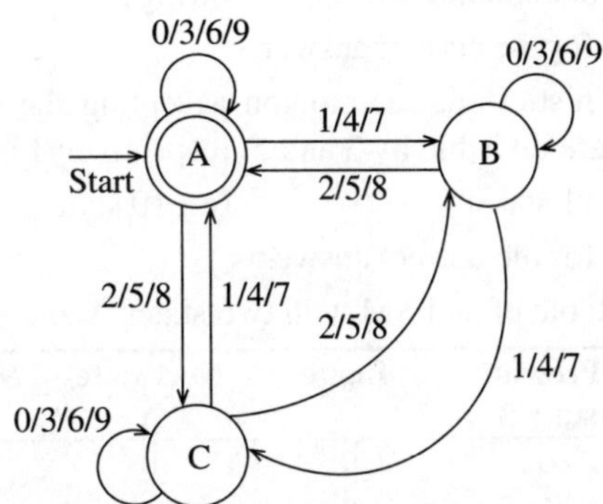

12. Design a DFA to accept strings over $\{a, b\}$ having at most one pair of *a*s or *b*s.
13. Design a DFA over $\{a, b, c\}$ so that the last two symbols are the same.
14. Design a DFA over $\{a, b, c\}$ so that every block of 4 contains at least two same symbols.
15. Design a DFA over $\{a, b\}$ containing exactly two *a*s but more than two *b*s.
16. Differentiate between Mealy machine and Moore machine. Describe the method of conversion from Mealy to Moore and Moore to Mealy.
17. What can be said about a regular language L over {a} whose minimal finite state automation has two states?

    (a) L must be $\{a^n \mid n \text{ is odd}\}$ (b) L must be $\{a^n \mid n \text{ is even}\}$

    (c) L must be $\{a^n \mid n \geq 0\}$ (d) Either L must be $\{a^n \mid n \text{ is odd}\}$ or L must be $\{a^n \mid n \text{ is even}\}$

    [GATE 2000]
18. Consider the following deterministic finite state automaton M.

    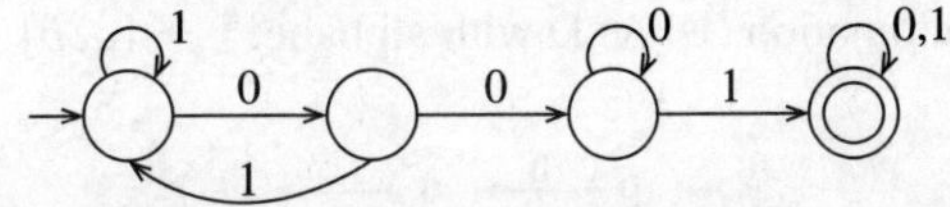

    Let S denote the set of seven bit binary strings in which the first, fourth and last bits are 1. The number of strings in S that are accepted by M is

    (a) 1 (b) 5 (c) 7 (d) 8

    Give a proper explanation for the correct answer. **[GATE 2003]**
19. Which language is recognised by the following finite state machine?

    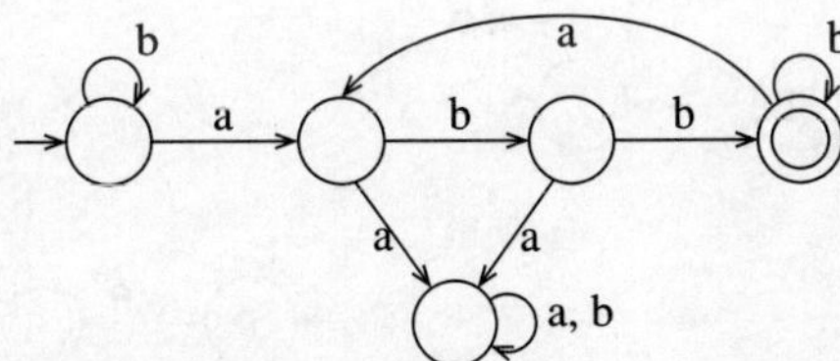

    (a) $\{w \in \{a, b\}^* \mid$ every *a* in *w* is followed by exactly two *b*s$\}$

    (b) $\{w \in \{a, b\}^* \mid$ every *a* in *w* is followed by at least two *b*s$\}$

    (c) $\{w \in \{a, b\}^* \mid$ *w* contains the substring '*abb*'$\}$

(d) $\{w \in \{a, b\}^* \mid w$ does not contain 'aa' as a substring$\}$

Give a proper explanation for the correct answer. **[GATE 2005]**

20. A minimum state deterministic finite automaton accepting the language $L = \{w \mid w \in \{0,1\}^*$, number of 0s and 1s in w are divisible by 3 and 5, respectively$\}$ has

(a) 15 states (b) 11 states (c) 10 states (d) 9 states

Give a proper explanation for the correct answer. **[GATE 2007]**

21. Given the following state table of an FSM with two states A and B, one input and one output:

Present state A	Present state B	Input	Next state A	Next state B	Output
0	0	0	0	0	1
0	1	0	1	0	0
1	0	0	0	1	0
1	1	0	1	0	0
0	0	1	0	1	0
0	1	1	0	0	1
1	0	1	0	1	1
1	1	1	0	0	1

If the initial state is A = 0, B = 0, what is the minimum length of an input stringthat will take the machine to state A = 0, B = 1 with output = 1?

(a) 3 (b) 4 (c) 5 (d) 6

Give a proper explanation for the correct answer. **[GATE 2009]**

22. Design an NDFA that accepts any binary string containing 00 or 11 as a substring.
23. Design a Moore machine to find the 2's complement of a binary number.
24. A deterministic finite automation (DFA) D with alphabet $\Sigma = \{a, b\}$ is given below:

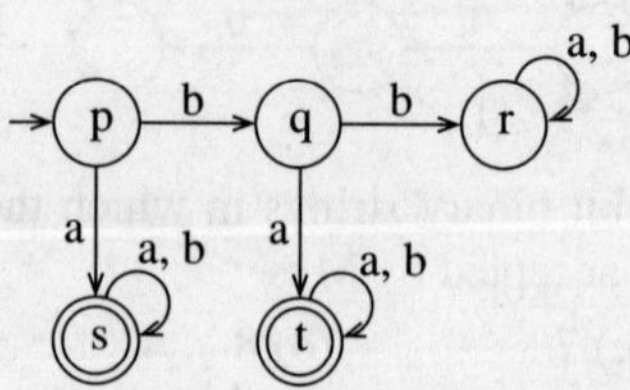

Which of the following finite state machines is a valid minimal DFA which accepts the same language as D? **[GATE 2011]**

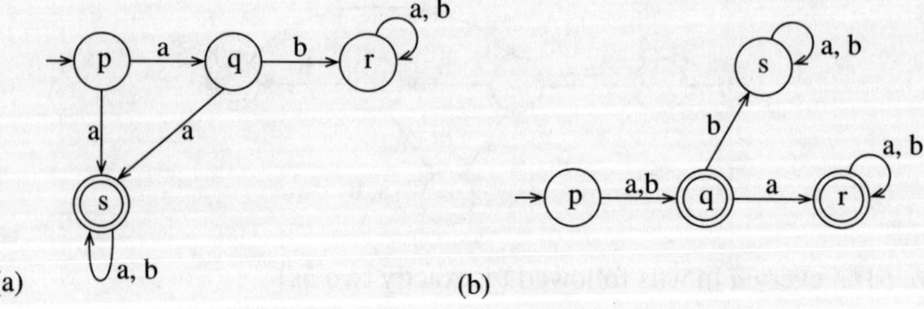

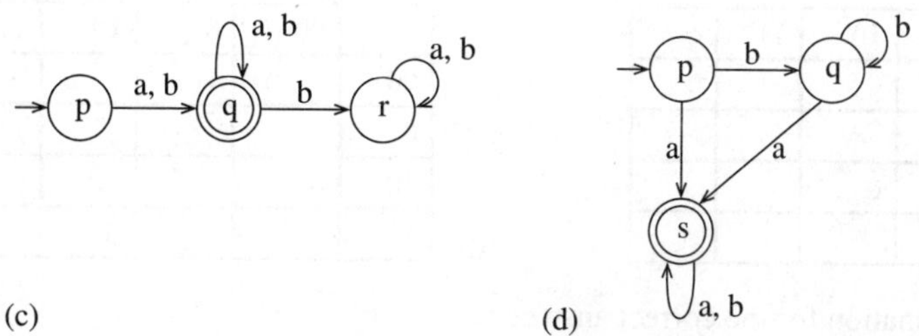

(c) (d)

25. Given the language L = $\{ab, aa, baa\}^*$, which of the following strings are in L*?

(1) abaabaaabaa (2) aaaabaaaa (3) baaaaabaaaab (4) baaaabaa

(a) 1, 2 and 3 (b) 2, 3 and 4 (c) 1, 2 and 4 (d) 1, 3 and 4

Give a proper explanation for the correct answer. **[GATE 2012, Rajasthan CS Lecturer Recruitment 2014, HP IT Lecturer Recruitment 2016]**

26. What is the complement of the language accepted by the NFA shown below? Assume $\Sigma = \{a\}$ and ε is the empty string.

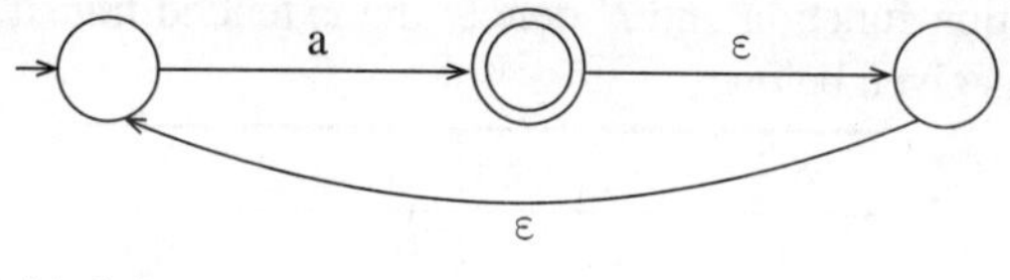

(a) $\varnothing$ (b) $\{\varepsilon\}$ (c) a^* (d) $\{a, \varepsilon\}$

Give a proper explanation for the correct answer. **[GATE 2012]**

27. Consider the set of strings on {0,1} in which, every substring of 3 symbols has at most two zeros. For example, 001110 and 011001 are in the language, but 100010 is not. All strings of length less than 3 are also in the language. A partially completed DFA that accepts this language is shown below.

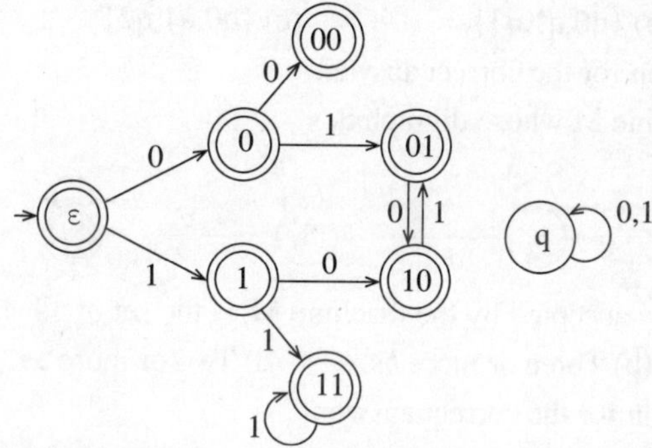

The missing arcs in the DFA are

(a)

	00	01	10	11	q
00	1	0			
01				1	
10	0				
11			0		

(b)

	00	01	10	11	q
00		0			1
01		1			
10				0	
11		0			

(c)

	00	01	10	11	q
00		1			0
01		1			
10			0		
11		0			

(d)

	00	01	10	11	q
00		1			0
01				1	
10	0				
11			0		

Give a proper explanation for the correct answer. **[GATE 2012]**

28.

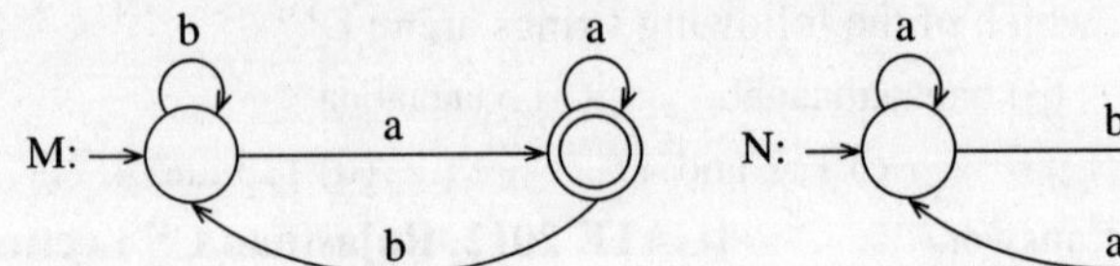

Consider the DFAs M and N given above. Find the number of states in a minimal DFA that accepts the language L(M) ∩ L(N). **[GATE 2015]**

29. Let δ denote the transition function and δ' denote the extended transition function of the $\in$ NFA whose transition table is given below:

δ	$\in$	a	b
$\rightarrow q_0$	$\{q_2\}$	$\{q_1\}$	$\{q_0\}$
q_1	$\{q_2\}$	$\{q_2\}$	$\{q_3\}$
q_2	$\{q_0\}$	$\varnothing$	$\varnothing$
q_3	$\varnothing$	$\varnothing$	$\{q_2\}$

Then $\delta'(q_2, \text{aba})$ is

(a) $\varnothing$ (b) {q0,q1,q3} (c) {q0,q1,q2} (d) {q0,q2,q3}

Give a proper explanation for the correct answer. **[GATE 2017]**

30. Consider a Moore machine M whose digraph is:

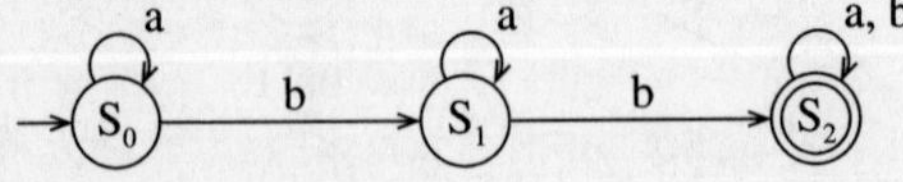

Then L(M), the language accepted by the machine M, is the set of all strings having:

(a) Two or more *b*s. (b) Three or more *b*s. (c) Two or more *a*s. (d) Three or more *a*s.

Give a proper explanation for the correct answer. **[UGC NET 2007]**

31. The minimum number of states of the non-deterministic finite automation that accepts the language $\{abab^n \mid n \geq 0\} \cup \{aba^n \mid n \geq 0\}$ is

(a) 3 (b) 4 (c) 5 (d) 6

Give a proper explanation for the correct answer. **[UGC NET 2012]**

32. Which of the following statements is true ?

(a) The model of deterministic finite state machine includes the model of non-deterministic finite state machine as a special case.

(b) The model of non-deterministic finite state machine includes the model of deterministic finite state machine as a special case.

(c) Non-deterministic finite state machines are not more powerful than deterministic finite state machines.

(d) None of these. **[Gujarat SET 2003]**

33. We require a four-state automaton to recognise the regular expression (*a*/*b*)* *abb*.

(a) Give an NFA for this purpose. **[GATE 2002, North Maharashtra University PhD Entrance 2004]**

(b) Give a DFA for this purpose. **[GATE 2002, North Maharashtra University PhD Entrance 2004, UGC NET June 2006]**

(c) Write the state transition for the string *'abababb'*. **[UGC NET 2006]**

34. Let A = $\{a, b\}$. Construct an automaton M such that L(M) will consist of those words in $w = \{a, b\}^*$ which begin with *a* and end in *b*. [UGC NET December 2006]

35. Explain why a finite state automaton cannot recognise a palindrome. **[UGC NET 2008]**

36. Design a finite state machine to check the divisibility by 7 and 11 of an input binary sequence with the least significant digit occurring first. **[UGC NET 2009]**

37. Design a DFA to accept strings containing exactly three 0s and at least two 1s over the alphabet $\{0, 1\}$. **[UGC NET 2009]**

38. Design NFA with the specified number of states recognising each of the following languages. The alphabet is $\{0, 1\}$. **[UGC NET 2010]**

(i) $\{W \mid W = x\,01 \mid$ where x has exactly one 1 and any number of 0s$\}$ (4 states)

(ii) $\{W \mid$ every odd position of W is a 1$\}$ (2 states)

(iii) $\{W \mid W$ contains either the substring 000 or substring 010$\}$ (4 states)

39. Prove whether the following statements are true or false. In the case of false, give a counterexample:

(a) Any *n*-state NFA, where all *n* states are final states, will accept all words in Σ^*.

(b) All regular languages can be accepted by a DFA containing an even number of states.

(c) For any NFA A, define the NFA A′ created by changing all final states of A to non-final and vice-versa accepts the complement of L(A). L(A) is the language accepted by A.

[Gujarat SET 2010]

40. For a binary string $x = a_0a_1...a_{n-1}$ define val(x) = decimal equivalent of the binary number represented by the string x. Design a finite automaton that accepts exactly the set of binary strings x such that val(x) is divisible by either 4 or 5. **[CMI PhD Entrance 2013]**

41. Construct a deterministic finite automaton accepting the following language:

$\{w \in \{0,1\}* : w$ has an equal number of 01s and 10s $\}$

For example, 101 is in the language because it contains one instance of 10 and one instance of 01 as well. **[ISI Kolkata JRF 2014]**

42. Indicate whether the following statements are true or false, providing a short explanation to substantiate your answers.

(i) A DFA with *n* states must accept at least one string of length greater than *n*.

(ii) A DFA that has *n* states and accepts an infinite language must accept at least one string *x* such that $2n < |x| < 3n$, where $|x|$ denotes the length of *x*.

(iii) If a language L is accepted by an NFA with *n* states then there is a DFA with no more than 2^n states accepting L **[CMI PhD Entrance 2010]**

43. Write a program that reads a transition table from a file and (a) tests whether the automaton is deterministic, (b) tests whether the language used by the automaton is void, and if not finds a word accepted by the language, (c) simulates the automaton on the given word (printing the transitions/states).

44. For a string $x = a_0a_1...a_{n-1}$ over the alphabet {0, 1, 2}, define val(x) to be the value of x interpreted as a ternary number, where a_0 is the most significant digit. Design a finite automaton that accepts exactly the set of strings $x \in \{0,1,2\}^*$ such that val(x) is divisible by 4.

[CMI PhD Entrance 2016]

45. Construct DFAs for the following languages:
(a) L = {w | $w \in \{a, b\}^*$, w has *baab* as a substring} **[GATE 2001]**
(b) L = {w | $w \in \{a, b\}^*$, w has odd number of *as* and an odd number of *bs*}

[GATE 2001, Gujarat SET 2008]

46. The following finite state machine accepts all binary strings in which the number of 1s and 0s are, respectively,

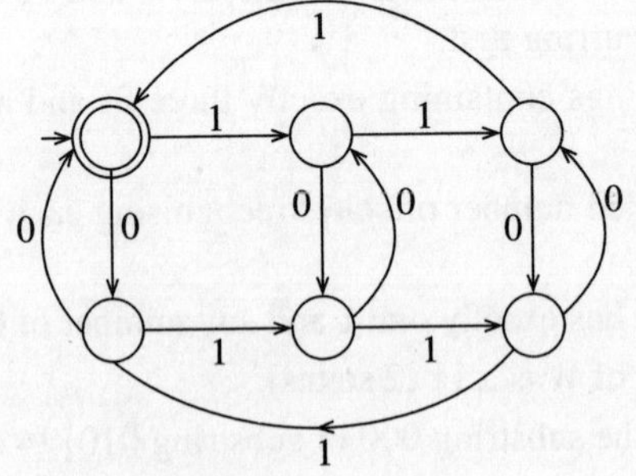

(a) Divisible by 3 and 2 (b) Odd and even
(c) Even and odd (d) Divisible by 2 and 3 **[GATE 2004]**

47. Consider the finite automaton in the following figure.

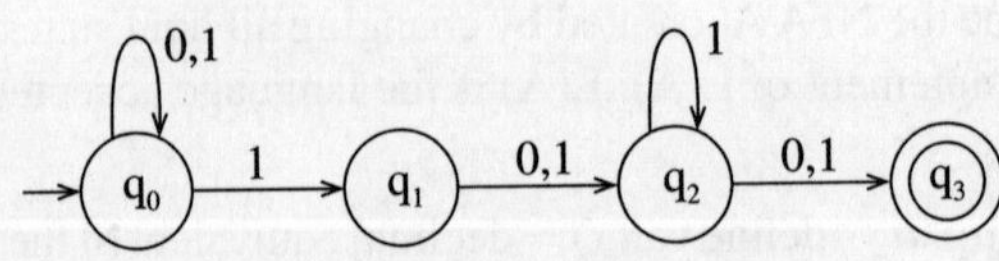

What is the set of reachable states for the input string 0011?
(a) {q_0, q_1, q_2} (b) {q_0, q_1} (c) {q_0, q_1, q_2, q_3} (d) {q_3} **[GATE 2014]**

48. If L1 and L2 are languages recognised by machines M1 and M2 with n and m states, respectively, then the machine M that recognises L1. L2 will contain the following number of states:
(a) n (b) m (c) n + m (d) nm **[West Bengal SET 2017]**

49. What does this NFA accept ?

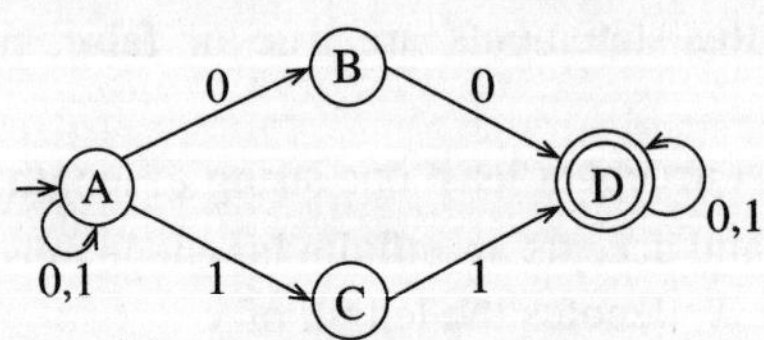

(a) 101010 (b) 0110 (c) Only 'A' (d) Both 'A' and 'B'

[Maharashtra SET 2018]

50. For finite automata, which one of the following statements is false?

(a) $\delta(q, xw) = \delta(\delta(q, x), w)$ (b) $\delta(q, xw) = \delta(q, wx)$

(c) $\delta(q, \varepsilon) = q$ (d) If $\delta(q, y) = \delta(q, x)$, then $\delta(q, xz) = \delta(q, yz)$

Give a proper proof or counterexample for each of the above. **[JNU M.Phil/PhD Entrance 2015]**

51. Construct a deterministic finite automaton (DFA) to recognise all strings w over $\Sigma = \{1, 2, 3\}$ such that the digits in w appear in non-decreasing order. For example, your DFA should accept 1123, but not 1232. **[ISI Kolkata JRF 2017]**

52. Suppose $L \subseteq \Sigma^*$ and for some positive integer n, there are n strings in Σ^*, any two of which are distinguishable with respect to L. Prove that every FA recognising L must have at least n states. **[UGC NET 2011]**

53. Let L be the language of all binary strings in which the third symbol from the right is '1'. Give a non-deterministic finite automaton that recognises L. How many states does the minimised equivalent deterministic finite automaton have? Justify your answer briefly. **[GATE 1991]**

54. Given two finite automata M1 and M2, outline an algorithm to decide if L(M1) $\subset$ L(M2). (*Note*: Strict subset). **[GATE 1999]**

55. The following deterministic finite automata recognises:

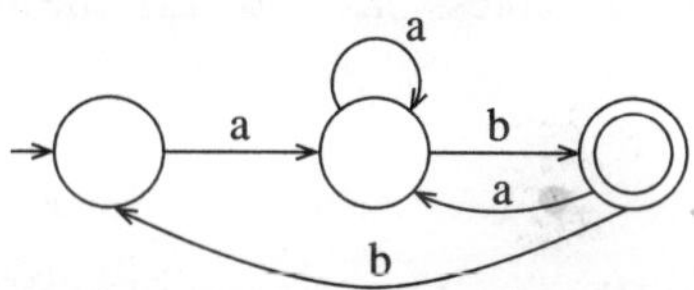

(a) Set of all strings containing 'ab' (b) Set of all strings containing 'aab'

(c) Set of all strings ending in 'abab' (d) None of the above **[UGC NET 2006]**

56. The number of states in the minimal state NFA where the transition diagram is given as follows is:

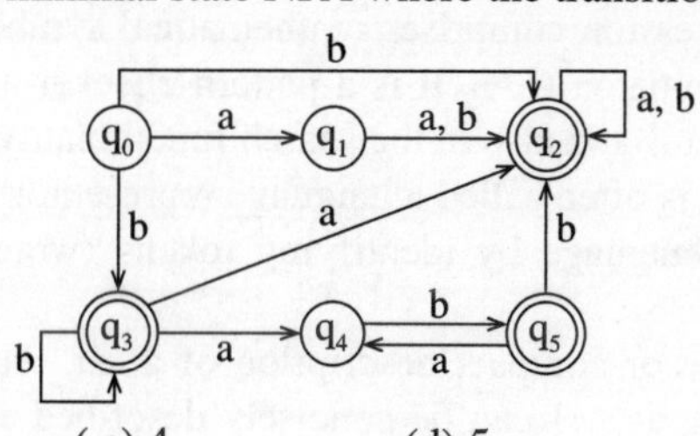

(a) 2 (b) 3 (c) 4 (d) 5 **[Maharashtra SET 2013]**

3 Regular Expressions

OBJECTIVES

At the end of this chapter, you will be able to

- Describe the mathematical modelling of languages in concise form
- Understand the operations on regular expressions
- Establish the correlation between regular languages and finite automata
- List the applications of regular expressions

3.1 INTRODUCTION

The concept of regular expression (RE) was first introduced by Kleene in the 1950s. It is a notation for language definition used in the study of formal languages. The notation may be used in natural languages as well as programming languages to describe their smallest pieces or units (namely, strings or words). They are also used in designing software.

By definition, a regular expression comprises mathematical symbols over Σ that describe a set of strings accepted by a finite automaton (FA). It is a pattern checker and is used in the command-line shells of various operating systems and also in the search functionality of text editors. In the context of programming languages, an RE is often called a language representor (scanner) as it helps to construct the sentences/statements of a language by identifying tokens (words) such as data type, operand, operator, etc.

RE is used to give a concise or compact description of a set, without listing all the elements in the set. For example, $S = \{\varepsilon, \text{a}, \text{aa}, \ldots\}$ can be concisely described as (a)*. It generates strings in a recursive manner. RE may be confused with arithmetic expressions. The main difference is that arithmetic expressions in mathematics give values, whereas regular expressions in automata theory produce sets of strings (language).

3.2 REGULAR EXPRESSIONS AND BASIC OPERATIONS

A regular expression is usually denoted by the symbols R or r. Let us first consider the base cases of RE before discussing the operations to be performed over them.

Base Cases

- Every single alphabet symbol is a regular expression; for example, $r = (\text{a})$, that is, $r = (\text{a}) = \{\text{a}\}$. So the language generated by r is denoted as $\text{L}(r) = \{\text{a}\}$, meaning that the language L consists of a single string '*a*'.

- The empty string (ε) is itself a regular expression; $r = (\varepsilon)$. So, $L(r) = \{\varepsilon\}$ [or $\{\Lambda\}$].
- Φ is a regular expression denoting the empty set, and $L(\Phi) = \Phi$. In reality, a regular expression (pattern) cannot describe an empty language (Φ). So, we would never want to use this pattern. However, FA can define an empty language. So, this case shows a lacuna between regular expression and FA.

Operators Used in RE

- The operators +, | and ∪ stand for union. They are binary operators representing the union operation.
- o (or juxtaposition operator) stands for concatenation. It is a binary operator.
- * represents closure and + (unary) represents positive closure. Both are unary operators.

Precedence of Operators

- * (closure) and + (positive closure)
- o (concatenation)
- + (union)

However, the parenthesis used for grouping REs is, by default, considered first. For example, $a(a + b)^* = a\ \{\varepsilon, a, b, aa, ab, ba, bb, ...\} = \{a, aa, ab, aab, ...\}$.

Here, * (closure) has higher priority than concatenation. Now, if concatenation had been given higher priority, then the resulting strings would be of even length; that is, {aa, ab}*. No odd-length strings such as *aba* and *abb* that begin with '*a*' can be generated.

Basic Operations: If *R* and *S* are two existing REs, then each of the operations, $R + S$ (union), *RS* (concatenation), R^* (closure) and S^*, forms a new RE. These operations are described in detail:

- **Union:** Let $R = \{a\}$ and $S = \{b\}$ be two REs over $\Sigma = \{a, b\}$. Then, $R_1 = (R + S) = (a \mid b) = (a + b) = (a \cup b)$ results in a new RE over Σ. The semantic interpretation of $R + S$ implies the union of all the strings generated by *R* and *S*, and so, $R + S = S + R$.
- **Concatenation:** Let R_1 and R_2 be two REs over Σ. The concatenation of R_1 followed by R_2 is written as: $R_1 \text{ o } R_2$ (or simply, R_1R_2), and it is a new RE over Σ that generates a set of strings such as $\{w = xy \mid x \in R_1 \text{ and } y \in R_2\}$. Similarly, we may construct $R_2 \text{ o } R_1$. However, $R_1 \text{ o } R_2 \neq R_2 \text{ o } R_1$ [for example, for two strings, *ab* and *ba*, $(ab)(ba) \neq (ba)(ab)$].
- **Closure:** Let $R = (a \mid b)$ be an RE over $\Sigma = \{a, b\}$. Then, $R_1 = (R)^* = (a \mid b)^*$ is a new RE obtained by applying closure on *R* (denoted as R^*) over {a, b}. In fact, R^* consists of all the strings that are the resultant of zero or more concatenations of the strings in *R*.

Formally, $(R)^* = R^0 \cup R^1 \cup R^2 \cup .... = (a \mid b)^* = (a \mid b)^0 \cup (a \mid b)^1 \cup (a \mid b)^2 \cup ... = \{\varepsilon, (a \mid b), (a \mid b) \text{ o } (a \mid b), (a \mid b) \text{ o } (a \mid b) \text{ o } (a \mid b), ...\}$, where 'o' is the concatenation operator. Thus, the set of strings (generated from this RE) includes: {ε, a, b, ab, ba, ...}.

A few simple REs are described below:

- $R = a^*$ gives {ε, a, aa, aaa, ...}, that is, zero or more occurrences of '*a*'. Hence, a* can generate strings {ε, a, aa, aaa, ...}.
- $R = a^+$ gives {a, aa, aaa, ...}, that is, one or more occurrences of '*a*'. Thus, a^+ can be expressed as aa*.
- $R = a \mid b^* = (a + b^*) = (r_1 + r_2)$, where r_1 and r_2 are two REs, and so, their union is also an RE. Here, $r_2 = b^*$ results in {ε, b, bb, bbb,...}. Thus, $R = a \mid b^* = (a + b^*) = (r_1 + r_2) = \{\varepsilon, a, b, bb, bbb,...\}$.

> **Note:** The set union $S = S_1 \cup S_2$ implies that any element $a \in S$ means $a \in S_1$ or $a \in S_2$ or both, where S, S_1 and S_2 are the sets. Further, RE is the mathematical expression that results in a set of

elements over Σ. Thus, the RE, $r = (a + b)$, is a set where both *a* and *b* are REs, and each generates a set. Here, the sets are {a} and {b}, respectively; that is, $r = (a + b) \equiv \{a\} \cup \{b\}$. Hence, the set *r* contains the elements *a* and *b*, which are in {a} or {b} or both. However, elements like *ab* and *ba* are not in *r*.

Regular languages and finite languages: A language that can be described by some regular expression is called a regular language. In other words, a language L over Σ is regular if there exists some RE over Σ for that language.

A language is finite if it contains a finite number of strings. Regular expression (*R*) for a finite language L with strings, say $\{w_1, _{...} w_n\}$ will be: $R = w_1 + \ldots + w_n$. So, every finite language is regular. This is discussed in detail in Section 3.5.

Regular expressions and finite automata: Regular expressions can be used to define the same languages that finite state automata describe – regular languages. In other words, every regular language is accepted by FA (DFA as well as NDFA). This is, in fact, known as the **equivalence property of regular language** (discussed later in this chapter). Interestingly, more than one FA can be designed for the same problem (P), but there exists one and only one minimum state FA. Likewise, several REs (each corresponding to an FA) may be described for P but all REs except one (which corresponds to the minimum FA) are redundant. Hence, the RE corresponding to the minimum FA is unique for each problem.

Note: If two or more REs generate the same language, these RE are considered equivalent. However, all the RE except one (the minimal one) are redundant.

3.3 EXAMPLES OF REGULAR EXPRESSIONS

EXAMPLE 3.1

Describe the following sets of strings by RE: (a){010}, (b) {0,1}, (c) {ε, ab}, (d) {1,11,111,...}, (e) the set of all strings of 0s and 1s ending in 00, (f) the set of all strings of 0s and 1s beginning with '0' and ending with '1', (g) the set of all strings over {a, b} that has only one '*a*', (h) the set of all strings over {0,1 } that has at most two zeros, (i) the set of all strings over {0,1 } that has at least three zeros.

Solution (a) As the set of strings contains only one string, its corresponding RE is $R = (010)$.

(b) Here, the set of strings contains only two simple strings, 0 and 1. Hence, the RE is $R = (0 + 1)$.

(c) The set contains only two strings, namely, ε and *ab*. So, the RE is $R = (\varepsilon + ab)$.

(d) The most appropriate RE for the given set is $R = 1(1)^* = 1^+$.

(e) The property of the given set states that each string in the set ends with 00. However, just before the end, any combination of '0' and '1' may appear in each valid string, and such combinations can be expressed as $(0 + 1)^*$. Hence, the set of strings can be described by the RE, $R = (0 + 1)^*00$.

(f) According to the given instruction, each valid string begins with '0' and ends with '1'. However, any substring of 0s and 1s may occur in between these two symbols. Obviously, such substrings can be described by $(0 + 1)^*$. Thus, the set of all valid strings can be expressed by an RE, $R = 0(0 + 1)^*1$.

Note: The RE for strings beginning with (that is, prefix) 1 and ending with (that is, suffix) 1 is $R = 1 + 1(1 + 0)^*1$. [Observe that the smallest string satisfying this property is '1' alone.]

(g) The problem states that each valid string contains exactly one *a*. However, any number of *b*s may appear before and after that *a* in the string. Such an expression can be written as: $R = b^*(a)b^*$.

(h) Given that each string in the set may have at most two zeros, the number of zeros in such a string may be 0 or 1 or 2. There is no restriction on the number of 1s before and after every 0. Thus, all these possibilities (namely, zero 0s, one 0 and two 0s) can be described by the RE, $R = 1^* + 1^*01^* + 1^*01^*01^*$.

We may note that the concept of OR logic is necessary here to view the three different possibilities, and '+' represents OR in the context of RE.

(i) Given that each string in the set has at least three zeros and the RE corresponding to this problem is $r = 1^*01^*01^*0(0 + 1)^*$. If we think of an alternative RE as: $(1 + 0)^*0(1 + 0)^*0(1 + 0)^*0(0 + 1)^*$, then some strings (though they are valid) may be generated multiple times. So, the latter RE is redundant.

EXAMPLE 3.2

Describe the following sets of strings by RE: (a) RE over {a, b} for generating strings of length 2; (b) RE over {a, b} for generating strings of length 3 or less; (c) RE for even-length words over {a, b}; (d) RE for odd-length words over {a, b}; (e) RE for strings ending in '1' and which do not contain the substring '00'; (f) RE for describing strings consisting of the substring '10'; (g) RE for accepting strings over {0, 1} ending with '1'; (h) RE for accepting strings consisting of the substring '10' preceded arbitrarily by many 1s; (i) RE of strings consisting of an even number of 1s over {0, 1}; (j) RE to generate strings of the form $\{0^k$: k is a multiple of 2 or 3$\}$.

Solution (a) The RE satisfying the given property for valid strings over {a, b} is: $R = (a + b)(a + b) = (a + b)^2$.

(b) $R = (\varepsilon + (a + b) + (a + b)(a + b) + (a + b)(a + b)(a + b)) = \varepsilon + (a + b) + (a + b)^2 + (a + b)^3$. This RE generates all possible strings of length 3 or less over {a, b}.

(c) As we need only even-length words over {a, b}, the desired RE is: $R = ((a + b)(a + b))^*$. It is important to note that $(a + b)^*(a + b)^*$ may not be the correct RE as it may generate odd-length strings as well; for example, *aba* is a valid string generated by this RE but it does not satisfy the constraints of the problem.

(d) An RE for generating odd-length strings over {a, b} is: $R = ((a + b)(a + b))^*(a + b) = (a + b)((a + b)(a + b))^*$. Here, the sub-expression $((a + b)(a + b))^*$ in R generates only even-length strings. When any even-length string resulting from this sub-expression is concatenated with (a + b), its length becomes odd. However, one may claim the RE for this purpose as: $R_1 = (a + b)^*(a + b)^*(a + b)$ or $(a + b)(a + b)^*(a + b)^*$. However, the sub-expression $(a + b)^*(a + b)^*$ of this RE may also generate an odd-length substring. Concatenating this odd-length substring with the substring generated from (a + b) results in even-length strings (which is not desired, according to the given instruction).

(e) One possible RE is $R = (1 + 01)^+$.

(f) The problem states that any valid string contains '10' as substring. Hence, an RE is $R = (0 + 1)^*10(0 + 1)^*$, as any combination of 0 and 1 may appear before and after the substring '10'.

Comment One may think that the substring '10' may be generated from the sub-expression $(1 + 0)^*$ (placed just before 10 in R), and so, the substring '10' appearing after the first $(0 + 1)^*$ is redundant. While it is true that the sub-expression: $(1 + 0)^*$ may generate '10', we may not ensure it. The reasons for its explicit presence after $(1 + 0)^*$ are given below.

- Firstly, we may not ensure the generation of '10' in each string from $(1 + 0)^*$; for example, ε, 0, 00, etc. are all generated for $(0 + 1)^*$ without '10' and may be wrongly treated as valid in the absence of '10'. $(1 + 0)^*$ also has the ability to generate a null string (ε). So, ε concatenated with '10' (if placed in the middle of R) results in a simple valid string with substring '10'. This is why $(0 + 1)^*$ is concatenated with '10'.
- On the other hand, $(1 + 0)^*$ appearing after '10'ensures that strings containing '10' (as substring) and ending with any combination of 0 and 1 are also valid.

Note: If the valid string contains exactly one substring '10', the RE will be $R = 0^*1^*(10)0^*1^*$.

(g) To satisfy the given constraint, one possible RE is $R = (0 + 1)^*1$. However, if one designs an RE such as $(0 + 1)^*11^*$ for this purpose, the last part (1^*) must be redundant.

(h) Assuming the meaning of the term 'many' as 'one or more', one possible RE for the given problem is: $R = (0 + 1)^*\ 1^+10(0 + 1)^*$. In fact, the sub-expression 1^+ helps generate many 1s before '10'.

(i) One possible RE for satisfying the given constraint is $R = (0^*10^*10^*)^*$. Here, each correct string consists of an even number of 1s but any number of 0s. The 0s may appear at any position in the string.

Comment If we design an RE, $R = (0^*10^*1)^*$ for this purpose, then strings such as 110, 1100,... may never be generated. However, such strings are undoubtedly valid according to the given instruction. Rejection of these valid strings can be detected easily if we let $(0^*10^*1) = \text{a}$ and express $R = (0^*10^*1)^*$ as $R = (\text{a})^* = \{\varepsilon, \text{a, aa, aaa,...}\}$.

(j) To satisfy the given properties for valid strings, the RE should be, $R = (00)^* + (000)^*$. Two distinct possibilities (based on the concept of OR) can occur here. However, one may want to design R as: $(00)^*(000)^*$. If so, then that RE will generate strings such as 00000, 0000000, etc. But, such strings are not valid as their length is not divisible by 2 or 3. Hence, the second choice is not correct. It is interesting to note that strings with length divisible by 2 AND 3 may be generated by $R = (000000)^*$, that is, taking closure of six 0s, and it is obtained from the LCM of 2 and 3.

EXAMPLE 3.3

(a) Give an RE for generating the set of all strings over {0, 1} with three consecutive 0s.

Solution An RE for solving this problem is, $R = (0 + 1)^*000(0 + 1)^*$.

(b) Let $L_1 = 10^*$ and $L_2 = 101^*$. Find $L_1 \cup L_2$, $L_1 \cap L_2$ and $L_1 - L_2$ in the form of RE.

Solution Here, $L_1 = 10^* = \{1, 10, 100,...\}$ and $L_2 = 101^* = \{10, 101, 1011,...\}$. Thus, the RE for $L_1 \cup L_2$ is $R = (10^* \cup 101^*)$. For $L_1 \cap L_2$, it is $R = (10)$, and for $L_1 - L_2$, it is $R = 1 + (100)0^*$.

EXAMPLE 3.4

Design an RE for the set of strings defined as $\{\text{a}^n \mid n \text{ is divisible by 2 or 3}, n > 0\}$ over {a}.

Solution To satisfy the given criterion, we need an RE, $R = (\text{aa})^+ + (\text{aaa})^+$.

EXAMPLE 3.5

(a) Design an RE to generate the set of all strings over {0, 1} containing exactly two 0s.

Solution Here, each valid string contains exactly two 0s. Thus, the RE for the given problem is $R = 1^*01^*01^*$, since any number of 1s may appear before and after each 0.

(b) Give the RE for the language $\text{L} = \{a^m b^n \mid m, n > 0\}$ over {a, b}.

Solution The RE for recognising the given language L is: $R = (\text{aa}^*\text{bb}^*) = (\text{a}^*\text{ab}^*\text{b}) = (\text{a}^+\text{b}^+)$. However, all these options are equivalent, as $RR^* = R^*R = R^+$ (see Section 3.4). This also means that there may be several REs for the same problem, but each generates the same set of strings.

(c) Give the RE for the language $\text{L} = \{a^m b^n \mid m, n > 0; (m + n) \text{ is even}\}$ over {a, b}.

Solution We know that odd + odd = even and even + even = even. So, we may design an RE for this problem as, $R = (\text{aa})^*(\text{bb})^* + \text{a}(\text{aa})^*\text{b}(\text{bb})^*$.

(d) Find the RE to describe strings over {a, b} whose third symbol from the right is *a*.

Solution To satisfy the given constraint, the most appropriate RE is: R = (a + b)*a(a + b)(a + b). The expression (especially, the sub-expression after '*a*') ensures that the third symbol from the right is '*a*'.

(e) Design an RE to identify valid identifiers of the C programming language.

Solution Each valid identifier in C starts with a letter, and any combination of letters and digits may appear after the starting letter. Hence, the desired RE is: R = letter (letter | digit)*, where letter = {'a'–'z', 'A'–'Z'} and digit = { 0, 1, 2, …, 9}.

(f) Give a short description in English of the language accepted by the RE, R = 0*10*10*10* over {0, 1}.

Solution The given RE (R) states that any string with exactly three 1s is valid.

(g) Give a short description in English of the language accepted by the RE, R = 0*(100*)*1* over {0, 1}.

Solution The given RE (R) states that any string that has no substring '110' is valid.

(h) Describe the RE: a*b* in English.

Solution The given RE states that any valid string consists of any number of *a*s followed by any number of *b*s but no *a* can appear after *b*.

(i) Describe the RE: (ab)* in English.

Solution The given RE states that all strings of *a* and *b* containing substring *ab* and its repetition (zero or more times) are valid.

3.4 ALGEBRAIC LAWS OF REGULAR EXPRESSIONS

Regular expressions may often be complex in structure and they may contain redundant parts. However, algebraic laws may be applied to simplify (or minimise) the REs. Before we discuss these algebraic laws, some basic mathematical concepts pertaining to these laws are presented below.

Basic Mathematical Concepts: Taking '*x*' as an element, the following terms are defined:

- An operator (*op*) is said to be idempotent if it behaves such that x *op* $x = x$.
- An identity (I) for an operator (*op*) is defined as: I *op* $x = x$. For instance, 0 is the identity for addition, and 1 is the identity for multiplication.
- An annihilator (A) for an operator (*op*) is defined as: A *op* x = A. For instance, A = 0 is the annihilator for multiplication.

Let us now discuss some important algebraic laws for regular expressions defined over alphabet Σ (the alphabet may be different for different REs).

(a) **Idempotent law:** If R is an RE, then $R + R = R$, but it is not necessary that $RR = R$. In other words, union (+) is an idempotent operation but concatenation (.) need not be. For instance, let R = a*b*. Now, $R + R = R$ but RR = a*b*a*b* $\neq R$. However, $RR = R$ if and only if each string $x \in R$ satisfies $x = x^R$.

(b) **Identity and annihilator:** Let R be an RE. Then,

(i) $\varepsilon R = R \varepsilon = R$. This law asserts that ε is the identity for concatenation.

(ii) $\Phi + R = R$. This states that Φ is the identity for union.

(iii) $\Phi R = \Phi$, where Φ is the null set (does not contain any string). Here, Φ is the annihilator for concatenation. Note that Φ in ΦR behaves like zero multiplication.

(c) **Commutative law:** Let P and Q be two REs. Then, $P + Q = Q + P$. Thus, the REs are commutative with respect to union. However, they are not commutative with respect to concatenation; for example, $PQ \neq QP$.

(d) **Associative laws:** Let P, Q and R be REs. Then,

(i) $(P + Q) + R = P + (Q + R)$ – associative with respect to union

(ii) $P(QR) = (PQ)R$ – associative with respect to concatenation

Thus, the REs are associative with respect to both union and concatenation.

(e) **Distributive laws:** Let P, Q and R be the REs. Then,

(i) $P(Q + R) = PQ + PR$ – (.) is left distributive over (+)

(ii) $(Q + R)\,P = QP + RP$ – (.) is right distributive over (+)

Hence, the REs hold distributive property over + and (.).

(f) **Laws involving closure property**

(i) Let R be an RE. Then, $(R^*)^* = R^*$. This law states that closure of an expression that is already closed does not change the language. For example, take the closure of R (R^*); it results in all possible strings over R. If we take the closure of R^* ($(R^*)^*$), this generates all possible strings that have already been generated from R^*. Thus, $(R^+)^* = R^*$ is also mathematically correct. However, $(RR)^* \neq R^*R^*$. To verify it, let $R = \{a\}$. Now, $(RR)^* = (aa)^* = \{\varepsilon, aa, aaaa, \ldots\}$, that is, a set that contains strings of an even number of *a*s. Again, R^*R^* $\{\varepsilon, a, aa, aaa, \ldots\}\{\varepsilon, a, aa, aaa, \ldots\} = \{\varepsilon, a, aa, aaa, aaaa, \ldots.\}$, that is, it is a set that contains strings of *a*s of any length.

(ii) $\varepsilon^* = \{\varepsilon, \varepsilon\varepsilon, \ldots\} = \varepsilon$

(iii) $\underline{RR^* = R^*R} = R^+$. The underlined expression states that any RE (R) commutates under concatenation with its own closure.

$R^* = \{\varepsilon, R, RR, \ldots\}$. Hence, $RR^* = R\,\{\varepsilon, R, RR, \ldots\} = \{R, RR, \ldots\} = R^+$. Further, $R^*R = \{\varepsilon, R, RR, \ldots\}\,R = \{R, RR, \ldots\} = R^+$. Thus, $RR^* = R^*R = R^+$.

(iv) $\varepsilon + RR^* = \varepsilon + R^+ = R^*$, since $RR^* = R^+$.

(v) $(P^*Q^*)^* = (P + Q)^* = (P^* + Q^*)^*$, where P and Q are two REs. To verify all these identities, we may choose suitable regular expressions, such as $P = a^*$, $Q = b^*$, etc.

(vi) $\varepsilon + R = R?$ Here, the $R?$ denotes zero or one copy of R.

Formally, zero copy of any RE is ε. Now, ε in ($\varepsilon + R$) simply results in zero copy, whereas the remaining R is used for the single copy. However, R^+ stands for one or more copies of R, and R^* represents zero or more copies of R.

> **Note:** Strictly speaking, an RE (R) is only an expression and not a language. In this respect, we should use L(R) for representing languages generated by R. However, we commonly use R to mean language too. Thus, if E and F denote REs, then
>
> (i) $L(E + F) = L(E) + L(F)$ (ii) $L(EF) = L(E)L(F)$ (iii) $L(E^*) = (L(E))^*$
>
> where L denotes language.

3.5 FINITE AND INFINITE LANGUAGES

We defined finite and infinite languages earlier in this chapter. These concepts are now described in detail.

Finite Language: Mathematically, it is the language whose strings can be explicitly listed in a finite amount of time/space. For example, L = {ab, ac, bb, bc} is a finite language over {a, b, c}. Another interesting example is L = {fixed-digit prime numbers}. The language L is regular as it is a finite language. However, the language of prime numbers of infinite size is not regular.

Why is a language of fixed digit (say, three-digit) prime numbers regular?
There is no RE that exists in a recursive manner to describe this problem. It is necessary here to specify all the three-digit prime numbers in a set; the RE will be, $R = (101 + 103 + ... + 997)$. This is, indeed, a finite specification of elements. Thus, it is a finite language for which an RE exists in a non-recursive manner. We may design an equivalent FA for the problem, and so, it is a regular language. However, a language consisting of all prime numbers is not regular, as we cannot obtain any pattern (or periodic pattern) for such a language. In fact, no mathematical formula has thus far been discovered (nor has it been proved that such a formula is possible) that can generate all the prime numbers.

Thus, every finite language can be analysed in this fashion, and hence, *every finite language is regular*.

Infinite Language: If the strings of a language cannot be listed explicitly, then that language is called an infinite language. For example, a language with an infinite number of stings of *a* with an RE, $R = a^*$ (here, a* is expressed in a recursive manner), is an infinite language. However, all infinite languages may not necessarily be regular. For example, an infinite language of prime numbers is not regular. To determine this, we may apply the concept of **equivalence class**.

Two strings *x* and *y* in Σ^* are **indistinguishable** with respect to a language L, if and only if, for any string *z* in Σ^*, both *xz* and *yz* are in L or both are not in L.

Let us take some examples to illustrate **indistinguishability** (or equivalence) of strings:

- Let L = {ab, ac, bb, bc}. Clearly, $\Sigma = \{a, b, c\}$. Let $x = a$, $y = b \in \Sigma^*$. Let $z = a$. It also belongs to Σ^*. Now, $xz = aa \notin L$ and $yz = ba \notin L$. Further, let $z = b$, then $xz = ab \in L$ and $yz = bb \in L$. So, the strings *a* and *b* are equivalent (indistinguishable) with respect to L. Likewise, we may check the other strings of Σ^*.
- Let us take another language $L = \{a^n \mid n \geq 0\}$ over $\Sigma = \{a\}$. Take two strings *a* and *aa* in $\{a\}^*$. These are indistinguishable with respect to the language L over alphabet {a} because aa^k and aaa^k are in the language $\{a^n \mid n \geq 0\}$ for any positive integer *k* (and $z = a^k$).

Let *x* and *y* be strings in Σ^*. We say that *x* and *y* are distinguishable by L if some string *z* exists in Σ^* whereby exactly one of the strings, *xz* or *yz,* is a member of L.

Let us take a language $L = \{a^n b^n \mid n \geq 1\}$ over $\Sigma = \{a, b\}$. With respect to this language, consider the strings: $x = a \in \Sigma^*$ and $y = aa \in \Sigma^*$. These are distinguishable because the string *ab* is in the language: $a^n b^n$, whereas the string *aab* is not in the language for $z = b$.

- A set of strings is pair-wise distinguishable with respect to a language L if every pair of strings in the set is distinguishable with respect to L.
- For any language, we will attempt to find a set of indistinguishable strings, and each such set is called the equivalence class. Further, if no *z* in Σ^* is found to insert the strings of a language L into equivalent classes, then there is one equivalence class for each string of L.

EQUIVALENCE CLASS AND REGULAR LANGUAGE

For a language L, if there exist many finite equivalence classes, then L is regular. In other words, there exists an equivalence relation (denoted as $\equiv_L$) for L. However, if there exists an infinite number of equivalence classes for L, then L is non-regular. For a non-regular language, there is one equivalence class for each string in Σ^*. An example of a non-regular language is $L = \{a^n b^n \mid n \geq\}$ over $\Sigma = \{a, b\}$. For this language, there exists an infinite number of equivalence classes (one for each string in Σ^*).

Lemma 1: If L is regular, then $\equiv_L$ (equivalence relation) has a finite number of equivalence classes.

Proof Let M be a DFA such that L = L(M). The lemma shows that $\equiv_L$ has at most as many equivalence classes as $\equiv_M$, which has a finite number of equivalence classes (equal to the number of states of M).

Claim As per Lemma 1, L = $\{0^n1^n \mid n \geq 0\}$ is not regular.

Proof Consider the sequence of strings $x_1, x_2, \ldots$ where $x_i = 0^i$ for $i \geq 1$. We now see that no two of these are equivalent to each other with respect to $\equiv_L$. Consider $x_i = 0^i$ and $x_j = 0^j$ for $i \neq j$. Let $z = 1^i$ and notice that $x^iz = 0^i1^i \in$ L but $x^jz = 0^j1^i \notin$ L. Therefore, no two of these strings are equivalent to each other for $z = 1^i$, and thus, L cannot be regular.

An advantage of this method for proving non-regularity of languages is that, unlike the pumping lemma, it is always guaranteed to work because the corollary above is a precise characterisation of regular languages.

Some important points for finite and infinite languages

1. *An infinite language can have finite automata.* This is possible if the strings of an infinite language (L) can be categorised/classified into a finite set of equivalence classes. Again, if it is possible, an FA can be built for the infinite language as well.
2. *If there are infinitely many strings that are pair-wise distinguishable with respect to L, then L cannot be regular.* If the strings in L are pair-wise distinct, we will be unable to find a finite number of subsets of equivalent classes, which makes the language non-regular.
3. *Any star (*)-free language (L) is a regular language.* It is important to note that the operator '*' is not allowed for such a language L. For example, r = a* is not star-free as the RE r contains *(closure). Accordingly, r derives an infinite number of strings. However, if a language L is star-free, then it contains only a finite number of elements (strings), which means that L is finite. Further, any finite language is regular. Thus, any star-free language is regular.

Finite Representation of Language: Not to be confused with finite language, the finite representation of a language shows the way to represent any language. Following are the essential criteria for the finite representation of a language:

- Σ (alphabet) is a finite set.
- Representation of each language is distinct so that no two languages have identical representation, that is, different languages have different representation.

EXAMPLE 3.6

(a) Consider the languages $L_1 = \Phi$ and $L_2 = \{a\}$. Find $L_1 L_2^* \cup L_1^*$.

Solution Here, $L_1 = \Phi$ and $L_2 = \{a\}$. Thus, the result of $L_1 L_2^*$ is Φ, as we know that Φ indicates empty language, and concatenation of Φ with any other language results in Φ. More specifically, Φ behaves like zero multiplication in arithmetic.

Again, $L_1^* = \Phi^* = \Phi^0 \cup \Phi^1 \cup \Phi^2 \cup \ldots = \{\varepsilon\} \cup \Phi = \{\varepsilon\}$. We know that the 0-th power of any language results in ε. Hence, $\Phi^0 = \{\varepsilon\}$. Further, union of any set with a null set gives the set itself. Thus, union of Φ and {ε} gives {ε}. Hence, $L_1L_2^* \cup L_1 = \Phi \cup \{\varepsilon\} = \{\varepsilon\}$.

(b) Under what circumstances is $L^+ = L^* - \{\varepsilon\}$?

Solution This is true if L does not contain ε (null string). An infinite number of languages satisfy this property.

(c) Give an example of a finite language whose closure is not regular.

Solution A language consisting of three-digit prime numbers is finite and regular. However, its closure is not regular as we may not find any pattern for this set.

(d) Simplify the regular expression: $R = \varepsilon + 1^*(011)^*(1^*(011)^*)^*$.

Solution Let $P = 1^*(011)^*$, where P is an RE. Now, we may write the given R as: $\varepsilon + PP^* = P^*$ (using law (*iv*) of closure property). Further, $P^* = (1^*(011)^*)^* = (P_1{}^*Q_1{}^*)^*$, letting $P_1 = (1)$ and $Q_1 = (011)$ as two REs.

$= (P_1 + Q_1)^*$ [using law (*v*) of closure] $= (1 + 011)^*$.

(e) Simplify the RE, $R = ((a + b)^*(c + d)^*)^+$.

Solution The given RE can be expressed as: $R = ((a + b)^*(c + d)^*)^* - \{\varepsilon\} = T - \{\varepsilon\}$, where $T = ((a + b)^*(c + d)^*)^*$. Now, T can be simplified as $T = (P^*Q^*)^*$, where $P = (a + b)^*$ and $Q = (c + d)^*$.

$= (P + Q)^* = \{\varepsilon, P, Q, PQ, QP, \ldots\}$

We may apply $P = (a + b)^*$ and $Q = (c + d)^*$ for listing the exact strings.

3.6 EQUIVALENCE OF FINITE AUTOMATON AND REGULAR EXPRESSION

It is true that every finite automaton has an equivalent regular expression, and vice versa. We will now discuss how to find the equivalent FA from a regular expression, and vice versa. First, we will discuss how to construct an FA (especially a DFA) from an RE.

3.6.1 Designing Equivalent DFA for Regular Expressions

In general, pattern checkers are expressed in the form of REs. Implementation of any pattern checker requires conversion to DFA and then to a state table, that is, RE → DFA → State table. Here, we will discuss two approaches for designing a DFA from an RE. In the first approach, we construct the DFA directly from the RE (direct method) and in the second approach, we map the RE to an NDFA and then convert it to DFA.

Approach 1: Direct Method: In this method, we take some elementary (base) regular expressions and their equivalent DFAs. The elementary REs help to design DFAs for more complex RE. It is important to note that even though we discuss a number of elementary building blocks, readers should be aware that the list presented here cannot be considered exhaustive, and in general, there is no (direct) universal strategy to convert RE to DFA. It requires a lot of observation and practice to master the art of conversion. However, it is possible to derive RE from any DFA or NDFA using Arden's theorem (discussed later).

We consider the following base cases to design a DFA from an RE.

Case 1 Consider the alphabet $\Sigma = \{a\}$ and a language $L = \{a^n \mid n \geq 0\}$. The RE for L is a*. The equivalent DFA is shown in Fig. 3.1.

Observe that the initial state is also the final state because the null string is also a valid string for the given language L. The self-loop represents the fact that there can be zero or more *a*s.

FIGURE 3.1 A DFA for a*

Case 2 Consider the alphabet $\Sigma = \{a\}$ and a language $L = \{a^n \mid n \geq 1\}$. The RE for L is a^+. The equivalent DFA is shown in Fig. 3.2.

Observe that at least two states are required here because the smallest possible string accepted by this language is of length 1. On accepting the first '*a*', there may be zero or more *as*, and hence, the self-loop is inserted at the final state.

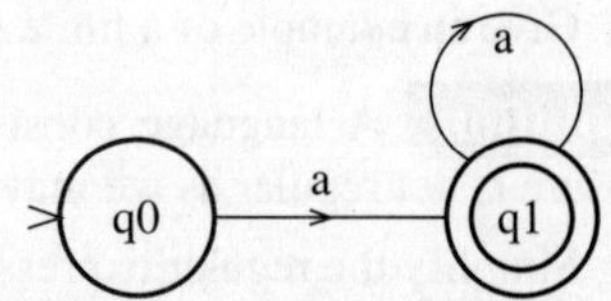

FIGURE 3.2 A DFA for a^+

> **Tip:** If the smallest possible string of a language has *n* number of symbols, the corresponding DFA must have at least $(n + 1)$ states.

Now, let us examine a few elementary cases for languages with more than one symbol.

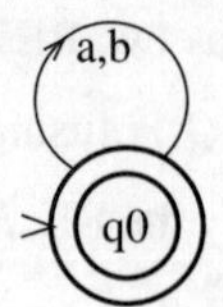

FIGURE 3.3 L = {any number of *as* and *bs* in any sequence over {a, b} including the null string}

Case 3 Consider the alphabet $\Sigma = \{a, b\}$ and RE = $(a + b)^*$. The language represented by this RE is L = {any number of *as* and *bs* in any sequence, including the null string}. The equivalent DFA is shown in Fig. 3.3.

Again, the initial state is also the final state because the null string is valid. '*a*' or '*b*' may be present in any sequence and any number of times; hence, the self-loop is inserted at the final state.

Case 4 Consider the alphabet $\Sigma = \{a, b\}$ and RE = $(a + b)^+$. The language represented by this RE is L = {any number of *as* and/or *bs* in any sequence, having at least one character}. The equivalent DFA is shown in Fig. 3.4.

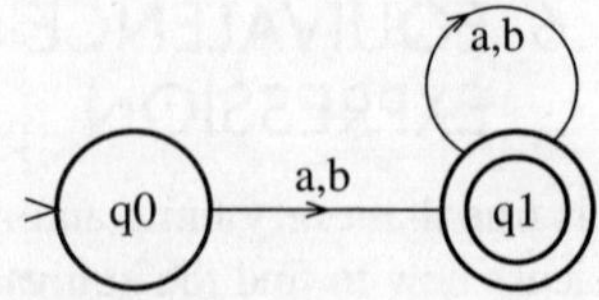

FIGURE 3.4 L = {any number of *as* and/or *bs* in any sequence over {a, b} excluding the null string}

The smallest string accepted by this DFA is of length 1. Hence, there are two states, as discussed in the earlier example. Also, on accepting the first character, the next character can be '*a*' or '*b*' – in any sequence – and occur any number of times; hence, the self-loop is inserted at the final state.

Case 5 Consider the alphabet $\Sigma = \{a, b\}$ and RE, $r = a^* + b^*$. The language specified by this RE is L = {any number of *as or bs*, including the null string}. The equivalent DFA is shown in Fig. 3.5.

We should examine this design very carefully.

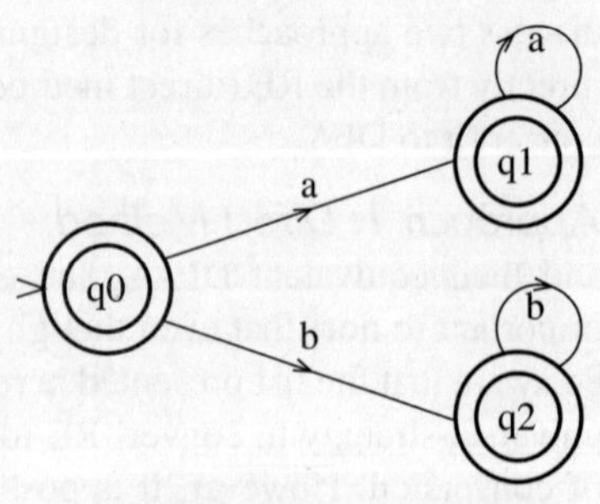

FIGURE 3.5 L = {any number of *as* or *bs*, including the null string}

- The initial state can also be the final state because the null string is valid and accepted by this automaton.
- A string accepted by this automaton can be made up of either *as* or *bs* but cannot contain combinations of *as* and *bs* – hence, we fork out two separate paths from the initial state: one for input symbol '*a*' and the other for input symbol '*b*'.
- Further, a string can have one or more symbols, and hence, there are self-loops in the final states.
- If '*b*' is encountered at q_1, the automaton must move to the dead state. Likewise, if '*a*' is encountered at q_2, it must move to the dead state.

This is a classic case where there are multiple final states.

> **Tip:** If an RE is actually the OR of multiple REs; that is, the expression is of the form $R = P_1 + P_2 + P_3 + \ldots + P_n$, where $P_1, P_2, P_3, \ldots, P_n$ are regular expressions, then we need to fork out *n* number of independent paths, one for each P_i.

Advantage

- Less effort is required to convert the RE to DFA.

Disadvantages

- No universal strategy exists to convert RE to DFA.
- It requires a lot of observation and practice. So, designing DFAs for complex REs like $\varepsilon + 1^*(011)^*(1^*(011)^*)^*$ is difficult.

To overcome the problems in the direct approach, Approach 2 may be used.

Approach 2: Mapping Regular Expressions to NDFA: We may adopt either of the following methods: (i) RE to NFA (without ε move), and then its conversion to DFA or (ii) RE to ε NFA and then DFA.

The second method is commonly preferred as combining FAs for RE for the concatenation, union and closure operations using ε transitions is easier than using NFA without ε move. More specifically, connecting adjacent parts in an FA is easier with ε transitions.

It is interesting to note that lexical analysis tools such as Lex/Flex use this method for implementation of regular expressions, and it is, indeed, performed in the sequence: RegExp ⇒ ε NFA ⇒ DFA ⇒ State table. This method is discussed below.

In practice, we refer to Thompson's construction algorithm (which includes some standard rules) for drawing ε NFA from RE. The rules are given below. This method converts regular expression to ε NFA; the resultant ε NFA is then converted into DFA by the subset construction approach. This approach is collectively named **Thompson's FA construction algorithm.**

Regular expression to NFA (without ε move) and then DFA: This strategy does not follow any standard rules for conversion; rather it depends on analysis of the problem as well as practice. Let us illustrate this method with the help of an example.

EXAMPLE 3.7

Design an equivalent DFA for the regular expression (a + b)*abb defined over the alphabet Σ = {a, b}.

Solution The language recognised by this RE can be described in English as, L = {set of all strings over Σ which end with *abb*}. The following steps may be adopted to draw an NFA from the given RE.

Step 1: To handle the ending with (suffix) *abb*, we need the final part of this automaton to be of the type shown in Fig. 3.6 (a).

FIGURE 3.6 (a) FA for abb

Step 2: The above is only a small fragment of this DFA. To consider the initial and final states of the FA, we may keep q_0 as the initial state and insert a self-loop over q_0 for input symbols '*a*' and '*b*'. This will handle the first part: (a + b)* as well.

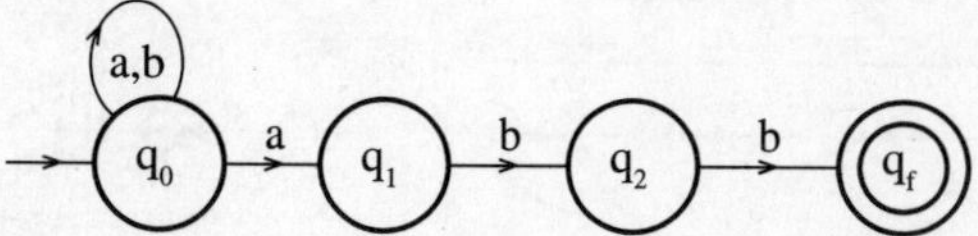

FIGURE 3.6 (b) An NDFA for (a + b)*abb

However, the FA (shown in Fig. 3.6 (b)) is an NDFA and not a DFA, since there is ambiguity about input symbol 'a' at state q_0. Now, following the conversion strategy (from NDFA to DFA), the equivalent DFA is presented in Fig. 3.6 (c).

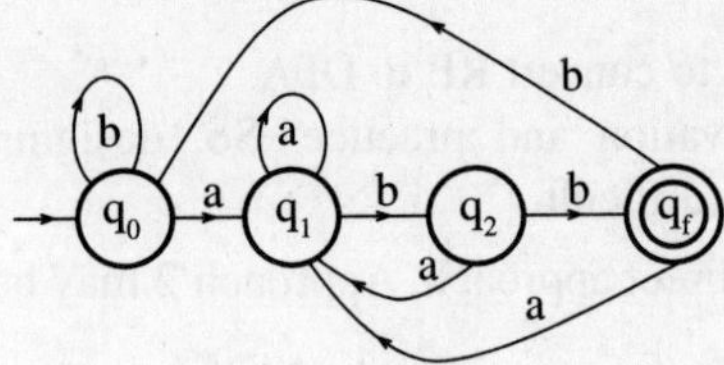

FIGURE 3.6 (c) A DFA for (a + b)*abb

Advantage and disadvantage: It is easier to draw the NDFA from the RE than drawing the DFA directly from the RE, as we need not show all the transitions explicitly. However, it is still problematic to design DFA from RE, as no standard rules are followed here.

Regular expression to ε NFA and then DFA: Thompson's FA construction algorithm: This is the most popular method for converting RE to DFA. The following rules are considered for constructing ε NFA from RE:

(i) Parallel path for union of expressions: Suppose $(r_1 + r_2)$ is an RE. The ε NFA for this union expression is shown in Figure 3.7 (a). An example for $r = (0 + 1)$ is also shown.

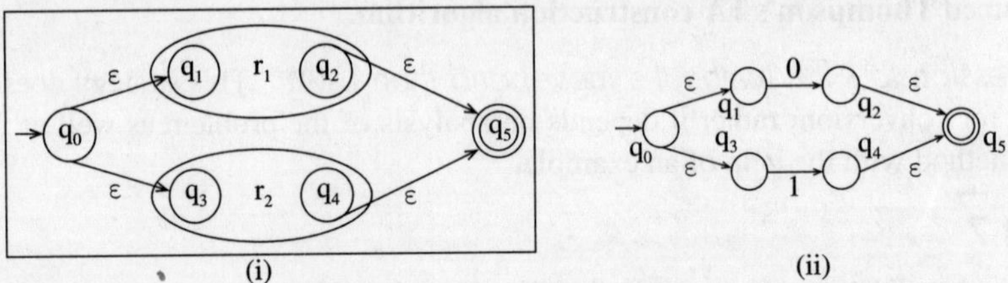

FIGURE 3.7 (a) (i) ε NFA for $(r_1 + r_2)$, (ii) ε NFA for (0 + 1): An example

(ii) Serial path for concatenation of expressions: Suppose $(r_1 r_2)$ is an RE. The ε NFA for this concatenation expression is shown in Fig. 3.7 (b). An example for $r = (01)$ is also presented.

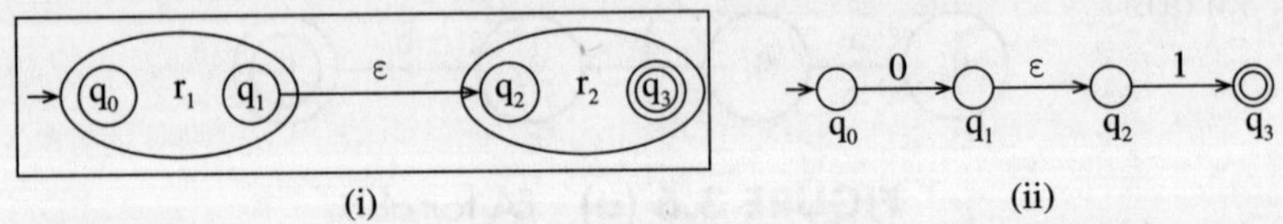

FIGURE 3.7 (b) (i) ε NFA for $(r_1 \cdot r_2)$, (ii) ε NFA for (0.1): An example

(iii) Path for closure of expression: Suppose $(r_1{*})$ is an RE. The ε NFA for it is shown in Fig. 3.7 (c). An example for $r = (0*)$ is also presented.

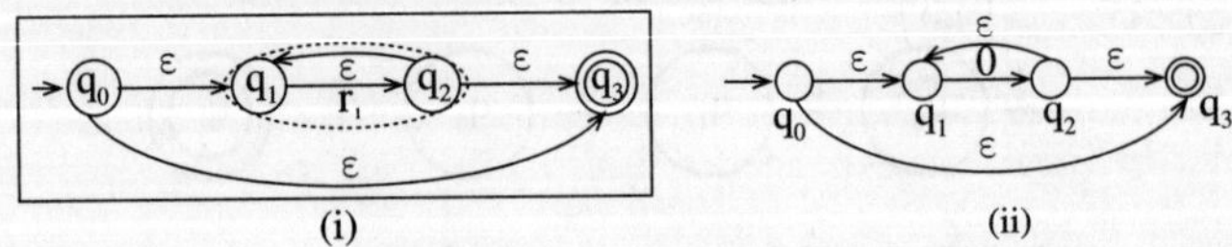

FIGURE 3.7 (c) (i) ε NFA for $r_1{*}$, (ii) ε NFA for (0*): An example

For generating null, it takes path: $q_0 - \varepsilon - q_2$, whereas it may take either of the paths: $q_0 - \varepsilon - q_1 - r_1 - q_2 - \varepsilon - q_3$ or $q_0 - \varepsilon - q_1 - r_1 - q_2 - \varepsilon - q_1 - r_1 - q_2 \ldots$ for generating one or more r_1. For example, ε NFA for 0* takes the path: $q_0 - \varepsilon - q_1 - 0 - q_2 - \varepsilon - q_1$ for generating one or more 0s. Interestingly, one may claim to connect q_0 to q_3 or q_1 to q_3 and so on through ε move to derive the null string. All these paths are valid but we are following some standard structures introduced by Thompson.

(iv) Path for positive closure: Suppose (r_1^+) is an RE. The ε NFA for this RE is shown in Fig. 3.7 (d). An example for $r = (0^+)$ is also presented.

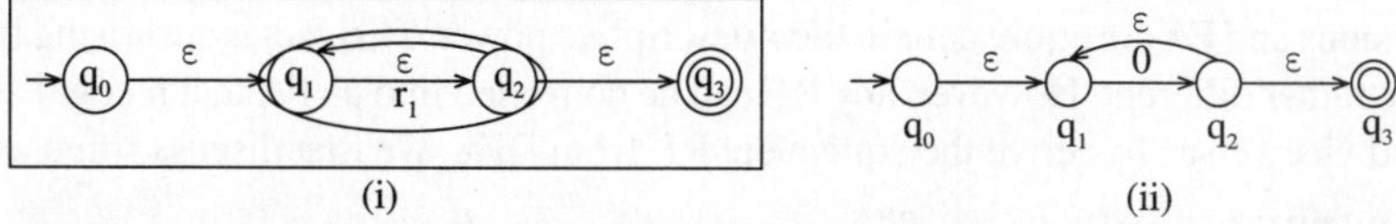

FIGURE 3.7 (d) (i) ε NFA for r_1^+, (ii) ε NFA for 0^+: An example

On the basis of these rules, an ε NFA for the RE, $R = (a + b)^*abb$ is shown in Fig. 3.7 (e) and its equivalent DFA is already shown in Fig. 3.6 (c).

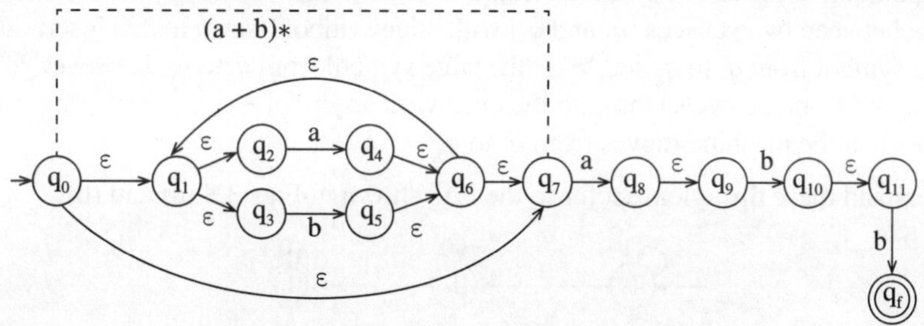

FIGURE 3.7 (e) ε NFA for (a + b)*abb

In particular, the ε NFA for the expression: (a + b)* can generate strings with any combination of *a*s and *b*s (including the null string).

It is important to note that designing the ε NFA often becomes time-consuming, so we may use a combination of NFA and DFA representations to maintain space and construction speed. Such a representation for Fig. 3.7 (e) is shown in Fig. 3.7 (f).

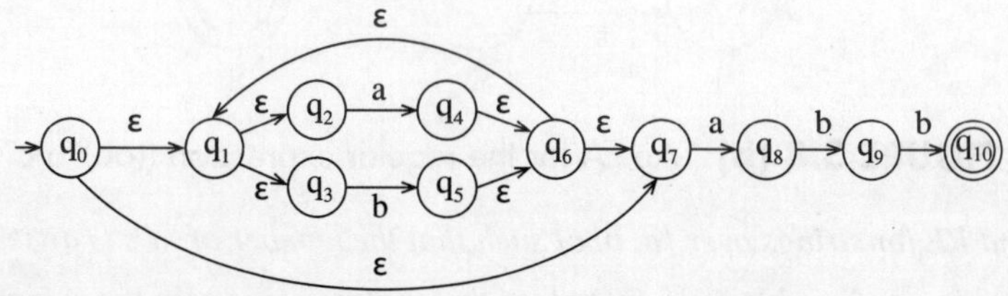

FIGURE 3.7 (f) A combination of ε NFA and DFA for (a + b)*abb

Note: It does not show the moves for '*a*' or '*b*' encountered at the final states in Fig. 3.7 (e). This is because it is an NFA and not a DFA. So it is not necessary to show them explicitly (through ε transitions from the final state to some other state). More specifically, the conversion strategy from ε NFA to DFA may explicitly show that part.

THEOREM 3.1

For every regular expression R, there exists a DFA (M) such that $L(R) = L(M)$.

Proof By induction on the structure of the regular expression (as shown above), we may design a non-deterministic FA. We may then find an equivalent DFA from the non-deterministic FA with ε transitions. The result follows from the equivalence of such automata with DFA.

3.6.2 Deriving Equivalent Regular Expressions from DFA

Regular expressions and FA are equivalent in their descriptive power. This fact is surprising because FA and RE appear to be rather different. However, any RE can be converted into an FA that recognises the language it describes, and vice versa. To derive the equivalent RE from DFA, we first discuss some useful tips.

Simple tips to find RE from DFA by applying the direct approach

- For a self-loop with the symbol '*a*', insert: a*. For example, let q be a state in a DFA. If there exists a self-loop labelled by an input symbol, say a, at q, then its corresponding RE is (a)*. However, if there exist two self-loops labelled by some input symbols, say a and b, respectively, at state q, then the corresponding RE is (a + b)*. Likewise, (a + b + c)* has three self-loops, and so on.
- For a cycle between two vertices (q_1 and q_2) with edge symbols, say a and b, insert (ab)*, where '*a*' is the edge symbol from q_1 to q_2 and '*b*' is the edge symbol from q_2 to q_1. However, to move from q_2 to q_3 (going out from the cycle) through the edge with a symbol, say c, insert ((ab)*a)c as one extra '*a*' is read when the machine moves from q_2 to q_3.

To better understand these tips, please refer to the FAs shown in Figs 3.8 (a) and (b).

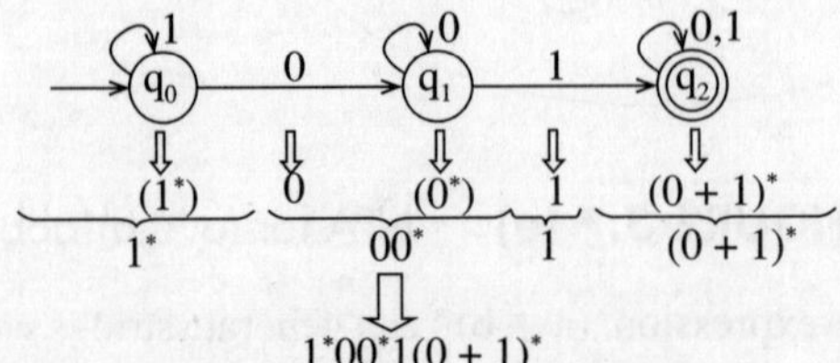

FIGURE 3.8 (a) An FA for the regular expression 1*00*1(0 + 1)*

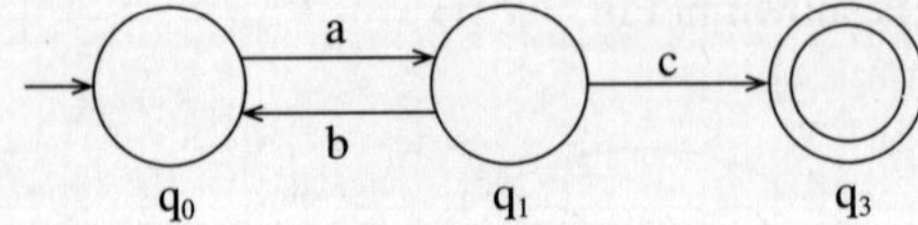

FIGURE 3.8 (b) An FA for the regular expression ((ab)*a)c

Simple tricks to find RE for strings over {a, b, c} such that the number of 'a's is divisible by n

1. As the number of *as* is divisible by n, first place the *as* at n positions in the expression.
2. Next, apply closure on the remaining symbols that appear after and before each a.
3. Finally, obtain closure of the entire expression.

For example, let $\Sigma = \{a, b, c\}$ and $n = 5$. So, RE R = ((b + c)*a(b + c)*a(b + c)*a(b + c)*a(b + c)* a(b + c)*)*. Likewise, if $\Sigma = \{a, b\}$ and $n = 2$ (even), then R = ((b)*a(b)*a(b)*)*.

Let us take another example. Suppose $\Sigma = \{a, b\}$ and n is not even, then R may be expressed as: $R = ((b)^*a(b)^*a(b)^*)^*a + a((b)^*a(b)^*a(b)^*)$, through concatenation of possible strings of even number of *a*s with the possible strings of odd number of *a*s and vice versa.

However, we may formulate the negation (complement) as follows: $R = \neg\ ((b^*ab^*ab^*)^*)$

Deriving RE from DFA is a complex task and depends on proper analysis and good practice. That is why we usually prefer the approach introduced by Arden.

ARDEN'S THEOREM

If P, Q and R are regular expressions, the equation $R = Q + RP$ has a unique solution $R = QP^*$.

Proof To prove the theorem, we may keep expanding the right-hand side (RHS) of the equation by using $R = Q + RP$ as follows:

$$\begin{aligned} R &= Q + RP = Q + (Q + RP)P \\ &= Q + QP + RPP \quad = Q + QP + (Q + RP)PP \\ &= Q + QP + QPP + RPPP \quad = Q + QP + QPP + (Q + RP)PPP \\ &= Q + QP + QPP + QPPP + RPPPP \\ &= \ldots.. \\ &= Q + QP + QPP + QPPP + QPPPP + QPPPPP + \ldots.\ \infty \text{ [applying } R = Q + RP \text{ infinitely]} \\ &= Q\{\ \varepsilon + P + PP + PPP + PPPP + PPPPP + \ldots.\ \infty\} \\ &= QP^* \text{ [since, } P + PP + PPP + \ldots = P^+ \text{ and } \varepsilon + P^+ = P^*] \end{aligned}$$

This proves the theorem.

Now, using this theorem, we shall attempt to derive regular expressions for a few simple DFA and NDFA.

Note: Each state in the state diagram represents an RE to apply Arden's theorem.

Applying Arden's Theorem

Assumptions

- The transition diagram must not have null transitions (if it exists, first remove it).
- It must have only one initial state.

Method

1. Derive equations in the following form for all the states of the DFA with n states. $q_i = \sum_{j=1}^{n} q_j R_{ji}$, ($i = 1, 2, \ldots n$), where R_{ij} represents the set of labels of edges from q_j *to* q_i. If no such edge exists, then $R_{ij} = \varnothing$. If q_i is the initial state, add ε (as the RE) to the equation associated with q_i.
2. Solve these equations to obtain the expression for the final state in terms of R_{ji}. If multiple final states exist, take their union to obtain the final RE.

EXAMPLE 3.8

Compute an equivalent regular expression for the DFA shown in Fig. 3.9.

Solution From Fig. 3.9, it is clear that q_0 is the initial as well as the final state. We can now express the state equation (using the technique discussed above) for q_0 as: $q_0 = q_0 a$.

Further, q_0 is the initial state, so it is expressed as: $q_0 = \varepsilon + q_0 a$.

Observe that it is of the form $R = Q + RP$, where $R = q_0$, $Q = \varepsilon$ and $P = a$.

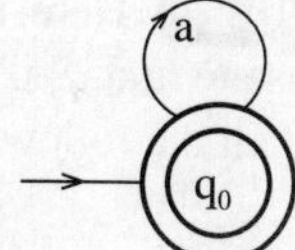

FIGURE 3.9 A DFA

Now, on applying Arden's theorem, we obtain the solution as:
$q_0 = \varepsilon\, a^* = a^*$.
Hence, the regular expression equivalent to this DFA is a* (as q_0 is the final state).

EXAMPLE 3.9

(a) Compute an equivalent regular expression for the DFA shown in Fig. 3.10 (a).

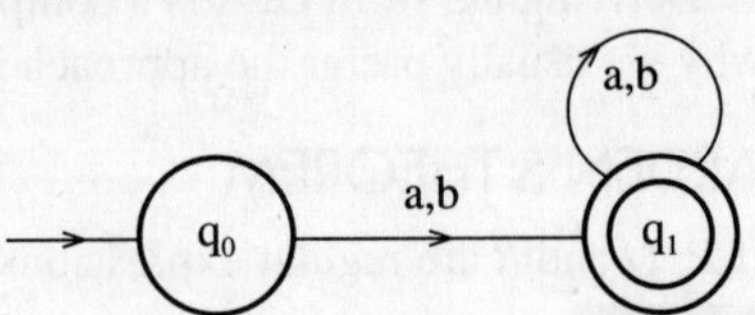

FIGURE 3.10 (a) A DFA over {a, b}

Solution From Fig. 3.10, we see that q_0 is the initial state and q_1 is the final state. Thus, there are only two states, say q_i ($i = 0, 1$). Now, the state equations are:

$i = 0$, $q_0 = q_0R_{00} + q_1R_{10} + \varepsilon = \varepsilon$, since $R_{00} = \varnothing$, $R_{10} = \varnothing$ and q_0 is the initial state.

$i = 1$, $q_1 = q_0R_{01} + q_1R_{11} = q_0(a + b) + q_1(a + b)$, since $R_{01} = (a + b)$ and $R_{11} = (a + b)$, that is, $q_1 = \varepsilon\,(a + b) + q_1(a + b)$ [substituting $q_0 = \varepsilon$] $= (a + b) + q_1(a + b)$.

This is also of the form $R = Q + RP$, where $R = q_1$, $P = (a + b)$ and $Q = (a + b)$. From Arden's theorem, we get,

$$q_1 = (a + b)(a + b)^*$$

Hence, the equivalent regular expression is $(a + b)^+$, since $q_1 = (a + b)^+$ and it is the final state.

(b) Compute an equivalent regular expression for the DFA shown in Fig. 3.10 (b).

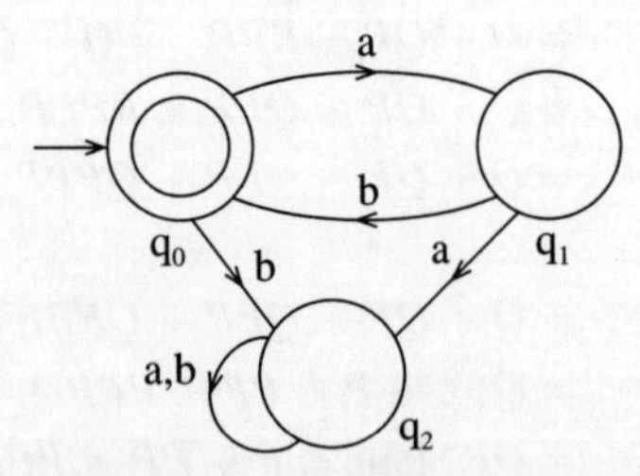

FIGURE 3.10 (b) A DFA over {a, b}

Solution Here, there are three states, namely q_0, q_1 and q_2, where q_0 is the initial as well as the final state and q_2 is the dead state. Thus, by applying Arden's theorem, we obtain the following equations:

$$q_0 = q_1 b + \varepsilon \quad \text{(i)}$$

$$q_1 = q_0 a \quad \text{(ii)}$$

$$q_2 = q_0 b + q_1 a + q_2.(a + b) \quad \text{(iii)}$$

Here, q_0 is the final state. So, the expression retained by this state is the final RE for the given DFA. However, both q_0 and q_1 do not depend on q_2. So, by substituting $q_1 = q_0 a$ in Equation (i), we get,

$q_0 = q_0\, ab + \varepsilon = \varepsilon + q_0 ab$.

Clearly, $R = q_0$, $Q = \varepsilon$ and $P = ab$. Hence, the unique solution is: $q_0 = (ab)^*$.

> **Note:** As soon as we observe that a specific state is a dead state, there is no need to compute the state equation for the same. Equation (iii) in this example is redundant.

(c) Compute an equivalent regular expression for the DFA shown in Fig. 3.10 (c).

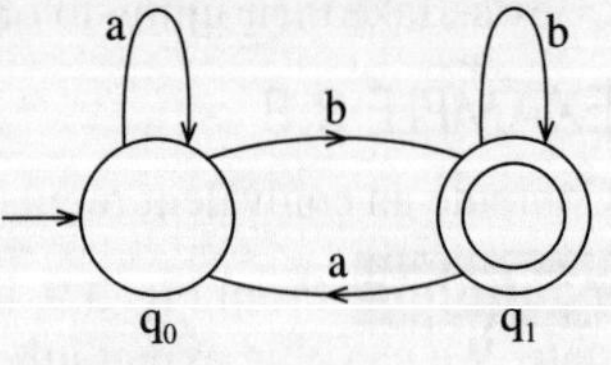

FIGURE 3.10 (c) A DFA over {a, b}

Solution Here, there are two states, namely q_0 and q_1, where q_0 is the initial state and q_2 is the final state. Thus, by applying Arden's theorem, we obtain the following equations:

$$q_0 = q_0 a + q_1 a + \varepsilon \quad \text{(i)}$$

$$q_1 = q_0 b + q_1 b \quad \text{(ii)}$$

Here, q_1 is the final state. So, the expression retained by this state is the final RE for the given DFA.

Now, by applying Arden's theorem on $q_1 = q_0 b + q_1 b$, we get

$$q_1 = q_0 bb^*, \text{ where } R = q_1, Q = q_0 b \text{ and } P = b$$

Further, by substituting $q_1 = q_0 bb^*$ in Equation (i), we get,

$$q_0 = q_0 a + q_0 bb^*a + \varepsilon = \varepsilon + q_0 (a + bb^*a) \text{ which gives}$$

$$q_0 = \varepsilon\,(a + bb^*a)^* \text{ by Arden's theorem.}$$

Hence, the resulting RE is $q_1 = q_0 bb^* = (a + bb^*a)^*bb^*$.

(d) Derive a regular expression for the state transition diagram shown in Fig. 3.10 (d).

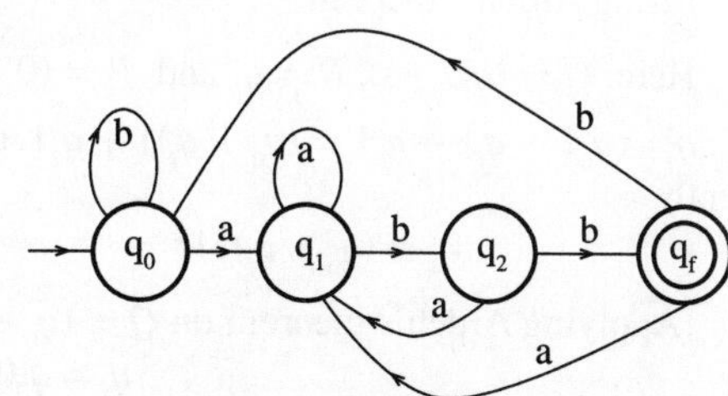

FIGURE 3.10 (d) A DFA

Solution From Fig. 3.10 (d), we obtain the following state equations:

$q_0 = q_f b + q_0 b + \varepsilon$, or equivalently

$q_0 = (q_f b + \varepsilon) + q_0 b = (q_f b + \varepsilon)b^*$ (i)

[using Arden's theorem, where $Q = (q_f b + \varepsilon)$ and $P = b$]

$q_1 = q_0 a + q_f a + q_2 a + q_1 a = (q_0 + q_f + q_2)a + q_1 a$,

or equivalently $q_1 = (q_0 + q_f + q_2)aa^*$ [using Arden's theorem]

or, even further simplified as $q_1 = (q_0 + q_f + q_2)a^+$ (ii)

$$q_2 = q_1 b \quad \text{(iii)}$$

$$q_f = q_2 b \quad \text{(iv)}$$

Using (iii) in (iv), we get $q_f = q_1 bb$ (v)

Using (iii) and (v) in (ii), we get $q_1 = q_0 a^+ + q_1 bba^+ + q_1 ba^+$

or, $q_1 = q_0 a^+ + q_1(bba^+ + ba^+)$, where $Q = q_0 a^+$ and $P = (bba^+ + ba^+)$

or, $q_1 = q_0 a^+(bba^+ + ba^+)^*$ [using Arden's theorem]

or, $q_1 = q_0 aa^*(bbaa^* + baa^*)^*$

or, $q_1 = q_0 aa^*\{bba^*a + ba^*a\}^*$

We are leaving a small part as an exercise for readers. Apply the laws of RE to simplify as

$$q_1 = q_0(a + b)^*a \quad \text{(vi)}$$

Use (vi) in (v) to get $q_f = q_0(a+b)^*abb$

$$= (q_f b + \varepsilon)b^*(a + b)^*abb \text{ [using Equation (i) for } q_0\text{]}$$

$$= b^*(a + b)^*abb + q_f b\, b^*(a + b)^*abb$$

Here, $Q = b^*(a + b)^*abb$ and $P = b\, b^*(a + b)^*abb$.

Using Arden's theorem, we get $q_f = b^*(a + b)^*abb\,(b\, b^*(a + b)^*abb)^*$. The simplified expression can be written as $(a + b)^*abb$.

(e) Compute the equivalent regular expression for the NDFA shown in Fig. 3.10 (e).

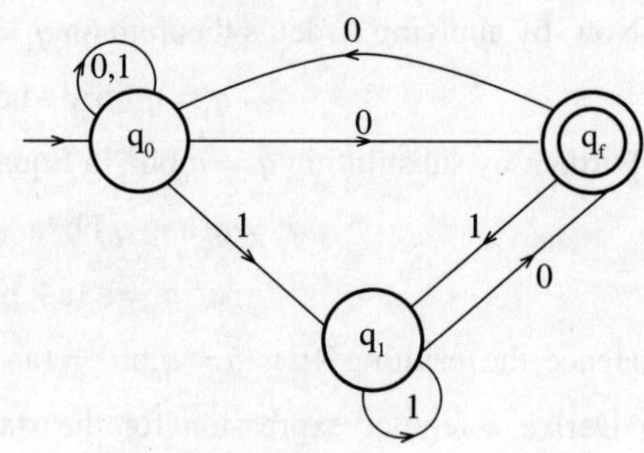

FIGURE 3.10 (e) An NDFA

Solution The initial state is q_0 and the final state is q_f. We need to express q_f as a function of q_0. Referring to Fig. 3.10 (e), the state equations are listed below.

$$q_0 = q_f0 + q_0(0 + 1) + \varepsilon, \text{ or equivalently}$$
$$q_0 = (q_f0 + \varepsilon) + q_0(0 + 1) = (q_f0 + \varepsilon)(0+1)^* \quad \text{(i)}$$

[using Arden's theorem]

Here, $Q = (q_f0 + \varepsilon)$, $R = q_0$ and $P = (0 + 1)$.

$q_1 = q_01 + q_f1 + q_11 = (q_0 + q_f)1 + q_11$, or equivalently

$$q_1 = (q_0 + q_f)11^* \quad \text{(ii)}$$

[Applying Arden's theorem on $Q = (q_0 + q_f)1$, $R = q_1$ and $P = 1$]

$$q_f = q_00 + q_10 = (q_0 + q_1)0 \quad \text{(iii)}$$

Using (i) in (ii), we get $\quad q_1 = \{q_f0(0 + 1)^* + q_f\}11^* \quad$ (iv)

Using (iv) in (iii), we get $\quad q_f = [q_0 + \{q_f0(0 + 1)^* + q_f\}11^*]0$

or, $q_f = q_00 + q_f\{0(0 + 1)^*11^*0 + 11^*0\}$

or, $q_f = q_00 + q_f\{0(0 + 1)^*1^+0 + 1^+0\}$

$= (q_f0 + \varepsilon)(0+1)^*0 + q_f\{0(0 + 1)^*1^+0 + 1^+0\}$ [using Equation (i) for q_0]

$= (0+1)^*0 + q_f(0(0+1)^*0 + \{0(0 + 1)^*1^+0 + 1^+0\})$

Using Arden's theorem, we get $q_f = (0+1)^*0(0(0+1)^*0 + \{0(0 + 1)^*1^+0 + 1^+0\})^*$

where $Q = (0+1)^*0$ and $P = (0(0+1)^*0 + \{0(0 + 1)^*1^+0 + 1^+0\})$.

Hence, the regular expression is $(0+1)^*0(0(0+1)^*0 + \{0(0 + 1)^*1^+0 + 1^+0\})^*$. It can be expressed as $0[\{0(0 + 1)^*1^+0 + 1^+0\}]^*$ in simplified form.

THEOREM 3.2

For every DFA (M), there exists a regular expression R, such that $L(M) = L(R)$.

Proof Let the states of M be $\{1, 2, ...n\}$. From the above discussion, we may consider that $R_{i,j}^{(k)}$ is the regular expression whose language is the set of labels of a path from i to j without visiting any state with a label larger than k.

Basis: $R_{i,j}^{(0)} = a_1 + a_2 + \cdots + a_n$, if $\delta(i, a_k) = j$ for $1 \le k \le n$

If $i = j$, then it also includes ε.

Induction: Compute $R_{i,j}^{(k)}$ using $R_{i,j}^{(k-1),s}$ as follows:

$$R_{i,j}^{(k)} = R_{i,j}^{(k-1)} + R_{i,k}^{(k-1)} \cdot \left(R_{k,k}^{(k-1)}\right) * R_{k,j}^{(k-1)}$$

Thus, the complete regular expression (R) can be obtained by applying the base case and the above-mentioned recursive relation.

Important points to find RE

- Arden's theorem is frequently used to simplify the set of state equations for any DFA. The RE (solution) held by the final state of the DFA is the RE of the DFA.
- If R is a regular expression such that it contains only symbols of the alphabet and no state symbols, then it is the equivalent regular expression for the given DFA.

- If there are multiple final states, say q_{f1} and q_{f2}, and the expressions for the final states are $q_{f1} = R_1$ and $q_{f2} = R_2$, then the regular expression for the automaton is $R_1 + R_2$.

EXAMPLE 3.10

Design a DFA for the RE, $R = (1)^*$ over {1}.

Solution Here, the set of generated strings is {ε, 1, 11, 111, ...}. Since ε (null string) is also a valid string, the initial state must be the final state of the designed DFA.

In addition, strings consisting of any number of 1s are also valid. Hence, the DFA for the given RE is as shown in Fig. 3.11.

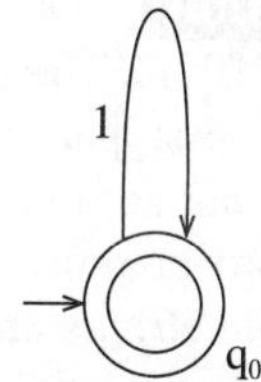

FIGURE 3.11 DFA for the RE: 1*

EXAMPLE 3.11

Design a minimum state finite automaton over {0, 1} in which the length of every string is divisible by 2 and 3.

Solution From the given constraints, it is clear that the length of any valid string must be a multiple of 6 (LCM of 2 and 3). So, the regular expression in its simplest form is given by ((0 + 1)(0 + 1)(0 + 1)(0 + 1)(0 + 1)(0 + 1))*, and the corresponding minimum state automaton is shown in Fig. 3.12.

FIGURE 3.12 DFA for ((0 + 1)(0 + 1)(0 + 1)(0 + 1)(0 + 1)(0 + 1))*

EXAMPLE 3.12

(a) Design a DFA that recognises all the strings accepted by the RE: (1 + 011)*.

Solution (a) The given RE, $r = (1 + 011)^*$ can be expressed as: (1*(011)*)* by applying the algebraic laws discussed earlier (more specifically, by applying (P*Q*)* = (P + Q)*).

Clearly, the null string is also valid. So, the initial state (say, q_0) of the proposed DFA must be the final state as well. Now, we insert a self-loop for 1* in the initial state. Next, we need two more states for (011)*. Finally, the state diagram of the proposed automaton is as shown in Fig. 3.13 (a).

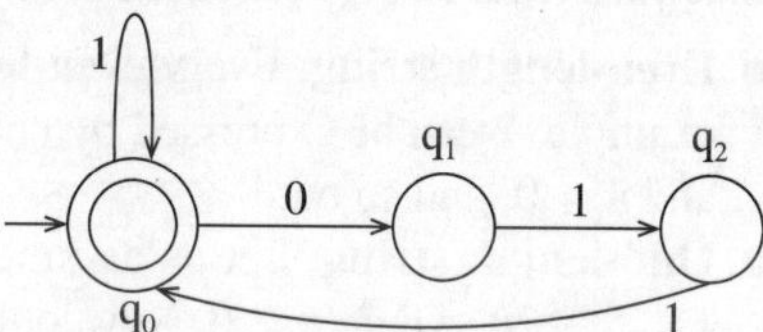

FIGURE 3.13 (a) DFA for (1 + 011)*

> **Note:** Let us now consider two REs: $r_1 = 011^*$ and $r_2 = (0 + 1)^*11^*$. The regular expression r_1 makes it clear that if a string starts with 1, the corresponding DFA moves to the dead state. Further, if 0 is encountered in the final state, then also it moves to the dead state. On the other hand, the regular expression r_2 states that the corresponding DFA starts with 0 or 1, and if 0 is received in the final state, it then moves to a non-final state, waiting for input 1 – it never moves to the dead state.

(b) Find the regular expression for the DFA shown in Fig. 3.13 (b).

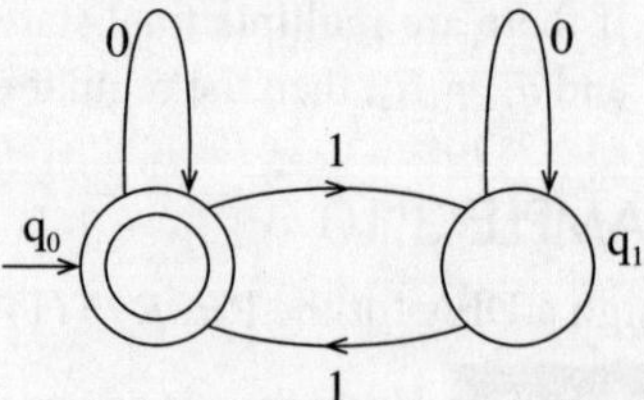

FIGURE 3.13 (b) A DFA over {0, 1}

Solution One possible RE for the given DFA is r = 0*(10*10*)*. The derivation is as follows.

The first part (0*) of r is required for any valid string containing zero or more 0s but no 1s, as a string with zero 1s and zero or more 0s is valid. On the other hand, the rest of the valid strings are generated by the second part of r, that is, (10*10*), and this part may be repeated zero or more times. Hence, concatenation with zero or more copies of (10*10*) preceded by 0* is the complete regular expression to generate all the valid strings for the given DFA. The complete regular expression is 0*10*10*.

3.7 ADDITIONAL TIPS FOR PROBLEMS ON FA AND RE

EXAMPLE 3.13

Simplify the RE, $R = (11 + 111)^*$.

Solution Analysing the given RE, $R = (11 + 111)^*$, it is clear that strings containing only 1s (except the string with a single 1) can be generated by this RE. Hence, it can be simplified as: $R = \{1\}^* - \{1\}$. Now, it is much easier to design an FA from the simplified version of the given RE, and that FA may possibly be the minimum state FA as well.

EXAMPLE 3.14

Simplify the RE, $R = (11 + 11111)^*$.

Solution The given RE simply states that each string consists of a single symbol, 1. However, the only consideration for any valid string is its length, and the length can be described by the expression: $(2m + 5n)$, where m and n are whole numbers.

Identifying valid strings based on m and n

- **Even-length string**: Every even-length string can be described by the expression: $(2m + 5n)$. For example, 2 can be expressed by choosing $m = 1$ and $n = 0$. Also, $4 = 2 \text{ x } 2 + 5 \text{ x } 0$, that is, $m = 2$ and $n = 0$, and so on.
- **Odd-length string**: Let us assume any odd number $k \geq 5$. Clearly, we can express k in the form $k = 5 + 2n$, where $n \geq 0$. Now, only two odd numbers, 1 and 3, are left to be analysed. However, it is true that neither of these can be expressed in the form: $(2m + 5n)$. Thus, every odd-length string except 1 and 3 can be expressed by $(2m + 5n)$. For example, 5 can be expressed by choosing $m = 0$ and $n = 1$. Likewise, $7 = 5 \times 1 + 2 \times 1$, where $m = 1$, $n = 1$, and so on.

Thus, the given RE can be reduced to the form: 1* – {1, 111}. It can be seen that reducing the given expression to this form makes designing the automata easier.

EXAMPLE 3.15

Find the correct option for the problem: A language consisting of all binary strings can be denoted by (i) (0 + 1)*, (ii) (0*1*)*, (iii) (0 + 1 + ε) (0 + 1)*(1 + 0 + ε)*, (iv) 1*(00*11*)*0* and (v) all of the above.

Solution Option (i) is correct. Option (ii) is also correct, as option (i) is true. This is because option

(i) derives option (ii) by applying the law of closure property on the RE: $((P + Q)^* = (P^*Q^*)^*)$. Option (iii) is a redundant version of option (i), and hence, is also true. Finally, option (iv) is an alternative form of option (i) used to generate binary strings. Hence, all the options are correct.

EXAMPLE 3.16

Let $L_1 = 10^*$ and $L_2 = 101^*$. Find $L_1 \cup L_2$, $L_1 \cap L_2$ and $L_1 - L_2$ (in form of RE).

Solution Given that $L_1 = 10^*$ and $L_2 = 101^*$. Hence, $L_1 = \{1, 10, 100, 100...\}$ and $L_2 = \{10, 101, 1011, 1011...\}$. Thus, $L_1 \cup L_2 = (10^* + 101^*) = 1(0^* + 01^*)$, $L_1 \cap L_2 = (10)$ and $L_1 - L_2 = (1) \cup (10^+)0$.

EXAMPLE 3.17

If the number of states in an FA is n, find the total number of strings with length up to $(n - 1)$ over an alphabet consisting of m symbols.

Solution The total number of strings (N) $= m^0 + m^1 + m^2 + ... + m^{(n-1)} = (m^n - 1)/(m - 1)$. However, the number of strings with length n is m^n (if repetition of symbols is allowed).

> **Note:** The number of strings with length 0 is m^0. Similarly, m^1 is the number of strings with length 1; m^2 is the number of strings with length 2; and so on.

EXAMPLE 3.18

Let $\Sigma = \{a, b, c, d, e\}$. Find the number of strings in Σ^* of length 4 such that no symbol is used more than once in a string.

Solution The number of strings is 120. If no symbol is repeated in a string, the number of ways to select four (as string length is 4) out of five symbols is ${}^5P_4 = 120$. In this case, we consider permutations within the group (or block) of length 4 out of the 5 given symbols. However, if repetition is allowed within the block, the answer will be $5^4 = 625$.

EXAMPLE 3.19

Find the minimum number of states required to construct a DFA that accepts strings of length n over Σ.

Solution The minimum number of states required here is $(n + 1)$, as the given problem focuses only on the length of the string and not on other criteria. However, the number is $(n + 2)$ as we need to include the dead state.

EXAMPLE 3.20

(a) If the length of a string is divisible by m and n, find the minimum number of states necessary to construct an FA.

Solution (a) The minimum number of states in an FA can be determined by finding LCM(m, n) (assuming ε is also valid).

Basic technique to reduce states in FA: Let us consider a DFA that satisfies two distinct properties: P_1 and P_2. For this purpose, we first design two distinct DFAs (M_1 for P_1 and M_2 for P_2) and then take the Cartesian product of both the sets of states (Q_1 for M_1 and Q_2 for M_2). Next, we perform intersection on M_1 and M_2 to obtain the FA, M, which satisfies both P_1 and P_2. Initially, the number of states in M will be the number of pairs of states (obtained from the Cartesian product). However, some of the states in M may not be reachable from the start state; these may be removed from the resulting machine.

If both the properties, P_1 and P_2, are of the same type, such as length of string (irrespective of any symbol) divisible by both *m* AND *n*, we may reduce the number of states by finding LCM(*m*, *n*) (assuming ε is also accepted). We concentrate on common multiples to reduce states. In the same way, if the length of the string is divisible by *m* OR *n*, then the minimum number of states in FA can be determined by finding LCM(*m*, *n*) (assuming ε is also a valid string). However, in the latter case, the number of final states will be more than in the AND case – it is around $(m + n - 1)$.

Properties such as 'the number of *a*s divisible by *m*' and 'the number of *b*s divisible by *n*' are treated as two different properties, as they are applied on two different symbols (one on *a* and the other on *b*), and we may not take the LCM of *m* and *n*.

(b) Find the minimum number of states to design a DFA for RE = (111 + 11111)*.

Solution The properties are of the same type. The common property here is string length, and the focus is on the divisibility of length by 3 or 5. Based on the idea of common multiples, we obtain $3 \times 5 =$ 15 states.

Any string expressed by the given RE will have as its length, $L = 3m + 5n$, where $m \geq 0$ and $n \geq 0$ and *m* and *n* are integers. The smallest possible non-zero positive value of L is 3. L = 4 is impossible. L = 5 is valid. L = 6 is also valid. L = 7 is impossible. L = 8 is possible with $m = n = 1$.

For $L \geq 9$, inserting $n = 0$ gives $3n$; that is, all the multiples of 3 or all integers that give zero on division by 3. Inserting $n = 2$ gives $3m + 10$, which represents all positive numbers ≥ 10 that give 1 as remainder on division by 3. Inserting $n = 1$ gives $3m + 5$, which represents all positive integers ≥ 5 that give 2 as remainder on division by 3.

Now, the possible values of remainders when a positive integer is divided by 3 are {0, 1, 2}. So, all integers ≥ 9 are represented by the formula $(3m + 5n)$ and this implies that all possible strings of length 9 and greater are possible.

However, the RE = (111 + 11111)* can be simplified as: = (111 + 11111)* = 1* – {1, 11, 1111, 1111111}, that is, all the strings of 1* (except 1, 11, 1111, 1111111) are valid, and only 8 states (instead of 15) are necessary to represent the DFA for it.

(c) Find the length of the shortest string not in the language over {a, b} generated by the RE, *R* = a*b*(ba)*a*.

Solution For the given RE, the shortest string (not belonging to the language) is *w* = *bab* with length 3, as the following are the 0-, 1-, 2- and 3-length strings generated by the given RE:

- ε is the 0-length string.
- Both *a* and *b* are unit-length strings.
- *ab*, *ba*, *aa*, *bb* are the 2-length strings.
- *aab*, *aba*, *abb*, *baa*, *aaa*, *bbb* (except *bab*) are the 3-length strings.

EXAMPLE 3.21

(a) If *l* (length of string) *mod n* = 0, find the minimum number of states required to construct an FA.

Solution The minimum number of states in the FA will be *n* (assuming that null string is also valid). Let $n = 2$ and the alphabet be Σ. A DFA satisfying the given property for $n = 2$ is shown in Fig. 3.14.

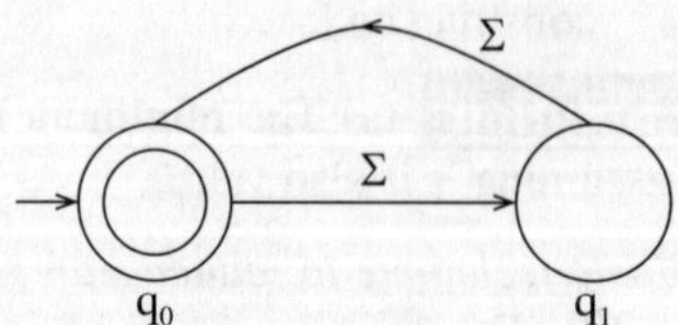

FIGURE 3.14 A DFA accepting string length divisible by n = 2

(b) If Σ consists of *m* number of symbols, find the number of strings with length *n*.

Solution The total number of strings (of length n) that can be generated over Σ is equal to m^n. If we design a DFA for this problem, then $(n + 1)$ states are necessary to generate any string of length n over Σ.

(c) *Prove or disprove*: Let L_1 and L_2 be two languages over Σ. If $L_1 \subseteq L_2$ and L_2 is not regular, then L_1 is not regular.

Solution False, as counterexamples exist. One such example is given below.

Counterexample: Assume that $\Sigma = \{0, 1\}$. Let L_1 = set of all strings over Σ such that each valid string has only 0s and no 1s at all. So, L_1 can be described by the RE, $R = 0^*$.

Further, let L_2 = set of all strings over $\Sigma = \{0, 1\}$ such that the number of 1s is either zero or a prime number. Unfortunately, it is not possible to design any RE for L_2. Formally, L_2 includes L_1. But, L_2 is not regular, whereas L_1 is regular. Although $L_1 \subseteq L_2$. Hence, the given statement is not valid.

(d) *Prove or disprove*: Let L_1 be a language over Σ. If L_1 is non-regular, then L'_1 (complement of L_1) is non-regular.

Solution Recall that if a language is regular, then its complement is also regular. However, the language here is non-regular and we need to prove or disprove that its complement is also non-regular. The given statement is not necessarily true, since counterexamples exist.

Counterexample: Let $\Sigma = \{0, 1\}$ and L_1 be a set of all strings over Σ such that string length is either odd or prime. We may not derive any RE for L_1.

On the other hand, L'_1 (complement of L_1) $= \Sigma^* - L_1$ is the set of all strings where string length is an even number (except 2, as 2 is prime and it is in L_1). Clearly, it is a regular language with an RE, $R = \varepsilon + (00)^+00$.

Formally, we may write an expression for L_1 and its complement as: $\varepsilon + 00(00)^+ + ? = \Sigma^*$. Now, the question is, what is the expression we need to use in place of (?) for L_1 to obtain Σ^*? However, there is no answer, though the RE for L'_1 exists and it is regular.

(e) Let A, B and C be three languages and AB = C, where A and C are regular. Prove that B is also regular.

Solution Here, AB = C, where A and C are regular. This means that concatenation of A and B results in C, which is regular. We know that concatenation of two regular languages produces a regular language. Here, A is regular, so B must also be regular.

3.8 PUMPING LEMMA FOR REGULAR LANGUAGES

The concept of pumping lemma is a good example of the **pigeon-hole principle**. This principle states that if n balls are to be placed in m boxes, then at least one box will have more than one ball if $n > m$.

The lemma is useful for disproving the regularity of a language in question. It was first proved by Dana Scott and Michael Rabin in 1959. However, the lemma was rediscovered shortly after by Yehoshua Bar-Hillel, Micha A. Perles and Eli Shamir in 1961, as a simplification of their pumping lemma for context-free languages.

We can say that a language is regular if it has a regular expression. Otherwise, it is not regular. However, just by looking at the structure of a language, it often becomes quite difficult to identify whether it is regular or not. So, we need to prove it in another way. The pumping lemma is a good tool for checking the regularity property of a language.

Statement of the lemma: Let $M = (Q, \Sigma, \delta, q_0, F)$ be a finite automaton with n states. Let L be a language accepted by M (L is regular) and a string $w \in L$, where $|w| \geq m$. Now, if $m \geq n$, we can break w into substrings x, y, z such that $w = xyz$, where $y \neq \varepsilon$ and $x(y)^iz$ is in L for $i \geq 0$. Here, $y \neq \varepsilon$ but $(y)^0 = \varepsilon$ when $i = 0$; then $w = xz$.

Proof Let $w = a_1a_2a_3 \ldots\ldots\ldots a_m$, $m \geq n$ be a string in L. Obviously, $|w|$ (length of w) $\geq n$ (number of states). Let us now apply transition function δ on $w = a_1a_2a_3 \ldots\ldots\ldots a_m$, starting from the start state q_0 of

M. Undoubtedly, the number of states required to process this string is $(m + 1)$ [as $\delta(q_0, a_1a_2a_3\ldots a_m) = q_i$, $i = 1, 2, 3, \ldots m$; that is, the set of necessary states $(Q_1) = \{q_0, q_1, q_2, \ldots q_m\}$].

However, there are only n distinct states in the FA. So, if $w \in L$, then at least two states of Q_1 must coincide to accommodate the strings within n states of the FA. In other words, there must exist pairs of repeated states in FA.

Let us take the first pair of repeated states: q_j and q_k $(q_j = q_k)$, where $0 \le j < k \le m$. Thus, the string w can be decomposed into three substrings: $(a_1a_2a_3\ldots a_j)(a_{j+1}\ldots a_k)(a_{k+1}\ldots\ldots a_m)$, namely, x, y and z such that the states q_j and q_k coincide for y. This implies that the machine M starts processing at state q_0. On reading the string x from q_0, it reaches q_j. Next, on receiving the string y at q_j, it reaches $q_j (= q_k)$; that is, it makes a loop for string y. So, q_j and q_k are the repeated states. Further, on reading the string z from q_j, M reaches the final state, q_f. Processing of the entire string w is shown in Fig. 3.15 (a).

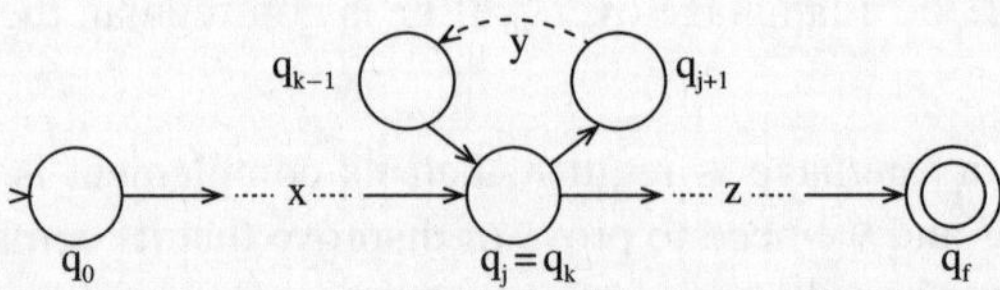

FIGURE 3.15 (a) A simple diagram for illustrating pumping lemma

An example of FA with two states is shown in Fig. 3.15 (b).

Though it is incomplete, even the complete diagram may not use more than two states. Let $w = aa$ be a string of length 2, recognised by FA. Now, to recognise a string of length 2, the number of states required is 3. Let us try to recognise as: $\delta(q_0, \text{aa}) = \delta((q_0, \text{a})\ \text{a}) = \delta(q_1, \text{a}) = q_2$, shown in Fig. 3.15 (c).

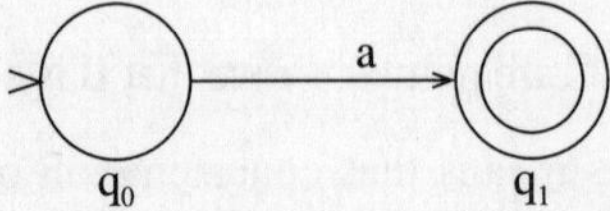

FIGURE 3.15 (b) An FA with two states

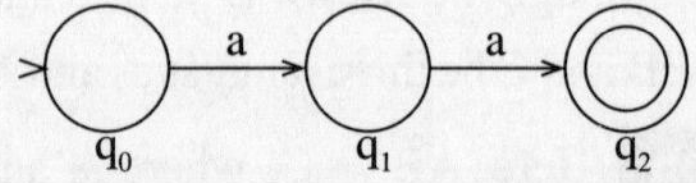

FIGURE 3.15 (c) Required FA with three states

Assume that the FA has only two states but the valid string requires three. This is possible if and only if q_2 moves back to any of the states q_0 and q_1 for input a, depending on the acceptance or rejection of the string (concept of repeating states). Another example is shown in Fig. 3.15 (d).

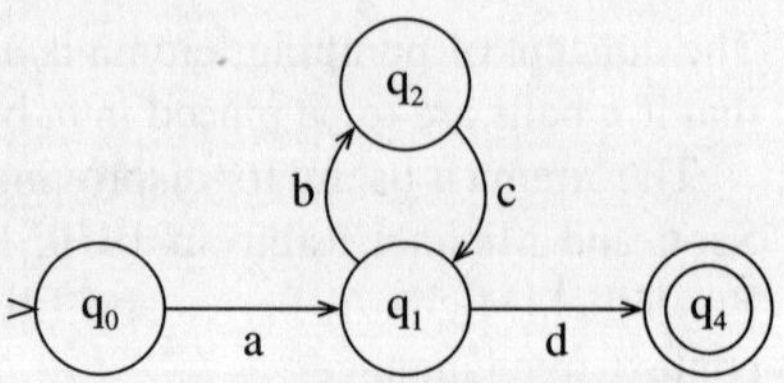

FIGURE 3.15 (d) An FA with four states

Here, $q_3 = q_1$ (from $q_2 - c \rightarrow q_3 = q_1$). Also, $x = a$, $y = bc$ and $z = d$ for $w = a(bc)d$. Clearly, the number of available states is 4, but the number of states required is 5. Likewise, $x = a$, $y = bcbc$ and $z = d$ for $w = a(bcbc)d$. Importantly, the other part of w (x and z, where $w = xyz$) can be decomposed in the same fashion (if possible).

Tips to apply pumping lemma

- First identify the constraints of the given language (L).
- To apply pumping lemma, select a string w according to the string format of L.
- Decompose the selected string w into three substrings, x, y and z, such that $y \neq \varepsilon$.

EXAMPLE 3.22

Show that L = palindrome over {a, b} is not regular.

Solution *Step 1*: Assume that the set L is regular and is recognised by an FA with n states. L must satisfy the principle of pumping lemma.

Step 2: Let us take a simple string: $w = a^nba^n$, where $|w| = (2n + 1) > n$. By pumping lemma, we can write $w = xyz$ as: $a^{n-j}a^jba^n$, where $x = a^{n-j}$, $y = a^j$ ($j > 0$, since $y \neq \varepsilon$) and $z = a^n$.

Step 3: By the principle of pumping lemma, xy^iz ($i \geq 0$) is also in L. Let $i = 0$. Then, $xy^iz = xy^0z = a^{n-j}(a^j)^0ba^n = a^{n-j}ba^n$.

As $j > 0$, $n - j \neq$ n. However, for a string to be a palindrome, $(n - j)$ must be equal to n. Hence, L is not regular, as a contradiction arises.

EXAMPLE 3.23

Prove that L= $\{a^n \mid n \text{ is prime}\}$ is not regular.

Solution *Step 1*: Assume that the given language (L) is regular. Therefore, it must satisfy the properties of pumping lemma.

Step 2: Let $w = a^p$ (of the form a^n) be a string in L. As per pumping lemma, the string w can be decomposed into three substrings as: $w = a^{m_1}a^ma^{m_2} = xyz$, where $x = a^{m_1}$, $y = a^m$ ($m > 0$, since $y \neq \varepsilon$), $z = a^{m_2}$ and $m_1 + m + m_2 = p$.

Step 3: Further, xy^iz ($i \geq 0$) also belongs to L (according to the principle of pumping lemma). Let us take $i > 0$. Now, $|xy^iz| = |xyz| + |y^{(i-1)}| = p + (i - 1)m$, since $|xyz| = m_1 + m + m_2$.

$= p + mp$ [letting $i = (p+1)$; as we may select any positive integer value of i]

$= p(1 + m) = k$ (say). Here, $m \neq 0$.

Clearly, the length (k) of the string: xy^iz (as computed above) is not a prime number, as two distinct factors, p and $(m + 1)$, are present in k.

Thus, the length of the selected string is not prime. However, it should be prime according to the constraints of the given language. Hence, a contradiction arises. So, the given language may not be regular.

We can take $i = 0$. Then, $w = xz$ and we can also show that w is not valid for L.

EXAMPLE 3.24

Prove that L = $\{0^i1^i \mid i \geq 0\}$ is not regular.

Solution *Step 1*: Assume that L is regular.

Step 2: Let $w = 0^i1^i$ be a string in L. According to the principle of pumping lemma, we may decompose w as: $0^i1^m1^p = xyz$, where $x = 0^i$, $y = 1^m$ ($m \neq 0$, since $y \neq \varepsilon$) and $z = 1^p$ such that $m + p = i$.

Step 3: Further, xy^kz ($k \geq 0$) also belongs to L (according to the lemma), and its length can be computed as follows.

Let us take $k \geq 1$. Then $|xy^kz| = |xyz| + |y^{k-1}| = 2i + m(k - 1)$, where $|xyz| = 2i$

$= 2i + m(1 + i - 1)$ [letting $k = (1 + i)$; since we may select any value of k]

$= 2i + mi = m(2 + i)$.

[Here, $m \neq 0$.]

$= 2j$ (say)

Therefore, $j = m(2 + i)/2$. However, this expression may also result in a fractional value of j, but the fractional value is not allowed according to the format of a string of L. Hence, a contradiction occurs. So, the given language is not regular.

Example of fractional value: Let $i = 3, m = 1$ and $p = 2$. So, $m + p = 3$.

Now, $j = m(2 + i)/2 = (1(2 + 3))/2 = 5/2 = 2.5$. Actually, j should be 3, but the expression gives 2.5. When $k = 0$, $|xz| = i + p$. Clearly, $p \neq i$ as $m \neq 0$ and $m + p = i$.

Note: Based on this example, we can prove that $L = \{0^i1^{2i} \mid i \geq 0\}$ is not regular. However, one may think of an RE, $r = 0^*(11)^*$ for this language, but this RE may also derive invalid strings such as 00011, 01111, etc. Hence, this RE is not correct.

EXAMPLE 3.25

(a) Prove that $L = \{ww \mid w \in \{a, b\}^*\}$ is not regular.

Solution Let L be a regular language and string $w_1 \in L$, where $w_1 = ww$. Assume that $|w| = \alpha$.

$\therefore |w_1| = |ww| = 2\alpha$

Further, $xy^kz \in L$ $(k \geq 0)$, as L is assumed to be regular.

Now, $|xy^kz| = |xyz| + |y^{k-1}|$

$= 2\alpha + (k - 1)m$ (assuming $y = a^m$, where $m \neq 0$, since $y \neq \varepsilon$)

$= 2\alpha + m \times m$ (letting $k = m + 1$)

$= 2\alpha + m^2 = 2\beta$ $(\beta \geq 0)$

According to the format of string (ww) specified in L, the length should always be even (odd + odd = even and even + even = even). Hence, β must always be a positive integer. However, according to the expression obtained for β, we cannot ensure that β is always an integer. For example, let $\alpha = 3$, $m = 3$.

So, $2\beta = 15$, that is, $\beta = 15/2$ (which is not admissible).

Hence, a contradiction occurs. Therefore, the given language (assumed to be regular) is not regular.

(b) Prove by pumping lemma that $L = \{a^m \mid m \geq 0\}$ is regular.

Solution *Step 1*: Assume that L is regular. So, it must satisfy the principle of pumping lemma.

Step 2: Let $w = a^n$ be in L. Now, we can decompose this w into three substrings: x, y and z as $x = a^i$, $y = a^j$ $(j \neq 0$, since $y \neq \varepsilon)$ and $z = a^k$ such that $i + j + k = n$.

Step 3: Further, xy^mz $(m \geq 0)$ also belongs to L (according to pumping lemma), and its length can be computed as follows.

Let $m > 0$. $|xy^mz| = |xyz| + |y^{m-1}| = n + (m - 1)j$, where $|xyz| = n$ and $j \neq 0$. Undoubtedly, for any values of $m > 0$, the string $a^{n+(m-1)j}$ is in L. Further, it is true for $m = 0$ as well.

3.9 CLOSURE PROPERTIES OF REGULAR LANGUAGES

Review of Terms

- **Closure property over a set:** Let a, b $\in S$ (set). If $a \oplus b \in S$ is satisfied, then S is closed with respect to the binary operator $\oplus$.
- '*' is a unary operator used for closure (Kleene star) with respect to a single entity.
- **Closure properties of languages:** Certain operations performed on the languages of certain

classes (such as types 0, 1, 2, 3, discussed in Chapter 4) produce other languages of that class; that is, type of derived language = type of base language. Here, the closure property behaves a little differently from closure on algebraic structures (where a set and an operator are used), though the idea is identical to that of the algebraic structure. In theory of computation, if a specific operator (such as +, *) is applied on two sets of the same type or on one set (language) to result in a new set of its base type, the operation is called a closed operation. On the other hand, in set theory, the closure property is applied on two elements of a set to obtain an element of that set.

In this section, some important closure properties of regular languages are discussed.

Closure Under Union

THEOREM 3.3

The union of two regular languages is also regular.

Proof Let L_1 and L_2 be two regular languages over an alphabet, say Σ. As they are regular, they must have regular expressions, say r_1 and r_2, respectively. We know that the union of two regular expressions results in another regular expression, that is, $r = r_1 + r_2$ is also an RE. Hence, the statement is proved.

Closure Under Concatenation

THEOREM 3.4

The concatenation of two regular languages is also regular.

Proof Let L_1 and L_2 be two regular languages over an alphabet, say Σ. As they are regular, they must have regular expressions, say r_1 and r_2, respectively. We know that the concatenation of two regular expressions results in another regular expression, that is, $r = r_1 . r_2$ is also an RE. Hence, the statement is correct.

THEOREM 3.5

The closure of any regular language is also regular.

Proof Let L be a regular language over an alphabet, say Σ. As L is regular, it must have a regular expression, say r. We know that closure of any regular language results in another regular expression, that is, $R = r^*$. Hence, the statement is true.

Note: We may prove this for positive closure as well.

Closure Under Complement

THEOREM 3.6

If L is a regular language over Σ, then $\Sigma^* - L$ (complement of L) is also regular.

Proof Given that L is regular. Let L be recognised by a DFA, say $M = (Q, \Sigma, \delta, q_0, F)$. Now, complement of M will be, $M' = (Q, \Sigma, \delta, q_0, F')$ such that $F' = Q - F$, that is, M and M′ differ only in their final states. More specifically, the set of non-final states of M will be the set of final states for M′ and the converse. In other words, if $w \notin L(M)$, then $w \in L(M')$ and the converse. Hence, DFA (M′) exists for the complement of L and M′ accepts the strings not validated by M, that is, $L(M') = \Sigma^* - L(M)$. Therefore, $\Sigma^* - L$ is also regular.

Closure Under Intersection

THEOREM 3.7

If X and Y are two regular sets over alphabet Σ, then $X \cap Y$ is also regular over Σ.

Proof Given that X and Y are regular. Hence, X' and Y' are regular.

By De Morgan's law for sets, we may write: $(X \cup Y)' = X' \cap Y'$. Now, applying De Morgan's law, $(X' \cup Y')'$ can be expressed as: $(X' \cup Y')' = (X')' \cap (Y')' = X \cap Y$ (since $(A')' = A$, where A is a set).

However, both X' and Y' are regular. Therefore, their union $(X' \cup Y')$ is also regular. Hence, $(X' \cup Y')'$ is also regular (as the complement of a regular language is also regular).

Finally, as $(X' \cup Y')'$ results in $X \cap Y$ and $(X' \cup Y')'$ is regular, $X \cap Y$ is also regular.

Closure Under Difference

THEOREM 3.8

If L_1 and L_2 are two regular languages, then $(L_1 - L_2)$ is also regular.

Proof Given that both L_1 and L_2 are regular. Now, $L_1 - L_2$ implies that element: $x \in L_1$ but $x \notin L_2$ Based on this concept, we may express $L_1 - L_2$ as follows. $L_1 - L_2 = L_1 \cap (L_2)'$, where $L'_2 = (S - L_2)$. Here, S denotes the universal set. Clearly, $(L_2)'$ is regular as L_2 is regular and we know that the complement of any language is also regular. Further, the intersection of any two regular languages is also regular.

Hence, $L_1 - L_2 = L_1 \cap (L_2)'$ is also regular.

Closure Under Reversal

THEOREM 3.9

A regular language is closed under reversal.

Proof Let L be a language and $M = (Q, \Sigma, \delta, q_0, F)$ be the FA for L. Now, we may construct $M^R = (Q, \Sigma, \delta^R, q_0, F)$ from M, where δ^R is the representation of δ with the orientation of the arcs reversed. In other words, if there is a path from q_0 (initial state) to q_f (final state) in M, then there exists a path from q_f to q_0 in M^R: $L(M^R) = L^R$, such that q_0 (initial state) in M is treated as the final state of M^R, and vice versa. The FA corresponding to L^R (reversal of the regular language L) can be constructed applying the following steps:

1. Invert the transitions.
2. Make the initial state as the final state and all final states as the initial states.
3. Add a new initial state to connect the initial states through ε transitions.
4. Finally, remove the ε transitions to convert it to DFA.

These steps are explained with a simple example.

Let $L = ab^* + ba$. The DFA corresponding to the given language L is shown in Fig. 3.16.

Here, q_0 is the initial state and both q_1 and q_2 are the final states. Now, the reverse of the language L is $L^R = b^*a + ab$, and its state diagram is shown in Fig. 3.17 by adding a new initial state (say s) and connecting it with other possible states through the ε move.

We can apply an ε transition elimination strategy to convert ε NFA to DFA.

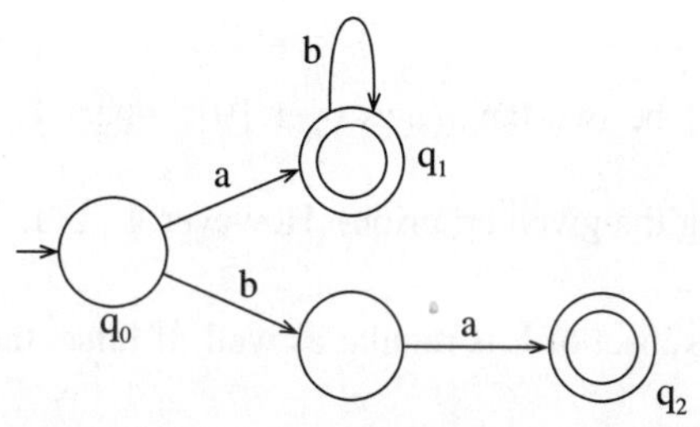

FIGURE 3.16 A DFA for L = ab* + ba

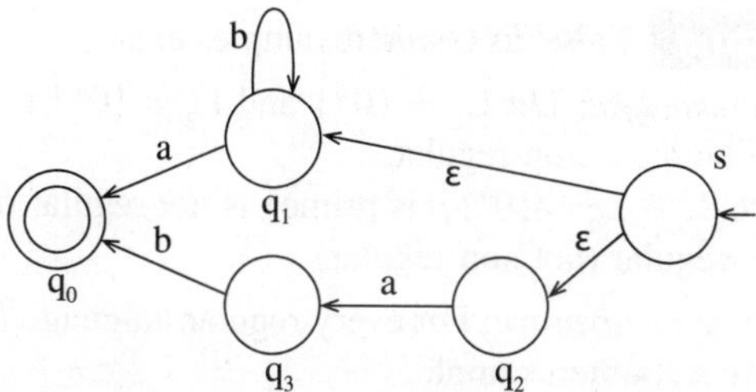

FIGURE 3.17 A DFA for L = b*a + ab

> **Note:** Other closure properties such as homomorphism and inverse homomorphism exist in the context of regular languages. These are not explained here.

EXAMPLE 3.26

(a) Prove that if $L_1, L_2, L_3, \ldots$ are all regular, then $\bigcup_{i-1}^{n} L_i$ is regular.

Solution Here, $L_1, L_2, L_3, \ldots$ are regular. We know that the union of two regular languages results in another regular language. So, $\bigcup_{i-1}^{n} L_i$ (finite union) is also regular.

(b) *Prove or disprove*: Infinite union of regular languages is regular.

Solution False, since counterexamples exist.

Counterexamples

- Let us take a finite language such as one for a single-digit prime number (say L_1), two-digit prime number (say L_2), and so on. Clearly, each of these languages is finite, and hence, regular. However, if we take an infinite union of these languages, that is, $\bigcup_{i-1}^{\infty} L_i$, the resulting language would not be regular. This is because the infinite union gives an infinite language of prime numbers of different digits, and we may not obtain a finite subset of equivalent classes of elements for this infinite language.
- Let $L_i = \{a^i b^i\}, i \geq 1$. Clearly, each $L_i = \{a^i b^i\}$ (for example, $L_1 = \{ab\}$, $L_2 = \{a^2b^2\}$, and so on) is regular. However, $\bigcup_{i-1}^{\infty} L_i = \{ab \cup a^2b^2 \cup a^3b^3 \cup \ldots\} = \{a^nb^n \mid n \geq 1\}$ is not regular.

Undoubtedly, the infinite union of regular languages results (in the second case) in an infinite language of the form:
$L = \{a^nb^n \mid n \geq 1\}$, and we may not find any finite subset of equivalence classes of the elements. It is, indeed, a context-free language. Recall that both $L = \{a^nb^n|n \geq 1\}$ and $L_1 = \{a^nb^n\}, n \geq 1$ are not equivalent. In fact, $L = \{a^nb^n \mid n \geq 1\}$ represents multiple strings of the form: $a^nb^n, n \geq 1$, whereas $L_1 = \{a^nb^n\}, n \geq 1$ represents only a single string of the form: $a^nb^n, n \geq 1$.

> **Note:** Let L be a regular language. $L^* = L^0 \cup L^1 \cup L^2 \cup \ldots$ (an infinite union of the powers of the language, L) results in a regular language, as the power is taken on the same language, L (and not on different languages).

(c) *Prove or disprove*: If L_1 is regular, L_2 is non-regular and $L_1 \cap L_2$ is non-regular; then $L_1 \cup L_2$ is non-regular.

Solution False, as counterexamples exist.

Counterexample: Let $L_1 = \{0^*\}$ and $L_2 = \{0^p \mid p \text{ is prime}\}$ be two languages over $\{0\}$, where L_1 is regular but L_2 is non-regular.

Here, $L_1 \cap L_2 = \{0^p \mid p \text{ is prime}\}$ is not regular (satisfying the given criterion). However, $L_1 \cup L_2 = \{0^*\}$ = regular (not non-regular).

(d) *Prove or disprove*: For every regular language L, every subset of L is regular as well. If false, then give a counterexample.

Solution False, as counterexamples exist.

Counterexample: Let $L = \{a^*b^*\}$ be a language over $\{a, b\}$. L is regular but it has a non-regular subset: $\{a^nb^n \mid n \geq 0)$.

(e) *Prove or disprove*: The intersection of any two non-regular languages is non-regular. If false, then give a counterexample.

Solution It is false. A simple counterexample is given below.

Let $L_1 = \{a^nb^n \mid n \geq 0\}$ and $L_2 = \{b^na^n \mid n \geq 0\}$ over $\{a, b\}$. Here, both L_1 and L_2 are non-regular. However, $L_1 \cap L_2 = \{\varepsilon\}$, and it is regular.

(f) Let s be a string over $(0 + 1)^*$, and let $n_0(s)$ denote the number of 0s in s, $n_1(s)$ the number of 1s in s. Is $L = \{s \in (0+1)^* \mid n_0(s) \text{ is a three-digit prime}\}$ regular?

Solution Given $L = \{s \in (0+1)^* \mid n_0(s) \text{ is a three-digit prime}\}$ is regular, as there exists a finite number of three-digit prime numbers that can be listed explicitly. So, we can design an FA for the given L.

(g) Let s be a string over $(0 + 1)^*$, and $n_0(s)$ denote the number of 0s in s and $n_1(s)$ the number of 1s in s. Is $L = \{s \in (0+1)^* \mid n_0(s) \bmod 6 = 0 \text{ and } n_1(s) \bmod 8 = 0\}$ regular?

Solution Given $L = \{s \in (0+1)^* \mid n_0(s) \bmod 6 = 0 \text{ and } n_1(s) \bmod 8 = 0\}$ is also regular. In fact, LCM(6, 8) = 24 states that are necessary to satisfy the conditions, $n_0(s) \bmod 6 = 0$ and $n_1(s) \bmod 8 = 0$, to design two distinct FAs. Accordingly, a total of 24 states will be present in the final automata. In particular, the final automata can be viewed as the product of two DFAs: one for $n_0(s) \bmod 6 = 0$ and the other for $n_1(s) \bmod 8 = 0$. More specifically, to satisfy both the properties, we take the Cartesian product of the states and apply the properties of the product scheme.

(h) Let n be a positive integer and $L = \{x \in \{0, 1\}^* \mid |x| = n \text{ and } n_0(x) = n_1(x)\}$, where $n_0(x)$ and $n_1(x)$ stand for the number of 0s and the number 1s in x. Is the language L regular?

Solution Yes, the language is regular as string length is finite. Likewise, it is true if $|n_0(x) - n_1(x)| = k$ $(k = 0, 1, 2)$.

Note: A problem like a palindrome of a finite-length string can be easily handled by the DFA (drawing all possible palindrome strings of given length over Σ).

(i) If $L_1 = \{a^n \mid n \geq 0\}$ and $L_2 = \{b^n \mid n \geq 0\}$, then justify that $L_1.L_2$ is regular.

Solution Both the languages L_1 and L_2 are regular and it is true that concatenation of two regular languages is regular. The resulting language looks like $\{a^nb^m \mid n, m \geq 0\}$ which is regular.

(j) Which of the following languages is regular?

(A) $\{ww^R \mid w \in \{0, 1\}^+\}$ (B) $\{ww^Rx \mid x, w \in \{0, 1\}^+\}$

(C) $\{wxw^R \mid x, w \in \{0, 1\}^+\}$ (D) $\{xww^R \mid x, w \in \{0, 1\}^+\}$

Solution Note that in each of these languages, string x can be anything, that is, $x = (0 + 1)^+$. So, the language $\{wxw^R \mid x, w \in \{0, 1\}^+\}$ can be expressed as a set of strings where the starting and ending symbols are the same. The regular expression for such language is simply $0(0 + 1)^+0 + 1(0 + 1)^+1$. All other languages are not regular and the same can be proved easily by applying pumping lemma. Proving this has been left as an exercise for the readers.

The correct option is (C).

Decision Problem and FA: The computational problems that FA can solve are limited to decision problems (where the output is yes/no based on the input). For example,

- Given the RE (r) and a string: x, does x belong to the language corresponding to r?
- Given an FA (say M), is there a string x that is accepted by M?
- Given two REs: r_1 and r_2, do they correspond to the same language?

EXAMPLE 3.27

Suggest an idea to test membership of a regular language.

Solution Let w be a string and L be a regular language. We need to test whether w belongs to L or not. For this purpose, we may perform the following basic steps.

Step 1: Design a DFA (say M) for the language L (since L is regular).

- It takes $O(mn)$ time, where m is the number of symbols in the alphabet, and n is the number of states in the DFA.

Step 2: Construct a state transition table corresponding to M using a two-dimensional array data structure.

Step 3: Process the given string: w, starting at the initial state (say q_0 of M) and consulting the transition table for each symbol of w.

- Each transition requires O(1) (constant time). Hence, to test whether w (with length, say n) is a member of L or not, it requires $O(n)$ time, once the DFA is built.

> *Exercise* Suggest an idea to check whether two regular languages L_1 and L_2 have at least one string in common.
> *Hint*: We may think about the product machine using the intersection approach.

EXAMPLE 3.28

Which of the following is true?

(A) Every subset of a regular set is regular. (B) Every finite subset of a non-regular set is regular.(C) The union of two non-regular sets is not regular. (D) Infinite union of finite sets is regular.

Solution Option (B) is correct.

Justification It is not necessarily true that every subset of a regular set is regular. For example, $r = a^*$. Certainly, one of its subsets may be taken as a string of *a*s such that the length of each string is a prime number. However, this subset is not regular. It may be noted that options (C) and (D) have already been explained.

3.10 NON-REGULAR LANGUAGES

Decision problems, such as odd number of 0s and even number of 1s in valid strings, the length of the valid string is divisible by a positive integer (say n), criteria such as at least/at most some measure (count), ordering of information, and so on, can be resolved by FA using the number of states in the FA. Importantly, satisfying criteria based on measure (m) = kn (in the form of arithmetic progression; that is, linear measure like L = $\{a^{kn} \mid k \neq n, n >= 0\}$ but not like a^{n^2} or a^{2^n} ···, which are non-linear expressions of n), can be resolved by FA (using loop or cycle with a finite number of states). Further, problems for matching of finite measure (requiring a finite amount of memory) can be solved by FA.

However, problems for matching of infinite measure (requiring an infinite amount of memory, like stack, to remember data) cannot be solved by FA. These come under non-regular languages. For example, checking balanced parenthesis, palindrome, equality in numbers and difference between two counts, cannot be solved by FA. Likewise, languages: $\{a^m b^n \mid m > n\}$ or $\{a^m b^n \mid n > m\}$ are not regular.

It is important to note that several constructs of programming languages may not be derived by FA. For instance, any program with the *main()* function in C language is valid if and only if the number of '{' is equal to the number of '}'. However, this structure may not be verified by any regular language, as it needs to remember how many '{' have been used, so that '}' must be checked to match the number of '{' and '}'. In fact, this requires an infinite number of distinct equivalence classes such as ({,}), ({{,}}), and so on. Hence, it is not regular. Likewise, a structure like L = $\{a^n b^n \mid n \geq 1\}$ is not regular.

Thus, these are the limitations of FA. For such cases, we need pushdown automata.

A non-regular language can be generally defined as a language that cannot be described by any regular expression. A few examples of interesting regular and non-regular languages are given below.

(i) $L_1 = \{a^n b^n \mid n < 10^{10}\}$
(ii) $L_2 = \{a^n b^n \mid n \geq 0\}$

The first language is finite, as n is bounded as $n \leq 10^{10}$, whereas the second is infinite (n is unbounded). Regardless, it is necessary to keep track of the number of *a*s to match with the number of *b*s for each case.

We know that the size of available memory in FA is fixed (limited) and no extra memory can be used. So it is simpler than PDA and TM with respect to transition rules, though the number of states for each type of automaton is finite. Thus, tracking the number of *a*s and *b*s for the first language (finite language) using FA is possible through distinct paths (transitions, possibly one for each string), where each transition requires some possible distinct states (equivalently memory). More specifically, memorising some input symbols is possible using state transitions. On the other hand, the second language (infinite language) may not be accepted by FA, as the number of strings is infinite (unbounded) but the size of memory in FA is limited (bounded). To overcome this limitation, we need a complex automaton (named PDA) that demands extra memory. Accordingly, the transition rules (functions) will be complex as compared to FA.

(iii) Balanced parentheses means that each opening parenthesis has a corresponding closing parenthesis and the pairs of parentheses are properly nested.
(iv) Balanced brackets means that each opening bracket ({) has a corresponding closing bracket (}) and the pairs of brackets are properly nested. Consider the following correctly balanced strings of brackets: { }. {{}}, { { } { } }. Such a problem cannot be satisfied by regular grammar; we need context-free grammar (CFG): $S \rightarrow \varepsilon$, $S \rightarrow \{S\}$, $S \rightarrow SS$. CGF is discussed in Chapter 4.

Balanced Parentheses and Similar Programming Constructs: Like balanced parentheses, there are numerous constructs in a typical programming language. These are usually parentheses themselves in expressions of all types, beginning and ending code blocks such as *begin* and *end* in PASCAL or curly braces in C. Certainly, the curly braces that appear in a C program must form a balanced sequence with '{' in place of the left parenthesis and '}' in place of the right parenthesis.

In truth, we need context-free languages to solve such problems and not regular languages (equivalently FAs). In fact, recognition of such languages requires a powerful machine called the pushdown automaton, which uses auxiliary memory (stack) to remember past information. However, regular language has the ability to identify tokens of any programming language on the basis of pattern matching. Similarly, a popular Unix command *grep* (global expression print) for finding pattern is based on RE.

Language Representor, Generator and Recogniser: In the context of programming languages, a language representor is a device or tool that determines whether a given string belongs to a language or not. Specifically, it recognises the strings or tokens of a language. Hence, FA (or equivalently RE) is a language representor. On the other hand, a language generator is a device/tool that checks whether a given sentence is syntactically correct with respect to a language. In other words, it assists in the formation of correct sentences in a language. This job is satisfied by a set of rules (known as grammar) and PDA assists its practical implementation. In fact, automata are known as language recognisers. Thus, automata and grammar are closely related.

3.11 APPLICATIONS OF FINITE AUTOMATA AND REGULAR EXPRESSIONS

Some applications of FA and RE are discussed below:

- FA recognises REs which are used to design the lexical analyser (a part of the compiler). The role of the lexical analyser is to identify the types of tokens, such as identifier, keywords, operators, numbers, and so on, of programming languages. Automata are more machine-like, whereas RE are more program syntax-like. In lexical analysis, implementation of REs is performed as: RegExp $\Rightarrow$ NFA $\Rightarrow$ DFA $\Rightarrow$ State tables.
- FA is used in designing digital circuits. The circuits of a real computer such as combinational (no memory at all) and sequential (very small amount of memory) are constructed based on the concept of FA. Examples of combinational circuits include encoders, pattern detectors (binary instruction matches for addition, multiplication) and circuits for performing complement operations, whereas the binary serial adder (with a small amount of memory to store the carry bit), parity bit generator and sequence detector fall under sequential circuits.
- FA are also used in biological modelling, image processing, speech processing, computer networking and designing cellular automata.

3.12 ADDITIONAL SOLVED PROBLEMS

PROBLEM 1

Given that a language L_1 is regular and the language $L = L_1 \cup L_2$ is regular. Is L_2 always regular? Prove your answer. **[GATE 1994, 1999]**

Solution Let us consider an alphabet $\Sigma = \{a\}$. Consider the language $L_1 = \{a^n \mid n \geq 2\}$ and $L_2 = \{a^m \mid m > 2$ and m is a prime number$\}$. We have already proved in this chapter that L_2 is not regular (using pumping lemma) but L_1 is obviously regular.

Now, a DFA accepting L_1 is shown in Fig. 3.18.

However, the language $L_1 \cup L_2 = \{a^n \mid n \geq 2\}$ is clearly regular. It is accepted by the DFA shown in Fig. 3.19. In fact, $L_2 \subseteq L_1$ and so the DFA for L_1 and $L_1 \cup L_2$ are identical.

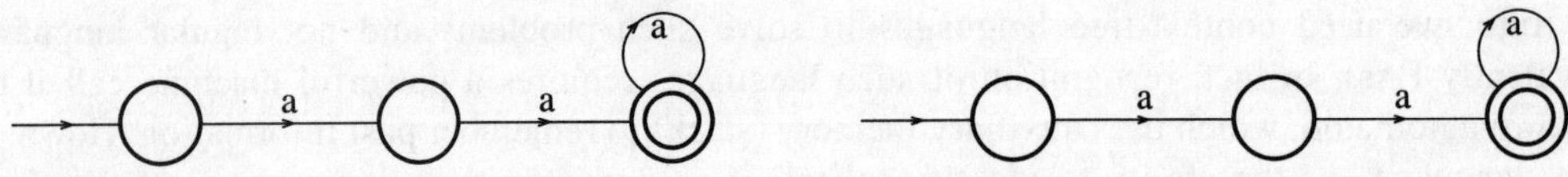

FIGURE 3.18 A DFA for $L_1 = \{a^n \mid n \geq 2\}$ **FIGURE 3.19** A DFA for $L_1 \cup L_2 = \{a^n \mid n \geq 2\}$

So, we have found a counterexample. Hence, L_2 need not be regular.

PROBLEM 2

Prove that $L = \{a^n \mid n \text{ is a prime number}\}$ defined over the alphabet $\Sigma = \{a\}$ is not regular.

Solution Thus far, no mathematical expression has been discovered (nor has it been proved that such an expression is possible) to always generate a prime number, and hence, it is impossible to come up with a regular expression or DFA to express this language.

PROBLEM 3

Prove that $L = \{a^m b^m \mid m \geq 0\}$ defined over the alphabet $\Sigma = \{a, b\}$ is not regular.

Solution The notion of RE and FA is completely stateless, that is, no state is aware of what was processed by the previous state. Consider the automaton shown in Fig. 3.20.

Can this automaton ensure that the state q_0 accepts the symbol a exactly n times, and after that, the state q_1 accepts the symbol b exactly $(n - 1)$ times? The answer is no because FA cannot remember what happened in the previous state(s).

To handle this problem, we need a system that can remember the previous states. One such automaton is PDA, which is capable of accepting specific categories of non-regular languages (discussed in later chapters).

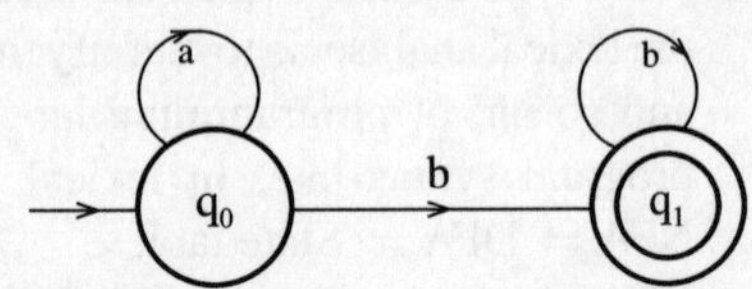

FIGURE 3.20 Incorrect DFA for $L = \{a^m b^m \mid m \geq 0\}$

PROBLEM 4

Which of the following identities of RE are true?
(a) $(r^*)^* = r^*$ (b) $(r^*s^*) = (r + s)^*$ (c) $(r + s)^* = r^* + s^*$ (d) $r^*s^* = r^* + s^*$ **[GATE 1992]**

Solution Only (a) is true. It is indeed one of the standard laws of regular expressions, already explained in this chapter. All the other expressions are invalid.

As a counterexample, let us take $\Sigma = \{a, b\}$, r = a and s = b. $(r + s)^* = (a + b)^*$ can generate strings such as 'abab', but the same string is not accepted by the regular expression a^*b^* and hence option (b) is invalid. Similarly, $(r + s)^* = (a + b)^*$ can generate strings such as 'abab', but the same string is not accepted by the regular expression $a^* + b^*$ and hence option (c) is invalid. $r^*s^* = a^*b^*$ can generate strings such as 'aabb', but the same string is not accepted by the regular expression $a^* + b^*$ and hence option (d) is invalid.

PROBLEM 5

In some programming languages, an identifier is permitted to be a letter followed by any number of letters or digits. If L and D denote the set of letters and digits, respectively, which of the following expressions defines an identifier?

(a) $(L \cup D)^+$ (b) $L(L \cup D)^*$ (c) $(LD)^*$ (d) $L(LD)^*$ **[GATE 1995]**

Solution The first character must be from set L, and then, any combination of letters and digits is allowed. Hence, only option (b) is valid.

PROBLEM 6

Which two of the following four regular expressions are equivalent? (ε is an empty string).

(i) $(00)^*(0 + \varepsilon)$ (ii) $(00)^*$ (iii) 0^* (iv) $0(00)^*$

(a) (i) and (ii) (b) (ii) and (iii) (c) (i) and (iii) (d) (i) and (iv) **[GATE 1996]**

Solution $(00)^*(0 + \varepsilon) = (00)^*0 + (00)^* \varepsilon = (00)^*0 + (00)^*$

$=$ {(strings of $(2n + 1)$ number of zeros) $\cup$ (strings of $2n$ number of zeros)}, where n is a natural number.

$=$ {(strings of *odd* number of zeros) $\cup$ (strings of *even* number of zeros, including null string)}

$=$ {strings made up of any number of zeros, including null string} $= 0^*$

Clearly, (i) and (iii) are equivalent. The correct option is (c).

Note that expression (ii), that is, $(00)^*$, accepts only those strings that have an *even* number of zeros, and hence, is not equivalent to (i) or (iii). Also, expression (iv), that is, $0(00)^*$, accepts only those strings that have an *odd* number of zeros, and hence, is not equivalent to (i) or (iii).

PROBLEM 7

Let $L \subseteq \Sigma^*$, where $\Sigma = \{a, b\}$. Which of the following is true?

(a) $L = \{x \mid x$ has an equal number of *as* and *bs*$\}$ is regular.

(b) $L = \{a^n b^n \mid n \geq 1\}$ is regular.

(c) $L = \{x \mid x$ has more *as* than *bs*$\}$ is regular.

(d) $L = \{a^m b^n \mid m, n \geq 1, m + n > 1\}$ is regular. **[GATE 1996]**

Solution We can apply pumping lemma to find out that (a), (b) and (c) are not regular. An easier method is to observe which languages need unlimited memory to identify valid strings. Clearly, any machine needs memory to determine whether an equal number of *as* and *bs* have been read to recognise (a) and (b). Similarly, a machine would also need memory to determine whether the number of *as* is greater than the number of *bs* to recognise (c). (d) is clearly regular; the language can be easily generated by the RE, a^+b^+. Therefore, the correct option is (d).

PROBLEM 8

Design an RE for $L = \{a^{2m}b^{2n} \mid m, n \geq 0\}$.

Solution An RE for the given language is $R = (aa)^*(bb)^*$.

PROBLEM 9

Design an RE with at most two *as* over {a, b}.

Solution An RE for the given language is $R = b^* + b^*ab^* + b^*ab^*ab^*$.

PROBLEM 10

Design an RE to accept strings with at least three consecutive 0s over {0, 1}.

Solution An RE for the given language is $R = (0 + 1)^*000(0 + 1)^*$.

PROBLEM 11

Design an RE to accept strings having no two consecutive 0s over {0, 1}.

Solution An RE for the given language is $R = (1 + 01)^*(0 + \varepsilon)$.

PROBLEM 12

Design an RE to accept strings where the third bit from the right is 0 over {0, 1}.

Solution An RE for the given language is $R = (0 + 1)^*0(0 + 1)(0 + 1)$.

PROBLEM 13

Design an RE to derive $L = \{0^m1^n \mid (m + n) \text{ is even}\}$.

Solution First, the DFA corresponding to the given language is as shown in Fig. 3.21.

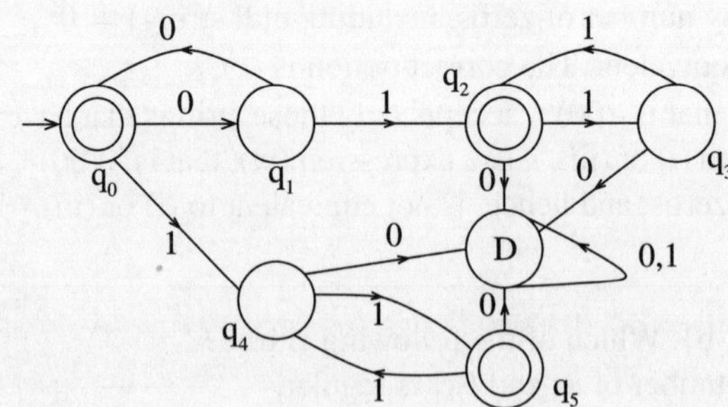

FIGURE 3.21 DFA for $L = \{0^m1^n \mid (m + n) \text{ is even}\}$

This is a problem of divisibility that can be managed by FA with the help of a number of states. The RE for the DFA is $r = \varepsilon + 11(11)^* + (00)^*01(11)^*$.

PROBLEM 14

Design an RE to derive $L = \{0^m1^n \mid mn \geq 3\}$.

Solution Again, it is identical to the problem of divisibility ($mn \geq 3$) and can be managed by FA with the help of a number of states. Here, $m, n \geq 1$, but the product of m and n is greater than or equal to 3. With these points in mind and concentrating on the format of any valid string of L, the RE is $r = 000^+1^+ + 0^+(111)^+ + (00)^+(11)^+$.

PROBLEM 15

Let c and d be two REs over Σ. If ε is not in the language corresponding to d, show that any RE, r, satisfying $r = c + rd$ corresponds to the language cd^*.

Solution Here, it is given that $r = c + rd$. Thus, $r = c + (c + rd)\, d = c + cd + rdd = c + cd + (c + rd)dd = c + cd + cdd + rdddd + \ldots$ (by substitution method)

$= c\,(\varepsilon + d + dd + ddd + \ldots \text{ to } \infty) = cd^*$.

This is a straightforward example of applying Arden's theorem.

PROBLEM 16

Find the complement of the language accepted by the NFA shown in Fig. 3.22.

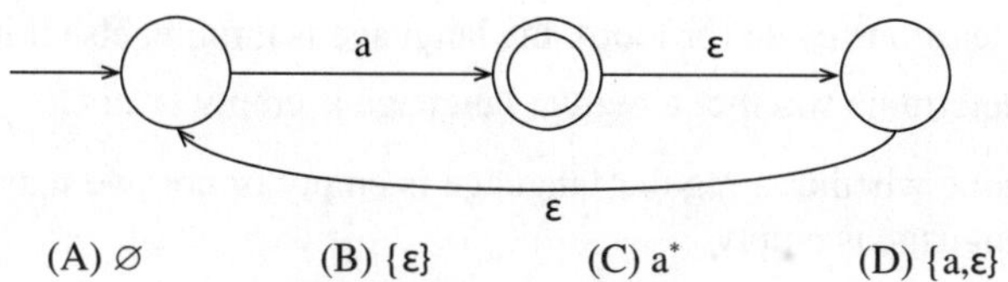

(A) ∅ (B) {ε} (C) a^* (D) {a,ε}

FIGURE 3.22 An NFA

Solution The given alphabet contains only one symbol 'a' and the given NFA accepts all strings with any number of occurrences of *a*. In other words, the NFA accepts a^+. Therefore, the complement of the language accepted by the automaton is simply an empty string.

PROBLEM 17

Is equality checking of two given regular languages (say, L_1 and L_2) over Σ solvable? If yes, then suggest an idea to solve it.

Solution Yes, it is solvable. Given that both L_1 and L_2 are regular.

Approach 1: If their symmetric difference gives Φ language, then both are equal; that is, if $L_1 - L_2 = L_1 \cap (L_2)' = \Phi$, otherwise not. Clearly, the set difference becomes null if both the sets have identical elements/strings. For this purpose, we can first draw two DFAs: one for L_1 and the other for L_2 over Σ. Next, we may draw their product machine by applying symmetric difference.

Approach 2: We may use the concept of intersection operation. Remember that if $L_1 \cap L_2 = L_1$, then it does not mean that $L_1 = L_2$. We need to check it as follows: If both $L_1 \cap L_2 = L_1$ and $L_1 \cup L_2 = L_1$, only then the two languages are equal. So, both union and intersection operations are required.

Approach 3: We can try to derive the regular expressions for both languages and then, using the algebraic laws of regular expressions, check if the two regular expressions are equivalent or not.

PROBLEM 18

Consider the following puzzle. On one side of a river, there are three humans, one big monkey, two small monkeys and one boat. Each of the humans and the big monkey is strong enough to row the boat. The boat can fit one or two bodies (regardless of size). If, at any time either side of the river, the monkeys outnumber the humans, the monkeys will eat the humans. How do you transport everyone to the other side of the river alive? Design an RE for this problem.

Solution Let Σ = {*b*, *s*, *h*}, where *b*, *s* and *h* denote big monkey, small monkey and human, respectively. Given that #b = 1, #s = 2, #h = 3, where # denotes the number. Now, based on the constraints specified in the problem, the RE, $r = (hs)(hh)(hb)(bs + hs)$.

We know that the order of appearance in concatenation operation is important. However, we have ignored this order, for example, $hs \equiv sh$; otherwise, we may use $(hs + sh)$. Also, single concatenation implies one trip of the boat but the sequence of concatenation is important to satisfy the given constraints. Further, one *h* of *hh* in *r* is the return of *h* to the earlier side of the river and the other is the left *h* to the returned side. Likewise, *h* in *hb* is the return of *h* to the earlier side and *b* is the left *b* to the return side.

PROBLEM 19

(a) Suggest an idea to decide whether a regular language is finite or infinite.

Solution To determine whether a regular language is finite of infinite, we may construct a DFA.

Now, if the DFA has at least one cycle (or loop), the language is infinite, else it is finite.

(b) Suggest an idea to determine whether a regular language is empty or not.

Solution To determine whether a regular language is empty or not, we may construct a DFA. If it has no final state, the language is empty.

PROBLEM 20

How do you formally prove that for every RE there exists an equivalent DFA?

Solution For formal proof of the statement "There exists an equivalent DFA for every RE", we may use the following steps.

Step 1: Construct an ε NFA for the given RE as we may easily design an equivalent ε NFA for an RE by applying Thompson's algorithm.

Step 2: For every ε NFA (or NFA), there is an equivalent DFA (discussed in Chapter 2).

So, the statement is proved.

PROBLEM 21

Consider the statement: For all languages L ⊆ {0, 1}*, if L* is regular, then L is regular. Is the above statement true? Justify your answer. **[ISI Kolkata JRF 2014]**

Solution We have a counterexample for this.

Let L = {w: |w| is a prime number}. Let there be a k state FSA accepting the language L. Let $w = a^n$ such that $|w| \geq k$. Let $w = xyz$, such that $|xy| \geq k$ and $|y| \geq 1$. Let $x = a^p$, $y = a^q$, $z = a^r$ $(p + q + r)$ = prime.

Consider the string xy^2z. $|xy^2z| = p + 2q + r$.

If $(p + q + r)$ is prime, there is no guarantee that $(p + 2q + r)$ will be prime. For example, $p = 2$, $q = 3$, $r = 6$. $p + q + r = 2 + 3 + 6 = 11$ = prime. But, $p + 2q + r = 2 + 6 + 6 = 14$ = not prime.

In general, if $(p + q + r)$ is prime, there is no guarantee that $(p + mq + r)$ is prime for any $m > 1$. So, there is a contradiction to the pumping lemma, and hence, L is not regular.

The sum of two prime numbers can be even or odd. In fact, all positive integers except 1 can be expressed as a sum of zero or more prime numbers (either unique or distinct primes). So, the language L* is defined as L* = {w: |w| ≠ 1, w ⊆ (0 + 1)*}.

It is indeed possible to design a DFA to accept this language, as shown in Fig. 3.23, and hence, L* is clearly regular.

So, L* is regular does not necessarily imply that L is regular.

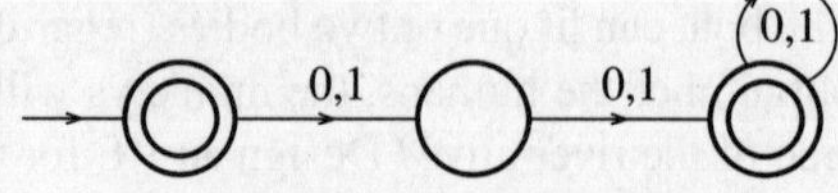

FIGURE 3.23 A DFA

PROBLEM 22

Find the shortest strings not present in the following:

(a) 0*1*0*1* (b) 0* + 1* + 1*0 + (1 + 0)*1 **[Gujarat SET 2010]**

Solution (a) 0*1*0*1*

- Empty string is accepted by this regular expression.
- Strings of length 1, that is, 0 and 1, are accepted by this regular expression.
- Strings of length 2, that is, 00, 01, 10 and 11, are accepted by this regular expression.

- Strings of length 3, that is, 000, 001, 010, 011, 100, 101, 110 and 111, are also accepted by this regular expression.

Now, observe strings of length 4.

0000 – accepted, 0001 – accepted, 0010 – accepted, 0011 – accepted, 0100 – accepted, 0101 – accepted, 0110 – accepted, 0111 – accepted, 1000 – accepted, 1001 – accepted, 1010 – not accepted.

Thus, the shortest string that is not valid is 1010.

(b) $0^* + 1^* + 1^*0 + (1 + 0)^*1$

- Empty string is accepted by this regular expression.
- Strings of length 1, that is, 0 and 1 are accepted by this regular expression.

Consider strings of length 2.

00 – accepted, 01 – accepted, 10 – accepted, 11 – accepted.

Consider strings of length 3.

000 – accepted, 001 – accepted, 010 – not accepted.

Thus, the shortest string (not valid) is 010.

PROBLEM 23

Consider the language $L = \{0^m1^n \mid m + n \text{ is even}\}$. *Prove or disprove*: L is regular.

[IIT Hyderabad PhD Entrance 2013]

Solution We shall prove that it is possible to draw a valid FA to accept the language L. Now, $(m + n)$ can be even in only two cases:

(i) Both m and n are odd numbers.

(ii) Both m and n are even numbers (inclusive of zero).

Case (i): Both m and n are odd numbers: The string 0^m1^n can be expressed as a concatenation of two strings 0^m and 1^n, such that both m and n are odd numbers. We can easily draw DFAs for 0^m and 1^n (m and n being odd), as shown in Fig. 3.24.

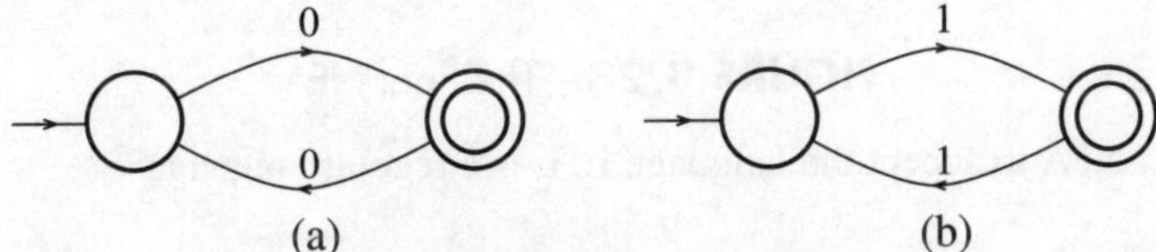

FIGURE 3.24 Two DFAs

Now, all we need to do is to concatenate these two DFAs as follows to generate the language $\{0^m1^n \mid m + n \text{ is even}, m = \text{odd and } n = \text{odd}\}$. The concatenation can be easily achieved by adding an ε move, as shown in Fig. 3.25.

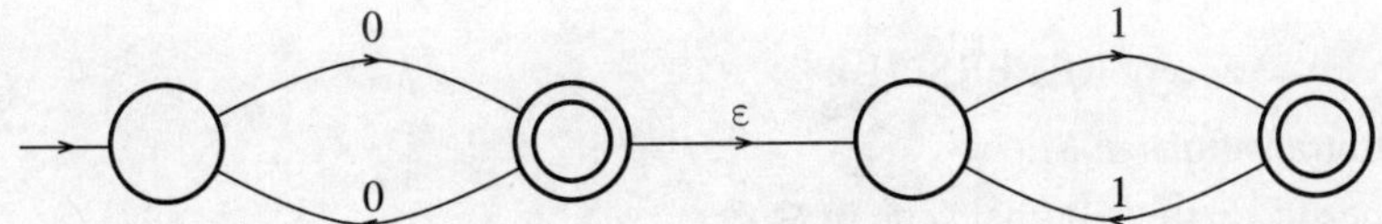

FIGURE 3.25 An ε NFA combining two DFAs (shown in Fig. 3.24)

We have achieved an NFA to represent the language $\{0^m1^n \mid m + n \text{ is even}, m = \text{odd and } n = \text{odd}\}$.

Case (ii): Both m and n are even numbers (inclusive of zero): The string 0^m1^n can be expressed as a concatenation of two strings, 0^m and 1^n, such that both m and n are even numbers. We can easily draw the DFAs for 0^m and 1^n (both m and n being even), as shown in Fig. 3.26.

FIGURE 3.26 Two DFAs

Now, all we need to do is to concatenate these two DFAs to generate the language L = $\{0^m1^n \mid m + n$ is even, $m =$ odd and $n =$ odd$\}$. The concatenation can be easily achieved by adding an ε move, as shown in Fig. 3.27.

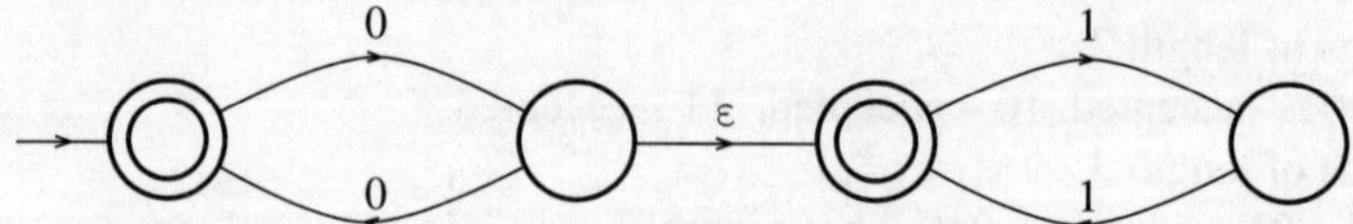

FIGURE 3.27 Concatenation of two DFAs (shown in Fig. 3.26)

We have achieved an NFA to represent the language $\{0^m1^n$: $m + n$ is even, $m =$ even and $n =$ even$\}$. The NFA to accept the language L is simply a union of the two NFAs designed for the two cases (shown in Fig. 3.28).

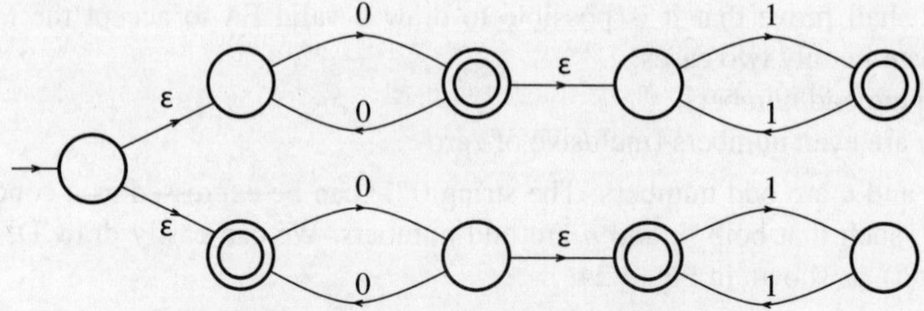

FIGURE 3.28 The final NFA

As there exists an NFA to accept the language L, L is a regular language.

PROBLEM 24

Let L be a language recognisable by a finite automaton. Determine the type of language realised by REVERSE(L) = {W such that W is the reverse of V, where V ∈ L}. Justify your answer.

[HCU PhD Entrance 2012]

Solution As L is a regular language, there exists a DFA, M, accepting the language, L.

To design an FA to accept REVERSE(L):

1. Reverse all transitions in M.
2. Convert the start state to a final state in M.
3. Convert all the final states to start states in M.
4. If there are multiple start states after these changes, introduce a new start state with ε transitions to these multiple start states.

Thus, there is a way to derive an NFA accepting REVERSE(L), and hence, REVERSE(L) is a regular language.

PROBLEM 25

For $\Sigma = \{a, b\}$, let us consider the regular language: $L = \{x \mid x = a^{2+3k} \text{ or } x = b^{10+12k}, k \geq 0\}$. Find the pumping length (the constant guaranteed by the pumping lemma) for L. **[GATE 2019]**

Solution The set x consists of $\{a^2, a^5, a^8, \ldots\}$. Here, the pumping length is 3; that is, to obtain another string, we can repeat the length 3.

The regular expression is R1 = aa(aaa)*

Or $\{b^{10}, b^{22}, b^{34}, b^{46}, \ldots\ldots\}$

Here, the pumping length is 12 and the regular expression is R2 = b^{10}(bbbbbbbbbbbb).

The possible pumping length combining both the cases is LCM(3,12) = 12.

3.13 RESEARCH DIRECTIONS

- Regular expressions have many applications in the field of 'search', which is itself a huge area for research. Every search engine, such as Google, Bing and DuckDuckGo, uses the concept of RE along with page-ranking algorithms to provide appropriate search results. Designing an efficient and foolproof algorithm for matching a string against a regular expression is a problem that has been widely researched – and is still open to optimisation.
- Several challenging problems involving regular expressions exist in the field of NLP (natural language processing). Mining of relevant text and summarisation of a lot of text written in natural language are a couple of areas where there is scope for research.
- Finding a generic algorithm to convert any regular expression to its equivalent DFA is also an open problem. It has not been solved, nor has it been proved till date that such an algorithm does not exist. If such an algorithm can be designed, the problem of efficient matching of string against RE will be solved easily.
- We propose another interesting problem here for readers to think about. Consider the following six variants of the same address of a fictitious person.

1.	2	3.
Mr Ambuj Kumar Flat W1109, West Wing XYZ Apartments PQRS Street 5 ABCD Nagar Hyderabad – 500018	Mr Ambuj Kumar Flat Number W1109 (West Wing) XYZ Apartment PQRS Street 5 ABCDNagar Hyderabad 500018	Ambuj Kumar W1109 West Wing XYZ Apartments PQRS Street 5 ABCD Nagar Hyderabad 500018
4.	**5.**	**6.**
Mr Ambuj Kumar W1109, West Wing, XYZ Apartments PQRS Street Number 5 ABCD Nagar Hyderabad 500018	Ambuj Kumar W1109 XYZ Apartments (West Wing) PQRS Street 5 ABCD Nagar Hyderabad - 500018	Mr Ambuj Kumar Flat W1109 XYZ Apartments (West Wing) PQRS Street Number 5 ABCD Nagar Hyderabad 500018

- These are six different ways (out of several dozen) of writing the address of a person. The human brain can identify that all six addresses are the same.

- Suppose the address written in cell labelled 1 is the address stored in some database. How can we design a foolproof algorithm that can accurately match any variation of the address (cell 2 to cell 6) against cell 1? The human brain can easily judge that all six addresses are equivalent, but how will general purpose software accomplish that? Can you think of a regular expression–based strategy to achieve it?

Project-based problems

1. Design a lexical analyser for the C compiler.
2. Design a tool to search for words in an English (or in any other natural language) dictionary using regular expressions.

EXERCISES

1. Design an RE for the following:
   (a) The set of all strings consisting of an odd number of 0s and an even number of 1s.
   (b) $L = \{b^m ab^n \mid m > 0, n > 0\}$.
   (c) The set of all strings over {a, b} in which the number of occurrences of 'a' is divisible by 3.
   (d) Any string whose length is a multiple of 5 over {0, 1, 2}.
   (e) Any sentence that begins with a capital letter and ends with a full stop.
   (f) A set of binary strings, each with exactly three 1s.
   (g) Obtain the regular expression for the language $L = \{a^{2n}b^{2m+1} \mid n, m \geq 0\}$.
2. *Prove or disprove*. If L_1 is regular and L_2 is non-regular, then $L_1 \cup L_2$ is non-regular.
3. *Prove or disprove*: If L_1 is regular and L_2 is non-regular, then $L_1 \cap L_2$ is regular.
4. By applying pumping lemma, prove or disprove $\{0^i1^j \mid i, j \geq 0\}$ is regular.
5. By applying pumping lemma, prove or disprove $\{xx^R \mid x \text{ in } \{0, 1\}^*\}$ is regular.
6. By applying pumping lemma, prove or disprove L = odd-length strings over {0, 1} with the middle symbol '0'} is regular.
7. By applying pumping lemma, prove or disprove that the following languages are regular.
   (a) $L_1 = \{a^ib^jc^k: i, j, k \geq 0, \text{ and } i + j = k\}$. (b) $L_2 = \{a^ib^jc^k: i, j, k \geq 0 \text{ and if } i = 1, \text{ then } j = k\}$.
8. Which one of the following regular expressions over {0, 1} denotes the set of all strings not containing 100 as a substring?
   (a) 0*(1 + 0)* (b) 0*1010* (c) 0*1*01 (d) 0(10 + 1)*
   Justify the answer. **[GATE 1997]**
9. An identifier in a programming language consists of up to six letters and digits, of which the first character must be a letter. Derive a regular expression for the identifier. **[GATE 1998]**
10. Which of the following statements is false?
    (a) Every finite subset of a non-regular set is regular.
    (b) Every subset of a regular set is regular.
    (c) Every finite subset of a regular set is regular.

(d) The intersection of two regular sets is regular.

Give a proper explanation for the correct answer. **[GATE 1998]**

11. Let S and T be languages over $\Sigma = \{a, b\}$ represented by the regular expressions $(a + b^*)^*$ and $(a + b)^*$, respectively. Which of the following is true?

(a) $S \subset T$ (b) $T \subset S$ (c) $S = T$ (d) $S \cap T = \Phi$

Give a proper explanation for the correct answer. **[GATE 2000]**

12. Match the following NFAs with their regular expressions:

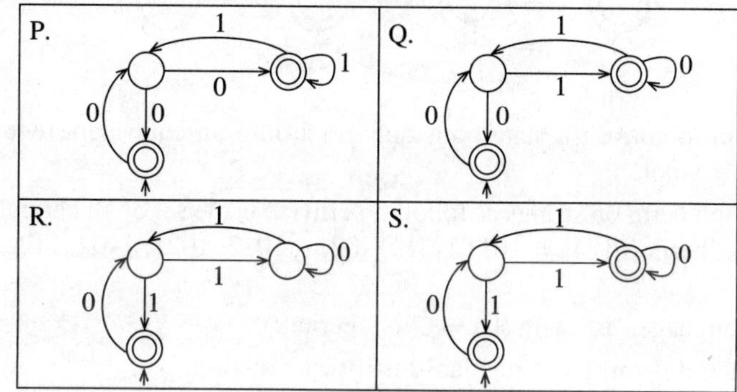

(1) $\varepsilon + 0(01^*1 + 00)^*01^*$ (2) $\varepsilon + 0(10^*1 + 00)^*0$

(3) $\varepsilon + 0(10^*1 + 10)^*1$ (4) $\varepsilon + 0(10^*1 + 10)^*10^*$

(a) P-2, Q-1, R-3, S-4 (b) P-1, Q-3, R-2, S-4

(c) P-1, Q-2, R-3, S-4 (d) P-3, Q-2, R-1, S-4 **[GATE 2008]**

13. Find the shortest string not in the language (over $\Sigma = \{a, b\}$) of the regular expression $a^*b^*(ba)^*a^*$.

14. Consider the following statements:

I. If all states of an NFA are accepting states, the language accepted by the NFA is Σ^*.

II. There exists a regular language A such that for all languages B, $A \cap B$ is regular.

Which one of the following is correct?

(a) Only I is true (b) Only II is true

(c) Both I and II are true (d) Both I and II are false

Give a proper explanation for the correct answer. **[GATE 2016]**

15. Which one of the following regular expressions represents the language: the set of all binary strings having two consecutive 0s and two consecutive 1s?

(a) $(0 + 1)^*0011(0 + 1)^* + (0 + 1)^*1100(0 + 1)^*$ (b) $(0 + 1)^*(00(0 + 1)^*11 + 11(0 + 1)^*00)(0 + 1)^*$

(c) $(0 + 1)^*00(0 + 1)^* + (0 + 1)^*11(0 + 1)^*$ (d) $00(0 + 1)^*11 + 11(0 + 1)^*00$

Give a proper explanation for the correct answer. **[GATE 2016]**

16. In each case, find a string of minimum length in (O | 1)* not in the language corresponding to the regular expression.

(a) $1^*(0\,1)^*0^*$ (b) $(0^* \mid 1^*)(0^* \mid 1^*)(0^* \mid 1^*)$ (c) $0^*(100^*)^*1^*$

17. Describe as simply as possible the language corresponding to the following regular expressions:

(a) $0^*1(1^*10^*1)^*0^*$ (b) $(1 \mid 0\,1)^*(0 \mid 0I)^*$

(c) $((0|1)^3)^*(\varepsilon \mid 0 \mid 1)$ (d) $(0|1)^*00$ (e) $(11 \mid 0)^*(00|1)^*$

18. Which of the following statements is true? Prove your answer.
    (a) The language $\{a^n \mid n \text{ is a prime number}\}$ is not regular.
    (b) Regular languages are closed under infinite union.
19. Is this language regular? $L = \{a^m b^n \mid m \geq 0, n \geq 0\}$. Justify your answer. **[UGC NET 2008]**
20. Derive the language accepted by the following automaton: **[UGC NET 2008]**

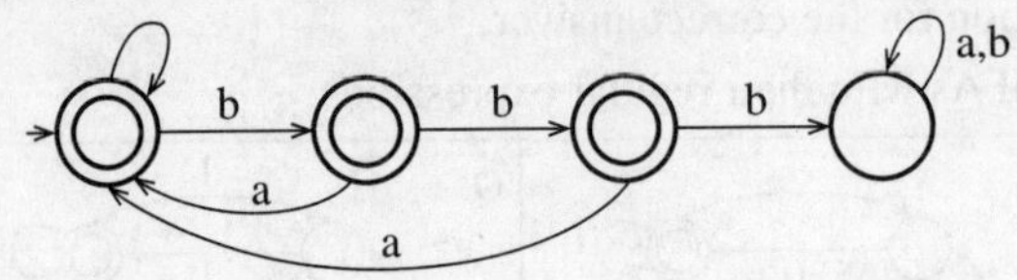

21. State an algorithm to convert a state transition graph into an equivalent regular expression, and illustrate with an example. **[UGC NET 2009]**
22. Define the operation perm on strings as follows: perm (x) is the set of all permutations of the letters of x. For example, perm (0121) = {0112, 0121, 0211, 1012, 1021, 1102, 1120, 1201, 1210, 2011, 2101, 2110}.
    Extend perm to languages as: perm (L) = $\cup_{x \in L}$ perm(x).
    If L is regular, should perm (L) be regular? Justify your answer. **[UGC NET 2010]**
23. Let $L \subseteq \{0, 1\}*$. Suppose L is regular and there is a non-deterministic automaton N which recognises L. Define the reverse of the language L to be the language $L^R = \{w \in \{0, 1\}* \mid \text{reverse}(w) \in L\}$; here, reverse(w) denotes the string w read in reverse. For example reverse(0001) = 1000.
    (a) Show that L^R is regular. How can you use N to construct an automaton to recognise L^R?
    (b) Show that the language $L{\cdot}L^R = \{x \in \{0, 1\}* \mid \exists y, z \text{ where } x = yz, y \in L, z \in L^R\}$ is regular.
    How can you use N to construct an automata for $L \cdot L^R$? **[CMI PhD Entrance 2011]**
24. Let $\Sigma = \{a, b\}$. Given a language $L \subseteq \Sigma^*$ and a word $w \in \Sigma^*$, define the languages: Extend (L,w) = $\{xw \mid x \in L\}$, Shrink(L,w) = $\{x \mid xw \in L\}$. Show that if L is regular, both Extend(L,w) and Shrink(L,w) are regular. **[CMI PhD Entrance 2015]**
25. Recall that a typical URL has the following form. It starts with a protocol specifier, followed by a colon (:) and two forward slashes (/), followed by a host name and a domain name. This is followed by an optional path specifier. Some sample URLs are given below.
    http://www.isical.ac.in/ http://www.isical.ac.in/deanoffice/results.html
    http://www.isical.ac.in/deanoffice/results/jrf/
    Assuming that A–Z, a–z and 0–9 are the only characters that can be used in a host / domain / file / directory name, write a regular expression for URLs. **[ISI Kolkata JRF 2011]**
26. Find the number of states in the minimum-sized DFA that accepts the language defined by the regular expression $(0 + 1)*(0 + 1)(0 + 1)*$.
27. Consider the following statements:
    I. If all states of an NFA are accepting states then the language accepted by the NFA is Σ^*.
    II. There exists a regular language A such that for all languages B, $A \cap B$ is regular.
    Which one of the following is correct?
    (a) Only I is true (b) Only II is true
    (c) Both I and II are true (d) Both I and II are false
    Give a proper explanation for the correct answer. **[GATE 2016]**

28. Which of the following language is regular?
(a) $L = \{a^n b^n \mid n \geq 1\}$ (b) $L = \{a^n b^m c^n d^m \mid n, m \geq 1\}$
(c) $L = \{a^n b^m \mid n, m \geq 1\}$ (d) $L = \{a^n b^m c^n \mid n, m \geq 1\}$
Give a proper explanation for the correct answer. **[UGC NET 2006]**

29. Consider the following languages:
$L1 = \{a^n b^l a^k \mid n + 1 + k > 5\}$ $L2 = \{a^n b^l a^k \mid n > 5, 1 > 3, k \leq 1\}$
Which of the following is true?
(a) L1 is regular language and L2 is not regular language.
(b) Both L1 and L2 are regular languages.
(c) Both L1 and L2 are not regular languages.
(d) L1 is not a regular language and L2 is a regular language.
Give a proper explanation for the correct answer. **[UGC 2013]**

30. Consider a language defined over the alphabet $\Sigma = \{0, 1\}$ as $L = \left\{0^{\left\lceil \frac{n}{2} \right\rceil} 1^n \mid n \geq 0\right\}$.
(a) 011 (b) 0111 (c) 0011 (d) 001111
Give a proper explanation for the correct answer. **UGC NET 2015]**

31. Which of the following regular expressions denotes a language comprising all possible strings over $\Sigma = \{a, b\}$ of length n, where n is a multiple of 3.
(a) (a + b + aa + bb + aba + bba)* (b) (aaa + bbb)*
(c) ((a + b)(a + b)(a + b))* (d) (aaa + ab + a) + (bbb + bb + a)
Give a proper explanation for the correct answer. **[Karnataka SET 2013, Gujarat SET 2016]**

32. For the regular expressions r, s and t, consider the following equalities (which may be true or false):
I. (r + s)* = (r* s*)* II. (rs + r)*r = r(sr + r)* III. (rs + r)* = r(sr + r)*
Which of these equalities are true?
(a) Only I (b) Only I and II (c) Only I and III (d) All of I, II and III
[TIFR PhD Entrance 2010, J&K SET 2013]

33. Which of the following is a valid word of the language (1*(01*01*)*) U (0*(10*10*)*)?
(a) 10110010101 (b) 1010101010 (c) 000101010000 (d) 001010001010
[Maharashtra SET 2013]

34. A regular expression that denotes a language comprising all possible strings of even length over the alphabet (0, 1) is:
(a) (0/1)(1/0)* (b) (0+1)*011 (c) (00|01|10|11)* (d) 0*1*
[JNU M.Phil/PhD Entrance 2011, Rajasthan CS Lecturer Recruitment 2014]

35. Consider the regular expression (a + b)(a + b)(a + b) … (a + b) [n times]. The minimum number of states in FA that recognise the language represented by this regular expression contains
(a) n states (b) $n + 1$ states (c) $n + 2$ states (d) $2n$ states
[Rajasthan CS Lecturer Recruitment 2014]

36. Let R_1 and R_2 be regular sets defined over the alphabet Z. Which of the following is true?
(a) R_1 U R_2 is regular (b) Z* – R_1 is regular (c) R_1* is not regular (d) Both (a) and (b)

Give a proper explanation for the correct answer.

[Pondicherry University PhD Entrance 2011, Uttarakhand SET 2012]

37. Let L1 and L2 be languages over an alphabet Σ such that L1 $\subseteq$ L2. Which of the following is true:
(a) If L2 is regular, then L1 must also be regular.
(b) If L1 is regular, then L2 must also be regular.
(c) Either both L1 and L2 are regular, or both are not regular.
(d) None of the above. **[CMI PhD Entrance 2015]**

38. How many strings of length less than 4 are present in the language described by the regular expression (x + y)*y(a + ab)*?
(a) 10 (b) 7 (c) 12 (d) 11 **[Pondicherry University PhD Entrance 2016]**

39. Consider the languages
L1 = $\{a^m b^n c^p \mid (m = n \text{ OR } n = p) \text{ AND } (m + n + p \geq 10)\}$
L2 = $\{a^m b^n c^p \mid (m = n \text{ OR } n = p) \text{ AND } (m + n + p \leq 10)\}$
Which of the following is true?
(a) L1 and L2 are both regular. (b) Neither L1 nor L2 is regular.
(c) L1 is regular and L2 is not regular. (d) L1 is not regular and L2 is regular.
(e) Both L1 and L2 are infinite.
Give a proper explanation for the correct answer. **[TIFR PhD Entrance 2015]**

40. Write a regular expression for all strings over {0, 1} in which the third and fourth symbols are the same as the first and second (respectively), the seventh and eighth symbols are the same as the fifth and sixth, and so on. Strings may be of any length. For example, 01011010101 is in the language.
Hint: One possible RE for this problem is $r = (0101)^+(0101 + 1010 + 1111 + 0000)^* + (1010)^+ (0101 + 1010 + 1111 + 0000)^* + (1111)^+ + (0000)^+$

41. Consider the regular language L = (111 + 11111)*. The minimum number of states in any DFA accepting this language is:
(a) 3 (b) 5 (c) 8 (d) 9
Draw the DFA and prove that it cannot be minimised any further.
[GATE 2006, Maharashtra SET 2015]

42. Which of the following languages is regular?
(a) $\{ww^R \mid w \in \{0, 1\}^+\}$ (b) $\{ww^Rx \mid x, w \in \{0, 1\}^+\}$
(c) $\{wxw^R \mid x, w \in \{0, 1\}^+\}$ (d) $\{xww^R \mid x, w \in \{0, 1\}^+\}$
Justify your answer with proper reasoning. **[GATE 2007]**

43. Let L1 = $\{w \in \{0, 1\}^* \mid$ w has at least as many occurrences of (110) as (011)$\}$.
Let L2 = $\{w \in \{0, 1\}^* \mid$ w has at least as many occurrence of (000) as (111)$\}$.
Which one of the following is true?
(a) L1 is regular but not L2 (b) L2 is regular but not L1
(c) Both L1 and L2 are regular (d) Neither L1 nor L2 are regular
Justify your answer. **[GATE 2014]**

44. The regular expression for the following DFA is

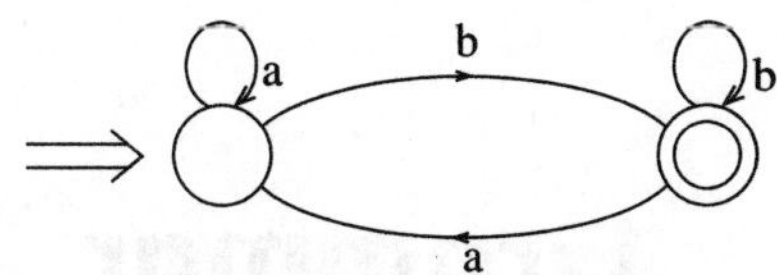

(a) ab*(b + aa*b)* (b) a*b(b + aa*b)* (c) a*b(b* + aa*b) (d) a*b(b* + aa*b)*

[UGC NET 2012]

45. Let P, Q and R be three languages. If P and R are regular and P.Q = R, then
   (a) Q has to be regular (b) Q cannot be regular
   (c) Q need not be regular (d) Q must be a CFL

   [Kerala SET 2010, JNU M.Phil/PhD Entrance 2014, Karnataka SET 2015]

46. Which of the following regular expressions is not equivalent to the regular expression (a + b + c)*?
   (a) (a* + b* + c*)* (b) (a*b*c*)* (c) (a*b* + c*)* (d) ((ab)* + c*)*
   Prove the correctness of your answer. **[Rajasthan SET 2011]**

47. Which pair of regular expressions is not a pair of equivalent expressions?
   (a) 1(01)* and (10)*1 (b) x(xx)* and (xx)*x (c) x^+ and x^*x^+ (d) (ab)* and a*b*
   Prove each case appropriately. **[Rajasthan CS Lecturer Recruitment 2014]**

48. A language is represented by the regular expression a*(a + ab + b). Which of the following strings does not belong to the regular set represented by the above expression?
   (a) aab (b) aba (c) abab (d) aaaaba **[Gujarat SET 2017]**

49. The regular expression (R + S)* S is equivalent to:
   (a) (R + S)* (b) (S + R)* (c) R*S* (d) (R + S)*
   Prove your answer. **[Maharashtra SET 2018]**

50. Let L be the set of strings over {0, 1} containing an unequal number of 0s and 1s. Prove that (i) L is not regular and (ii) L^2 is regular. **[ISI Kolkata JRF 2011]**

51. State whether L = $\{a^n b a^m b a^l \mid l, m, n \geq 1\}$ is regular or not. Justify your claim. **[UGC NET 2007]**

52. Show that the language L = $\{0^n 1 0^n \mid n = 0, 1, 2, \ldots\}$ is not a finite state language. **[UGC NET 2009]**

4 Grammar

OBJECTIVES

At the end of this chapter, you will be able to

- Understand the mathematical notations (formal grammar) used for expressing a language
- Describe the classification of grammars and languages
- Understand the relationship between regular grammar, regular expression and finite automata
- Understand ambiguity in grammar
- List the applications of grammar

4.1 INTRODUCTION

In linguistics, grammar is a set of structural rules governing the composition of clauses, phrases and words in a natural language. It is a powerful fundamental tool for describing languages. A **grammar** consists of a set of rules by which valid sentences (strings in formal languages) are constructed. A grammar is said to be **formal** if its rules are expressed concisely by formal notations. A brief discussion on formal grammar was presented in Chapter 1. We will examine it in detail in this chapter.

Every language has its own distinct grammar to describe its syntactical construction. For example, English has a grammar. Let us try to understand the importance of grammar by taking a simple English sentence: "*Ram is a boy*" is a grammatically correct English sentence. But why is it correct? Why not "*boy a is Ram*"?

The answer is that the sentence "*Ram is a boy*" is constructed following the rules (syntax) of English grammar. However, "*boy a Ram is*" does not satisfy any of the proper rules of that grammar. Here, we refer to some common rules for constructing correct English sentences. These rules are informally expressed as follows: *Every sentence starts with an article followed by the subject or simply the subject without any article. It consists of verbs, objects, articles, and so on.*

To derive the sentence "*Ram is a boy*", we may express the rules in a formal way:

$S_{en} \rightarrow S_{ub}$ V O // Non-terminals: S_{en} stands for sentence, S_{ub} for subject, V for verb phrase and O for object.

$S_{ub} \rightarrow$ A N | N // Non-terminals: A stands for article, N for noun phrase.

$O \rightarrow$ N | A N

$A \rightarrow$ a | an | the // These are terminals (that is, terminal strings).

$N \rightarrow A_{dj}$ N // Non-terminal: A_{dj} stands for adjective.

N → Ram | boy | ... // These are terminals (terminal strings).

V → is | run | am | ...

A_{dj} → good | bad | ...

Here, all the terminal strings (words) are generated over the alphabet set: {a–z, A–Z}. The generation (also known as derivation) of the sentence "*Ram is a boy*" using a sequence of steps that apply these formal rules is shown below:

$S_{en} \Rightarrow S_{ub}$ V O ⇒ N V O Ram V O ⇒ Ram is A N ⇒ Ram is a N ⇒ Ram is a boy.

That is, $S_{en} \overset{*}{\Rightarrow} S_{ub}$ V O ⇒ Ram is a boy

Note: Any derivation involves the application of productions. When the Head (LHS) of the rule is replaced by its RHS (Body), the derivation step-count is one. We use the notation: → for grammar rules/productions; ⇒ to mean one-step derivation (the production is applied once). This operator is transitive. The notation: $\overset{*}{\Rightarrow}$ means derivation comprising zero or more steps. Thus, a *k*-step derivation means *k*-consecutive single-step derivations. Now, if *k* = 0, this means that no rules are applied, so the string does not change (discussion on derivations is presented below).

Alternatively, we may formally express the rules as:

(i) <Noun> <Verb> <Article> <Noun>

(ii) <Noun> <Verb> <Adverb>

(iii) <Noun> <Verb> <Adjective> <Adverb>..., and so on.

The rules are in **Backus–Naur Form (BNF)**, which is the formal mathematical way of specifying context-free grammar. For example, in the sentence, "*Ram is a boy*", "*Ram*" is substituted as noun, "is" as verb, "*a*" as article and "*boy*" as noun.

Clearly, *Rule (i)* (as specified above) is satisfied. On the other hand, the sentence "*boy a is Ram*" does not match any proper rule of English grammar in the same manner (word-by-word substitution). Further, "*A Ram is boy*" is grammatically correct (according to the rules of English grammar) but does not make semantic sense. Thus, grammars are a tool for syntax, not semantics.

It is important to note that the grammar of any language (whether programming or natural) can be represented formally, and it must have a distinct start symbol. In order to generate a string from a formal grammar, we begin at its start symbol and then apply the rules successively. In practice, any particular sequence of legal choices made during this process yields one particular string in the language, called the **derivation** of that string. In computer science, this process is often known as **parsing** (viewed in the form of a tree called the derivation or parse tree). Interestingly, the same approach is applied for every type of language.

The branch of mathematics that deals with the properties of formal grammars and languages is called **Formal Language Theory**.

Formal languages vs. programming or natural languages: In formal language, strings and sentences are treated identically, whereas they are different in the context of programming or natural languages. For example, let L = $\{a^n b^n \mid n \geq 1\}$ over {a, b}. It is a formal language. Here, each ab, a^2b^2, ... is a string as well as a sentence. On the other hand, let us consider a natural language, say English. "*Ram is a boy*" is a correct sentence in this language. Each word of this sentence is a word or string. However, "*Ram is a boy*" is a string/sentence with respect to the formal version of English. In the same way, "*int x;*" is a sentence with respect to C language, and "*int*" and "*x*" are strings/tokens.

4.2 FORMAL DEFINITION

We know that grammar provides a set of rules for generating the sentences/statements of a language through substitution of strings (words), and such a description of grammar is informal. However, in the late 1950s, American linguist Noam Chomsky begun to study the formal (mathematical) representation of languages, known as formal grammar, that actually describes the syntax but not the meaning. Chomsky proposed a hierarchy that categorised formal grammars into four classes (known as the Chomsky hierarchy) based on increasing expressing power of representation of languages. These are well-suited to four different types of automata, namely, finite automata (FA), pushdown automata (PDA), linear bounded automata (LBA) and Turing machine (TM). In fact, in computer science, formal languages (especially regular and context-free) are used to model computer languages, whereas in linguistics, they are used for the scientific study of natural languages. In the classic formalisation of generative grammars, a grammar $G = (V_N, \Sigma, P, S)$ is a four-tuple mathematical model, as described below:

- V_N is a non-empty finite set of non-terminal symbols or variables (that can be replaced/expanded to a sequence of symbols). The symbols are normally denoted by capital letters, such as A, B, C, etc.
- Σ is a non-empty finite set of terminal symbols (which cannot be expanded further; each terminal is an actual word/token in a language) and $V_N \cap \Sigma = \Phi$. The terminal symbols are normally denoted by lower-case letters, such as *a*, *b*, *c*, etc. In particular, any string (terminal string) of a programming language takes symbols of Σ only. One may note that $V_N \cup \Sigma$ is often called the **complete alphabet** of the grammar.
- S denotes the start symbol of the grammar from where the parsing/derivation process begins, and $S \in V_N$.
- P is a finite set of production rules. A production rule is a grammar rule that describes how symbols should be replaced/exchanged. Each rule has the form: LHS (left-hand side) $\rightarrow$ RHS (right-hand side); equivalently, Head $\rightarrow$ Body. Formally, each rule can be expressed as: $x \rightarrow y$, where $x \in (V_N \cup \Sigma)^* V_N (V_N \cup \Sigma)^*$ and $y \in (V_N \cup \Sigma)^*$. Here, $\cup$ stands for set union and * is the Kleene star operator. This expression implies that the LHS of a rule must have at least one non-terminal symbol, whereas the RHS can be any combination of non-terminal and terminal symbols (it can even be null). Thus, a **production rule** is one that produces or derives a set of words to create a correct sentence.

Note: If $x \rightarrow y$ is a production, it does not necessarily mean that $y \rightarrow x$ is also a production.

Recursive Rule: If the same non-terminal symbol appears on both the LHS and the RHS of a rule (assuming that the LHS contains only the non-terminal), it is called a recursive rule. For example, $S \rightarrow aSb$ is a recursive rule.

In addition, when multiple rules have the same LHS, we may simply use the symbol '|' (OR) on the RHS to distinguish between these rules. For example, we can write the rules: $S \rightarrow aSb$ and $S \rightarrow c$ as: $S \rightarrow aSb \mid c$. This notation is sometimes referred to as **compressed/compact notation**.

If the recursive non-terminal is on the left of the RHS, for example, $A \rightarrow u \mid Av$, we call the production **left-recursive**. Similarly, we can define a **right-recursive** production as $A \rightarrow u \mid vA$.

A grammar is said to be **recursive** if its rule set contains at least one recursive rule, otherwise it is non-recursive. A non-recursive grammar can produce only a finite language; and every finite language can be produced by a non-recursive grammar. For example, a straight-line grammar (SLG) produces exactly one word. Thus, the language generated from SLG is finite.

Derivation or Generation of Strings: Derivation is a generic term; it is the process of generating (deriving) the target string of a language by substituting a sequence of rules of the grammar of that language. The language of a formal grammar $G = (V_N, \Sigma, P, S)$ is denoted as L(G). Conceptually, L(G) is defined as all the strings over Σ that can be generated from the start symbol S by applying the production rules in P.

During generation of a string from the rules, the RHS contains a combination of terminal and non-terminal symbols in every step. Each such combination is known as the **sentential form** of the string. When the target string is derived, the form is termed a **sentence**.

For example, let us assume a grammar (G) with $V_N = \{S\}$, $\Sigma = \{a, b\}$, S is the start symbol and P (the set of rules) = {S → aSb, S → ba}. Here, in P, there are two rules associated with S: (i) S → aSb and (ii) S → ba.

Now, we start with S and choose a rule associated with it. If we select *Rule* (*i*), we obtain: *aSb* on the RHS as a sentential form at this stage. However, if we again choose *Rule* (*i*) and replace *S* with *aSb*, we obtain *aaSbb* as the current sentential form. In fact, this process can be repeated many times and finally *Rule* (*ii*) is applied in the sentential form to generate sentences (strings): for example, *aaSbb* is a sentential form and *aababb* (as a sentence) is generated, where *ba* is replaced by *S*. The derivation of this string can be expressed as: $S \Rightarrow aSb \Rightarrow aaSbb \Rightarrow aababb$. In short, it is $S \overset{*}{\Rightarrow} aababb$, where $\overset{*}{\Rightarrow}$ denotes applying zero or more steps to generate the target string. This process generates a set of strings: {ba, abab, aababb, aaababbb, …} from grammar G. It is, indeed, L(G).

In fact, the derived language can be expressed as $L(G) = \{(a)^n ba(b)^n \mid n \geq 0\}$.

Types of Derivation: In context-free or lower languages, two derivations, namely leftmost and rightmost, are used to derive strings.

- **Leftmost derivation (LMD):** A derivation $A \overset{*}{\Rightarrow} w$ is called the leftmost derivation if, at every step, a production rule is applied only to the leftmost non-terminal in the sentential form. Here, $\overset{*}{\Rightarrow}$ is the general notation for derivation and it denotes multiple derivation steps, and A and *w* are, respectively, the start symbol of G and the target string.
- **Rightmost derivation (RMD):** A derivation $A \overset{*}{\Rightarrow} w$ is called the rightmost derivation if we apply a production rule only to the rightmost non-terminal in every step.

To better understand the two types of derivation, consider the rules: S → 0B | 1A, A → 0 | 0S | 1AA, B → 1 | 1S | 0BB and a target string: *w* = 00110101. For leftmost derivation, to generate *w* = 00110101, we follow: S ⇒ 0B ⇒ 00BB ⇒ 00(1)B … ⇒ 00110101. Here, in every step, only the leftmost non-terminal is expanded to derive *w* = 00110101. The rightmost non-terminal is never expanded in any stage of leftmost derivation.

For rightmost derivation, to generate *w* = 00110101, we follow: S ⇒ 0B ⇒ 00BB ⇒ 00B(1S) … ⇒ 00110101. In every step, the rightmost non-terminal is expanded to derive 00110101, but the leftmost non-terminal is never expanded .

Note: In context-free grammar (CFG), derivation of a string from a grammar G is possible if and only if there exists a corresponding derivation tree. In addition, during formation of the tree, either leftmost or rightmost derivation is followed, but not both (leftmost derivation in one step and rightmost derivation in another, or vice versa). If leftmost derivation is considered in the context of a compiler, it simply implies that the parser scans the input from left to right. On the other hand, a parser scans input from right to left when rightmost derivation is adopted. Almost all parsers follow left-to-right scanning.

4.3 CHOMSKY HIERARCHY OF GRAMMARS

Noam Chomsky classified the generative grammars into four categories on the basis of decreasing expressive power (that is, based on the structure of the rules), now known as the Chomsky hierarchy. These are Type 0 (phrase structured or unrestricted), Type 1 (context-sensitive), Type 2 (context-free) and Type 3 (regular) grammars. The main difference among these types of grammars is that they possess increasingly stricter production rules and can express fewer formal languages. In this context, an interesting point is that an equivalent regular grammar exists for every regular expression. Further, machines such as TM, LBA, PDA and FA are designed based on the structure of the rules of grammar. The types of grammars are described below.

Type 0 (Phrase Structured or Unrestricted) Grammar: This is phrase structured grammar without any restrictions. The production rules are of the form: $\alpha \rightarrow \beta$, where $\alpha \in (V_N \cup \Sigma)^* V_N (V_N \cup \Sigma)^*$ and $\beta \in (V_N \cup \Sigma)^*$. In fact, the only restriction is that α contains at least one non-terminal symbol; no other restriction is applied. Such a type of grammar is recognised by TM.

Consider a rule set of a grammar G as: P = {S → MaP, P → RP, P → ε, aR → Raa, MR → M, M → ε}.

The rules in this set (P) follow the structure of Type 0 grammar. The set P generates the language $L = \{a^{2^n} | n \geq 0\}$. It is true that L may not be generated by other types of grammar.

Let us consider a few **derivations** with respect to this grammar.

$w = a$ $(n = 0)$; S ⇒ MaP ⇒ Ma(ε) ⇒ εa = a = a^{2^0} [for M and P, ε is substituted]

$w = aa$; S ⇒ Ma(P) ⇒ M(aR)P ⇒ MRaa(P) ⇒ (MR)aa ε ⇒ (M)aa ⇒ ε aa = aa = a^{2^1}

$w = aaaa$; S ⇒ Ma(P) ⇒ MaR(P) ⇒ MaRRP ⇒ MaRR ε ⇒ M(aR)R

⇒ MRa(aR) ⇒ MRaRaa ⇒ MR(aR)aa ⇒ MRRaaaa

⇒ MRaaaa ⇒ Maaaa ⇒ ε aaaa = aaaa = a^{2^2}

Type 1 (Context-sensitive) Grammar: The most general form of each rule of Type 1 grammar is: $\alpha \rightarrow \beta$ with $|\beta| \geq |\alpha|$, and at least one symbol of α is the non-terminal symbol, that is, $|\alpha| \geq 1$. Hence, $|\beta|$ may not be 0 ($\beta \neq \varepsilon$), as $|\beta| \geq |\alpha|$. However, on considering the presence of the null string in a language, a rule like: A → ε is allowed, subject to the condition that A does not appear on the RHS of any rule in the rule set. Thus, the characteristics of CSG state that every CSG is a subset of Type 0 grammar, but not vice versa.

Importantly, any production rule of a Type 1 grammar in **normal form** is: $\Phi A \psi \rightarrow \Phi \beta \psi$, where $A \in V_N$, $(\Phi, \psi) \in (V_N \cup \Sigma)^*$, $\beta \in (V_N \cup \Sigma)^+$ (β may not be null). Here, Φ and ψ are called **contexts**. As Φ and ψ remain unchanged, $\Phi A \psi \rightarrow \Phi \beta \psi$, where β is the replacement for A. In a Type 1 production, the non-terminal A cannot be erased. In particular, if $\beta = \varepsilon$, the length of the RHS of the production is less than the length of the LHS of the production (not allowed in Type 1 production). However, a rule like A → ε may be allowed (special case) subject to the condition that A does not appear on the RHS of any production. If it does appear, then $|\beta| \geq |\alpha|$ is violated. The language generated by such a grammar is called a **context-sensitive language (CSL)**, and it is recognised by LBA.

For example, G = ({S, B, C, T, X}, {a, b, c}, P, S), where V_N = {S, B, C, T, X}, Σ = {a, b, c} and the rule set (P) is:

S → aSBC (left and right contexts of S (Φ and ψ) are ε, and β = aSBC)

bB → bb (left and right contexts of B are b and ε, and β = b)

S → aBC (left and right contexts of S are ε and β = aBC)

bC → bc (left and right contexts of C are, respectively, b and ε, and β = c)

CB → CX (left context of B = C but its right context = ε, and B substitutes X)

CX → BX (right context of C = X but its left context = ε, and C substitutes B)

BX → BC (left context of X = B but its right context = ε, and X substitutes C)

cC → cc (left and right contexts of C are, respectively, c and ε, and β = c)

aB → ab (left and right contexts of B are, respectively, a and ε, and β = b)

Tip: To find the contexts (left and right), identify the substitution part.

Note: The subset: {CB → CX, CX → BX, BX → BC} is equivalent to the rule: CB → BC. However, if the single rule: CB → BC (instead of the subset) is considered, it does not satisfy the normal form of a Type 1 grammar.

Derivation of a string from the rule set: Let $w = abc$. To derive this, we may apply:
S ⇒ aBC ⇒ (aB)C ⇒ abC ⇒ a (bC) ⇒ abc.

Let $w = aabbcc$. Now, S ⇒ aSBC ⇒ a(aBC)BC ⇒ aaB(CB)C ⇒ aaB(CX)C ⇒ aaB(BX)C ⇒ aaB(BC)C
⇒ a(aB)BCC ⇒ a(ab)BCC ⇒ aa(bB)CC ⇒ aabbCC
⇒ aab(bC)C ⇒ aab(bc)C ⇒ aabb(cC) ⇒ aabb(cc)

It is important to note that we may not substitute BC with BX (the LHS of the rule: BX → BC, when aaB(CB)C is found as the sentential form) and so on to finally obtain CB. If we do so, this shows reduction not derivation. However, the process we have adopted is derivation, not reduction. Truly, the language $L = \{a^n b^n c^n \mid n \geq 1\}$ cannot be generated by Type 2 or Type 3 grammars. Further, we can generate language $L = \{a^n b^n a^n b^n \mid n \geq 1\}$ by modifying the above-mentioned rule set.

Another interesting example of Type 1 language is: $L_{ww} = \{ww \mid w \in \{a, b\}^*\}$. In particular, the 'define/use' structure in the C function prototype can be captured by this language. One possible rule set for this language is {S → FM, F → FaA | FbB, Aa → aA, Ab → bA, Ba → aB, Bb → bB, AM → Ma, BM → Mb, F → ε, M → ε}.

In another example, the grammar G = ({S, A, B, C, a, b, c}, {a, b, c}, P, S), where P (the set of rules) = {S → ABC, S → ABCS, AB → BA, AC → CA, BC → CB, BA → AB, CA → AC, CB → BC, A → a, B → b, C → c}. In fact, the language generated by this grammar is $L = \{w \in \{a, b, c\}^+ \mid \#(a) = \#(b) = \#(c)\}$. Undoubtedly, this is not context-free but context-sensitive.

Context-sensitivity in natural languages: An example of a natural language is *a* versus *an* in English, where *a* and *an* are the determiners (denoted as DET). In English, these determiners are used as follows:

- *a* appears before consonants; for example, a banana, a large apple
- *an* appears before vowels; for example, an apple, an exceptionally large banana

Thus, the context-sensitive rules (schematic only) for this purpose will be:

- DET [c-word] → a [c-word], where [c-word] means word starts with consonant and DET substitutes *a*.
- DET [v-word] → an [v-word], where [v-word] means word starts with vowel and DET substitutes *an*.

KEY POINTS

- The class of CSL is large when compared to CFLs and regular languages.
- CSLs are accepted by space-bounded TM. These are widely used in natural language processing. It is interesting to note that most Indian languages are CSL, whereas most European languages are CFL.
- With CSGs, we need to think in terms of generic derivation but not derivation trees or parse trees.
- The main problem associated with CSGs is that there are no known practical methods for constructing parsers from them, as no specific derivation strategy (such as leftmost or rightmost) may be adopted.

Type 2 (Context-free) Grammar (CFG): In this type of grammar, the LHS of each production rule consists of only one symbol and it is a non-terminal, whereas the RHS takes the form: $(V_N \cup \Sigma)^*$. In other words, each rule takes the form: $A \rightarrow (V_N \cup \Sigma)^*$, where $A \in V_N$. This is simpler than the previous types of grammar, and many languages can be generated by this type. The language generated by such a grammar is called a context-free language (CFL), and every CFL is recognised by a PDA. Grammars, specifically CFGs, are used in programming languages to describe the syntax of the language.

The structure of a CFG states that every CFG is a CSG, but the converse is not true. One very well-known example of a CFL is $L = \{a^n b^n \mid n \geq 1\}$. To generate $L = \{a^n b^n \mid n \geq 1\}$, we may design a CFG $(G) = (V_N, \Sigma, P, S)$, where $V_N = \{S\}$, $\Sigma = \{a, b\}$, $P = \{S \rightarrow aSb, S \rightarrow ab\}$ and S is the start symbol of this grammar. Derivation of the language is shown below.

$S \Rightarrow aSb \Rightarrow a(aSb)b$

$\Rightarrow aaSbb \Rightarrow aa(aSb)bb$

....

$\Rightarrow a^n b^n$ (after repetition of the n-th step)

It is important to note that every string w in L(G) of a CFG has at least one leftmost and at least one rightmost derivation.

Note: This grammar (replacing 'a' with '(' and 'b' with ')') generates a construct such as $(^n)^n$ which may not be nested. To generate a nested construct with balanced parenthesis, we may consider the rule set $\{S \rightarrow aSb|SS, S \rightarrow ab\}$. Further, the rule set $\{S \rightarrow aSb, S_1 \rightarrow SS_1 \mid \varepsilon\}$ is capable of generating the language $L_1 = \{a^n b^n a^m b^m ...\} = L^* = L^0 \cup L^1 \cup L^2 \cup ...$, where $L = \{a^n b^n \mid n \geq 1\}$. Nested parenthesis is not allowed here.

Types of CFGs: CFGs may be of two types: deterministic (DCFG) and non-deterministic (NCFG). If a CFG is accepted by deterministic pushdown automata (DPDA), it is a DCFG; otherwise it is an NCFG.

The most basic characteristic of a DCFG is that productions such as $A \rightarrow a$ and $A \rightarrow b$ are not allowed, whereas these are allowed in NCFG. However, it may not be the prime criterion. For example, the rule set: $S \rightarrow aSa \mid bSb \mid \varepsilon$ (for even-length palindromes over $\{a, b\}$) meets the above criterion but is still not DCFG.

Interestingly, every DCFG is unambiguous, but the converse is not true. For example, $S \rightarrow aSa \mid bSb \mid \varepsilon$ is unambiguous but is not DCFG. Undoubtedly, DCFLs are of practical interest as they can be parsed in linear time. In fact, DCFG is the most important and useful type of grammar in designing the parser module of any compiler. Most compilers are based on this type of grammar.

It is interesting to note that the Sanskrit grammar defined by Panini (probably 400 BCE) is accepted as the first CFG for a natural language. In Sanskrit, sentences can be generated iteratively from scratch, using the grammar.

Note: PDA accepts CFG. A CFL can be recognised in $O(n^3)$ time. That is, for every CFL, a machine can be built that takes a string as input and determines in $O(n^3)$ time if it is a member of the language, where n is the length of the string.

Type 3 (Regular) Grammar: The LHS of each rule in regular grammar contains only one symbol, and it is a non-terminal. The RHS is restricted. In particular, the RHS may be the empty string or a single terminal symbol or a single terminal symbol followed by a non-terminal symbol, or vice versa.

Formally, the normal form of rules is expressed as: $A \rightarrow \varepsilon, A \rightarrow a, A \rightarrow aB, A \rightarrow Ba$, where $A, B \in V_N$ and $a \in \Sigma$. Note that a rule like: $A \rightarrow \varepsilon$ is allowed if A does not appear on the RHS of any production. The format of regular grammar indicates that every regular grammar is a CFG, but the converse is not true.

Categories: In computer science, a linear grammar is a CFG that has at most one non-terminal on the RHS of each production. For example, S → aSb | ε. Thus, regular grammars are a special type of linear grammar, and they may be of two types: **right-linear** and **left-linear**.

Rules of the form: A → aB are called **right-linear**, whereas those of the form: A → Ba are **left-linear**. However, in a regular grammar, there should not be any mix up of these two types. If present, the rule must be transformed to one type (left or right). For parsing, left-linear rules must be converted into right-linear rules.

Note: Linearity and its type are taken into account in terms of non-terminals on the RHS of the rule to the left (called left-linear) or right (called right-linear).

Extended form: Sometimes, we may express the production rules of a regular grammar in a more formal way as: A → wN or A → w or A → ε, where A and N are non-terminals and *w* is a string of terminal symbols ($w \in \Sigma^+$). For example, A → abcN is a rule of a regular grammar, where *abc* is a string of terminal symbols. Of course, the rule: A → abcN can be expressed as a collection of multiple rules, each of which has the form: A → aB, such as $A \rightarrow aA_1$, $A_1 \rightarrow bA_2$, $A_2 \rightarrow cN$, where A_1 and A_2 are the new non-terminal symbols to be included in the existing set of non-terminal symbols. The outcomes are, indeed, based on the idea of the Greibach Normal Form (GNF) (discussed later).

KEY POINT

- When more restrictions are imposed on rules, it results in simpler rules.

THEOREM 4.1

All regular languages are necessarily DCFLs.

Proof It is true that any regular language can be represented by a DFA. Further, any DFA can be interpreted by a DPDA, where the stack remains empty. Hence, any regular language is a DCFL.

Note: The language generated by a regular grammar is called a regular language (RL), and it is recognised by FA in linear time. In practice, regular grammars are expressed using RE. However, it is often difficult to design RE for some regular languages, but regular grammar makes it easier to describe such regular languages.

Context-free Grammars and Regular Grammars in Programming Languages: The tokens of any programming language may be finite or infinite sets generated by patterns. We already know that a finite set/language is regular and can be described by regular expression. Also, each infinite set/language satisfying some pattern is regular. So, RE is sufficient for identifying tokens of any programming language. Thus, RE (equivalently regular grammar) is powerful enough to model the morphology (words) of any natural language as well as the lexical analyser of any computer language. However, regular languages are not powerful enough to model syntax. For example, one valid feature of any C program is the *main*() function, which consists of '{' and '}' such that #({) = #(}) (assuming that { appears before }). This structure may not be verified by any regular language, as it needs to remember how many '{' have been used in the program, so the opening and closing parentheses must be examined to ensure that they match. Hence, it is not regular. In particular, this construct can be fulfilled by a set of CFG rules: {S → SS, S → {S}, S → ε}. Likewise, $L = \{a^n b^n \mid n \geq 1\}$. This format follows constructs such as $(^n)^n$, for example, (((a + b)*c – d)*e). For all such constructs, we need CFGs.

Context Sensitivity in Programming Languages: It may come as a surprise that most programming languages are not context-free. More explicitly, though most arithmetic expressions are generated by CFG, not all features/constructs of programming languages such as C, Fortran, and so on, can be captured by CFG. For instance:

- In C programming language, we can declare a function prototype that can introduce an arbitrary number of arguments. Later, when the function is defined, the same argument list must appear in the same order. This 'define/use' structure is identical to the language: $L_{ww} = \{ww \mid w \in \{0, 1\}^*\}$, which is a CSL. According to the language derivation capability of grammars, derivation of this define/use construct is not possible by CFG (or by PDA). However, CFG is always preferred for designing compilers as no practical methods exist for constructing parsers from CSGs but it is easier using CFG. That is why the derivation of CSL features (power of LBA) in compilers is often managed by combining CFG and the symbol table (through information from *id*). It is important that we may not use stack here, as the last symbol of the first w (in L_{ww}) pushed onto the stack should not be matched with the first symbol of the second w in L_{ww}. However, it is true that most constructs of the C language (or other languages) can be recognised by CFGs (equivalently by using stack).
- Further, $L = \{wcw \mid w \in \{a, b\}^*\}$ is neither regular nor context-free. It is a special case of the CSL feature mentioned earlier. An example of a similar construct is the *identifier* that is declared and checked for its definition before its use.
- Typing rules; for example, the type of LHS expression to the assignment/equal to operator conforms to the type of the RHS expression.
- Access constraints (for example, use of public vs private methods in Java).

Note: We do not know any efficient parsing algorithm for context-sensitive languages. However, algorithms may be designed for parsing of CSGs. Actually, compilers first parse the context-free features. Parsing of context-sensitive features, such as type checking, is carried out in a separate stage after parsing in a syntactic manner (for example, checking that the variables used are defined in the symbol table, otherwise it generates an error). This is performed using a syntax-directed scheme.

Critical Analysis of Context-sensitive, Context-free and Regular Grammars: Recall that the language $L = \{a^n b^n c^n \mid n \geq 1\}$ is not a CFL. However, $L = \{a^n b^n \mid n \geq 1\}$ is context-free. In particular, $L = \{a^n b^n \mid n \geq 1\}$ can be easily generated by Type 2 grammar (CFG), as the corresponding machine (PDA) uses one stack to remember the number of *as* that can be matched with the number of *bs*. In other words, it works in a store-and-match fashion, whereas FA has finite measuring and matching capabilities that cannot be extended for infiniteness (as no stack is used). On the other hand, generation of language: $L = \{a^n b^n c^n \mid n \geq 1\}$ is not possible by CFG (and its corresponding PDA), as the machine is allowed to use only one stack. The stack may be used for matching the substring: $a^n b^n$ but problems occur while matching substring: $b^n c^n$. So, we need to extend the power of PDA. It is then named LBA.

Simple block diagram: On the basis of the strength or controlling power of the four types of grammars, the relation among them is shown in Fig. 4.1.

From Fig. 4.1, it is clear that Type 0 grammar includes Type 1 grammar, whereas Type 1 includes Type 2, and Type 2 includes Type 3 grammar. Thus, types 1, 2 and 3 fall under Type 0 grammar, but the converse is not true. In other words, regular languages are CFLs, CFLs are CSLs and CSLs are unrestricted languages, but the converse is not true.

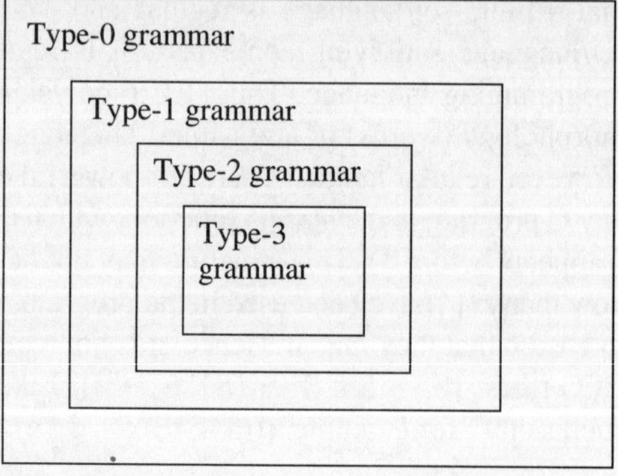

FIGURE 4.1 Chomsky hierarchy of grammars

KEY POINT

- The rule set of a grammar G for generating a language L must be designed in such a way that derivation begins from the start symbol of G and every accepted string (satisfying the properties specified) is generated.

EXAMPLE 4.1

Find the language generated by grammar $G = (V_N, \Sigma, P, S)$, where $V_N = \{S\}$, $\Sigma = \{a, b\}$, $P = \{S \to aS, S \to bS, S \to a, S \to b\}$, and S is the start symbol of G.

Solution The smallest possible strings generated by this grammar are '*a*' and '*b*'. Also, the rules $S \to aS$, $S \to bS$ together imply that any symbol in a string generated by this grammar can be followed by '*a*' or '*b*'. Hence, it is clear that the set is capable of generating all strings (except ε) over {a, b}. Thus, $L(G) = \{a, b\}^* - \{\varepsilon\} = \{a, b\}^+$.

EXAMPLE 4.2

(a) Find the language generated by the grammar with rules: $S \to aAa$, $A \to aAa$, $A \to a$, where S is the start symbol.

Solution From the given grammar, we obtain $S \Rightarrow aAa \Rightarrow a(aAa)a\ldots \Rightarrow a^{n-1}a\,a^{n-1}$ (substituting A with '*a*' in the *n*-th derivation step).

$= a^{2(n-1)+1} = a^{2n-1}$ ($n \geq 2$, where *n* is the number of derivation steps).

Hence, the language $L(G) = \{a^{2n-1} \mid (n \geq 2)\}$.

Note: The rule set $\{S \to aSa, S \to \varepsilon\}$ generates the language $L = \{a^{2n} \mid n \geq 0\}$, whereas the rule set $\{S \to aSa, S \to a\}$ generates the language $L = \{a^{2n-1} \mid n \geq 1\}$.

(b) Find the language generated by the grammar (G) with rules: $S \to aSbb \mid a$ over {a, b}, where S is the start symbol.

Solution For 0 derivation (that is, direct derivation), $w = a$: $S \Rightarrow a$.

For 1 derivation, $w = aabb$: $S \Rightarrow aSbb \Rightarrow a(a)bb = aabb$.

For 2 derivation, $w = aaabbbb$: $S \Rightarrow aSbb \Rightarrow a(aSbb)bb = aa(a)bbbb$.

Thus, a language L generated by the given grammar G is $L(G) = \{a^{n+1}b^{2n} \mid n \geq 0\}$.

EXAMPLE 4.3

(a) Consider the grammar $G = (V_N, \Sigma, P, S)$, where $V_N = \{S\}$, $\Sigma = \{a, b\}$, $P = \{S \to aSb, S \to \varepsilon\}$, S is the start symbol of G. Generate the language.

Solution The given grammar can generate the language: $L(G) = \{a^n b^n \mid n \geq 0\}$. The value of *n* may be zero, as the grammar can derive a null string ε by applying the rule: $S \to \varepsilon$.

Comment: This grammar generates the language: $L = \{a^n b^n \mid n \geq 0\}$; that is, the language contains strings with an equal number of *as* and *bs* (which appear after the *as*). Certainly, no '*a*' appears after '*b*' (strings like aba^2b^2 and aba^2b^2ab are not correct with respect to this language). However, in order to derive a language consisting of such strings, one possible rule set is: $P_1 = \{S \to SS, S \to aSb, S \to \varepsilon\}$.

(b) Construct a grammar that generates strings with an equal number of *as* and *bs*.

Solution In such a language, the positions of '*a*' and '*b*' are unpredictable. The rule sets, $P = \{S \to aSb, S \to \varepsilon\}$ and $P_1 = \{S \to SS, S \to aSb, S \to \varepsilon\}$ cannot generate strings with an equal number of *as* and *bs*. Four possible rule sets are presented below to solve the given problem.

- $P_1 = \{S \rightarrow aB, S \rightarrow bA, A \rightarrow a \mid aS \mid bAA, B \rightarrow b \mid bS \mid aBB\}$.
- $P_2 = \{S \rightarrow aSb \mid bSa \mid SS \mid \varepsilon\}$. Let $w = aababb$ be the string to be derived. The derivation process will be: $S \rightarrow aSb \rightarrow a(SS)b \rightarrow a(aSb)Sb \rightarrow aa(\varepsilon)bSb \rightarrow aab(aSb)b \rightarrow aaba(\varepsilon)bb = aababb$.
- Another possible rule set is: $P_3 = \{S \rightarrow ab \mid ba \mid aSb \mid bSa \mid SS\}$.
- One very concise rule set is: $P_4 = \{S \rightarrow aSbS \mid bSaS \mid \varepsilon\}$.

EXAMPLE 4.4

Find the language generated by the grammar (G) with rule: $S \rightarrow SS$, where S is the start symbol.

Solution $L(G) = \Phi$, since the grammar G consists of only one production rule: $S \rightarrow SS$ with no terminal symbols on the RHS (the rule cannot derive any terminal string).

EXAMPLE 4.5

Let L be the set of all palindromes[1] over {a, b}. Construct G for L.

Solution Let S be the start symbol of G. To construct a grammar for this language, the following palindrome patterns can be considered over {a, b}:

(a) ε (null) is a palindrome with length zero.
(b) *a*, *b* are palindromes (as these are strings with a single symbol).
(c) a *x* a, b *x* b are palindromes, where *x* is a string over {a, b} and is itself a palindrome.

Now, to derive ε, '*a*' and '*b*', the simple rules: $S \rightarrow \varepsilon$, $S \rightarrow a$ and $S \rightarrow b$ are necessary. On the other hand, the rules: $S \rightarrow aSa$ and $S \rightarrow bSb$ are necessary for generating palindromes like: a *x* a and b *x* b.

EXAMPLE 4.6

Construct a grammar (G) for $L = \{WcW^T \mid W \in \{a, b\}^*\}$, W^T is the reverse of *W*.

Solution Clearly, each string of the form: WcW^T has two halves: the first half which appears before terminal '*c*', and the second half which appears after the terminal symbol '*c*'. Further, the second half is exactly the reverse of the first half. Keeping these points in mind, we may frame the production rules: $S \rightarrow aSa$, $S \rightarrow bSb$. In fact, on applying these rules, the first half and its reverse in WcW^T are easily derived.

In order to generate a '*c*' in the middle of these two halves, a simple rule: $S \rightarrow c$ is essential. Thus, the complete rule set is: $P = \{S \rightarrow aSa, S \rightarrow bSb, S \rightarrow c\}$.

Validation of the rule set: Let $w = abcba$. $S \Rightarrow aSb \Rightarrow a(bSb)a$ (replacing S with bSb)
$\Rightarrow abSba \Rightarrow ab(c)ba$ (replacing S with '*c*')
Here, $W = ab$ appears before '*c*' and $W^T = ba$ after '*c*'.

EXAMPLE 4.7

Construct a grammar to generate strings that do not contain two consecutive 0s.

Solution Here, the restriction is that two consecutive 0s should not appear in a valid string. To satisfy this, we concentrate on the following:

[1] A palindrome is a word that reads the same forward and backwards. Examples are noon, radar, etc.

- Rules such as S → 0A, A → 1A are essential to prevent generation of 0s at any two consecutive places in a valid string.
- Any number of 1s may be present in any part of the string. To fulfil this requirement, the sub-rule sets: {S → 1S |1} and {A → 1A | 1} are necessary.
- Finally, the simple string '0' is also valid, and to derive it, we require the rule: S → 0. Therefore, the complete rule set of this grammar will be: {S → 0A |1S |0 |1, A → 1A |1S |1}.

EXAMPLE 4.8

Construct a grammar to generate strings with balanced parentheses; that is, each left parenthesis must have a matching right parenthesis.

Solution To satisfy the property of the given problem, the rule set of the grammar should be: {S → SS, S → (S), S → ε}. For example, the string: (( ) ( )) has balanced parentheses, and it can be generated from the following derivations performed in sequence:

$$S \Rightarrow (\,S\,) \Rightarrow (SS) \Rightarrow (\,(S)\,S\,) \Rightarrow (\,(\varepsilon)\,S\,) \Rightarrow (\,(\,)\,(\,S\,)\,) \Rightarrow (\,(\,)\,(\varepsilon)) = (\,(\,)\,(\,)\,)$$

EXAMPLE 4.9

Construct a grammar (G) for L = $\{W = xaay \mid x, y \in \{a, b\}^*\}$.

Solution Here, both x and y are substrings over {a, b}, with any combination of '*a*' and '*b*'. Also, in between x and y, one fixed substring: '*aa*' must be present in every correct string of the language L(G). In order to fulfil these properties, one possible rule set (P) is: {S → AaaA, A → aA | bA | ε}. In particular, the sub-rule set: {A → aA | bA | ε} is responsible for generating (a + b)*, which are the substrings x and y.

EXAMPLE 4.10

Construct a grammar G for L = $\{a^n b a^m \mid m, n \geq 1\}$.

Solution To simplify the problem, the valid string may be decomposed into three substrings, namely, first, second and third.

(i) The first substring contains any number of *as*, whereas (ii) the second substring has a single '*b*', which appears just after the first substring. (*iii*) Finally, the third substring contains any number of *as*, but they appear just after the '*b*' (after the second substring).

To derive the first, second and third substrings, we can consider rule subsets such as (S → A, A → aA), (A → bB) and (B → aB, B → a). However, this would be problematic because when S → A and A → bB are taken in sequence, the first substring would not contain any *a*, which is not permissible according to the given criteria.

Hence, we need to modify the first subset as: {S → aA, A → aA} to generate at least one '*a*' in the first substring of a valid string. However, the remaining subsets are correct. Thus, the complete rule set of the grammar will be: P = {S → aA, A → aA | bB, B → aB | a}.

EXAMPLE 4.11

(a) Construct a grammar for L(G) = $\{a^j b^n c^n \mid j \geq 0, n \geq 1\}$.

Solution Clearly, every valid string in L(G) can be viewed as two parts (in sequence). (i) The first part has the substring: a^j which can be generated by the rule: S → aS. (ii) The second part consists of $b^n c^n$, $n \geq 1$, and this part can be easily derived from rules such as: A → bAc | bc (as $n \geq 1$).

Further, in a^j, $j \geq 0$; so the first part of the string can also be ignored (a valid string may not have the first part at all). Therefore, we can construct the complete rule set by merging the above subsets of rules as: P = {S → aS | A, A → bAc | bc}.

(b) Give a CFG for the language L = {$a^n b^m a^n$| $n, m \geq 1$}.

Solution The rule set {S → aAa, A → aAa | bB, B → bB | ε} can generate the given language. The rules: S → aAa, A → aAa will generate an equal number of *a*s appearing before and after any number of *b*s. On the other hand, any number of *b*s can be generated by the rules: A → bB, B → bB | ε.

(c) Find a CFG for the regular expression r = (a + b)*abb.

Solution The rule set {S → AB, A → aA | bA | ε, B → abb} can generate the given language. This grammar is, indeed, a CFG. On the other hand, a rule set of regular grammar corresponding to the given RE is: {S → aS | bS | aB, B → bC, C → b}.

EXAMPLE 4.12

(a) Construct a grammar for L(G) = {$0^m 1^n$ | $m + n \geq 1$}.

Solution The basic properties of every correct string in the given L(G) are:

- All the 0s in a valid string appear before all the 1s.
- The sum of the number of 0s and 1s is always greater than or equal to 1.

The following possibilities may occur for the second property, and may be satisfied by the following rules ensuring the first property as well:

- Any correct string may have one or more 0s but no 1s at all (as it fulfils the condition $(m + n) \geq 1$). Clearly, such strings can be generated by the rules: S → 0S, S → 0.
- Any correct string may have one or more 1s but no 0s at all (as it also fulfils the condition $(m + n) \geq 1$). Such strings can be generated through the rules: S → A, A → 1A, A → 1, since no '0' appears after '1'.

Thus, to satisfy these cases, one possible complete rule set of the grammar could be:

P = {S → 0S |A | 0, A → 1A | 1}.

(b) Design the CFG for L = {$a^n b^m$ | $n > m$}.

Solution We may not design regular grammar for this language as it is necessary to remember the number of *a*s and the number of *b*s in the derived string, and that is not possible by FA. However, we may design a CFG for this language. One possible CFG rule set is P = {S → aS | aT, T → aTb | ε}.

(c) Construct a context-free grammar to accept the language L over Σ = {0, 1} such as: L = {w | w starts and ends with the same symbol}.

Solution On analysing the characteristics of the language, we may derive the RE, R = 1(0 + 1)*1 + 0(1 + 0)* + 0 +1 for this language. To ensure that the start and end symbols are identical in valid strings, one possible rule set is {S → 0A0 | 1A1|0 |1, A → 0A | 1A | ε}. However, the grammar is not regular; rather it is a CFG because of the presence of the rule: S → 0A0 | 1A1.

However, one possible regular grammar for the language is:
S → 0A | 1C |0 |1, A → 0A | 1A| 0B, B → ε C → 0C | 1C | 1D, D → ε.

(d) Find a CFG for the language L = {w | |w| is odd and its middle symbol is 0} over {0, 1}.

Solution In order to satisfy the given properties of valid strings, one possible rule set is: $\{S \to 0 \mid 0S0 \mid 0S1 \mid 1S0 \mid 1S1\}$.

(e) Design a grammar for $L = \{a^n b^m \mid n, m \geq 0\} = (a^*b^*)$ over {a, b}.

Solution On analysing the format of valid strings of L, it is clear that no *a* will appear after *b* and there is no restriction on the number of *a*s and *b*s. To satisfy the criteria, one possible rule set for this language is: $\{S \to aS \mid bB \mid \varepsilon, B \to bB \mid \varepsilon\}$.

So, the designed rule set is a rule set of regular grammar. It may be noted that we may not design any regular rule set for $L = \{a^n b^n \mid n \geq 0\}$ over {a, b}, since regular grammar (equivalently FA) does not have the ability to match the number of *a*s with the number of *b*s. This is because regular grammar is a more restricted grammar than CFG. However, CFG (equivalently PDA) has such capability, and one possible context-free rule set is: $\{S \to aSb \mid \varepsilon\}$.

EXAMPLE 4.13

(a) Design a grammar to generate strings of *a*s and *b*s such that every string has $2k$ ($k = 0, 1, 2, 3, \ldots$) number of *b*s.

Solution The rule set to generate such strings is: $\{S \to a \mid aS \mid SbSbS \mid \varepsilon\}$.

(b) Design a CFG for the language $L = \{a^n b^{n-2} \mid n \geq 2\}$.

Solution Every string in L can be expressed as $a^2(a^{n-2}b^{n-2})$. Clearly, two extra *a*s in comparison to the number of *b*s are present in every valid string. To generate the extra *a*s, rules such as $S \to aA$, $A \to aB$ are required. On the other hand, the rules $B \to aBb$, $B \to \varepsilon$ are essential for deriving $a^{n-2}b^{n-2}$.

Hence, the complete rule set is: $\{S \to aA, A \to aB, B \to aBb, B \to \varepsilon\}$.

[Another rule set that may be considered is $\{S \to aaA, A \to aAb, A \to \varepsilon\}$.]

Note: While designing the rule set for deriving any language, we must try to reduce the number of rules in the set in order to reduce the time taken by the parser for implementation.

(c) Design a regular grammar for the language $L = \{(ab)^n \mid n \geq 1\}$ over {a, b}.

Solution One possible set of rules of a regular grammar for the given language is: $\{S \to aA, A \to bS \mid b\}$. Another rule set is: $\{S \to AB, A \to aB, B \to bA \mid b\}$. This set contains four rules, whereas the earlier one contained three. So, the first is better. Further, the first rule: $S \to AB$ in the second set does not follow the format of the rules of regular grammar.

(d) Design a CFG for the language $L = \{w \mid w \in \{a, b\}^+$ and *w* does not contain *ba* as substring}.

Solution Here, the valid strings are of the form: {a, aa, aaa, …, b, bb, bbb, .., abb……, aa…bb….}, but not *ba* or *baa, ba*…. Therefore, the appropriate rule set is: $\{S \to aS \mid Sb \mid b \mid a\}$.

KEY POINT

- All type 1, 2 and 3 grammars are Type 0 grammars, since the latter is the superset of all other types. For example, $L = \{a^i b^i c^i \mid i > 0\}$ is CSL (Type 1 language), and there exists a Type 1 grammar. Obviously, it is a special case of a Type 0 grammar. Thus, every Type 1 grammar is a Type 0 grammar but the converse is not true. For example, $L = \{a^{2^n} \mid n \geq 0\}$ has no Type 1 grammar but it has Type 0 grammar.

EXAMPLE 4.14

Design a CFG for the language $L = \{a^i b^j c^k \mid i, j, k \geq 0 \text{ and } (i = j) \text{ or } (j = k)\}$.

Solution The grammar is given below. We achieve an equal number of *as* and *bs* by growing inside out, as captured by the rule M → a M b. We achieve zero or more *cs* by the rule: C → c C | ε. Most CFGs are designed using such idioms.

Thus, the complete rule set is: {S → M C | A N, M → a M b | ε, N → b N c | ε, C → c C | ε, A → a A | ε}, where S is the start symbol (non-terminal) of the grammar.

EXAMPLE 4.15

Give a CFG for L = {*w* ∈ {0, 1}* | *w* contains at least three 1s}.

Solution Let G = (V, Σ, P, S) be the CFG with V = {S, X}, Σ = {0, 1} and P = {S → X1X1X1X, X → 0X | 1X | ε}. It should be noted that the rule set with the start symbol X essentially derives (0 + 1)*.

EXAMPLE 4.16

(a) Find the CFL for the language $L = \{a^i b^j c^k \mid j = i + k\}$.

Solution To generate the given language, we may consider the following cases.

(i) $n_c(w) = 0$ (no *c* appears in *w*), but $n_a(w) = n_b(w)$ and no *a* appears after *b*. Here, $n_x(w)$ stands for the number of *x* in the string *w*. To satisfy this constraint, the possible rule subset is: $S_1 \rightarrow a\, S_1 b \mid \varepsilon$.

(ii) $n_a(w) = 0$ (no *b* appears in *w*) but $n_b(w) = n_c(w)$ and no *b* appears after *c*. To satisfy this constraint, the possible rule set is: $S_2 \rightarrow b\, S_2 c \mid \varepsilon$.

(iii) $n_a(w) + n_c(w) = n_b(w)$. This states that the number of (*as* + *cs*) = the number of *bs*. The possible final rule set is $S_0 \rightarrow aS_1bS_2 \mid S_1bS_2c \mid \varepsilon$.

(b) Give a CFG for $L = \{a^i b^j c^k \mid i, j, k \geq 0, i + j = k\}$ over {a, b, c}.

Solution Let G = (V, Σ, R, S) be the CFG with V = {S, X}, Σ = {a, b, c} and R = {S → aSc| X, X → bXc | ε}.

EXAMPLE 4.17

Give a CFG for $L = \{a^m b^n c^p d^q \mid m + n = p + q\}$ over {a, b, c, d}.

Solution Let G = (V, Σ, R, S) be the CFG with V = {S, T, X, Y, Z}, Σ = {a, b, c, d} and R = {S → T | X | Y | Z, T → aTd | X | Y | ε, X → aXc | Z | ε, Y → bYd | Z | ε, Z → bZc | ε}.

EXAMPLE 4.18

Design a simple set of CFG rules for checking the validity of the C declaration statement.

Solution A simple set of CFG rules is given below.

D → T L; *// D stands for declaration, T stands for type, L is used for variables. They are all non-terminal symbols.*

T → char | int | float | double

L → L, id

L → id // x, y, ..

Sentence: int x, y;

4.4 STRATEGIES TO SIMPLIFY CONTEXT-FREE GRAMMARS

Elimination of Useless Symbols

Non-terminal: X is a useless symbol for a grammar $G = (V_N, \Sigma, P, S)$ if it does not directly/indirectly generate any terminal symbol. Further, if it is not possible to reach a particular non-terminal (variable) Y from the start symbol of G, then Y is also treated as a useless symbol. To reduce the derivation steps, useless symbols in the grammar are removed.

EXAMPLE 4.19

Eliminate the useless symbols (if any) from the rule set: $S \rightarrow AB \mid a, A \rightarrow b$.

Solution Here, B is a useless symbol as it does not derive any terminal. So, if B is eliminated from G, the rule: $S \rightarrow AB$ is also eliminated from G. Hence, the revised rule set is: $S \rightarrow a, A \rightarrow b$.

Further, A derives a terminal symbol but it is not reachable from S. So, A is also useless, and the rule $A \rightarrow b$ is eliminated from the revised rule set. Thus, the final rule set contains only $S \rightarrow a$. Therefore, the grammar generates the language $L = \{a\}$.

Elimination of Null Productions

Null (ε) *production*: A production of the form: $A \rightarrow \varepsilon$ is known as an ε production of $G = (V_N, \Sigma, P, S)$, and it is undesirable in G, unless an empty string is derived from the start symbol of the grammar.

Suppose that a language generated from a grammar G does not contain any null string and the grammar consists of an ε-production. Then, such a production is unnecessary and should be removed from G. In fact, it causes extra steps in derivation if it is not removed from G.

Nullable variables: A nullable variable for a grammar $G = (V_N, \Sigma, P, S)$ is defined as follows:

If $A \rightarrow \varepsilon$ is a production in P, then A is a nullable variable.

If $A \rightarrow B_1B_2 \ldots. B_n$ is a production in P and all $B_1, B_2, \ldots, B_n$ are nullable variables, then A is also nullable.

For example, $B \rightarrow \varepsilon, C \rightarrow \varepsilon, A \rightarrow BC$. Here, the nullable variables are A, B and C, whereas $B \rightarrow \varepsilon$ and $C \rightarrow \varepsilon$ are null productions, but $A \rightarrow BC$ is not a null production (though A is a nullable variable).

Elimination: To eliminate any null production from a grammar G, we may apply the following strategy: For every production: $A \rightarrow \alpha$ in P (old rule set) of grammar G, add the rules to the new rule set: P_1 deleting one or more nullable variables from α. If no deletion is possible from α, add $A \rightarrow \alpha$ directly to P_1. Again, if deletion gives null production, discard the resulting null production.

EXAMPLE 4.20

Consider the rule set (P) of a grammar (G) as: $S \rightarrow ABa,\ A \rightarrow B \mid a,\ B \rightarrow b \mid \varepsilon$. Eliminate the null production from this grammar.

Solution Here, the nullable variables are A and B. The resulting productions (after eliminating ε productions) in the new rule set (P_1) are shown in the last column of Table 4.1.

TABLE 4.1 Resulting rules eliminating ε productions

S. No	Productions in P	Resulting productions in P_1
1.	$S \rightarrow ABa$	$S \rightarrow Aa \mid Ba \mid a$
2.	$A \rightarrow B \mid a$	$A \rightarrow a$
3.	$B \rightarrow b \mid \varepsilon$	$B \rightarrow b$

Unit Production and Its Elimination: A production of the form: $A \rightarrow B$ is called unit production. Here, both A and B are non-terminals (variables). If such a production exists in a rule set, it must be removed; otherwise extra derivation steps will be required. However, unit productions are often useful in cases such as for creating unambiguous grammar from ambiguous grammar and deriving correct strings satisfying the language constraints.

The following steps are applied to remove unit productions from a rule set (P) of grammar (G).

Step 1: Remove all the productions of the form: A → A (a special case of unit production) from P.

Step 2: Find all the other unit productions of P and apply Step 3.

Step 3: Construct dependency graph(s) (if possible) for the variables (Fig. 4.2 (a)).

Step 4: Modify the non-unit productions from the dependency graph(s) to remove the unit productions (as explained below).

From Fig. 4.2 (a), edge: B → C implies that C can be obtained from B. More explicitly, all non-unit productions generated by C can also be generated from B. Likewise, edge: A → B implies that all non-unit productions generated by B can also be generated from A.

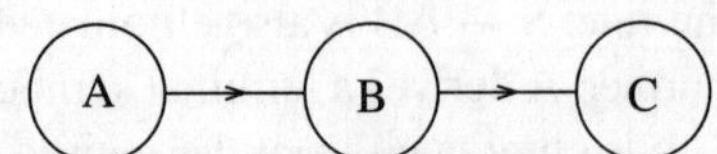

FIGURE 4.2 (a) A dependency graph for the variables A, B and C

Finally, A → B → C, that is, A → C, implies that all non-unit productions generated by B and C can be generated from A.

EXAMPLE 4.21

Eliminate all unit productions from the rule set given below.

$$P = \{S \rightarrow AB, A \rightarrow a, B \rightarrow C \mid b, C \rightarrow D, D \rightarrow E \mid bC, E \rightarrow d \mid Ab\}$$

Solution The non-unit productions in the rule set are shown below.

$$S \rightarrow AB, A \rightarrow a, B \rightarrow b, D \rightarrow bC, E \rightarrow d \mid Ab$$

The unit productions in the rule set are shown below.

$$B \rightarrow C, C \rightarrow D, D \rightarrow E$$

The dependency graph for these unit productions is shown in Fig. 4.2 (b).

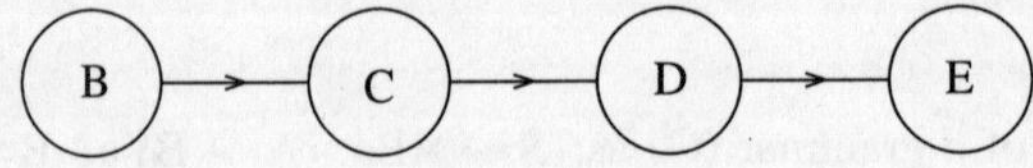

FIGURE 4.2 (b) A dependency graph for the variables B, C, D and E

From Fig. 4.2 (b), $D \rightarrow E$. So, the non-unit productions from E can be generated from D too. Here, the non-unit productions from E are $E \rightarrow d \mid Ab$. So, these can be generated from D; that is, add $D \rightarrow d \mid Ab$ to the set of rules. Thus, the resulting D productions are: $D \rightarrow d \mid Ab$, $D \rightarrow bc$ (self).

Again, from the dependency graph: we obtain $C \rightarrow D$, $D \rightarrow E$. This implies that $C \rightarrow E$. So, the non-unit productions from E can be generated from C. Hence, $C \rightarrow d \mid Ab$ is added as a C production to the rule set. Further, $C \rightarrow D$. So, the non-unit productions from D can be generated from C. Therefore, $C \rightarrow bC$ (since $C \rightarrow D$) is added as a C production. Thus, the resulting C productions are: $C \rightarrow bC$ $C \rightarrow d \mid Ab$.

Finally, from the dependency graph, we get, $B \xrightarrow{*} C$, $C \xrightarrow{*} D$ and $D \xrightarrow{*} E$. This implies that $B \xrightarrow{*} E$. So, the non-unit productions from C, D and E can also be generated from B. Thus, the resulting B productions are: $B \to bc$, $B \to d \mid Ab$, $B \to b$ (self).

Hence, the resulting rule set

$$P_1 = \left\{ S \to AB,\ A \to a,\ B \to b \left| d \right| Ab \left| bC,\ C \to bC \right| d \left| Ab,\ D \to bC \right| d \mid Ab,\ E \to d \mid Ab \right\},$$

where S is the start symbol.

4.5 CONVERSION OF FINITE AUTOMATA TO REGULAR GRAMMAR AND REGULAR EXPRESSION TO REGULAR GRAMMAR

In this section, we discuss the strategies used to convert FA and RE to regular grammar.

4.5.1 Designing Right-Linear Regular Grammar (RLRG) from DFA

Regular languages are recognised by DFAs. Hence, there must exist a regular grammar for each DFA. To generate a regular grammar for a given DFA, the following steps are adopted.

Step 1: Introduce one distinct non-terminal symbol for each state of the given DFA.

Step 2: Add the rule(s) for each edge of the DFA as follows:

2.1: $A_i \to aA_j$ for $\delta(q_i, a) = q_j$. Here, A_i and A_j are the non-terminal symbols introduced for q_i and q_j. Note that q_i and q_j may be the same state as well.

2.2: If $q_i \in F$, add rule: $A_i \to \varepsilon$ to the rule set.

Based on this method, we may generate a rule set (P) for the given DFA (shown in Fig. 4.3 (a)).

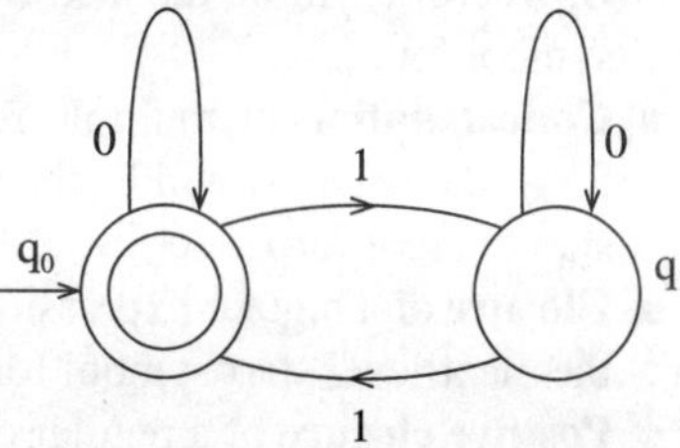

FIGURE 4.3 (a) A DFA to accept strings with an even number of 1s

$P = \{A_0 \to \varepsilon, A_0 \to 0A_0, A_0 \to 0, A_0 \to 1A_1, A_1 \to 0A_1, A_1 \to 1A_0, A_1 \to 1\}$, where A_0 and A_1 are the non-terminal symbols introduced for states q_0 and q_1, respectively (with A_0 as the start symbol of the derived grammar).

Another example is given in Fig. 4.3 (b).

The resulting rule set derived from the state graph (shown in Fig. 4.3 (b)) is: $P = \{A_0 \to \varepsilon, A_0 \to aA_1, A_1 \to b, A_1 \to bA_0, A_0 \to bA_2, A_1 \to aA_2, A_2 \to aA_2, A_2 \to bA_2\}$, where A_0, A_1 and A_2 are used for q_0, q_1 and q_2, respectively. One significant point is that the non-terminal symbol A_2 (introduced for q_2) does not directly/indirectly derive any terminal string. Hence, this is a useless variable. Accordingly, $A_0 \to bA_2$, $A_1 \to aA_2$, $A_2 \to aA_2$, $A_2 \to bA_2$ are useless and we can discard them from the rule set.

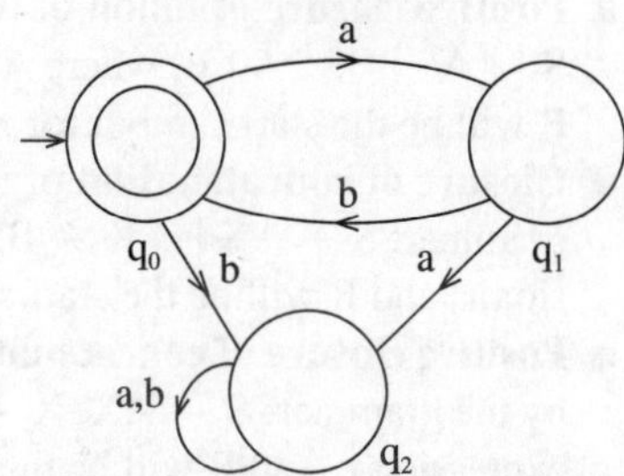

FIGURE 4.3 (b) A DFA for $L = \{(ab)^n \mid n \geq 0\}$

Note: There is a unique minimal DFA for every regular language. In fact, any regular language may have multiple sets of rules but only one (among these) is the minimum set. Hence, the DFA with the minimum set of rules is the minimal FA.

4.5.2 Designing Left-Linear Regular Grammar (LLRG) from DFA

To find LLRG from DFA, we can use the following steps:

1. First, perform reversal of the DFA (as discussed in Chapter 3).
2. Find the left-linear rules.
3. Convert them directly to right-linear rules, for example, $A_i \rightarrow aA_j$ to $A_i \rightarrow A_ja$.

Generating Regular Grammar from Regular Expression: As discussed in Chapter 3, regular expressions generate regular languages. We have already seen that Type 3 (or regular) grammars generate regular languages. So, logically it must be possible to generate a grammar directly from a regular expression. Here, we discuss the technique for converting a regular expression to a regular grammar.

- A regular expression r, made up of **only terminal symbols** and no operator, can be directly written as: S → r; for example, simple expressions such as *a*, *ab* and *abc* can be expressed by a single rule grammar: S → a, S → ab, S → abc.
- The union of two regular expressions implies a **bifurcation/split** of the first grammar rule. Suppose a regular expression is expressed as $r_1 + r_2$. In this case, the grammar will be of the form: S → A | B, where A will be the next-level start symbol for the regular expression r_1 and B will be the start symbol for r_2.
- **Concatenation** of multiple regular expressions such as $r_1.r_2$ can be expressed by the grammar: S → AB, where A will be the next-level start symbol for the regular expression r_1 and B will be the start symbol for r_2.
- **Closure** of a regular expression, r^*, can be expressed by the grammar: S → AS | ε, where A will be the next-level start symbol for the regular expression r.
- **Positive closure** of a regular expression, r^+, can be expressed by the grammar: S → AS | A, where A will be the next-level start symbol for the regular expression r.
- **Closure of union** of two regular expressions $(r_1 + r_2)^*$ can be expressed by the grammar: S → AS | BS | ε, where A will be the next-level start symbol for the regular expression r_1 and B will be the start symbol for r_2.
- **Positive closure of union** of two regular expressions $(r_1 + r_2)^+$ can be expressed by the grammar: S → AS | BS | A | B, where A will be the next-level start symbol for the regular expression r_1 and B will be the start symbol for r_2.
- **Closure of concatenation** of multiple regular expressions such as $(r_1.r_2)^*$ can be expressed by the grammar: S → AX | ε, X → BS, where A will be the next-level start symbol for the regular expression r_1 and B will be the start symbol for r_2.
- **Positive closure of concatenation** of multiple regular expressions such as $(r_1.r_2)^+$ can be expressed by the grammar: S → AX, X → BS | B, where A will be the next-level start symbol for the regular expression r_1 and B will be the start symbol for r_2.
- In many cases, observing the regular expression and minimising it using the standard algebraic laws of regular expressions makes it easy to derive the grammar.

For all practical purposes, these rules are sufficient to generate a grammar from any valid regular expression.

Please refer to **Annexure I** for converting a right-linear regular grammar to NFA.

EXAMPLE 4.22

(a) Give an equivalent grammar for the RE: a* + b*.

Solution This is a union of two regular expressions a* and b*. The first rules are: S → A | B, where further rules are needed for A to generate a*, and for B to generate b*. To generate a* with A as the start symbol, the required rules are A → aA | ε. Similarly, to generate b* with B as the start symbol, the required rules are B → bB | ε.

The complete grammar is: S → A | B, A → aA | ε, B → bB | ε. This grammar is in GNF (discussed later in this chapter).

(b) Give the equivalent grammar for the RE: $(a + b^*a)^+c^*$.

Solution This is a concatenation of two regular expressions $(a + b^*a)^+$ and c*. The first rule is: S → PQ, where P will derive the regular expression $(a + b^*a)^+$ and Q will derive the regular expression c*. To generate c* with Q as the start symbol, the required rules are: Q → cQ | ε.

The regular expression $(a + b^*a)^+$ is a positive closure of the regular expression (a + b*a). We need to add a new set of rules: P → XP | X, where X will derive (a + b*a).

(a + b*a) is logically equivalent to b*a (applying the laws of regular expressions). b*a is again a concatenation of two regular expressions b* and a. So, we add the rule: X → YZ, where Y will derive b* and Z will derive the symbol *a*. We add the rule: Z → a. To generate b* with Y as the start symbol, the required rules are Y → bY | ε.

The complete grammar is: S → PQ, P → XP | X, X → YZ, Y → bY | ε, Z → a, Q → cQ | ε.

This grammar is in Chomsky Normal Form (CNF) (discussed later in this chapter).

Note: Readers may have noticed that the symbol Z is redundant. We could have directly written the rule as X → Ya and got rid of the non-terminal symbol Z. However, if we do so, the grammar will not be in CNF or GNF.

(c) Give the equivalent grammar for the RE: (00)*(0 + ε)(11)*.

Solution We need to apply a bit of observation here and simplify the RE. Here, (00)*(0 + ε) represents a sequence of an odd number of 0s, while (11)* represents a sequence of an even number of 1s.

The language specified by this RE is $\{0^{2m+1}1^{2n}$, where $m \geq 0$, $n \geq 0\}$. Using observation, we can write the grammar as: S → AB, A → 00A | 0, B → 11B | ε.

Note: It is possible to derive an NFA or DFA from an RE. Hence, readers may also try to convert RE to DFA/NFA and then apply the technique of generating left-linear or right-linear grammar from it.

4.6 DERIVATION/PARSE TREE

Though the term derivation is applicable to all four types of languages, it is usually assumed to be appropriate for context-free language. The derivation process of a string can be described in the form of a tree. In other words, it is the graphical representation of the derivation sequence of the sentential forms. It helps us to find out if a given string/statement is syntactically correct or not. It also helps us to understand the semantics, that is, we can determine how an expression (derived from the expression grammar) should be evaluated using the tree. It is important to note that we consider the derivation tree only for CFG.

A derivation tree for a CFG, (G) = (V_N, Σ, P, S), is one that satisfies the following **properties**:

- Every vertex has a label that represents a non-terminal (variable) or a terminal or ε.
- The root has label S (start symbol of the grammar).
- The label of each leaf vertex is either a terminal symbol or ε.

- The label of each internal (interior) vertex is a non-terminal symbol. If an interior node is labelled A, its children are labelled $X_1, X_2, \ldots, X_n$ from the left, if and only if rule: $A \to X_1 X_2 \ldots X_n$ exists, where each X_i ($i = 1, \ldots n$) may be a terminal or a non-terminal.

So, at every internal vertex, the rule corresponding to the vertex label is applied; that is, the vertex label represented by the LHS of the rule is expanded. Thus, every non-terminal vertex represents the LHS of a rule.

A **path** in a parse tree is a sequence of distinct nodes from the root to the leaf. The number of line segments (arcs) in a path is its **length** and the height of a parse tree is the **length of the longest path** in the tree.

KEY POINTS

- The children of a node in a tree are ordered from left to right. The same strategy is followed here as well. Any parse tree conventionally follows leftmost derivation (scanning of the string is from left to right).
- The concept of parse tree and rightmost derivation is that it is essentially the mirror image of left-most derivation. More specifically, we need to scan the string from right to left in this case.
- For every parse tree, there is a unique leftmost and a unique rightmost derivation.
- The concatenation of the labels of the leaves from left to right gives the derived string (w), and the left-to-right order shows the pre-order traversal.

Note: An input statement (a string in formal version) is valid with respect to a given grammar G if it is appropriately placed in a derivation tree that follows the rules of G.

Derivation Tree for CFG: CFG is considered in the derivation tree (parse tree) because a node of the tree may not represent multiple symbols (taken from sets of terminal or non-terminal symbols). If it is a non-terminal, it is the LHS of any rule. In fact, the LHS of each rule in CFG has only one symbol, and it is strictly a non-terminal symbol. More precisely, the parse tree for CFG is transitive but it is not true for CSG.

Further, every regular grammar is CFG; so the derivation tree is applicable for regular grammar too.

Let us consider the rules of a grammar G as: $S \to aS \mid ab$. Suppose we need to derive the string $w = aab = a^2b$. Now, the derivation: $S \Rightarrow aS \Rightarrow aaS \Rightarrow aab$ can be represented as shown in Fig. 4.4 (a). We call such a picture, a parse tree.

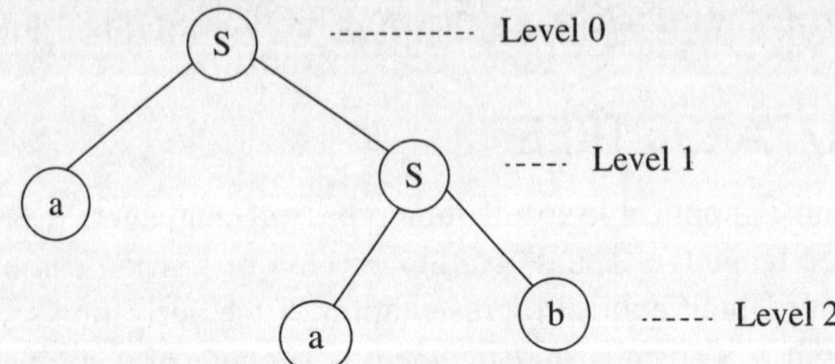

FIGURE 4.4 (a) A derivation tree for $w = aab$

Here, S is the start symbol of G, and it is placed at the root. The current input of the string $w = aab$ is 'a'. At label 0, the root: S is expanded as aS to derive the input: 'a'. However, you can expand S as: $S \to ab$, but the second input: 'a' of aab would not match in that case. If it is expanded in this way, backtracking is necessary to derive the target string.

S is now at level 1 of the tree, which is an internal node (when root S is expanded as aS), and the rule: $S \to ab$ associated with S is expanded to derive the symbols 'a' and 'b'. Thus, terminal symbols 'a', 'a'

and '*b*' are placed at the leaf nodes. Lastly, concatenating the labels of the leaves from left to right, we obtain the derived string $w = aab$. This is called the **yield** of the parse tree.

Note: If rightmost derivation is used here to construct the tree, the adopted derivation does not generate a structurally different tree.

Let us consider another grammar (G) with rules: G → E, E → E *Op* E | *id*, *Op* → + | – | * and $w = x - 2 * y$. Two parse trees (one following leftmost derivation and the other following rightmost derivation) are shown in Fig. 4.4 (b) for *w*. The given grammar is, indeed, an expression grammar, where E stands for expression (the generic expression) and *Op* stands for operator.

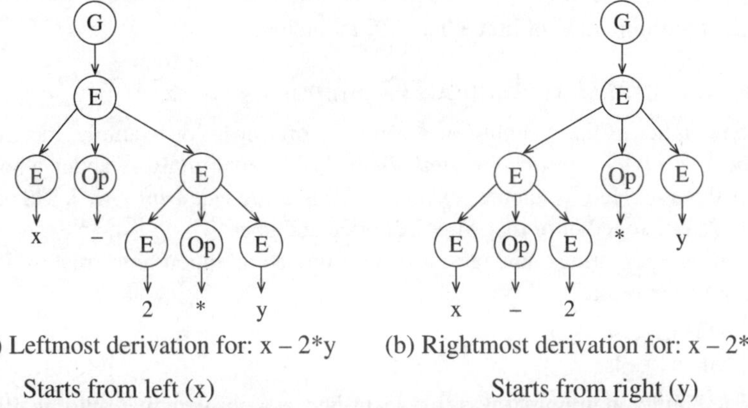

(a) Leftmost derivation for: x – 2*y (b) Rightmost derivation for: x – 2*y

Starts from left (x) Starts from right (y)

FIGURE 4.4 (b) Leftmost and rightmost derivations

Note: Here, the leftmost and rightmost derivations result in two structurally different trees for the same string.

4.6.1 Parsing Techniques

Parsing or syntax analysis is the process of analysing (or inferring) a string of symbols of a language – more specifically, a stream of tokens/words with respect to a programming/natural language – conforming to the rules of a formal grammar. Two approaches are usually adopted in this respect: recursive inference and derivation. The more conventional **recursive inference** uses rules from the body to the head (reduction), whereas **derivation** uses the reverse strategy (head to body, beginning at the start symbol of the grammar). Thus, derivation is a strategy for parsing strings where rules are used to produce/derive the string.

From a programming language point of view, a **parser** for a grammar G is a program that takes *w* (a string) as input and produces as output either a parse tree for *w* if it is a sentence of G or an error message, indicating that *w* is not a sentence of G. Two types of parsers are commonly used to construct compilers: top-down and bottom-up .

The **top-down parser** attempts to construct the derivation tree (top-down parse tree) of an input string, starting from the root (labelled by the start symbol of the grammar) and moving towards the bottom. Here, the non-terminal symbols at intermediate nodes are replaced (expanded) by the RHS of the rules associated with the respective symbols.

In **bottom-up parsing**, analysis of the string starts at the bottom and moves towards the root. A kind of reduction technique (instead of expansion) is followed here. This process can also be represented by a tree structure, termed the bottom-up parse tree.

In this book, we will concentrate on derivation strategies for parsing.

Problems in Derivation: A list of the issues commonly encountered by the parser during derivation/parsing is given below:

- *Conflict in derivation due to ambiguous grammar*: It causes conflict in the meaning of the sentence/string.
  - The solution is conversion of ambiguous grammar to unambiguous grammar.
- *Backtracking due to the presence of a common prefix in rules*: It leads to increase in the time taken to parse the string.
  - It can be avoided by applying left-factoring strategy.
- *Falling into an infinite loop due to the presence of immediate left-recursive rules in the rule set*:
  - The solution is elimination of immediate left-recursion.

4.6.2 Ambiguous and Unambiguous Grammar

Ambiguity is the conflict in the meaning of a sentence or word. For example, the English sentence: "*John gave a bath to his dog wearing a pink T-shirt*". The ambiguity is about who is wearing the T-shirt – John or the dog. Here is another example: "*Amina ate rice with ghee*". This could mean that Amina and ghee ate rice together or that Amina ate rice and ghee.

In the context of programming languages, a very common ambiguous construct is "if *b* then if *b* then *s* else *s*". This can be viewed as:

- If *b* then (if *b* then *s*, else *s*) or
- If *b* then (if *b* then *s*), else *s*

The concept of ambiguity in grammar was first identified as a problem by Cantor in 1962 and inherent ambiguity by M Gross in 1964.

A CFG grammar (G) is said to be **ambiguous** if, for at least one string: *w*, there exists:

- More than one LMD, or
- More than one RMD, or equivalently
- More than one distinct (structurally different) parse tree that follows the same derivation (either LMD or RMD).

Note: In Computer Science, we will talk about ambiguity of CFG. In the case of ambiguous grammar, if multiple parse trees are generated for a string (*w*) following the same derivation (either LMD or RMD), then multiple parse trees are generated for *w* following the two different derivations. However, LMD is usually adopted for checking ambiguity of grammar. Ambiguity causes problems to the parser, because the parser becomes confused as to which parse tree it needs to generate. On the other hand, an **unambiguous grammar** is a grammar for which every valid string has a unique leftmost derivation (equivalently unique rightmost derivation). In particular, for unambiguous grammar, we can have different LMD and RMD for a string, but we cannot have different leftmost and rightmost derivation trees for it. For example, consider the unambiguous grammar with rule set: $\{S \rightarrow AB, A \rightarrow a, B \rightarrow b\}$ and w = ab. LMD for w: $S \Rightarrow AB \Rightarrow aB \Rightarrow ab$; RMD for w: $S \Rightarrow AB \Rightarrow Ab \Rightarrow ab$. Truly, LMD and RMD vary here, although derivation trees are identical if drawn. Thus, it should be noted that the existence of multiple derivations (LMD and RMD) of the same string does not suffice to indicate that the grammar is ambiguous; only multiple leftmost/rightmost derivations (or, equivalently, multiple parse trees) indicate ambiguity.

Many languages admit both ambiguous and unambiguous grammars, while some admit only ambiguous grammars. The only language without ambiguous grammars is the empty language.

EXAMPLE 4.23

Let G = ({S}, {a, b, *, +}, P, S), P = {S → S + S | S * S | a | b}. Here '*' represents the multiplication operator. Prove that this grammar is ambiguous.

Solution The given grammar is, indeed, an expression grammar. Suppose we need to generate $w = a + a * b$ from this grammar. Here, we obtain more than one parse tree (Figs 4.5 (a) and (b)) for w.

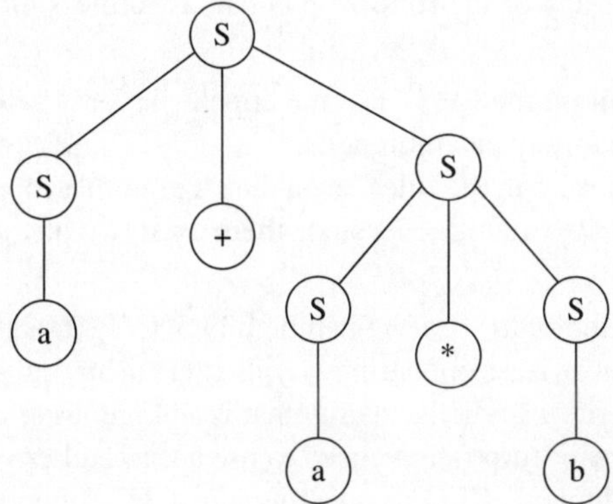

FIGURE 4.5 (a) A parse tree corresponding to $a + a * b$

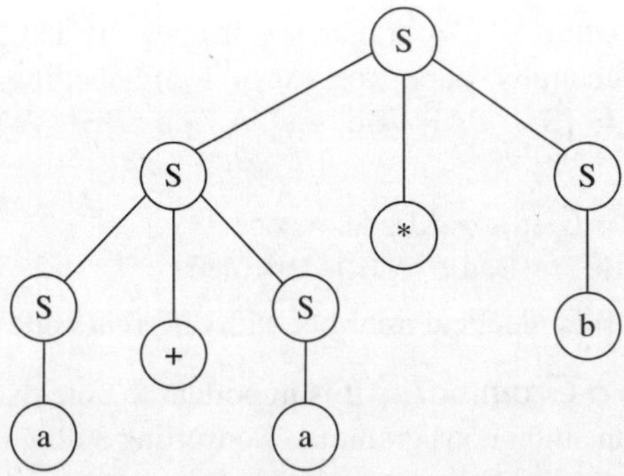

FIGURE 4.5 (b) Another parse tree corresponding to $a + a * b$

Inherently Ambiguous Grammar: In the above examples, the ambiguity arose from the grammar, but it could be removed by finding an equivalent unambiguous grammar. In some instances, however, this is not possible because the ambiguity is in the language itself. So, if there exists a language L, for which there is no unambiguous grammar, the language is called an **inherently ambiguous language**, and the grammar from which the language is derived is called an **inherently ambiguous grammar**.

Language L = $\{a^n b^n c^m d^m \mid m, n \geq 1\}$ is an example of an inherently ambiguous language. One possible rule set for this language is: S → AB | C, A → aAb | ab, B → cBd | cd, C → aCd | aDd, D → bDc | bc. The set is of ambiguous grammar. It can be easily verified by taking string $w = abcd$. Two different parse trees can be constructed to derive this string.

Disadvantages of Ambiguous Grammar: One major demerit is that due to the presence of ambiguous grammar in the compiler, the parser can select any one out of multiple possible parse trees. Accordingly, undesired output may be obtained for any expression (string); that is, it causes non-determinism in meaning.

For instance, in Example 4.23, if the values of 'a' and 'b' are chosen as 3 and 4, respectively, and the parser selects the tree shown in Fig. 4.5 (a), the exact value, 15, is computed for the expression: $3 + 3 * 4$ (the lowermost sub-tree derives the sub-expression: $a * b$, which is evaluated first). On the other hand, if the parse tree shown in Fig. 4.5 (b) is selected by the parser for the same expression, it gives undesired output, 36, for $3 + 3 * 4$ (the lowermost sub-tree here represents the sub-expression: $a + b$, which will be evaluated first).

Usually, ambiguity arises if any of the following occurs regardless of any grammars (expression or others):

- If any non-terminal present on the LHS of a rule appears at least twice on the RHS of the same or different rules, the grammar may be ambiguous.
- If no such occurrence is present in the rules, even then a grammar may be ambiguous. For example, S → aS | X, X → aX | a is also ambiguous, since there exist two distinct parse trees for a string, say $w = aaaa$.

Thus, the most important criterion for verifying ambiguity of CFG (G) is to check whether the grammar derives multiple parse trees for at least one string (w) in L(G) using the same derivation. However, no standard algorithm exists for verifying whether a grammar is ambiguous or not (to find the string for which multiple parse trees exist). For this purpose, we need to use a trial-and-error approach. However, it is not always possible to convert ambiguous CFG to unambiguous CFG. Importantly, for every unambiguous grammar, there is an equivalent ambiguous grammar, but the converse is not true. Let us consider a rule set of unambiguous CFG: {S → aSb | ab}. Its equivalent ambiguous grammar is: { S → aSb | ab | T, T → ab}.

Structurally Ambiguous Grammar: A grammar is structurally ambiguous if it generates the same string twice with the same (structurally) parse tree, except for relabelling of nodes.

For example, the rule set P = {S → dAB | Be | Ae|, A → e | Bed | AAd, B → d | Bed | Ae}, where Σ = {d, e}. Let $w = dede$.

- The first LMD is: S ⇒Ae ⇒(Bed)e ⇒(d)ede = dede
- The second LMD is: S ⇒Be ⇒ (Bed)e ⇒ (d)ede = dede

These two derivations construct identical trees but with different node labels.

Removing Ambiguity from a Grammar: It is important to note that there is no standard method for removing ambiguity from an ambiguous grammar. Converting ambiguous to unambiguous grammar requires practice (trial and error), and so it is unsolvable (please see Chapter 7). We may note that the grammars (especially expression grammars) for real-world programming languages are often ambiguous due to absence of specification of associativity of the operators or precedence of the operators or both. Ambiguity in expression grammar may be removed by employing the following methods:

- Explicitly mentioning the priority and associativity of the operators, but not changing the rule set
- Transforming the rules into another set of rules

Note: For ambiguous grammars other than expression grammars, making it unambiguous depends on analysis. For example, the ambiguous grammar: S → aS | X, X → aX | a, can be made unambiguous as: S → aS | a. Also, converting ambiguous expression grammar into the equivalent unambiguous grammar usually increases the size of the original rule set. However, the size of the rule set in unambiguous grammar (for grammar other than expression grammar) may also decrease in comparison to the size of the rule set in ambiguous grammar.

Removing associativity: The concept of associativity comes into play when the same operator or different operators with same precedence appear on both sides of an operand in an expression.

Therefore, taking a decision about the order of evaluation becomes difficult. Also, undesired results may be obtained. For example, ($num_1 - num_2 + num_3$) (say, 7 – 5 + 3): here, the problem is whether ($num_1 - num_2$) or ($num_2 + num_3$) will be evaluated first.

Let us take an ambiguous grammar without explicitly specifying the associativity of the operators: {E → E + E | E – E | num}. Here, precedence of '+' and '–' is the same. Non-terminal: E stands for expression.

Let w = 7(num) – 5(num) + 3(num).

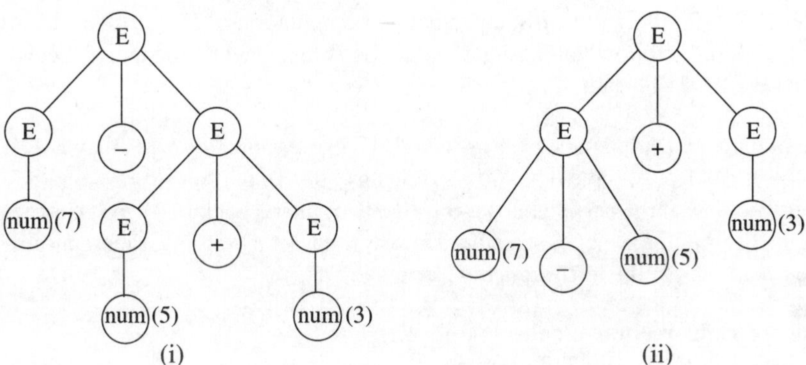

FIGURE 4.5 (c) Ambiguous grammar and two parse trees using leftmost derivation for 7 – 5 + 3

Obviously, two parse trees are generated to validate the same expression, but one is correct while the other is not, from the evaluation point of view. This occurs because of the ambiguous grammar in which associativity is absent.

The first tree gives an undesired result (7 – (3 + 5)) = –1), whereas the second gives the correct output (7 – 5) + 3 = 5, as the lowermost sub-trees are evaluated first. Placing the expression at the lower level of the derivation tree means placing it at the top of the stack. Hence, it will be reduced first by the bottom-up parser and will therefore be evaluated earlier as compared to other sub-expressions placed in the lower part of the stack. Interestingly, if the expression contains only operator: +, all the derivation trees generated by this grammar would give the correct result, but the trees (with respect to the grammar) would still be problematic (conflicting) to the parser (which needs to choose one tree).

The correct parse tree may be deterministically selected if the grammar here is restructured based on left-associativity. More specifically, the issue of associativity can be resolved by the use of the **left-recursive rule** (for maintaining left-associativity). Such recursive rules explicitly define the associativity of the operators. For instance, the rules: E → E + E | E – E | num (where associativity is not defined) may be restructured for this purpose as: {E → E + T | E – T | T, T → num} or {E → E + num | E – num | num} (based on left-associativity). Here, T (for term) is the new non-terminal symbol introduced in the set.

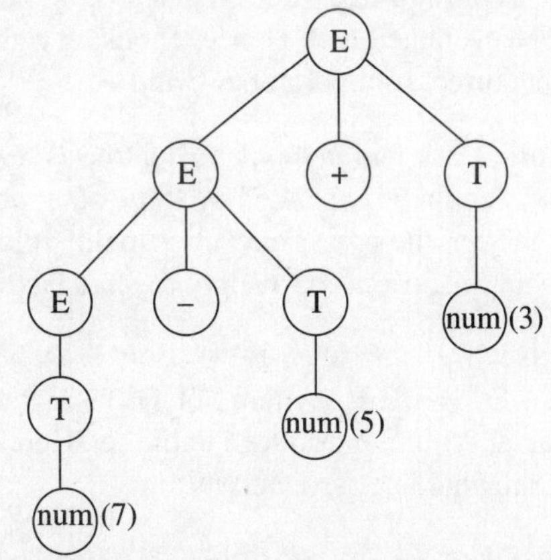

FIGURE 4.5 (d) Left-recursive grammar and leftmost derivation tree for 7 – 5 + 3

Derivation of expression from left-recursive grammar: Two points are followed in this regard:

- Scanning of expression takes place from left to right.
- Leftmost derivation is adopted.

The rules will be expanded only on the left side of the parse tree following left-associativity. The expression: 7 – 5 + 3 can be derived easily by applying the rules specified above. The derivation is illustrated using a derivation tree (shown in Fig. 4.5 (d)).

The expression is 7 – 5 + 3. Here, the first input is 7, the next input is '–', and the third is 5. In Fig. 4.5 (d). 7, '–' and 5 are placed at the lower levels of the tree, when compared to the '+' and 3 appearing after 5. So, (7 – 5) will be evaluated before computation of (5 + 3). This clearly means that the operators: '+' and '–' follow left-associativity and it is managed by the left-recursive rule:

$E \rightarrow E + T \mid E - T \mid T, T \rightarrow num.$

Note: If we initially expand the tree as: $E \rightarrow E - T$, the expression: 7 – 5 + 3 may not be validated, as T cannot be expanded by some other non-terminals to derive the sub-expression: (5 + 3), which appears after 7 –. Now, if the parser chooses this rule first, then it backtracks to select the alternative path in order to derive the exact expression. However, backtracking is a time-consuming job. To avoid it, we need to apply the **left-factoring strategy**.

Let us take the **right-recursive rules** that give right-associativity: $E \rightarrow T + E \mid T - E \mid T, T \rightarrow num.$

Derivation of expression from right-recursive grammar: In this connection, the following basic points are considered:

- Scanning of expression takes place from left to right
- Leftmost derivation is followed.

The derivation tree for the expression: 7 – 5 + 3 is shown in Fig. 4.5 (e). It is interesting to note that following right-associativity means evaluating the right sub-expression, (5 + 3) first, before evaluating (7 – 5).

Undoubtedly, an incorrect result will again be obtained here, as the sub-tree: (5 + 3) placed at the lowermost level will be evaluated first. Hence, choosing right-associativity is not correct for operators: + and –.

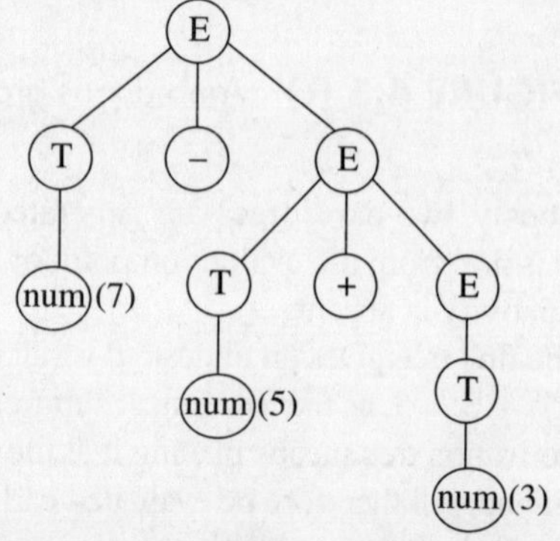

FIGURE 4.5 (e) Right-recursive grammar and rightmost derivation tree for 7 – 5 + 3

Note: If we initially expand the tree as: $E \rightarrow T + E$, the expression: 7 – 5 + 3 may not be validated, as T cannot be expanded by some other non-terminals to derive (7 – 5), which is desired before +3. However, the parser may choose this rule first and then backtrack to select the alternative path. To avoid this, we need to apply the left-factoring strategy.

Likewise, the rule: $E \rightarrow E * E$ (where associativity is not defined) may be restructured as: $\{E \rightarrow E * F \mid F, F \rightarrow num\}$ or $\{E \rightarrow E * num \mid num\}$ (based on left-associativity). F is the new non-terminal symbol introduced in the set. Here also, the rules are expanded only on the left side of the parse tree showing left-associativity.

Note: If we take the left-recursive rules for expression grammar, then left-associativity for the operator in the rules with the same precedence is adopted, and so the left sub-expression of an operator is evaluated first, as it is placed in the lowermost part of the derivation tree. Likewise, it is true for the right-recursive rules too.

Removing the Problem of Precedence: The concept of precedence of operators in an expression is important when operators with different precedence appear in that expression. In such a case, it is essential to maintain the order of evaluation according to the precedence level of the operators; otherwise, the value computed from the expression may be incorrect.

From the compiler's perspective, to overcome the issue of precedence of operators, we define precedence level-wise in the rules, where the rules associated with higher precedence appear at lower level in the rule set. Accordingly, the parse tree is built by placing the higher precedence operators at the lower level of the tree, where it will be evaluated before the operators placed at the upper level of the tree.

For example, the rule set (P) = {E → E + E | E * E | num} can be restructured by explicitly defining associativity and precedence as:

P_1 (new rule set) = {E → E + T (left associativity), E → T,
T → T * F (left associativity but precedence of '*' is higher than that of '+'),
T → F, F → num}

Here, T stands for term, F denotes factor (a multiplicative part) and num denotes the terminal (number).

Note that an expression may contain only one type of operator. Then the unit productions: E → T and T → F handle the operation.

Conversion from one set to another aims to preserve priority as well as the associativity of the operations, but the number of rules increases in the new set.

Let us attempt to consider the exponent operator (denoted as ↑) in the previous expression grammar. We know that exponent follows right-associativity in the expression. For example, in 2^{2^3}, 2^3 is first evaluated (2^{2^3} is equivalent to 2^8), then 2^8 is evaluated. So, in order to evaluate such an expression, we may update the above-presented expression grammar as:

E → E + T | E – T (left associative), E → T, T → T * F (left associativity but precedence of '*' is higher than that of '+' and '–'), T → F, F → G ↑ F (here, G ↑ F results in right associativity, as F → G ↑ F is a right-recursive rule), F → G, G → num.

A more complete rule set of expression grammar can be shown as:

E → E + T | E – T (left-associativity), E → T, T → T * F |T / F (left-associativity but precedence of '*' and '/' is higher than that of '+' and '–'), T → F, F → G ↑ F (here, G ↑ F results in right-associativity, as F → G ↑ F is a right-recursive rule), F → G, G → num.

Note: Instead of taking num (number) as terminal, we may take id (identifier) as the terminal token that can represent both the value (number) and the variable, for example, G → id (instead of G → num).

Validation of Postfix Expression: Thus far, we have shown the grammar for infix expressions and its validation. Now, we discuss the grammar for postfix expressions.

A grammar (G) with a rule set (P) = {E → EE+ | EE* | id} for checking the validity of postfix expression is not ambiguous, as multiple parse trees for a valid postfix expression may not be derived following either LMD or RMD. For example, let us take a valid postfix expression: 3 4 5 * +.

To derive it, we may follow the RMD as:

E ⇒ EE+ ⇒ E(EE*)+ ⇒ EE(id = 5)*+ ⇒ E(id = 4)5*+ ⇒ (id = 3)4 5*+ = 3 4 5*+.

Surely, we may not derive multiple RMDs for 3 4 5 * + and for any other string too. As multiple RMD trees for any postfix string is possible, it will be true for LMD trees as well.

EXAMPLE 4.24

(a) Consider the rule set of the Boolean grammar: P = {$E_b \rightarrow E_b$ OR E_b | E_b AND E_b| NOT E_b | True | False}. Find its equivalent unambiguous grammar.

Solution Here, the associativity and precedence of the operators have not been defined, and the given grammar is, indeed, ambiguous. Now, based on left-associativity for the Boolean operators OR and AND and right-associativity for NOT, and maintaining precedence of the operators in ascending order as: OR, AND, NOT, we may restructure the rules as:

E → E OR F (OR as left-associativity), E → F, F → F AND G (AND has left-associativity but precedence of AND is higher than that of OR), F → G, G → NOT G (NOT has right-associativity and higher precedence than AND), G → True | False.

(b) Consider the following grammar and check the associativity and precedence of the operators: P = {E → E*F | F + E | F, F → F – F | id}.

Solution Here, E → E * F means that '*' follows left-associativity, whereas E → F + E means that '+' follows right-associativity. Again, from E → E * F and E → F + E, we may say that '+' and '*' have the same level of precedence. Obviously, precedence of '+' and '*' is not explicitly defined. Further, the rule: F → F – F reports that associativity for '–' is undefined. Hence, the given grammar is ambiguous.

For string: id + id * id, the rules: E → E * F | F + E | F, F → id result in two parse trees following leftmost derivation:

1. E ⇒ E * F ⇒ F + E * F ⇒ id + E * F ⇒ id + F * F ⇒ id + id * F ⇒ id + id * id (since E ⇒ F and F ⇒ id).
2. E ⇒ F + E ⇒ id + E ⇒ id + E * F ⇒ id + F * F ⇒ id + id * F ⇒ id + id * id (since E ⇒ F and F ⇒id).

Of greater interest, let us consider another rule set: P = {E → E – F | F + E | F, F → id}. Here, '–' is left-associative and '+' is right-associative.

For the string: id + id – id and the rules: E ⇒ E – F | F + E | F, F ⇒ id, two parse trees following leftmost derivation are derived:

1. E ⇒ E – F ⇒ F + E – F ⇒ id +E – F ⇒ id + F – F ⇒ id + id – F ⇒ id + id – id (since E ⇒ F and F⇒ id).
2. E ⇒ F + E ⇒ id + E ⇒ id + E – F ⇒ id + F – F ⇒id + id – F ⇒ id + id – id (since E ⇒ F and F ⇒ id).

Thus, the rule set is ambiguous. In particular, we may conclude that in a rule set, if rules to derive expressions consisting of operators with the same precedence contain different associativity, the rule set is ambiguous.

(c) Let G = ({R}, {a, b, c, *, +}, P, R), P = {R → R + R | RR | RR* | a | b | c}. Prove that the given grammar is ambiguous and make it unambiguous.

Solution The given grammar represents all the operations of RE (*R*), where '+' denotes union and '*' denotes closure. Also, *RR* implies concatenation of *R*. Let us take an RE, *r* = a + bc. This RE may be derived in two ways (following LMD):

(i) *R*⇒*R* + *R*⇒a + *R*⇒a + *RR*⇒a + b*R*⇒a + bc
(ii) *R*⇒*RR*⇒*R* + *RR*⇒a + *RR*⇒a + b*R*⇒a + bc

Thus, two parse trees may be constructed to derive the same RE, $r = a + bc$. Hence, the grammar is ambiguous. To make it unambiguous, we use left-associativity and priority of the operator (denoted as o), as follows: + < o < *, and finally, we may consider one possible unambiguous rule set:

$R \to R + R_1 \mid R_1, R_1 \to R_1 + R_2 \mid R_2, R_2 \to R_2{*} \mid a \mid b \mid c.$

KEY POINTS

- Left-recursion in a rule results in left-associativity.
- Right-recursion results in right-associativity.
- Precedence of operators in an expression is maintained by specification of the level of the rules in the grammar.
- In a parse or syntax tree, if an operator (O_1) appears lower in the tree than another operator (O_2), then the precedence of O_1 is higher than that of O_2.
- In the case of associativity of operators with the same precedence, if a sub-tree representing a sub-expression on the left side of an operator (say, O) in an expression appears lower in the tree than the subtree representing the sub-expression on the right of the operator, then O is left-associative (right-associative in the converse case).

Deterministic and Non-deterministic Grammar: In deterministic grammar, there cannot be multiple options to derive a string from a non-terminal, whereas a non-deterministic grammar may attempt many options for the same non-terminal to derive a string. The options usually occur due to the presence of a common prefix (left factor) appearing on the RHS of the rules associated with the same non-terminal used on the LHS in CFG. This leads to backtracking during parsing, which is time-consuming. This issue can be avoided by using the left-factoring strategy, explained below.

Left-factoring strategy: Let $A \to \alpha\beta|\alpha\gamma$ be the rules in a rule set, P. Here, α is the common prefix that appears in both the rules. Unfortunately, the presence of such a common left-factor in the rules may lead to the parser requiring more time for processing, as it results in **backtracking**. To avoid this, we may modify the rules as: $A \to \alpha C$, $C \to \beta|\gamma$.

For example, the rules: A → ab | ad (with common prefix: a) can be modified as: A → aC, C → b | d, where C is the new non-terminal introduced in G.

KEY POINTS

- Backtracking leads to several extra parsing steps. More specifically, it causes non-determinism in parsing.
- Backtracking may be resolved by left-factoring.
- Ambiguity causes non-determinism in meaning. It does not hamper string derivation. Ambiguity may not be resolved by left-factoring.

EXAMPLE 4.25

(a) Apply the left-factoring strategy on the rule set: A → abB | adD, B → b, D → d.

Solution Here, $\alpha = a$. Thus, the given rules may be modified as: A → aC, C → bB | dD, B → b, D → d, where C is the new non-terminal symbol introduced in G.

(b) Apply the left-factoring strategy on the rule set: S → aSSbS | aSaSb |abb | b.

Solution Here, we may initiate left-factoring with 'a' as the common prefix for the rules: S → aSSbS | aSaSb |abb or 'aS' as the common prefix for the rules: S → aSSbS | aSaSb. Both the cases provide the same result.

Let us take 'aS' as the common prefix and apply left-factoring. We get:
{S → aSS′, S → abb, S′ → SbS | aSb}, introducing the new symbol 'S′'. However, in the rule subset: {S → aSS′, S → abb}, the common prefix 'a' is again present on the RHS. So, we can further restructure this subset as:
{S → aS″, S″ → SS′, S″ → bb}, introducing the new symbol 'S″'. Thus, the resulting rule set is:
{S → aS″, S″ → SS′, S″ → bb, S′ → SbS | aSb}.

Note: There may exist indirect (hidden) left-factors in the rules of a grammar, and it may cause backtracking during derivation. One such example is: S → aA | Bb, A → aA | b, B → aB | b, where the rule S → Bb shows a common left-factor with the rule: S → aA (as B → aB is present in the rule set).

Eliminating immediate left-recursion: If a non-terminal A appears on both the LHS and the RHS of a rule (*r*) and the RHS begins with A, then the rule (*r*) is called the **immediate left-recursive rule** (or left-recursive); for example, $A \to Aa$.

Unfortunately, most top-down parsers do not work properly for a left-recursive rule like: $A \to A\alpha|\beta$, as left-recursion causes the parser to fall into an infinite loop. A practical example is give below.

Let us take a simple rule set: $\{A \to bB,\ B \to Ba|a\}$. In order to generate the language: L = {$ba^n \mid n \geq 1$} (with strings: ba, ba^2, ba^3, …}, the parser may fall into an **infinite loop**. This is because at any instant of parsing, the parser may not know whether the required *as* have been derived up to this point, as *a* appears after the non-terminal B. So, it will expand A repeatedly and will fall into an infinite loop.

Therefore, it is necessary to eliminate left-recursion from immediate left-recursive rules like: $A \to A\alpha|\beta$. More explicitly, the rules: $A \to A\alpha|\beta$ generate language: βa^* but direct provision of a rule like $A \to \beta a^*$ is not permissible in order to resolve the issue of infinite loop, since $A \to \beta a^*$ is not the format of a valid production rule. So, we need to restructure the rules as follows:

- First, identify the rules that do not start with the non-terminal symbol appearing on the LHS and where the non-terminal has an immediate left-recursive rule. For $A \to A\alpha|\beta$, it is: $A \to \beta$.
- Introduce a new non-terminal symbol and remove the left-recursion as: A → βA′, A′ → α A′ | ε, where A′ is the new non-terminal symbol introduced in the set of non-terminals. In particular, rule: A′ → αA′ is responsible for deriving the number of αs. So, the modified rule set is equivalent to the given left-recursive set. Importantly, if we substitute A in place of A′, the target string will not be generated; that is, any number of βs may be generated (which is not expected). Also, β may appear after α, which is not desired.

EXAMPLE 4.26

(a) Apply the immediate left-recursion elimination strategy on the rule set: E → E + E | a.

Solution Here, E → E + E. So, $\alpha = +E$. Further, E → a. So, β = a. Thus, the modified rules (after eliminating immediate left-recursion) are: E → aE′, E′ → +E′ | ε.

Note: Indirect (hidden) left-recursion may be present in the grammar and it may cause an infinite loop during derivation. One such example is: S → Aa | a, A → Sa | b, where the non-terminal: A on the RHS of the rule S → Aa equivalently creates immediate left-recursion in the rule set.

(b) Apply elimination of immediate left-recursion strategy on the rule set: A → Aa | b.

Solution Here, α = a and β = b. Thus, A → bA′, A′ → aA′ | ε.

4.7 DERIVATION OF NORMAL FORMS OF CONTEXT-FREE GRAMMAR

Thus far we have seen that the RHS of any production rule of CFG does not follow any specific format; that is, CFG places no restrictions on the RHS of rules. Two normal forms are introduced in this respect: Chomsky Normal Form (CNF) and Greibach Normal Form (GNF).

Chomsky Normal Form: A CFG is in CNF if every production is of the form: $A \rightarrow BC$ or $A \rightarrow c$, where A, B and C are arbitrary variables (non-terminal symbols) and 'c' is an arbitrary terminal symbol.

The key advantage of CNF is that the parse trees are always binary. If w is a string of length n, there will be exactly $(2n - 1)$ nodes labelled by variables (non-terminals). So, the derivation of a string of length n contains exactly $2n - 1$ steps (except for the special case, $w = \varepsilon$). Thus, we can determine if a string is in a language by carrying out an exhaustive search of all the derivations.

It is important to note that any efficient method for membership testing of string (w) in a CFL (L) generally takes exponential time. If the grammar is in CNF, its takes $O(2^n)$ time, where n is the length of w and the RHS of each rule contains at most two symbols – the number of parse trees possible will be 2^n to test whether $w \in L$ or not On the other hand, if the RHS consists of three symbols, then the number of parse trees will be 3^n. As $3^n > 2^n$, CNF is preferred in most cases.

This concept may be understood with the help of examples. Let $|w| = 2$, $\Sigma = \{0, 1\}$ and $|w| = 3$, $\Sigma = \{0, 1, 2\}$. In the case of $\Sigma = \{0, 1\}$, the possible strings are, respectively, 00, 01, 10 and 11 (N (no. of strings) $= 2^2 = 4$). On the other hand, if $\Sigma = \{0, 1, 2\}$, the possible strings are 00, 11, 22, 01, 10, 02, 20, 12 and 21 ($N = 3^2 = 9$).

Conversion of CFG to CNF can be accomplished using the following steps:

Input: $G = (V_N, \Sigma, P, S)$: a CFG; Output: G' (the CNF of G).

1. Remove all ε productions from G.
2. Eliminate all unit productions from G.
3. Split the RHS of each production into smaller productions (based on the rule format of CNF) if $|RHS| > 2$, introducing a number of new non-terminal symbols in the set of non-terminals. Thus, the number of non-terminals and rules of the original grammar increases and $G' = (V'_N, \Sigma, P', S)$, where V'_N is the new set of non-terminals and P' is the new set of rules (each in CNF).

Note: $S \rightarrow \varepsilon$ is not excluded from CNF, where S is the start symbol of the grammar. This implies that the language contains the null string as well.

Illustration of the conversion strategy: Let $P = \{S \rightarrow aSb \mid c\}$. Clearly, $V_N = \{S\}$, $\Sigma = \{a, b, c\}$, $P = \{S \rightarrow aSb \mid c\}$, where S is the start symbol of G. Here, $S \rightarrow c$ directly matches the format of CNF. However, the rule: $\Sigma \rightarrow aSb$ does not follow the format and length of the RHS of the production (aSb is greater than 2). So, we must modify it by replacing 'a' and 'b' with 'A' and 'B' on the RHS of $S \rightarrow aSb$. The conversion of $S \rightarrow aSb$ would be: $\Sigma \rightarrow ASB$, where $A \rightarrow a$ and $B \rightarrow b$.

Next, since the length of the RHS of each rule is at most 2, $S \rightarrow ASB$ is split as: $S \rightarrow AA_1$ and $A_1 \rightarrow SB$, where A_1 is a newly introduced variable in the grammar. Thus, the new rule set $(P') = \{S \rightarrow AA_1, A_1 \rightarrow SB, S \rightarrow c, A \rightarrow a, B \rightarrow b\}$. Also, $V'_N = \{S, A, B, A_1\}$.

Clearly, the number of rules (n) in the original set = 2, whereas the number of rules in the new rule set (P_1) is $5 = 2^2 + 1 = n^2 + c$, where c is constant. Thus, the number of rules in the new set can be bounded as $O(n^2)$. Also, derivation of a string of size m takes $2m - 1$ steps (refer to Theorem 4.2, which appears later in this chapter).

This is illustrated by the following test cases.

Test case 1: Let w = acb. Thus, $|w| = m = 3$.

$$
\begin{aligned}
S &\Rightarrow AA_1 && (1)\\
&\Rightarrow a\,A_1 && (2)\\
&\Rightarrow aSB && (3)\\
&\Rightarrow acB && (4)\\
&\Rightarrow acb && (5)
\end{aligned}
$$

Number of derivation steps = 5 = $2m - 1$ = 2 x 3 – 1

Test case 2: Let w = aacbb. $|w| = m = 5$.

$$
\begin{aligned}
S &\Rightarrow AA_1 && (1)\\
&\Rightarrow a\,A_1 && (2)\\
&\Rightarrow aSB && (3)\\
&\Rightarrow a(AA_1)B && (4)\\
&\Rightarrow aaA_1B && (5)\\
&\Rightarrow aaSBB && (6)\\
&\Rightarrow aacBB && (7)\\
&\Rightarrow aacbB && (8)\\
&\Rightarrow aacbb && (9)
\end{aligned}
$$

Thus, the number of derivation steps = 9 = $2m - 1$ = 2 x 5 – 1.

Note: Any CFG can be converted to an equivalent CNF.

EXAMPLE 4.27

(a) Convert the rule set: P = {E → E + T | T, T → T * F | F, F → id} to CNF.

Solution After eliminating unit productions from P, we get:

P_1 = {E → E + T, E → id, T → T * F, T → id, F → id}.

Take the rule: E → E + T. It can be rewritten as: E → E A T, A → +. Here, |E A T| > 2. So, we can split it as: E → E E_1, E_1 → A T.

Take T → T * F. It can be rewritten as: T → T M F, M → *. Here, |TMF| > 2. So, we can split it as: T → T T_1, T_1 → M F.

(b) Convert the rule set: S → S(S) to CNF.

Solution Here, S → S(S) can be expressed as: S → SASB, where A → (and B →). Now, |SASB| > 2. So, we need to split it further to convert it to the form: C → DE. The splitting is carried out as:

S → SS_1, S_1 → AS_2, S_2 → SB. Here, A, B, S_1 and S_2 are the new non-terminals introduced in the grammar.

Greibach Normal Form (GNF): A context-free grammar is in GNF if the RHS of each rule has one terminal followed by zero or more non-terminals. The structure is shown below.

$$A \rightarrow aX$$

One terminal symbol ↗ ↖ Zero or more non-terminal symbols

Advantages: The parsing process using PDA becomes easier; on encountering the current input symbol, the parser can easily decide the rule to be parsed (as the rules are of the form: A → aX, where 'a' is the terminal and 'X' may be zero or more non-terminals). Further, deriving a string: w from a grammar (G) in GNF takes only one step for each symbol. So the number of derivation steps = $|w|$. Thus, the string can be parsed quickly.

Limitation: It is hard to find the GNF of any CFG, whereas it is very easy to find the CNF.

An example of a grammar in GNF: S → aB | bA, A → a | aS | bAA, B → b | bS | aBB.

Each RHS consists of exactly one terminal followed by zero or more non-terminals.

Let us derive this string: w = aababb, $|w| = 6$. Now, the derivation of w is as follows:

$$S \Rightarrow aB \Rightarrow aaBB \Rightarrow aabSb \Rightarrow aabaBB \Rightarrow aababB \Rightarrow aababb$$

$$|aababb| = 6.$$

The number of derivation steps = 6.

Some examples of grammars not in GNF are shown below:

Terminal at the end is not allowed

S → aBc ✗

B → b (it is valid, as the number of non-terminals after b is zero)

P = {S → aA, A → B, B → C, C → a}. This grammar is not in GNF.

Derivation of the string: aa : $S \Rightarrow aA \Rightarrow aB \Rightarrow aC \Rightarrow aa; |aa| = 2$

Here, the number of derivation steps = 4

To convert CFG into its equivalent GNF, we often consider the following lemma.

Lemma: Deleting non-terminals that appear as the first symbols on the RHS of rules is performed as follows.

Let $A \to B\alpha$, $B \to \gamma_1 X | \gamma_2 Y$, where γ_1 and γ_2 may be terminals or non-terminals. Now, the reformatted rules are: $A \to \gamma_1 X\alpha | \gamma_2 Y\alpha$, $B \to \gamma_1 X | \gamma_2 Y$.

To convert CFG to GNF:

1. Remove the ε productions from the rule set.
2. Remove the unit productions from the rule set.
3. Restructure the rules in GNF.

Note: The variables may be renamed first to keep track of them (if introduced).

Consider the grammar given below and find its equivalent GNF.

G not in GNF	convert	G in GNF
S → aBc	⟹	S → aBC
B → b		B → b
		C → c

EXAMPLE 4.28

(a) Find the GNF for P = {S → Ab, A → aA | c}.

Solution The variables are first renamed and the rules are re-written as: $A_1 \to A_2b$, $A_2 \to aA_2 | c$. The rules: $A_2 \to aA_2 | c$ are in GNF. Now, consider the rule: $A_1 \to A_2b$. Here, variable A_2 appears at the beginning of the RHS. So, we reformat it as: $A_1 \to aA_2b | cb$ [since $A_2 \to aA_2 | c$]. Both the rules: $A_1 \to aA_2b | cb$ are not in GNF. So, we need to reformat them as: $A_1 \to aA_2A_3 | cA_3$, $A_3 \to b$.

Hence, the resulting rule set is: $P_1 = \{A_1 \to aA_2A_3 | cA_3, A_3 \to b, A_2 \to aA_2 | c\}$

(b) Find the GNF for P = {S → ab | aS | aaS} of grammar G.

Solution Let us first convert the given rules to CNF as: $S \rightarrow B_aB_b \mid B_aS \mid B_aB_aS$, where B_a and B_b are the new non-terminals (variables) introduced in the original grammar G and $B_a \rightarrow a$, $B_b \rightarrow b$.

Now, we may rename the variables S, B_a and B_b as A_1, A_2 and A_3, respectively; the modified rules would be: $A_1 \rightarrow A_2A_3 \mid A_2A_1 \mid A_2A_2A_1$, $A_2 \rightarrow a$, $A_3 \rightarrow b$.

To obtain the rule structure of GNF: $A \rightarrow aX$, where 'a' is the terminal and 'X' may be zero or more non-terminals, we may restructure the rules as: $A_1 \rightarrow a\,A_3 \mid a\,A_1 \mid aA_2A_1$, $A_2 \rightarrow a$, $A_3 \rightarrow b$.

THEOREM 4.2

If G is a CFG in CNF, then for any string w in L(G) of length $n \geq 1$, exactly $2n - 1$ steps are required for any derivation of w.

Proof We begin with a single non-terminal (S) and form the string by making substitutions according to the rules. Each rule of the form $A \rightarrow BC$ adds a new non-terminal, and each rule of the form $A \rightarrow a$ converts one non-terminal into a terminal. As it takes $n - 1$ steps to grow one string from the original non-terminal to n non-terminals, and n steps to convert each of those non-terminals into terminals, it takes a total of $2n - 1$ steps to derive a string of length n.

THEOREM 4.3

Let G be a CFG in CNF that contains b variables. If G generates some string using a derivation with at least 2^b steps, then L(G) is infinite.

Proof In the largest possible derivation, starting with the start variable (S) of G, we replace it with two other variables, each of which is replaced with two other variables, and so on. Since no variable can be repeated along any branch of the tree without creating a cycle, the tree will be of depth: b, and it will contain $2^b - 1$ nodes. Now, if a derivation requires 2^b or more steps, the grammar must contain a cycle (it must have the recursive rule) and L(G) must be infinite.

4.8 PUMPING LEMMA FOR CONTEXT-FREE LANGUAGE

In regular languages, the key point is that any sufficiently long string (in the form of repetitions) will visit the same state twice. In context-free languages, we note that any sufficiently large syntax tree (in the form of repetitions) will have a downward path that visits the same non-terminal twice. We can then 'pump in' extra copies of the relevant sub-tree and remain within the language. This is shown in Fig. 4.6. Here, S is the start symbol of the grammar G, and P is the non-terminal for which part of the syntax tree is repeated.

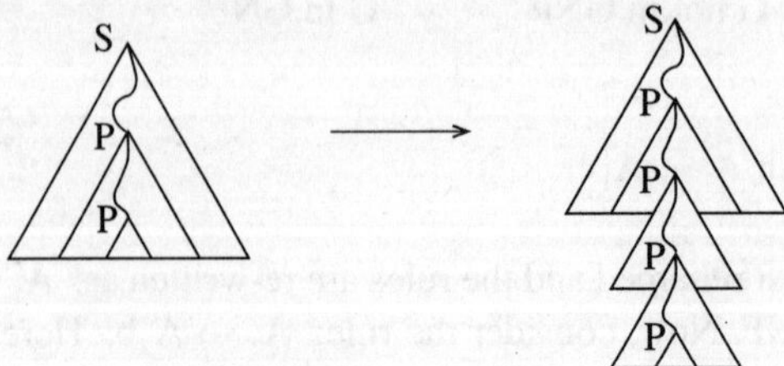

FIGURE 4.6 Pumping of syntax tree

Statement: Let G be a CFG. There exists a constant k such that any string w in L(G) of length greater than k can be rewritten as $w = uvxyz$, in such a way that either v or y is non-empty and uv^nxy^nz is in L(G) for $n \geq 0$.

Proof Let G = (V_N, Σ, R, S). There is a number k such that any terminal string w in L(G) of length greater than k has a derivation of the form:

$S \overset{*}{\rightarrow} uAz \overset{*}{\rightarrow} uvAyz \overset{*}{\rightarrow} uvxyz$, where $u, v, x, y, z \in \Sigma^*$, $A \in (V_N - \Sigma)$ and either v or y is non-empty. Also, $A \rightarrow vAy \mid x$.

Clearly, the derivation: $A \overset{*}{\rightarrow} vAy$ can be repeated any number of times to obtain strings of the form: uv^nxy^nz, $n \geq 0$. Now, we will prove that both v and y may not be empty. To do so, let p be the largest number of symbols on the RHS of any rule in G; that is, $p = \max \{|\alpha| : A \rightarrow \alpha \text{ is a rule in G}\}$.

For any $m \geq 1$, a parse tree (T) of height m can have at most p^m leaves. Now, if T is a parse tree with string of length greater than p^m, then T has some path of length greater than m. Let $m = |V_N - \Sigma|$, $k = p^m$ and suppose w has length exceeding k. Let T be a parse tree with root labelled S and with yield w. Then T has at least one path of length greater than $|V_N - \Sigma| + 1$.

To explain this discussion graphically, consider a parse tree T′ whose root is a node labelled A on this path and its leaves are the leaves of T, except that some subsequent node labelled A on the path is a leaf (Fig. 4.7).

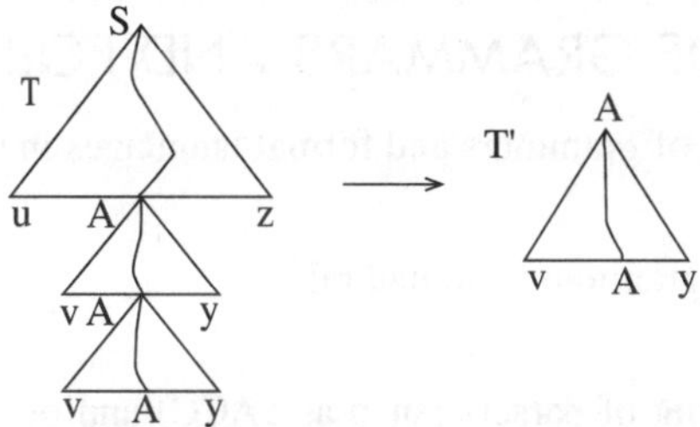

FIGURE 4.7 Pumping of syntax tree

The parse tree T′ has a root labelled A and yield of the form vAy for some $v, y \in \Sigma^*$.
Suppose $v = y = \varepsilon$. The tree T′ (except the node A) can be removed from T without affecting the tree T as a whole; this makes it impossible to obtain a parse tree that derives string $w = uvxyz$ and uv^nxy^nz, $n \geq 0$. Thus, both v and y cannot be ε. Hence, the lemma is proved.

EXAMPLE 4.29

Show that $L = \{a^p \mid p \text{ is prime}\}$ is not CFL.

Solution Assume that L is CFL and $a^p \in L$. Let k be any natural number for the language L. Again, let p be any prime number greater than k.

As L is CFL, we can write $w = a^p$ as $uvxyz$ and $uv^ixy^iz \in L$, for $i \geq 0$. When $i = 0$, $uv^ixy^iz = uxz \in L$. Let $|uxz| = q$ and $|vy| = r$.

Now, $uv^qxy^qz = |uxz| + |(vy)^q| = q + qr = q(1 + r)$; since we are allowed to take any value of i, and so, we have chosen $i = q$.

However, according to the constraints of the given language, $q(1 + r)$ should be prime, but it is not prime here, as the two factors q and $(1 + r)$ exist. Hence, a contradiction occurs. Therefore, $L = \{a^p \mid p \text{ is prime}\}$ is not CFL.

EXAMPLE 4.30

Show that $L = \{a^nb^nc^n \mid n \geq 1\}$ is not CFL.

Solution *Step 1*: Assume that the language L is CFL. Let n be a natural number obtained by using pumping lemma.

Step 2: Let $w = a^nb^nc^n$. So, $|w| = 3n > n$. According to the pumping lemma for CFL, we may write $w = uvxyz$, where $|vy| \geq 1$.

Step 3: $w = uvxyz = a^nb^nc^n$. As $1 \leq |vy| \leq n$ [$|vxy| \leq n$, so $|vy| \leq n$], v or y cannot contain all the three symbols, a, b and c. So, v or y will be one of the following forms:

1. Contain only a and b, that is, in the form a^ib^j
2. Contain only b and c, that is, in the form b^ic^j
3. Contain only repetitions of a, b or c.

Let us take the value of i as 2. So, v^2 or y^2 will be in the form: $a^ib^ja^ib^j$ (as v is a string, aba is not equal to a^2b or ba^2) or $b^jc^kb^jc^k$. Thus, uv^2xy^2z cannot be in the form $a^mb^mc^m$ So, uv^2xy^2z does not belong to L.

If v or y contain repetitions of a, b or c, then v or x will be of the form a^j, b^k or c^m Let us take the value of i as 0. So, $uv^ixy^iz = uxz$. In this string, the number of occurrences of one of the other two symbols in uvy is less than n. So, uv^2xy^2z does not belong to L.

4.9 APPLICATIONS OF GRAMMARS AND FORMAL LANGUAGES

There are numerous applications of grammars and formal languages in the field of computer science:

- Designing automata
- Constructing languages (programming and natural)
- Processing text

More specifically, the designing of parsers (such as YACC) and markup languages (such as XML, HTML, etc.) are based on the concept of CFGs.

Programming languages: The concept of formal grammars and languages is very useful for describing any programming language, such as Pascal, C, C++, etc. In markup languages like HTML and XML, the action (semantic action) is set for a block of matching texts.

4.10 ADDITIONAL SOLVED PROBLEMS

PROBLEM 1

Design a regular grammar for $L = \{a^nb^n \mid n \leq 3\}$.

Solution The language $L = \{a^nb^n \mid n \geq 0\}$ is a CFL. However, $L = \{a^nb^n \mid n \leq 3\}$ is a finite language. We know that every finite language is a regular language. Hence, $L = \{a^nb^n \mid n \leq 3\}$ is also a regular language, and we may design a regular grammar for it.

A simple rule set (P) based on the structure of regular grammar is:

$P = \{S \to aA, A \to b, A \to aB, B \to bD, D \to b, B \to aE, E \to bF, F \to bG, G \to b\}$.

Actually, the rules: $S \to aA$ and $A \to b$ derive string 'ab', while the rules: $S \to aA$, $A \to aB$, $B \to bD$ and $D \to b$ are responsible for deriving string 'a^2b^2', and so on.

Likewise, we may design a regular grammar for specific values of n of the language $L = \{a^nb^n\}$. The number of rules would increase here.

PROBLEM 2

(a) Design a Type 0 grammar for $L = \{a^nb^n \mid n \geq 1\}$ over $\{a, b\}$.

Solution The Type 0 rule set for L = {a^nb^n | $n \geq 1$} is: P ={S → aAb, aA → aaAb, A → ε}. This is a Type 0 grammar (not Type 1), since A appears on the RHS of S → aAb and A derives null string also.

Derivation of w = aabb: S → aAb → (aaAb)b → aa(ε)bb = aabb.

However, restructuring the rule set P as {S → aAb, aA → aaAb, A → B, B → ε}, we obtain Type 1 (context-sensitive) grammar for L.

(b) Design a CFG for the language L = {$a^mb^nc^{m-n}$ | $m \geq n \geq 0$}.

Solution Split the words in this language into sections: a^{m-n}, followed by $a^n b^n$, followed by c^{m-n}. Notice that $(a^{m-n})(a^n b^n)(c^{m-n})$ gives $a^mb^nc^{m-n}$.

Now, use different non-terminals to generate the first and third substrings (simultaneously) and to generate the second substring.

S → aSc |T, T → aTb | ε [Here, b will not appear after c].

PROBLEM 3

Let A and B be two sets of words (strings) from Σ* derived from some alphabet of symbols Σ. Suppose B is a subset of A. Which of the following statements must always be true of A and B?

(i) If A is finite, then B is finite.
(ii) If A is regular, then B is regular.
(iii) If A is an RE, then B is an RE.
(iv) If A is context-free, then B is context-free.
(v) If A is infinite, then B is infinite.

Solution Only (i) is correct. Certainly, if A is finite, then any subset of A is also finite. Accordingly, B is finite. Regarding option (ii), the set A may be regular, but B may not necessarily be regular. For example, A is a set of all strings over Σ for which an RE exists and it is Σ*. Let B be a set of prime-length strings over Σ. Definitely, set B is a subset of A but it has no RE and hence is not regular. Therefore, option (ii) may not be correct.

Likewise, option (iii) may not be true. Here, the example provided for option (ii) may be used to prove option (iii) as incorrect. Further, the same example may be used by taking a context-free language, instead of a regular language, to prove option (iv) as false. Certainly, there does not exist a context-free language for a prime-length string.

Finally, option (v) may also not be true. Since B is a subset of A, we may construct a finite B.

PROBLEM 4

Draw a state transition diagram to accept a string with not more than three *as* over Σ = {a, b}, and then generate the grammar from that diagram.

Solution The state diagram for recognising the given language is shown in Fig. 4.8.

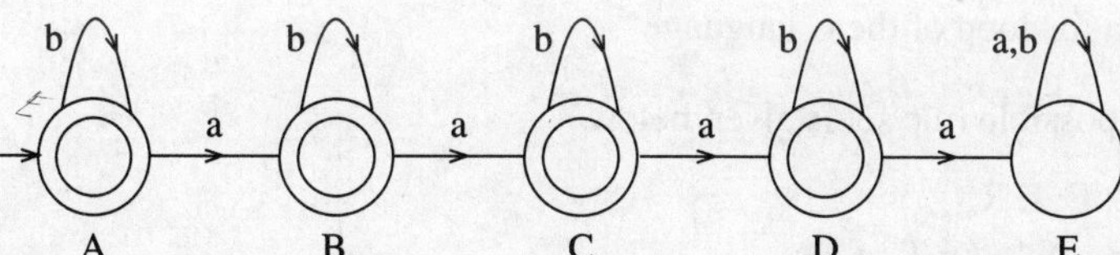

FIGURE 4.8 A state diagram for accepting strings with not more than three *as*

Here, all the states: A, B, C and D are accepting states. State A is the initial state of the machine, whereas E is the dead state. From Fig. 4.6, we can generate a regular grammar with rules:

A → bA | aB | ε, B → bB | aC | ε, C → bC | aD | ε, D → bD | aE | ε, E → aE | bE.
Note that the non-terminal E is useless, as it is not capable of deriving a terminal string. This is because E is the dead state.

PROBLEM 5

Find the regular grammar for the state transition diagram shown in Fig. 4.9 (a).

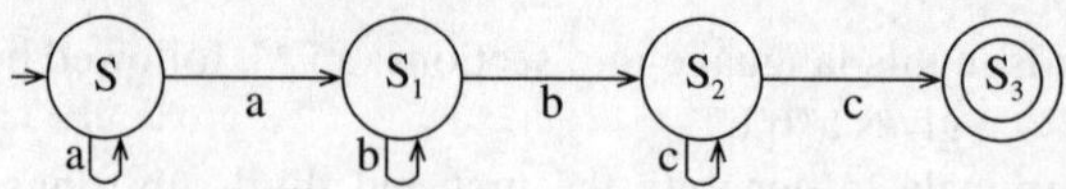

FIGURE 4.9 (a) An NFA over {a, b}

Solution The given FA is NDFA. Now, based on the conversion strategy (discussed in Section 4.5.1), the regular rule set is:
$S \to aS, S \to aS_1, S_1 \to bS_1, S_1 \to bS_2, S_2 \to cS_2, S_2 \to cS_3, S_3 \to \varepsilon$.
For an edge connected through ε in NDFA, we may derive the unit rule, as shown in Fig. 4.9 (b).

The derived rule for A – ε – B is A → B. If B becomes the final state, then add B → ε.

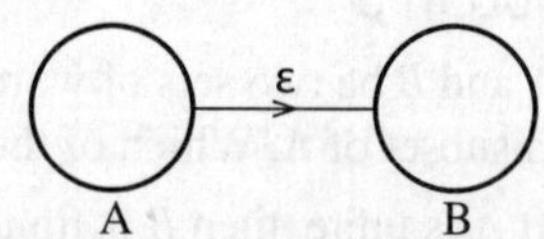

FIGURE 4.9 (b) An ε NFA

PROBLEM 6

Design a simple CFG for valid C assignment statement. Also, modify it for multiple assignment statement.

Solution For validity checking of C assignment statement, a simple set of rules is presented as:
S → id = E; E → E + T |T, T → T * F | F, F → id.
Here, only two operations, addition and multiplication, are taken into account. We may add more operators as desired. Priority is given to '*' (multiplication operator) as compared to '+'. Also, left-associativity is followed for both cases.

Now, we may restructure this grammar for validating multiple assignment as:
S → id = M, M → id = M | E, E → E + T | T, T → T * F | F, F → id.

Certainly, for a multiple assignment statement such as $a = b = c + d$; (valid in C), right-associativity is followed and this must be managed by the right-recursive rule for the assignment construct. Hence, the right recursive rule: M → id = M is used in the above rule set.

PROBLEM 7

Design a CFG for a for-loop of the C language.

Solution One possible rule set is given below.

$L \to \text{for } L_p\ E;\ E;\ E\ R_p$
$E \to id = N\ B_{op}\ N \mid id = T \mid S \mid \varepsilon,$
$S \to T,\ T \to T \text{ op } F \mid id,\ F \to num \mid \varepsilon,\ N \to num,$
$op \to B_{op} \mid U_{op},\ B_{op} \to + \mid - \mid * \mid /,\ U_{op} \to\ ++ \mid --,$
$L_p \to (,\ R_p \to),$

Note: E (expression) may be null or non-null. So, the for-loop may have a total of $2^3 = 8$ options (including the structure consisting of *null; null; null*, or *for*(; ;)).

PROBLEM 8

Deciding whether a CFL by a CFG is finite or not.

Solution To verify finiteness, we may draw a directed graph (in the form of dependency graph) on the basis of the rule set of the given grammar. If the dependency graph (using only non-terminals) contains no loop, the given language is finite, else it is infinite. Let us take an example of a CFG and draw the dependency graph based on the non-terminals and the rules present in the grammar to check if the language generated from the grammar is finite or infinite. Let the rule set of the CFG be:
P = {S → AB, A → aCb | a, B → bB | bb, C → cBS}.

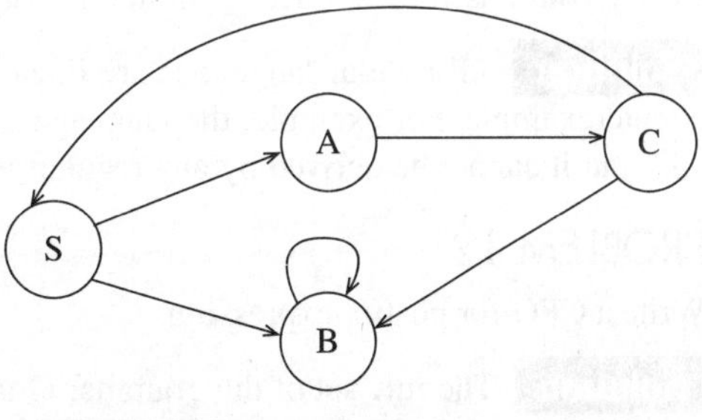

FIGURE 4.10 Dependency graph for the given CFG

Here, the non-terminals present in the grammar are S, A, B and C. Now, based on the non-terminals and the rules present in *P*, we obtain the dependency graph shown in Fig. 4.10.

As the dependency graph has loops/cycles, the language generated from this grammar is infinite.

Tip to draw dependency graph: If A → BCD is a rule, then connect one directed edge from A to each of B, C and D, where A, B, C and D are non-terminals.

Note: According to the decision algorithm, if S (start symbol of G) is found to be useless, then L (G) is empty.

PROBLEM 9

Prove or disprove that the union of two non-regular languages is non-regular.

Solution The union of two non-regular languages is not necessarily non-regular; it may be regular. We shall show this using counterexamples.
Suppose the first language $L_1 = \{a^m b^n \mid m, n \geq 0 \text{ and } m = n\}$. Language L_1 is a CFL but is not regular. Let us take the second language $L_2 = \{a^m b^n \mid m, n \geq 0 \text{ and } m \neq n\}$. L_2 is again a CFL but not regular. Now, the union of L_1 and L_2 is $L = \{a^m b^n \mid m, n \geq 0\}$; that is, there is no restriction over *n* and *m* values and the regular expression for L is a*b*.

Hence, $L = L_1 \cup L_2$ results in a regular language.

PROBLEM 10

Prove or disprove that the intersection of two non-regular languages is non-regular.

Solution The intersection of two non-regular languages is not necessarily non-regular; that is, it may be regular. We shall show this using counterexamples.
Suppose, the first language $L_1 = \{a^n b^n \mid n \geq 1\}$. Language L_1 is a CFL but is not regular. Let us take the second language $L_2 = \{b^m a^m \mid m \geq 1\}$. L_2 is again a CFL but not regular. Now, the intersection of L_1 and L_2 is $L = \{a^n b^n \mid n \geq 1\} \cap \{b^m a^m \mid m \geq 1\} = \Phi$, which is regular.

Further, let the languages be $L_1 = \{a^n b^n \mid n \geq 0\}$ and $L_2 = \{b^m a^m \mid m \geq 0\}$.

Then, $L = \{a^n b^n \mid n \geq 1\} \cap \{b^m a^m \mid m \geq 1\} = \{\varepsilon\}$, which is also regular. Hence, the intersection of two non-regular languages is not necessarily non-regular.

PROBLEM 11

Justify your answer: "A linear grammar is not necessarily equivalent to a regular grammar".

Solution All regular languages are linear but the converse is not true. We shall show this using a counterexample. For example, the language $L = \{a^n b^n \mid n \geq 0\}$ with $R = \{S \rightarrow aSb, S \rightarrow \varepsilon\}$ is a linear CFL but it cannot be derived by any regular grammar. Hence, the statement is not correct.

PROBLEM 12

Write a CFG for postfix expression.

Solution The rule set of the grammar G is $\{E \rightarrow E\,E\text{ op}, E \rightarrow \text{num}, \text{op} \rightarrow + \mid - \mid * \mid /\}$.

A postfix expression is a parenthesis-free one.

This is often called reverse Polish notation, whereas prefix is called Polish notation, which means parenthesis-free notation. It was introduced by the Polish mathematician Jan L in 1920.

PROBLEM 13

Explain whether determining if a string (w) belongs to CFL (L) is a solvable problem or not.

Solution *Approach 1*: Whether a string (w) belongs to a CFL is solvable though it may take exponential time (an NP problem). For an FA, M, this membership problem can be solved easily, but it is not difficult in the case of PDA. We need to check if an FA, M, accepts a string x, run M with input and see what happens. In the case of PDAs, a solution cannot be simple because non-determinism cannot always be eliminated (as PDAs are inherently non-deterministic). If we think of a PDA, M, as making a move once a second, and choosing arbitrarily, then "what happens" may be different from one run to the next. A sequence of moves resulting in the input x being rejected does not mean that there is no other sequence leading to acceptance. From the specifications for a PDA, we can formulate some backtracking algorithm that will be guaranteed to answer the question. Simpler to describe (though probably inefficient to carry out) is the adoption of CNF of CFG and then construction of a derivation tree that may take exponential time.

Approach 2: Given a string w in L(G), we can parse it by systematically constructing all possible (say, leftmost) derivations and checking whether any of them match w. Specifically, we start Round 1 by viewing all productions of the form $S \rightarrow x$, finding all the x values that can be derived from S in one step. If it does not result in a match with w, we move to the next round, in which we apply all applicable productions to the leftmost variable of every x that can be derived from S in two steps. Again, if none of these sentential forms gives w, we proceed to Round 3, and so on.

This gives us a set of sentential forms, some of which may lead to w. On each subsequent round, we again take all the leftmost variables and apply all possible productions. Thus, the method will eventually produce a leftmost derivation of w. This is, indeed, exhaustive search parsing or brute force parsing.

For example, consider the grammar (G): $S \rightarrow SS \mid aSb \mid bSa \mid \varepsilon$ and the string $w = aabb$.

Round 1 gives

$S \Rightarrow SS$

$S \Rightarrow aSb$

$S \Rightarrow bSa$

$S \Rightarrow \varepsilon$

The last two of these can be removed from further consideration for obvious reasons (since the string does not start with *b* and the string is not null). Round 2 then yields the sentential forms:

$S \Rightarrow SS \Rightarrow SSS$

$S \Rightarrow SS \Rightarrow aSbS$

$S \Rightarrow SS \Rightarrow bSaS$

$S \Rightarrow SS \Rightarrow S$

These are obtained by replacing the leftmost S in sentential form 1 with all applicable substitutes. Similarly, from sentential form 2, we obtain the additional sentential forms:

$S \Rightarrow aSb \Rightarrow aSSb$

$S \Rightarrow aSb \Rightarrow aaSbb$

$S \Rightarrow aSb \Rightarrow abSab$

$S \Rightarrow aSb \rightarrow ab$

Again, several of these can be removed from contention. In the next round, we find the actual target string from the sequence: $S \rightarrow aSb \rightarrow aaSbb \rightarrow aabb$. Therefore, *aabb* is in the language generated by the grammar under consideration.

Exhaustive search parsing has serious flaws. The most obvious is its tediousness; it is not to be used where efficient parsing is required. Though this search method gives a theoretical guarantee that parsing can always be performed, its practical usefulness is limited because the number of sentential forms generated may be excessively large. In fact, the total number of sentential forms:

$$M = |P| + |P|^2 + \ldots + |P|^{2|w|}$$
$$= O(P^{2|w|+1}),$$

where |P| is the number of rules in the rule set and $|w|$ gives the length of the string w to be checked. This indicates that the work for exhaustive search parsing may grow exponentially with the length of the string, making the cost of the method prohibitive.

PROBLEM 14

Explain why checking the equality of two CFLs is an unsolvable problem.

Solution Here, we cannot achieve a proof analogy to the DFA case, where the concept of set difference may be applied. In particular, in the case of DFA, we may apply the concept of production automata for the set difference operation for two regular languages, say L_1 and L_2. If the resulting automaton has no final state, then L_1 and L_2 are equal. However, applying such a strategy over two PDAs may not be correct, as the PDAs are by definition NPDA.

PROBLEM 15

Consider the parse tree shown in Fig. 4.11 for the expression: a#bcd#e#f involving two binary operators: $ and #.

Which one of the following is correct for the given parse tree?

$ has higher precedence and is left-associative; # is right-associative

has higher precedence and is left-associative; $ is right -associative

$ has higher precedence and is left-associative; # is left-associative

has higher precedence and is right-associative; $ is left-associative

Solution In a parse or syntax tree, if an operator (O_1) appears lower than another operator (O_2), then O_1 has higher precedence than O_2. Thus, the precedence of $ is higher than that of #. In the case of associativity of operator(s) with the same precedence, if the subtree representing a sub-expression on the LHS of an operator (say, O) in an expression appears lower in the tree than the subtree representing a sub-expression on the RHS of the operator O, then O is left-associative (it is right-associative for the reverse case).

Thus, $ is left-associative (concluded from: bcd) but # is right-associative (concluded from a#bcd#e#f).

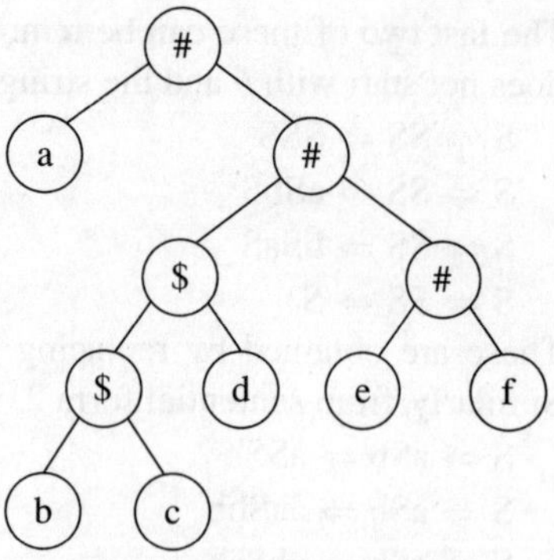

FIGURE 4.11 A parse tree for the expression: a#bcd#e#f

PROBLEM 16

(a) Which of the languages spoken in the world correspond to the three notations – infix, prefix and postfix?

Solution In natural languages, sentences that follow the rule format S (subject), V (verb), O (object) are considered as using an infix-like format. Natural languages like English and Chinese come under this structure. On the other hand, examples of postfix-like languages that follow the SOV format are Turkish and Japanese, while examples of prefix-like languages that follow the VSO format are Arabic and Celtic.

(b) Which programming languages correspond to these three notations – infix, prefix and postfix?

Solution In programming languages, the format: *operand op operand* is said to be of infix form. Most conventional languages such as C, C++, Java, Python, etc. are examples of infix languages. On the other hand, postfix (*operand operand op*) languages are very rare. One example is PostScript (which supports a syntax like 23 34 +). Lastly, Lisp-like functional language is an example of a prefix language.

(c) Given a grammar and a string, how do you know if the string is ambiguous?

Solution First, convert the grammar to CNF. If multiple parses can be found for this string, it is ambiguous.

PROBLEM 17

Let L denotes the language generated by the grammar $S \rightarrow 0S0|00$. Which of the following is true?

(a) $L = 0^+$
(b) L is regular but not 0^+
(c) L is context-free but not regular
(d) L is not context-free

Solution Interestingly, the given grammar is CFG and it generates the language (L):
$S \Rightarrow 0S0 \Rightarrow 00S00 \Rightarrow 000S000 \Rightarrow \ldots\ 0^kS0^k \Rightarrow 0^k000^k = 0^{2(k+1)}$, where the smallest possible string derivable from S is 00. Now, L can be expressed in form of a regular expression as $(00)^+ \neq 0^+$.

Obviously, it is a regular language. Further, every regular language is by default context-free as well. Although, the given grammar is not in form of CFG, it can be converted to the form of regular grammar, as the language is regular. Hence, the correct option is (b).

PROBLEM 18

The language $\{a^m b^n c^{m+n} \mid m, n \geq 1\}$ is **[GATE 2004]**

(a) Regular (b) Context-free but not regular
(c) Context-sensitive but not context-free (d) Type 0 but not context-sensitive

Solution Apply pumping lemma for regular languages. Let $s = xyz$, where $|y| \geq 1$, $|xy| \leq$ p. For $s \in$ L, xy^kz must be a member of L for all $k \geq 0$. Let $x = a^m$, $y = b^n$, $z = c^{m+n}$, such that $m + n \leq$ p.

$xy^pz = a^m b^{nk} c^{m+n} \in$ L. So, L is not regular, as pumping lemma is violated if $k > 1$. Option (a) is thus eliminated.

Apply pumping lemma for context-free languages. Let $s = uvwxy$, such that $|vwx| \leq$ p, $|vx| \geq 1$. Then, $uv^nwx^ny \in$ L for all $n \geq 0$.

Let $u = a^p$, $v = a^{(m-p)}b^q$, $w = b^{(m-q)}$, $x = c^r$, $y = c^{(n+m-r)}$. Clearly, $uv^2wx^2y \in$ L. So, L is not context-free, as pumping lemma is violated for $n = 2$. Option (b) is also eliminated.

Clearly, one stack is sufficient to determine the acceptance or rejection of $a^m b^n c^{m+n}$. So, a PDA can be designed for this language. Thus, this language is context-sensitive. The correct option is (c).

PROBLEM 19

Which of the following features cannot be captured by context-free grammars? **[GATE 1994]**

(a) Syntax of if-then-else statements (b) Syntax of recursive procedures
(c) Whether a variable has been declared before its use (d) Variable names of arbitrary length

Solution CFGs are used to show syntactic rules while designing the compiler, and these rules do not check for meaningful things such as if a variable has been declared before its use or not. Such things are meant to be handled by the semantic analysis phase (requires the power of CSG). To check whether a variable has been declared before its use, we need to parse strings such as $\{wcw \mid w \in (a \mid b)*\}$, which cannot be done using CFG. The correct option is (c).

PROBLEM 20

Which of the following definitions generates the same language as L, where L = $\{x^ny^n$ such that $n \geq 1\}$?

I. $E \rightarrow xEy \mid xy$ II. $xy \mid (x^+xyy^+)$ III. x^+y^+

(a) Only I (b) I and II (c) II and III (d) II only

[GATE 1995, Punjab Tech University PhD Entrance 2010, HCU PhD Entrance 2015]

Solution $E \rightarrow xEy \mid xy$ can be expanded as:

$E \Rightarrow xEy \Rightarrow xxEyy \Rightarrow xxxEyyy \Rightarrow \ldots. \Rightarrow x^{(n-1)}Ey^{(n-1)} \Rightarrow x^ny^n$ (replacing E with xy).

$xy \mid (x^+xyy^+)$ and x^+y^+ can also derive strings of the form 'xxyyy', which is clearly not of the form x^ny^n. So, options (b), (c) and (d) are eliminated. The correct option is (a).

PROBLEM 21

Give the grammar for a language of palindromes over the alphabet {a, b, c, d, e}.

[Gujarat SET 2011]

Solution The smallest possible (palindrome) strings accepted by the language are ε, a, b, c, d, e. This defines one rule of the grammar as $S \rightarrow a \mid b \mid c \mid d \mid e \mid \varepsilon$.

Also, any palindrome string that starts with 'a' must end with 'a'. This needs the rule: $S \rightarrow aSa$. Any palindrome string that starts with 'b' must end with 'b'. This needs the rule: $S \rightarrow bSb$. Any palindrome string that starts with 'c' must end with 'c'. This needs the rule: S → cSc. Any palindrome string that starts with 'd' must end with 'd'. This needs the rule: S → dSd. Any palindrome string that starts with 'e' must end with 'e'. This needs the rule: S → eSe.

Hence, the complete grammar is: S → aSa | bSb | cSc | dSd | eSe | a | b | c | d | e | ε. It is clearly a context-free grammar.

PROBLEM 22

Consider the grammar G: S → SbS | a.

(a) Does the string *abababa* belong to the language generated by the grammar G?

(b) Is this grammar ambiguous? Justify your answers. **[HCU PhD Entrance 2010]**

Solution (a) To derive the string *abababa*, proceed by leftmost derivation as: S ⇒ SbS ⇒ abS ⇒ abSbS ⇒ ababS ⇒ ababSbS ⇒ abababS ⇒ abababa. Yes, the given string belongs to L(G).

(b) Consider the derivation of another string *aba* by this grammar. Two different derivations are possible.

Leftmost derivation: S ⇒ SbS ⇒ abS ⇒ aba.

Rightmost derivation: S ⇒ SbS ⇒ Sba ⇒ aba

As more than one derivation is possible for the same string, the grammar is ambiguous.

PROBLEM 23

Give a grammar G in Chomsky Normal Form to produce the language L = $\{a^{2n}b^n \mid n > 0\}$. **[Gujarat SET 2008]**

Solution We first derive an intuitively visible grammar to generate the language L = $\{a^{2n}b^n \mid n > 0\}$, and then attempt to convert it to CNF. The smallest string accepted by this grammar is *aab*. So, one rule is S → aab. For generating larger strings, the rule is S → aaSb. The complete grammar is S → aaSb | aab.

Every rule in CNF is of the type A → BC or A → c, where A, B, C are non-terminal symbols and c is a single terminal symbol. For this purpose, we introduce two new rules as A → a and B → b. Now, the rule set becomes {S → AASB | AAB, A → a, B → b}.

Introduce one more rule as: X → AA so that the rule: S → AAB can be modified to obey the rules of CNF. Now, the rule set becomes {S → XSB | XB, X → AA, A → a, B → b}.

The rule S → XSB is still not in CNF. We modify this rule by introducing a new rule as: Y → SB. Now, our rule set is transformed as {S → XY | XB, X → AA, Y → SB, A → a, B → b}. This rule set is in Chomsky Normal Form.

PROBLEM 24

If L is a language recognisable by an FA, the language REVERSE(L) = {W | W is the reverse of W, where W ∈ L} is a:

(a) Context-free language (b) Context-sensitive language
(c) Regular language (d) Recursively enumerable language **[Rajasthan SET 2012]**

Solution L is a regular language, as it is accepted by a finite automaton. Also, an FA can be designed for the language obtained by reversing all the strings in L. Hence, REVERSE(L) is a regular language. The correct option is (c).

PROBLEM 25

Remove left recursion from the grammar: $S \rightarrow Sa| \ Sb \ | \ a \ | \ b$ **[GATE 2001]**

Solution To eliminate left recursion from the rule $S \rightarrow Sa \mid a$, the modification is $S \rightarrow aX$, $X \rightarrow aX \mid \varepsilon$, where X is the new non-terminal introduced. Similarly, the rules $S \rightarrow Sb \mid b$ are modified as $S \rightarrow bY, Y \rightarrow bY \mid \varepsilon$.

Thus, the modified grammar is: $S \rightarrow aX \mid bY, X \rightarrow aX \mid \varepsilon, Y \rightarrow bY \mid \varepsilon$.

PROBLEM 26

Let L be a language over the alphabet {1, 2, 3, (,)} generated by the following grammar (with start symbol S, and non-terminals {A, B, C}): $S \rightarrow A\ B\ C$, $A \rightarrow ($, $B \rightarrow 1\ B \mid 2\ B \mid 3\ B$, $B \rightarrow 1 \mid 2 \mid 3$, $C \rightarrow)$. Then, which of the following is true? **[TIFR PhD Entrance 2017]**

(a) L is finite
(b) L is regular
(c) L contains only strings of even length
(d) L is context-free but not regular

Solution The rules: $B \rightarrow 1\ B \mid 2\ B \mid 3\ B \mid 1 \mid 2 \mid 3$ define the regular expression $(1 + 2 + 3)^*$. So, the set of strings defined by this grammar is given by $((1 + 2 + 3)^*)$, which is a valid regular expression. So, the language L is regular. The correct option is (b).

PROBLEM 27

Find CFGs for the following languages:

(i) $L = \{a^n b^i \mid i \in \{n, 3n, 5n\}, n \geq 0\}$

(ii) $L = \{wa^n b^n w^R \mid w \in \{a, b\}^*, n \geq 0\}$.

Solution The context-free rule sets for the given languages are:

(i) $\{S \rightarrow aSb \mid aSbbb \mid aSbbbbb \mid \varepsilon\}$

(ii) $\{S \rightarrow aSa \mid bSb \mid S_1, S_1 \rightarrow aS_1b \mid \varepsilon\}$

4.11 RESEARCH DIRECTIONS

There are several problems for extensive research in the area of formal grammar. Let us re-visit the structure of sentences in natural languages, as was discussed in the introduction to this chapter.

Examine these English sentences:

This is a book. This chapter is very interesting. I am a student of computer science.

Based on English grammar, the structure of these sentences can be visualised as:

[This = *Pronoun*] [is = *Verb*] [a = *Article*] [book = *Noun*].

[This = *Pronoun*] [chapter = *Noun*] [is = *Verb*] [very = *Adverb*] [interesting = *Adjective*].

[I = *Pronoun*] [am = *Verb*] [a = *Article*] [student = *Noun*] [of = *Preposition*] [computer science = *Noun*]

What does this interpretation mean? It means that it is possible to express the rules of the grammar of the English language in the form of formal grammar.

Let us use these symbols. S = Start symbol, P = Pronoun, V = Verb, A = Article, N = Noun, E = Adverb, D = Adjective, R = Preposition

In the context of the sentences discussed above, we may devise the following rules:

$S \rightarrow PVAN$

S → PNVED

S → PVANRN

However, it is not that simple. English is a very context-sensitive language. The same word may acquire multiple meanings and can also act as different parts of speech, depending on the context in which it is used.

- Run fast – here the word "run" is a verb.
- They have added one more run to the total – here the word "run" is a noun, if this sentence is mentioned in the context of a game of cricket or baseball.

Thus, we need to design a system that can determine whether an English sentence is grammatically and contextually correct or not. When you attempt it, you will come to appreciate the complexity of something that seems so natural and routine.

Project-based problems

1. Design a grammar for your native language.
2. Design a grammar for the C programming language.

EXERCISES

1. Let P = {S → AAA | B, S → aA | B, B S → AAA | B → ε}. Find the useless rules and eliminate them. Also eliminate the ε production and unit production and convert the grammar to CNF.
2. Show that S → SS | (S) | ε is an ambiguous grammar.
3. Apply left-factoring for the rule sets:

   (a)P= $\{S \to S0S1S|\,01\}$, (b)P= $\{S \to (L)|\,x, L \to L,S|\,S\}$ (c)P= S→ $\{aSSbS|\,aSaSb|\,abb|\,b\}$
4. Convert the rule set P = $\{S \to AA|\,a,\ A \to SS|\,b\}$ to GNF. (Deleting A from S → AA causes immediate left-recursion, which must be removed).
5. Convert the rule set P = $\{E \leftrightarrow E+T|\,T,\ T \to T*F|\,F, F \to id\}$ to GNF.
6. If G is a context-free grammar and *w* is a string of length *l* in L(G), how long is a derivation of *w* in G, if G is in Chomsky Normal Form? Justify the answer.

   (a) $2l$ (b) $2l + 1$ (c) $2l - 1$ (d) l
7. Context-free languages are:

   (a) Closed under union (b) Closed under complement

   (c) Closed under intersection (d) Closed under Kleene closure
8. Convert the grammar (G): S → YXY, Y → bba | ε, X → Y | Y c to Chomsky Normal Form.
9. Convert the grammar (G): S → A, A → a B a | a, B → b A b | b to GNF.
10. (a) Prove that, for any context-free grammar (which does not have null production), there is an equivalent grammar in Chomsky Normal Form.

    (b) Prove that, for any context-free grammar (which does not have null production), there is an equivalent grammar in Greibach Normal Form.

    Hint: Apply conversion strategies for CNF and GNF.
11. How can we decide whether or not two different CFGs define the same language?

12. How can we tell whether or not the complement of a given CFL is also CFL?
13. Given two CFGs, how can we tell whether or not they have a word in common?
14. Given a CFG, how can we tell whether or not there are any words that it does not generate?
15. Write the grammar for the language $\{a^nb^n \mid n \text{ is a positive integer}\}$. **[Gujarat SET 2011]**
16. For a string $x \in \{0, 1\}^*$, let x^R denote the string obtained by reversing the string x, and let x be the string obtained from x by changing all 0s to 1s and 1s to 0s. Show that the set $\{x \mid x^R = x\}$ is context-free. **[IIT Chennai PhD Entrance 2014]**
17. Write a grammar for the prefix notation (Polish) arithmetic expression.
18. Give a context-free grammar G that generates $L = \{0^i1^j0^k \mid i + k = j\}$. Prove that L = L(G). **[ISI Kolkata JRF 2013]**
19. Construct a context-free grammar (CFG) to generate the following language: $L = \{a^nb^mc^{n+m}: n, m \text{ are integers, and } n \geq 1, m \geq 1\}$. **[ISI Kolkata JRF 2015]**
20. Eliminate all left recursions from the following grammar: S → AB | Sc, A → BA | a, B → SB | Ab | b. **[JNU PhD Fellowship Exam 2016]**
21. Consider the grammar S → a | Sa | b | bS. Show that $L(G) = \{b^ma^n \mid m, n \geq 0\}$. **[North Maharashtra University PhD Entrance 2004]**
22. Convert the following grammar to CNF: S → Aba, A → aab, B → Ac **[UGC NET 2006]**
23. Convert the following grammar to GNF: S → AB, A → aA | bB | b, B → b **[UGC NET 2006]**
24. Consider the grammar: S → Aa|b, A → Ac|Sd| ε. Construct an equivalent grammar with no left recursion and with minimum number of production rules. **[GATE 1998]**
25. Which of the following languages is generated by the grammar, S → aS | bS | ε?
    (a) $\{a^nb^m \mid n, m \geq 0\}$ (b) $\{w \to \{a,b\}^* \mid w$ has equal number of *a*s and *b*s$\}$
    (c) $\{a^n \mid n \geq 0\} \cup \{b^n \mid n \geq 0\} \cup \{a^nb^n \mid n \geq 0\}$ (d) $\{a,b\}^*$ **[GATE 2016]**
26. Which of the following grammars is free from left recursion?

    (a) S → AB
    A → Aa | *b*
    B → c

    (b) S → Ab | B*b* | c
    A → Bd | ε
    B → e

    (c) S → Aa | B
    A → Bb | Sc | ε
    B → d

    (d) S → Aa | B*b* | c
    A → Bd | ε
    b → Ae | ε

    [GATE 2016]
27. A student wrote two context-free grammars, G1 and G2, to generate a single C-like array declaration. The dimension of the array is at least one. For example, int a[10][3]; The grammars use D as the start symbol and have six terminal symbols, int ; id [ ] num.

Grammar G1	*Grammar G2*
D → int *L*;	*D* → int *L*;
L → id [*E*	*L* → id *E*
E → num]	*E* → *E* [num]
E → num] [*E*	*E* → [num]

    Which of the grammars correctly generates the declaration mentioned above?
    (a) Both G1 and G2 (b) Only G1 (c) Only G2 (d) Neither G1 nor G2 **[GATE 2016]**

28. The grammar {S → aSa, A → bB, B → bB, B → C} is of: **[Karnataka SET 2014]**
    (a) Type 3 (b) Type 2 but not Type 3 (c) Type 1 but not Type 2 (d) Type 0 but not Type 1
29. Use pumping lemma to prove that this language is not regular: L = {}.
30. Write CFG to accept the language $\{a^n b^n c^m d^m \mid n \geq 0, m \geq 0\}$ **[UGC NET 2008]**
31. Find a GNF grammar equivalent to the following context-free grammar:
    S → AA | 0, A → SS|1 **[UGC NET 2010]**
32. Which of the following grammars generates the language $L = \{ww^R \mid w \to \{0, 1\}^*\}$?
    (a) S → 0S1 | 1S0 | ε (b) S → 0S0 | 1S1 | ε
    (c) S → 0S1 | 1S0 | 0S0 | 1S1 | ε (d) S → 0S1 | 1S0 | 0S0 | 1S1
    [JNU M.Phil/PhD Entrance 2013, HP SET 2015]
33. Let G be a grammar with productions S → AACD, A → aAb | ε, C → aC | a, D → aDa | bDb | ε. Convert the grammar G to Chomsky Normal Form. **[UGC NET 2011]**
34. Consider the grammar {S → ABSc|Abc, BA → AB, Bb → bb, Ab → ab, Aa → aa}. Which of the following strings can be derived by this grammar?
    (a) abc (b) aab (c) abcc (d) abbc
    [Pondicherry University PhD Entrance 2013, Karnataka SET 2015]
35. Given the following context-free grammar with production rules: S → aS | A | C, A → a, B → aa, C → aCb. Eliminate useless symbols and productions, and find the equivalent grammar.
36. Given the languages: $L1 = \{a^n b^m \mid n \leq m\}$, $L2 = \{a^n b^m \mid n \neq m\}$. Which of the following is correct? (a) L1 is regular and L2 is not regular. (b) L1 is not regular and L2 is regular.
    (c) Both L1 and L2 are regular. (d) Both L1 and L2 are not regular. **[HP SET 2014]**
37. Which of the following strings is not generated by the grammar: S → SaSbS | ε?
    (a) aabb (b) abab (c) aababb (d) aaabb **[Karnataka SET 2013]**

ANNEXURE I

Converting a Right-Linear Regular Grammar to NFA

Regular grammar defines a regular language, and NFA also defines a regular language. So, there has to be a clear equivalence between regular grammars and NFA. Designing an NFA from the right linear form of regular grammar is very easy. If the regular grammar is not specified in the right linear form, then that grammar must first be converted to the right linear form.

Recall that right linear regular grammar has rules of the form: A → *w*N or A → *w* or A → ε, where $w \in \Sigma^*$ and $N \in V_N$.

> **Note:** It should be observed that any regular grammar expressed in Greibach Normal Form (GNF) is obviously a right linear regular grammar. This illustrates one more important feature of GNF – ease of conversion to NFA.

This enables a straightforward technique to derive an NFA for the grammar. We list the building blocks below:

The start symbol S is always the initial state of the NFA.

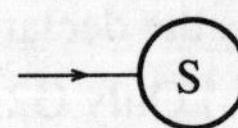

FIGURE A.1 NFA representing the start symbol S as the start state

- $A \to x$, where $x \in \Sigma$ and $A \in V_N$. This implies that on encountering the terminal symbol: x at state A, the NFA reaches a final state.

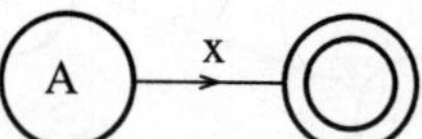

FIGURE A.2 NFA for rule: $A \to x$

- $A \to xB$, where $x \in \Sigma$ and $A, B \in V_N$. This implies that on encountering the terminal symbol x at state A, the NFA reaches another state B.

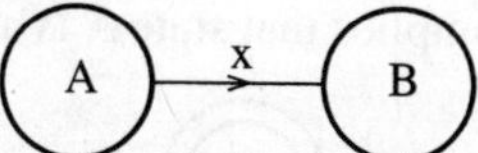

FIGURE A.3 NFA for rule: $A \to xB$

- $A \to xyz$, where $x, y, z \in \Sigma$ and $A \in V_N$. This defines the following linear chain, ending in a final state.

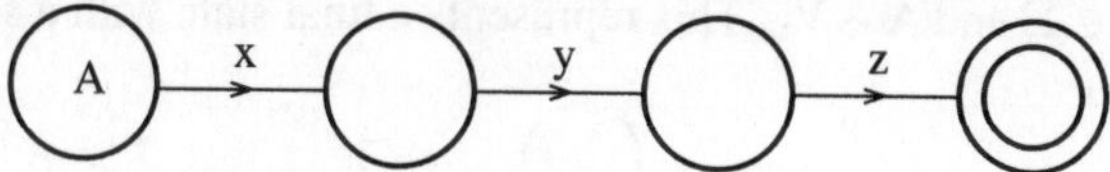

FIGURE A.4 NFA for rule: $A \to xyz$

- $A \to xA \mid x$, where $x \in \Sigma$ and $A \in V_N$. This defines a self-loop along with a final state.

FIGURE A.5 (a) NFA for rules: $A \to xA \mid x$

This structure clearly defines an NFA, as there are two possible state transitions for input symbol x from state A. It should be intuitively evident to the readers that this structure is logically equivalent to the following deterministic state transition.

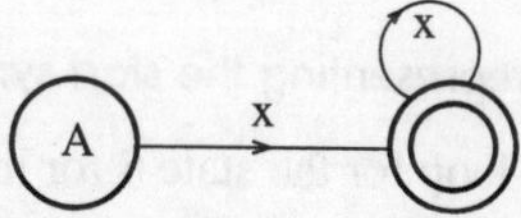

FIGURE A.5 (b) NFA for rules: $A \to xA \mid x$

- $A \to xB \mid xC$, where $x \in \Sigma$ and $A, B, C \in V_N$. This is also a simple NFA transition.

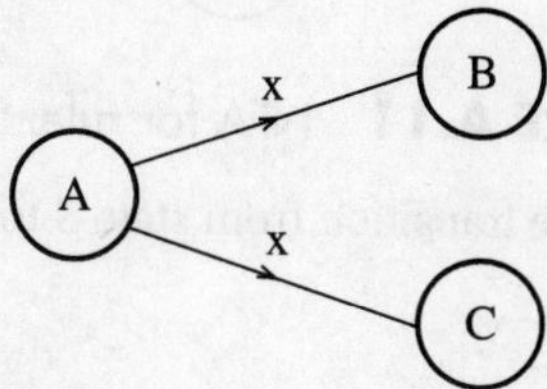

FIGURE A.6 NFA for rules: $A \to xB \mid xC$

- $A \rightarrow xB \mid yC$, where $x, y \in \Sigma$ and $A, B, C \in V_N$. This is also a simple state transition.

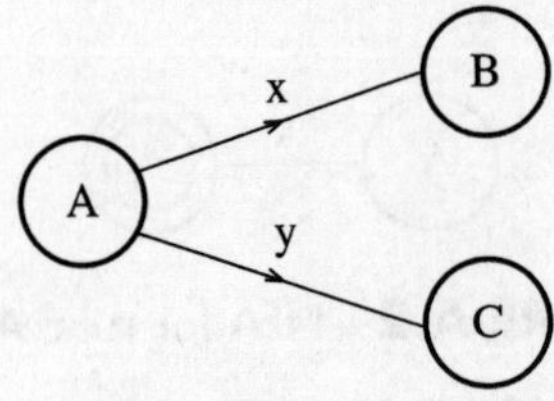

FIGURE A.7 NFA for rules: $A \rightarrow xB \mid yC$

- $A \rightarrow \varepsilon$, where $A \in V_N$. This simply implies that state A is a final state.

FIGURE A.8 NFA for rules: $A \rightarrow xA \mid \varepsilon$

- $A \rightarrow xA \mid \varepsilon$, where $x \in \Sigma$ and $A \in V_N$. This represents a final state with a self-loop for symbol x.

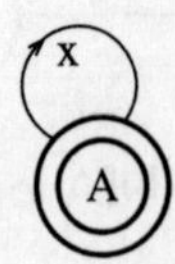

FIGURE A.9 NFA for rules: $A \rightarrow xA \mid \varepsilon$

Keeping these rules in mind, we explain the technique of generating an NFA from a right linear regular grammar using an example.

Generate an NFA for the grammar: $S \rightarrow aS$, $S \rightarrow bA$, $A \rightarrow \varepsilon$, $A \rightarrow cA$.

Step 1: Define the initial state S for the start symbol.

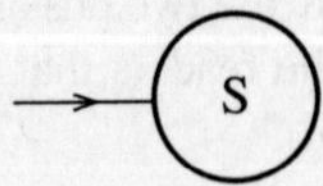

FIGURE A.10 NFA representing the start symbol S as the start state

Step 2: For the rule: $S \rightarrow aS$, add a self-loop for the state S for input symbol a.

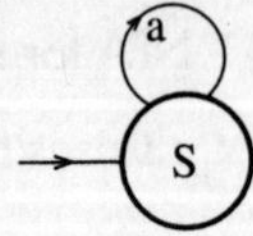

FIGURE A.11 NFA for rule: $S \rightarrow aS$

Step 3: For the rule $S \rightarrow bA$, add a state transition from state S to state A for the input symbol b.

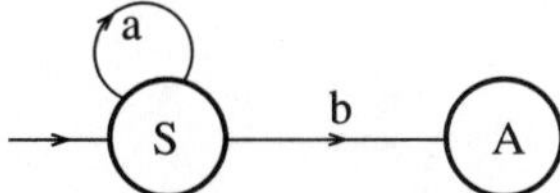

FIGURE A.12 Extension of NFA shown in Fig. A.11 for rule: S → *b*A

Step 4: The rule A → ε implies that the state A is a final state. Mark the state A as a final state.

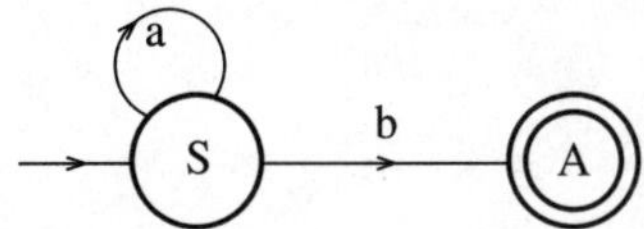

FIGURE A.13 Extension of NFA shown in Fig. A.12 for rule: A → ε

Step 5: For the rule A → *c*A, add a self-loop for the state A for input symbol *c*.

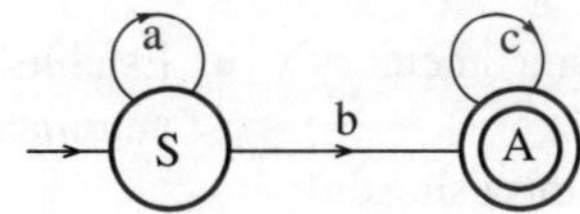

FIGURE A.14 Extension of NFA shown in Fig. A.13 for rule: A → *c*A

This completes construction of the NFA.

Note: The NFA generated in this case is indeed deterministic, without any dead states. The regular expression corresponding to this grammar or NFA is a*bc*.

5 Pushdown Automata

OBJECTIVES

At the end of this chapter, you will be able to

- Explain the limitations of constant memory automata and how to overcome them
- Express stack-based parsing algorithms using the formal notation of pushdown automata
- Establish a correlation between pushdown automata and context-free languages

5.1 INTRODUCTION

The idea of pushdown automata (PDA) has been attributed to Oettinger (in 1961) and to Schutzenberger (in 1963), while the concept of deterministic pushdown automata (DPDA) was first introduced by Fischer in 1963.

You may recall that decision problems such as odd number of 0s and even number of 1s in valid strings, length of valid strings divisible by a positive integer (say, *n*), properties such as at least/most some measure (count), and so on, can be resolved by using the finite number of states of finite automata (FA). To solve such problems, an FA reaches a particular state that satisfies a specified count and then repeats that path (if required). Further, ordering of symbols (for example, the appearance of *a*s before *b*s, strings beginning with *a* and ending with *b*) can be easily managed by state control or state change in FA. State control can also be used to solve problems with finite measures (that is, counts requiring a finite amount of memory) and matching.

However, problems that satisfy infinite measures and matching (requiring an almost infinite amount of memory, such as stack, to remember data) cannot be solved by FA. The limitation of FA is that it cannot handle languages that require some form of unbounded memory – something that can be implemented easily by a single (unbounded) integer register. More specifically, infinite languages that satisfy constraints such as equal number of *a*s and *b*s, *n* number of *a*s followed by *n* number of *b*s, checking for palindrome and the well-formedness of parentheses, verifying a particular pattern (P_1) *n* times in comparison to another pattern (P_2), and so on, cannot be resolved by simple FA; such languages have infinitely many distinguishable strings and require machines to have an infinite number of states. In short, if we need to compare two or more properties that occur in an infinite language, it cannot be satisfied by FA. To solve such problems, the automaton requires extra memory to remember some contents for future processing. In other words, it is necessary for the machine to memorise the path through which it proceeds. To do so, we may use a special type of FA in which stack (an auxiliary memory that is both readable and writeable) is present. Such a type of FA is termed **pushdown automata (PDA)**. In PDA, the stack can store an infinite amount of information for future processing and it plays an important role in recognising some non-regular sets. Of course, all regular sets must be recognised by PDA. The bigger application of PDA is to design the parser part of the compiler (mainly shift-reduce parser). It may be noted that the above-mentioned constraints of strings can also be satisfied by more powerful automata, such as linear bounded automata (LBA) or Turing machine (TM), where the tape heads are read/write heads and can move backwards too.

FA and PDA: Expressed concisely, PDA = FA + STACK. Here, FA and STACK can be viewed as the CPU (with limited memory; in-built chip) and RAM (primary memory), respectively. Hence, such automata are identical to deterministic finite automata (DFA) or non-deterministic finite automata (NFA), except that they carry memory in the form of a stack in addition. As more memory is added, its power is greater than that of FA; that is, it can solve more complex problems. It may be noted here that the behaviour of the reading head of PDA is the same as of that in FA. However, the transition function (δ) of PDA depends on the symbol at the top of the stack, which specifies how the stack is to be changed at each transition.

In summary, PDA differ from FA in two ways:

- The size of the input tape in PDA is infinite, whereas the size of the tape in FA is assumed to be fixed and the input is marked by two end markers or by the two ends of the input tape.
- A PDA uses the top of the stack to decide which transition to use, that is, it can manipulate the stack as part of performing a transition.

However, the main **drawback** of PDA is that, at any given moment, it can remember only the stack-top symbol; so matching is possible only with the stack-top symbol. The stack cannot count numbers in the way a counter does, though it does help in counting through matching. This is explained below with an example.

A farmer has some goats. He cannot count the exact number of goats he has. He employs a boy to look after the goats. Every day, the boy goes to the farmer's house in the morning and the farmer hands him the goats one by one, putting one stone for each goat in a bag. When the boy comes back with the goats in the evening, the farmer counts the goats by taking out one stone for each goat. Here, the bag works like a stack in PDA. The general model of PDA is shown in Fig. 5.1.

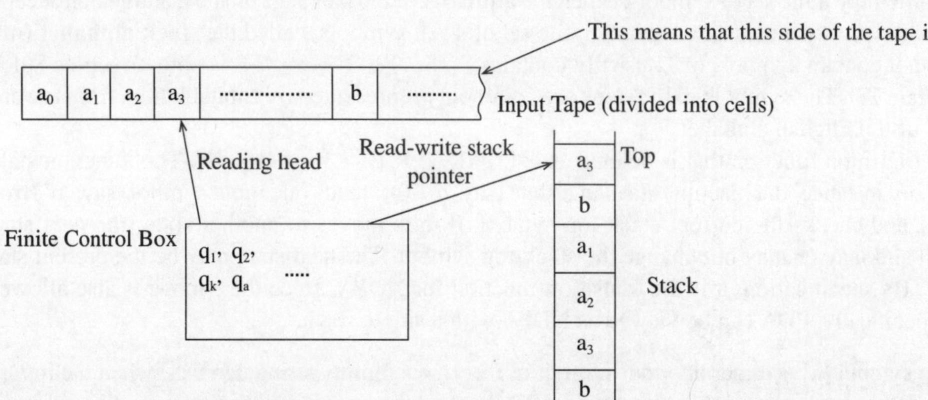

FIGURE 5.1 A schematic representation of PDA

Figure 5.1 shows that a PDA has the following **components**:

- A one-way read-only **input tape**: Input data is stored on the tape. The symbols in the tape cannot be erased. The size of the tape is infinite but it is bounded on one side.
- A **read head** (as the input tape is read-only): The head is unidirectional; it can move forward but not backwards.
- A **stack**: The stack is a potentially infinite, ordered collection of initially blank cells. A read-write **stack pointer** connected with the stack processes the stack contents. The pointer can move up or down across the stack. Here, data may be erased.
- A **finite control box (FCB)**: This helps the reading head to sense the symbol written at any position on the tape. It also guides the stack and changes the state, if necessary.

This type of automaton can recognise context-free languages (CFLs) as such recognition requires more powerful machines than FA. After all, a PDA accepts input by consuming it and entering into an accepting state. We call this approach **acceptance by final state**. However, another equivalent approach, called **acceptance by empty stack**, exists for PDA. Both the approaches are explained later in the chapter.

It may be noted here that, *for every context-free grammar, there exists a pushdown automaton such that the language generated by the grammar is identical to the language recognised by the automaton, and vice versa.*

Note: Referring to a real computer, the computational power of PDA may be viewed as equal to the computational power of CPU with RAM (main memory, though it is in practical terms finite but theoretically unbounded, like the stack). Interestingly, PDA is often called a **zero-address hypothetical machine**. This is because all the operations are maintained through the stack-top (auxiliary memory), and so the stack does not need to remember other addresses for insertion or deletion operations.

5.2 FORMAL DEFINITION

A PDA is formally defined as a 7-tuple, $M = (Q, \Sigma, \Gamma, \delta, q_o, Z_0, F)$, where

- Q is a non-empty finite set of states.
- Σ is a non-empty finite set of symbols, called the **alphabet** of the language, that the automaton accepts.
- Γ (a Greek letter) denotes a non-empty finite set of stack symbols, called the **stack alphabet** (may or may not contain symbols of Σ; usually contains Σ). So, $\Sigma \subset \Gamma$ but $\Sigma \subseteq \Gamma$ is not correct (as Σ does not contain Z_0). The symbols in the stack are commonly represented by capital letters from towards the end of the English alphabet.
- δ is the transition function that is defined as $\delta: Q \times (\Sigma \cup \{\varepsilon\}) \times \Gamma \rightarrow Q \times \Gamma^*$. This function states that, at any instance, the machine stays in a state (say, $q_i \in Q$), reads one input symbol (say, 'a') from $\Sigma \cup \{\varepsilon\}$ and checks the current stack-top symbol. It then moves to another state (the next state, $q_j \in Q$) and may or may not change the stack-top symbol. The next state may be the present state q_i itself. By specification, it is the transition function for NDPA, since the ε move is also allowed. More specifically, PDA can be viewed as NDFA with a single stack.

Note: The symbol 'a' is either an input symbol in Σ or $a = \varepsilon$ (empty string). In fact, when the input is processed, the null string (ε) is assumed by PDA to end the process. Other markers like '#' and blank space (denoted by B or $\not{b}$) are not used here to end the input.

- q_0 is the start state, that is, the state from which the PDA starts processing; $q_0 \in Q$.
- Z_0 is the initial stack symbol. It is the end marker of the stack and used to indicate an underflow situation (no data left for processing; the stack is empty).

Importance of Z_0: When the stack becomes empty, it is difficult for the machine to take decisions about moves such as push and pop (these operations depend on the present input symbol, present state and stack-top symbol). That is why special symbols like Z_0 or '\$' are encountered to move to the accepting state.

- $F \subseteq Q$ (a subset of Q) is a set of final states. (Here, Z_0 is not in Σ but it is in Γ. Thus, it is not necessary that all the symbols of Γ are to be taken only from Σ).

Interestingly, the contents of Γ^* in δ: $Q \times (\Sigma \cup \{\varepsilon\}) \times \Gamma \to Q \times \Gamma^*$ play an important role in representing stack operations such as push, pop and read. Here, the symbol ε is usually taken to indicate the end of processing.

The transition configuration can be generalised as follows: $(p, a, x) = (q, \alpha)$. Here, p denotes the present state of the PDA; a denotes the current input read by the machine; x is the current stack-top content; q is the next state; and α represents the updated string of stack symbols (such as $\alpha = a\,x,\ldots$, assuming that a is at the top). It may be noted that x may not be null unless the concept of empty stack is taken into account.

Now, according to the generalised form of transition, it is clear that a number of operations are possible in PDA based on the input symbol, the stack-top symbol and the present state specified in the generalised form. These parameters guide the operations of the PDA. The three important operations, namely, push, pop and read, performed by a PDA are explained below.

- **Push**: To represent push, α in $(p, a, x) = (q, \alpha)$ is formally shown as ax, that is, $(p, a, x) = (q, ax)$. Specifically, $a \neq \varepsilon$, $\alpha \neq \varepsilon$ and $x \neq \varepsilon$ (except in an empty stack). In other words, the transition: $(p, a, x) = (q, \alpha)$ implies that the machine changes state from p to q and α becomes: $\alpha = ax\ldots$, that is, the earlier stack-top symbol was x and currently it is 'a'. More explicitly, input: 'a' is pushed onto the stack-top of PDA. However, depending on the nature of the problem, the state need not be changed; the machine may remain in the same state; that is, $p = q$.
- **Pop**: To represent pop, α in $(p, a, x) = (q, \alpha)$ is formally shown as ε, that is, $(p, \text{a}, x) = (q, \varepsilon)$. This implies that input 'a' is read and the present stack-top is deleted; ε replaces x (stack-top).
  Note that the input 'a' may be ε as well. This means that the stack-top may be deleted without consuming any input. For example, if the machine is at p and sees stack-top 'b' ($x = b$), it reads no input and deterministically performs a pop operation. This is shown in Example 5.2 (b). The state changes here from p to q but it may remain in the same state as well. Likewise, it may be true for the read (instead of the pop) operation, and x remains unchanged.
- **Read**: For the read operation, α in $(p, a, x) = (q, \alpha)$ is formally shown as x, that is, $(p, a, x) = (q, x)$. In other words, one read operation means that the stack-top remains unchanged. However, the state may change from p to q. As the stack-top is not changed, neither push nor pop is performed. This is why it is named the read operation. Note that the input 'a' may be ε here. In practice, $a = \varepsilon$ arises when an empty string is also valid (without consuming any input, the machine moves to the final state or it is acceptable in the case of NPDA). This operation is used to resolve several problems.

Note: If transitions for the dead state are not explicitly specified, it simply means that the PDA moves to the dead state for such case(s). Accordingly, the machine crashes for such moves.

Instantaneous Description (ID): This description shows how a PDA processes a string to accept or reject it. More specifically, it shows how the PDA goes from configuration to configuration to process an input. For an ID of PDA, we will use the state relations (moves) of the designed PDA.

State diagram for representing PDA: For diagrammatic representation of the operations push, pop and read, the following edge labels may be considered:

- $<x, z> \mid xz$ for push, where x represents the input symbol, z is the present stack-top symbol and x in xz denotes the latest stack-top.
- $<x, z> \mid z$ for read, where x is the input symbol, z is the present stack-top symbol and z (after the pipe symbol) denotes the stack-top after the read operation, that is, the stack-top remains unchanged.
- $<x, z> \mid \varepsilon$ for pop, where x is the input symbol, z is the present stack-top symbol and ε signifies that the stack-top is removed through the pop operation.
- The edge label for the read operation for null string in a non-empty stack (stack with top symbol, say Z_0) is $< \varepsilon, Z_0> \mid Z_0$.
- The edge label for the read operation for null string in an empty stack is $< \varepsilon, \varepsilon > \mid \varepsilon$.

State diagram of FA vs state diagram of PDA: The state diagram of PDA is somewhat different from that of FA:

- Stack is used in PDA, whereas there is no such concept in FA. As stack is used in PDA, the number of possible combinations using state, stack and input symbols increases. Accordingly, the number of possible transitions also increases.
- ε move is acceptable in DPDA but not in DFA (as the chance of non-determinism for the ε move cannot be avoided in DFA). The main reason for this is the absence of stack.

Transitions of PDA: The transition function stated above for PDA is by definition for a non-deterministic PDA, as it allows the ε transition. In the context of FA, we have seen that ε transition (explicit ε move) always causes non-deterministic behaviour (due to consideration of implicit ε move). However, PDA may control this behaviour for ε transition with the help of the stack-top content (not present in FA). Actually, a DPDA in the current state must examine the top of the stack and decide whether to make the ε transition or to read the (real) input for a move.

Note: In transitions, the state plays an important role by fulfilling the restrictions specified in the problem. States other than the initial or final states are used in transitions to satisfy some important properties of languages. For example, let us consider two problems: (i) $L = \{a^n b^n \mid n \geq 0\}$, (ii) $L = \{a^n b^n \mid n \geq 1\}$. The problems are identical except for the value of n. The second problem ensures that the value of n is at least 1, whereas n is 0 for the first problem. Now, the presence of at least one '*a*' and one '*b*' in a string may be validated by state control.

Ways to Accept a Language by PDA: Two ways are used by PDA to accept a language:

- While accepting a string, the PDA moves to a final state and does not erase the end marker of the stack, Z_0. In other words, the language is accepted by the final state.
- The PDA uses an empty or null stack (where the end marker of the stack is erased); the final state is not necessary for accepting strings. The language here is simply accepted by an empty stack.

Let us explain these two cases with the help of transitions. Suppose a PDA (M) accepts a string (w) if it is at state q_i, and sees input ε and stack-top Z_0. Then, the transition for the first case can be expressed as: $(q_i, \varepsilon, Z_0) = (q_f, Z_0)$. This transition can be represented using an empty stack as: $(q_i, \varepsilon, Z_0) = (q_i, \varepsilon)$.

Utility of Empty Stack: The main purpose of taking an empty stack is to use fewer number of states in PDA. However, a recognition run (transition) need not end at the point of recognition if the stack is empty. In other words, null moves may permit further transitions, and such transitions do not affect recognition of the input. Thus, if the stack becomes empty, a null string may be allowed to read with a null stack to proceed further.

5.2.1 Types of Pushdown Automata

It is well-accepted that if any language is accepted by a PDA, it is called a CFL. Two types of PDA are taken into account for CFLs: deterministic pushdown automata (DPDA) and non-deterministic pushdown automata (NPDA). A PDA is a DPDA if the following two conditions are satisfied:

- $\delta(q, a, X)$ is either empty[1] (Φ) or singleton for every $a \in \Sigma$.
- If $\delta(q, \varepsilon, X) \neq \Phi$, then $\delta(q, a, X) = \Phi$ for every $a \in \Sigma$. Normally, the ε transition in DPDA is taken as the last transition and is used to end the processing of input.

Now, if any one of these two conditions fails, the PDA is called an NPDA. However, in the case of NPDA, it may be assumed that the machine guesses the right transition to perform the right operation for solving the problem. The non-deterministic case usually occurs due to an ε move, and some special symbols (like $, #) may be chosen as the right end marker for the input string to make the machine a DPDA. Actually, while storing data in the tape, a marker like $, #, may be appended when no real input is left to be stored.

Deterministic Context-free Language (DCFL) and Non-deterministic Context-free Language (NCFL): Based on the language acceptance capability of PDA, CFLs are categorised as deterministic context-free languages (DCFL) and non-deterministic context-free languages (NCFL). A CFL accepted by DPDA is called a DCFL, whereas a CFL accepted by NPDA (strictly not by DPDA) is known as NCFL (or simply CFL). In this respect, a simple block diagram is shown in Fig. 5.2.

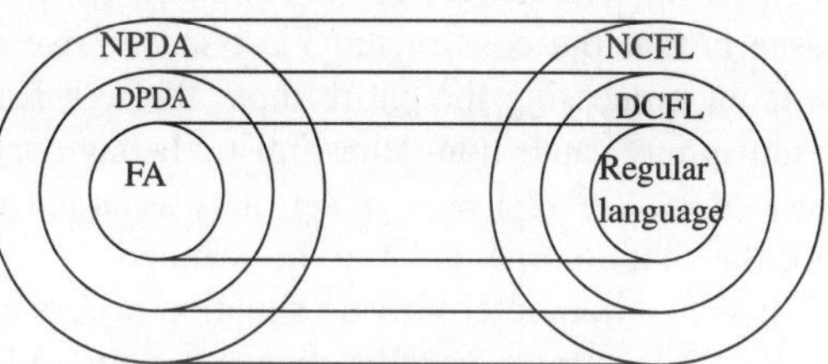

FIGURE 5.2 Mapping between PDA and CFL

From Fig. 5.2, we may easily claim that NPDA is more powerful than DPDA. The concept of non-determinism for PDA is made clear through the state diagram shown in Fig. 5.3 (a)).

DPDA AND REGULAR LANGUAGES

The class of languages that DPDA accepts sits between regular languages and CFLs (NCFL). Thus, DPDAs accept all regular languages. Truly, a DPDA includes a stack, but in order to simulate a regular language, it does not use the stack. The stack remains inactive.

In particular, if A is an FA for accepting the language L, then $\delta_M(q, a, z) = \{(p, z)\}$ of DPDA (M) for all p, q belong to Q (set of states) such that $\delta_A(q, a) = p$.

Note: A CFL accepted by a DPDA is called a DCFL.

[1] An empty (Φ) transition means a transition that is not specified. A singleton transition results in only one pair of outcomes (unique), where $p \in Q$, $a \in \Sigma$ or $a = \varepsilon$ and $X \in \Gamma$. More specifically, it does not result in multiple states or multiple operations like push, pop, read or accept with multiple states.

EXAMPLE 5.1

Design a PDA for the language L = {a^n | $n \geq 0$} and check if it is DPDA or NPDA.

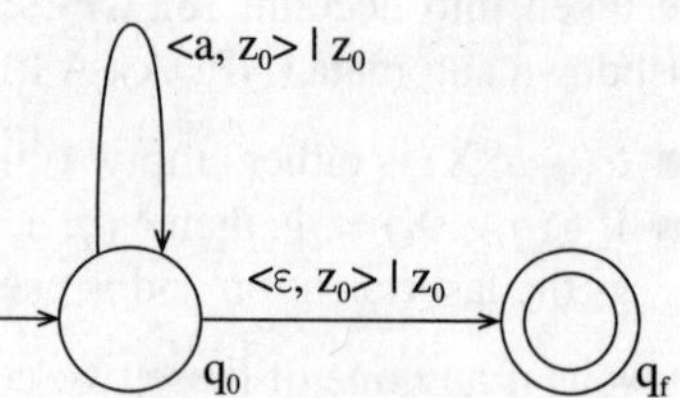

FIGURE 5.3 (a) An NPDA for L = {a^n | $n \geq 0$}

Solution Let M = (Q, Σ, Γ, δ, q_0, Z_0, F) be the formal model of the proposed PDA for solving the given problem, where Q = {q_0}, Σ = {a, b}, Γ = {Z_0} and F = {q_0}. The transitions are depicted in Fig. 5.3 (a).

Here, only two transitions are required to solve the problem:

1. $\delta(q_0, a, Z_0) = (q_0, Z_0)$ // Only *a*s are read but not pushed, that is, a simple read operation
2. $\delta(q_0, \varepsilon, Z_0) = (q_f, Z_0)$ // Moving to the final state when no input is left

It is, indeed, an NPDA, as condition (2) fails. In reality, the issue is that the constructed PDA may move to the final state without processing the entire input through Rule 2; that is, the ε move may cause non-deterministic behaviour. However, in the case of $\delta(q_0, \varepsilon, Z_0) = (q_f, Z_0)$, it is assumed that the machine guesses that no input is left to be scanned.

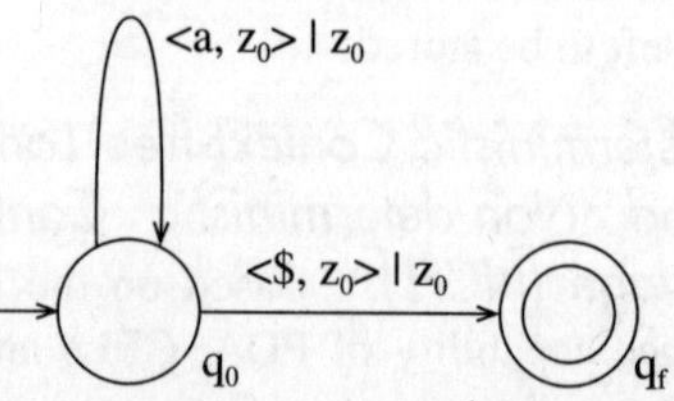

FIGURE 5.3 (b) A DPDA for L = {a^n | $n \geq 0$}

If such a non-deterministic situation occurs in PDA (as shown in Fig. 5.3 (a)), we usually choose a right end marker (such as $) for the input string to avoid it. By so doing, the NPDA is converted to DPDA, as shown in Fig. 5.3 (b).

Now, the transitions are:

1. $\delta(q_0, a, Z_0) = (q_0, Z_0)$
2. $\delta(q_0, \$, Z_0) = (q_0, Z_0)$

For this problem, to make the machine a DPDA by using $, no ε move is considered. Simultaneously, the conditions for DPDA are satisfied. However, conversion from NPDA to DPDA depends on the problem to be solved. This may be observed in the problems discussed later in this chapter.

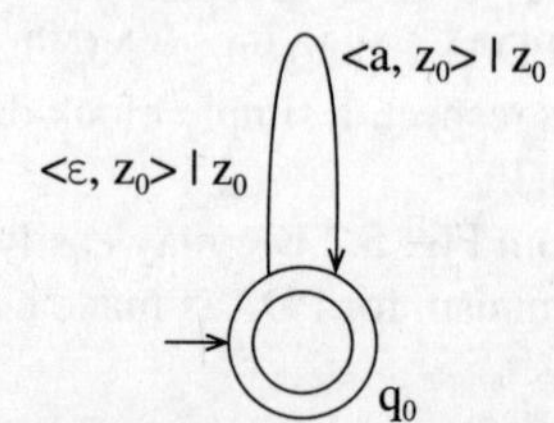

FIGURE 5.3 (c) A PDA for L = {a^n | $n \geq 0$}

Note: In pictorial representations of PDA, we usually do not consider the initial state as the final state for accepting a null string (allowed in FA, assuming that FA moves to the final state without consuming any input). This is because we need to show the stack contents over the edges for the PDA (not required in FA). In particular, if the initial state is taken as the final state for PDA to show the acceptance of the null string, then the stack content is ignored. Hence, this representation is not usually accepted in PDA. In practice, one more state is taken to represent the final state, labelling the edge with the null string and the stack-top contents (say, Z_0). This practice is also followed in TM, where movement of the read-write head is necessary to show the transition explicitly. However, many authors manage this case as required. In this book, separate states (for PDA as well as TM) are considered for managing this situation. Accordingly, the PDA shown in Fig. 5.3 (c) for L = {a^n | $n \geq 0$} is not allowed.

EXAMPLE 5.2

Design a PDA to accept the language $L = \{WcW^R \mid W \in \{a, b\}^*\}$.

Solution To design a PDA to solve the problem, we may concentrate on the following key points:

- *Push mode*: Before reading input 'c', the machine pushes the contents of the string onto the stack.
- *Read mode*: On reading 'c', it switches its state.
- *Pop mode*: After the read operation, the machine reads the remaining symbols of input, ensuring that each new read symbol is identical to the symbol popped out from the stack.
- The machine accepts the processed string if Z_0 is left in the stack and no input remains to be scanned (or the stack becomes empty in an empty stack case). Otherwise, the string is rejected.

Formal model using Z_0 as the initial stack-top symbol: Let the PDA, $M = (Q, \Sigma, \Gamma, \delta, q_0, Z_0, F)$, where $Q = \{q_0, q_1, q_f\}$, $\Sigma = \{a, b, c\}$ and $\Gamma = \{a, b, Z_0\}$. q_0 is the start state, Z_0 is the initial stack symbol, $F = \{q_f\}$. δ is defined as follows to solve the problem.

Push operation

- *State is q_0, present input is 'a' or 'b' and stack-top is Z_0.*

1. $\delta(q_0, a, Z_0) = (q_0, aZ_0) \Rightarrow$ The string: aZ_0 shows that the earlier stack-top content was Z_0 but it is currently '*a*' (by pushing onto the stack); that is, Z_0 is just below '*a*'.
2. $\delta(q_0, b, Z_0) = (q_0, bZ_0) \Rightarrow$ The string: bZ_0 implies that the earlier content of the stack-top was Z_0 but currently it is '*b*'; that is, Z_0 is just below '*b*'.

- *State is q_0, present input is 'a' or 'b' and stack-top is 'a' or 'b'.*

3. $\delta(q_0, a, a) = (q_0, aa) \Rightarrow$ '*aa*' implies that the earlier content of the stack-top was '*a*' but currently it is '*a*'. In other words, the current '*a*' is just above the earlier '*a*'.
4. $\delta(q_0, b, a) = (q_0, ba) \Rightarrow$ '*ba*' implies that the earlier content of the stack-top was '*a*' but currently it is '*b*'; that is, '*b*' is just above the '*a*'.
5. $\delta(q_0, a, b) = (q_0, ab) \Rightarrow$ '*ab*' implies that the earlier stack-top symbol was '*a*'.
6. $\delta(q_0, b, b) = (q_0, bb) \Rightarrow$ '*bb*' implies that the earlier content of the stack-top was '*b*' but currently also it is '*b*'. In other words, the current '*b*' is just above the earlier '*b*'.

(*Transitions 1–6 continue until input 'c' is read, and state change is not required*).

Read operation: The state is q_0, input is '*c*', stack-top is '*a*' or '*b*' or Z_0.

7. $\delta(q_0, c, b) = (q_1, b)$
8. $\delta(q_0, c, a) = (q_1, a)$
9. $\delta(q_0, c, Z_0) = (q_1, Z_0)$ [One may note that $\delta(q_0, c, Z_0) = (q_f, Z_0)$ is not correct, as it is necessary to view the substring after '*c*', that is, to check whether it is null or not (for example, $w = cab$ is not valid). That is why state change is essential. Further, Transitions 7, 8 and 9 are all simple read operations. To avoid pushing of *a*s and *b*s after the read operation, a state change is necessary].

Pop operation: The state is q_1, the present input is '*a*' and the stack-top is '*a*', or the present input is '*b*' and the stack-top is '*b*'.

10. $\delta(q_1, b, b) = (q_1, \varepsilon)$
11. $\delta(q_1, a, a) = (q_1, \varepsilon)$.

Moving to the final state: The state is q_1, no input remains to be scanned and the stack-top is Z_0.

12. $\delta(q_1, \varepsilon, Z_0) = (q_f, Z_0)$. When no input remains to be read (it is ε) and the stack-top is Z_0 (initial stack symbol), the machine moves to the final state (q_f) and the processed string is valid.

- With respect to the empty stack, Transition 12 may be expressed as: $\delta(q_1, \varepsilon, Z_0) = (q_1, \varepsilon)$.
- The designed PDA is deterministic, as both the conditions (mentioned earlier) are satisfied.

The state diagram corresponding to the transitions specified above for the problem: $L = \{WcW^R \mid W \in \{a, b\}^*\}$ is shown in Fig. 5.4.

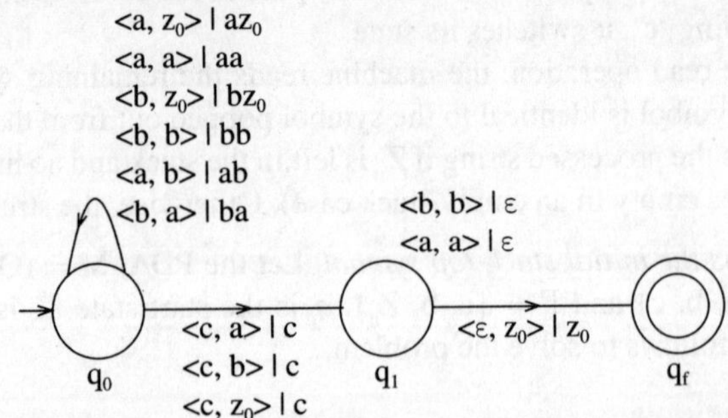

FIGURE 5.4 A PDA for $L = \{WcW^R \mid W \in \{a,b\}^*\}$

Example to demonstrate the working of the designed PDA: Let $w = abcba$. According to Fig. 5.4, the machine starts to process w at q_0. It first reads '*a*', performs a push operation and moves to q_0. Next, it reads '*b*', performs a push operation and moves again to q_0. As soon as '*c*' is encountered, it performs a read operation and moves to q_1. Here, it executes pop operation for two simultaneous inputs: '*b*' and '*a*'. When no input is left to be scanned (input is ε) and the stack-top is Z_0, it moves to the accepting state.

Instantaneous description of the proposed model

Acceptance of string: Let $w = abcba$. The instantaneous description of the model for the string: $w = abcba$ is shown below, using the move relation: $\vdash$. As mentioned earlier, the RHS of $\vdash$ states that the output configuration is obtained by applying a specific transition.

Initial configuration: $(q_0, \text{abcba}, Z_0) \vdash (q_0, \text{bcba}, aZ_0)$ by Transition rule 1.

$\vdash (q_0, \text{cba}, baZ_0)$ by Transition rule 2 [In this configuration, the LHS of $\vdash$ uses (q_0, bcba, aZ_0), which is not shown here but the same is assumed; this practice is followed in subsequent cases as well].

$\vdash (q_1, \text{ba}, baZ_0)$ by Transition rule 7
$\vdash (q_1, \text{a}, aZ_0)$ by Transition rule 10
$\vdash (q_1, \varepsilon, Z_0)$ by Transition rule 11
$\vdash (q_f, Z_0)$ by Transition rule 12

Thus, $(q_0, \text{abcba}, Z_0) \overset{*}{\vdash} (q_f, Z_0)$, where $\overset{*}{\vdash}$ denotes zero or more move relations.

Rejection of string: Let $w = abcbab$.

$(q_0, \text{abcbab}, Z_0) \vdash (q_0, \text{bcbab}, aZ_0)$ by Transition rule 1
$\vdash (q_0, \text{cbab}, baZ_0)$ by Transition rule 2
$\vdash (q_1, \text{bab}, baZ_0)$ by Transition rule 7
$\vdash (q_1, \text{ab}, aZ_0)$ by Transition rule 10
$\vdash (q_1, \text{b}, Z_0)$ by Transition rule 11

For (q_1, b, Z_0), no transition is specified; hence the given string is rejected.

EXAMPLE 5.3

(a) Design a PDA to recognise the language, $L = \{a^n b^n \mid n \geq 0\}$.

Solution *Logic*: The basic steps to solve this problem are outlined below:

1. First, push all the *as* of the input string onto the stack (memory) of the proposed PDA (M).
2. If '*b*' is the current input and '*a*' is the stack-top symbol, pop (deletion of stack-top) is performed. For each such pair, the operation continues.
3. If no input remains to be read and the stack-top of M reaches Z_0 (initial stack-top symbol), the string is accepted by the designed PDA.

Modelling of the machine: Let $M = (Q, \Sigma, \Gamma, \delta, q_0, Z_0, F)$, where $Q = \{q_0, q_1, q_f\}$, $\Sigma = \{a, b\}$, $\Gamma = \{a, Z_0\}$ (as the input symbol: '*a*' only will be shifted to the stack).

Here, q_0 is the start state, Z_0 is the initial stack symbol and $F = \{q_f\}$, δ is defined to solve the problem.

- *Push operation*: For the following transitions (rules), the push operation is performed at state q_0 when input is '*a*' and stack-top is either Z_0 or '*a*'.

  Rule 1:

  - $\delta(q_0, a, Z_0) = (q_0, aZ_0) \Rightarrow$ The string aZ_0 indicates that the present input '*a*' is pushed onto the stack and Z_0 is just below '*a*'.
  - $\delta(q_0, a, a) = (q_0, aa) \Rightarrow$ The string '*aa*' indicates that the current input '*a*' is pushed onto the stack, whereas the earlier stack-top '*a*' is just below the current '*a*'. This transition continues till input '*a*' is found.

- *Pop operation*: For the following transitions (rules), the pop operation is performed.

  Rule 2: At state q_0, if the input is '*b*' and stack-top is '*a*', a pop operation is performed. The equivalent transition is: $\delta(q_0, b, a) = (q_1, \varepsilon) \Rightarrow$ The string ε replaces '*a*' (top-stack), that is, when input is '*b*' and the stack-top symbol is '*a*', stack-top '*a*' is removed from the stack.

  Rule 3: In state q_1, if the input is '*b*' and the stack-top is '*a*', a pop operation is performed as: $\delta(q_1, b, a) = (q_1, \varepsilon) \Rightarrow$ String ε replaces '*a*' (top-stack). More explicitly, when the input is '*b*' and the stack-top symbol is '*a*', the symbol '*a*' is removed from the stack. This transition continues till input '*b*' and stack-top '*a*' are found.

  Rule 4: In state q_1, if the input is ε and the stack-top is Z_0, the PDA moves to the final state: q_f and the transition is: $\delta(q_1, \varepsilon, Z_0) = (q_f, Z_0) \Rightarrow$ This indicates that the processed string is valid.

  Rule 5: In state q_0, if the input is ε and the stack-top is Z_0, the machine moves to the final state q_f and the transition is: $\delta(q_0, \varepsilon, Z_0) = (q_f, Z_0)$. This transition is necessary for accepting a null string.

  Rules for moving to dead state

  * Rule 6: $\delta(q_0, b, Z_0) = (D, Z_0)$. Here, *D* stands for dead state.

  * Rule 7: $\delta(q_1, a, a) = (D, a)$. Here, *D* denotes dead state.

  Is the designed PDA a DPDA or NPDA? The presence of Rule 1: $\delta(q_0, a, Z_0)$ and Rule 5: $\delta(q_0, \varepsilon, Z_0)$ in the proposed model indicates that the PDA does not meet the second condition for DPDA (as specified above). Hence, it is an NPDA.

Note: As it is an NPDA (designed for Example 5.3 (a)), the machine may validate an incorrect string (for example, $w = aab$) on moving to the final state by reading the ε string initially (when stack-top $= Z_0$).

Fundamentally, no push should be permissible after the pop operation for the given language. That is why state changes while applying Rule 2. More explicitly, if the state is not changed (if the PDA remains in the same state), the input '*a*' appearing after '*b*' is also processed and pushed onto the stack. To avoid this situation, the state is changed.

> **Note:** In PDA, it is assumed that the stack initially contains Z_0 (marker). However, if Z_0 is not present there, then we may initially add an extra (push) transition: $\delta(q_0, \varepsilon, \varepsilon) = (q_0, Z_0)$. Assuming that the stack is initially empty, we may denote the initial state as the final state to accept a null string for problems that contain a null string. Some authors accept this concept.

Recognising $L = \{a^n b^n \mid n \geq 0\}$ by empty stack PDA: With respect to the empty stack, Rule 4 would be $\delta(q_1, \varepsilon, Z_0) = (q_1, \varepsilon)$ and Rule 5 would be: $\delta(q_0, \varepsilon, Z_0) = (q_0, \varepsilon)$.

Analysis of the transitions: For this example, the machine moves to the dead state for Rules 6 and 7. Though these are explicitly specified here, they may be ignored. Further,

- If state q_1 is considered as the final state too, it simply means that the designed PDA moves to the final state by removing just a single '*a*' from the stack (a string like *aab* is also treated as valid).
- If $n \geq 1$ for $L = \{a^n b^n\}$ is considered, Rule 5 is not allowed (since null string is not valid).

ID of the proposed model

Acceptance of string: Let $w = aabb$. The instantaneous description of the proposed model for this string is presented below, using the move relation $\vdash$.

Initial configuration: $(q_0, aabb, Z_0) \vdash (q_0, abb, aZ_0)$ by Transition rule 1
$\vdash (q_0, bb, aaZ_0)$ by Transition rule 1
$\vdash (q_1, b, aZ_0)$ by Transition rule 2
$\vdash (q_1, \varepsilon, Z_0)$ by Transition rule 3
$\vdash (q_f, Z_0)$ by Transition rule 4

Hence, $(q_0, aabb, Z_0) \overset{*}{\vdash} (q_f, Z_0)$, where $\overset{*}{\vdash}$ denotes zero or more move relations.

Rejection of string: Let $w = aabbb$.
$(q_0, aabbb, Z_0) \vdash (q_0, abbb, aZ_0)$ by Transition rule 1
$\vdash (q_0, bbb, aaZ_0)$ by Transition rule 1
$\vdash (q_1, bb, aZ_0)$ by Transition rule 2
$\vdash (q_1, b, Z_0)$ by Transition rule 3
However, for (q_1, b, Z_0), no such transition is specified. Hence, the given string is not valid.

(b) Design a PDA to recognise the language, $L = \{a^n b^{2n} \mid n \geq 1\}$.

Solution The transitions of a PDA (M) for solving the given problem are presented below:

1. $\delta(q_0, a, Z_0) = (q_0, aaZ_0)$ // Two *as* are pushed for a single '*a*' read by M.
2. $\delta(q_0, a, a) = (q_0, aaa)$ // Two *as* are pushed for a single '*a*' read by M.
3. $\delta(q_0, b, a) = (q_1, \varepsilon)$ // Pop operation
4. $\delta(q_1, b, a) = (q_1, \varepsilon)$ //Pop operation
5. $\delta(q_1, \varepsilon, a) = (q_f, \varepsilon)$ // Pop operation

Note: Based on a small variation of the idea presented in Example 5.3 (a), a language like $L = \{a^m b^n \mid n = 2m + 1, m \geq 0\}$ can be recognised by a PDA. Here, each valid string can be expressed as: $a^m bb^{2m}$. For the new machine M, for each push operation, two *as* must be shifted onto the stack. When all the *as* are consumed, only one read operation (neither push nor pop) is performed for the first '*b*'. Next, the machine applies pop operations as per the usual, that is, for each matching pair: (b, a). Now, can we design a PDA for $L = \{a^n b^n c^n \mid n \geq 0\}$ by shifting two *as* for each '*a*' onto the stack? The answer is no. This is because we may not be able to track #*as* = #*bs* = #*cs* (along with the order of appearance of *as*, *bs* and *cs*) by shifting two *as* for each '*a*' onto the stack. If so, then invalid strings like $w = a^2bc^3$, $w = a^2b^3c$, etc. will be validated.

L = {a^nb^m | $n, m \geq 0$} AND THE FOLLOWING CONSTRAINTS

- $n = m$ is not regular. Firstly, the language is not finite. Secondly, the number of *as* and the number of *bs* should match for each string with $n = m$. However, it may not be satisfied by FA, as no stack is used in FA.
- Likewise, $n = 2m$; $n \neq m$; $n \geq m$, etc. may not be verified by FA due to the absence of stack in FA, and the language is infinite.

(c) Design a PDA to accept the language, L = {$a^nb^ma^n$ | $n \geq 0, m \geq 1$}.

Solution The basic idea to solve this problem is to shift (push) the input *as* onto the stack of the PDA. When symbol '*b*' is read, the machine changes state once and continues to perform the read operation in the same state until '*a*' is read or ε is found. The state then changes, as no push operation is to be allowed after a read operation.

Next, pop is performed by changing the state if '*a*' is found. If the state is not changed, the read operation may continue to be performed after the pop operation, which is not allowed (for example, string like *w* = *aabbaba*). Finally, if no input is left to be read and the stack-top reaches Z_0 (the initial stack-top symbol), the machine moves to the final state to accept the string.

Let M = (Q, Σ, Γ, δ, q_o, Z_0, F), where Q = {q_0, q_1, q_2, q_f}, Σ = {a, b} and Γ = {a, Z_0} (as only the input symbol '*a*' is to be shifted onto the stack). q_0 is the start state, Z_0 is the initial stack symbol and F = {q_f}.

Transition functions: δ is defined to solve the given problem, considering the following possible cases of valid strings.

Case I: Presence of as *to the left and right in valid string*

1. $\delta(q_0, a, Z_0) = (q_0, aZ_0) \Rightarrow$ Push
2. $\delta(q_0, a, a) = (q_0, aa) \Rightarrow$ Push (for more *as*)
3. $\delta(q_0, b, a) = (q_1, a) \Rightarrow$ Read (No push is to be allowed after read. To maintain this, only the state is changed).
4. $\delta(q_1, b, a) = (q_1, a) \Rightarrow$ Read (for more *bs*)
5. $\delta(q_1, a, a) = (q_2, \varepsilon) \Rightarrow$ Pop (when both the stack-top and the present input is '*a*') [No read is to be allowed after pop. To maintain this, only the state is changed. If the state is not changed, invalid strings like a^3baba^2 will also be accepted].
6. $\delta(q_2, a, a) = (q_2, \varepsilon) \Rightarrow$ Pop (for more *as*)
7. $\delta(q_2, \varepsilon, Z_0) = (q_f, Z_0) \Rightarrow$ Accept

Case II: No a *to the left and right of* b *in valid string*

1. $\delta(q_0, b, Z_0) = (q_1, Z_0)$ [Read operation. The state changes but it is not yet the final state, as the decision is pending until ε is found].
2. $\delta(q_1, b, Z_0) = (q_1, Z_0)$ (For more *bs*)
3. $\delta(q_1, \varepsilon, Z_0) = (q_f, Z_0)$

With respect to empty stack, Transition 10 would be: $\delta(q_1, \varepsilon, Z_0) = (q_1, \varepsilon)$. [Contemplate designing a PDA for L = {$a^nb^ma^n$ | $n \geq 0, m \geq 0$}].

(d) Design a PDA to accept the language, $L = \{a^n b^m \mid n \neq m\}$.

Solution Let $M = (Q, \Sigma, \Gamma, \delta, q_o, Z_0, F)$, where $Q = \{q_0, q_1, q_f\}$, $\Sigma = \{a, b\}$, $\Gamma = \{a, Z_0\}$ (as only the input symbol '*a*' needs to be shifted to the stack). q_0 is the start state, Z_0 is the initial stack symbol and $F = \{q_f\}$. δ is defined as follows:

Case I: Only as *but no* bs *in valid strings*

- $\delta(q_0, a, Z_0) = (q_0, aZ_0) \Rightarrow$ Push
- $\delta(q_0, a, a) = (q_0, aa) \Rightarrow$ Push
- $\delta(q_0, \varepsilon, a) = (q_f, a) \Rightarrow$ Accept

Case II: Only bs *but no* as *in valid strings*

- $\delta(q_0, b, Z_0) = (q_1, Z_0) \Rightarrow$ Read (but the decision is pending until the whole string is fully processed. Hence, read is performed and the state changes, as no push is allowed after read).
- $\delta(q_1, b, Z_0) = (q_1, Z_0) \Rightarrow$ Read
- $\delta(q_1, \varepsilon, Z_0) = (q_f, Z_0) \Rightarrow$ Accept

Case III: Both as *and* bs *may be present in valid strings*

- $\delta(q_0, b, a) = (q_2, \varepsilon) \Rightarrow$ Pop from state q_0 (The state changes as push after pop is not allowed).
- $\delta(q_2, b, a) = (q_2, \varepsilon) \Rightarrow$ Pop (For more *as* to pop)

Case III (a): The number of as *is fewer than the number of* bs.

- $\delta(q_2, b, Z_0) = (q_3, Z_0) \Rightarrow$ Read (All the *as* are popped out (stack becomes empty) and *b* appears in the string. However, the decision is pending until the string is fully processed. The state changes to avoid the possibility of an equal number of *as* and *bs*).
- $\delta(q_3, b, Z_0) = (q_3, Z_0) \Rightarrow$ Read (For more *bs*)
- $\delta(q_3, \varepsilon, Z_0) = (q_f, Z_0) \Rightarrow$ Accept

Case III (b): The number of bs *is fewer than the number of* as.

- $\delta(q_2, \varepsilon, a) = (q_f, a) \Rightarrow$ Accept (The string is fully processed but the number of *as* is more than the number of *bs*, as the stack-top is '*a*').

EXAMPLE 5.4

(a) Design a PDA to accept strings in which the number of *as* is equal to the number of *bs*. $\Sigma = \{a, b\}$.

Solution *Logic*: To design a PDA (M) to solve the given problem, we use the following logic: If the stack-top symbol and the current input symbol are, respectively, '*a*' and '*b*' or vice versa, then the pop operation must be performed; else the push operation (except for null input and stack-top = Z_0).

The proposed mathematical model: Let $M = (Q, \Sigma, \Gamma, \delta, q_o, Z_0, F)$, where $Q = \{q_0, q_f\}$, $\Sigma = \{a, b\}$, $\Gamma = \{a, b, Z_0\}$. The state q_0 represents the start state of the machine, Z_0 stands for the initial stack symbol and $F = \{q_f\}$. The transition function δ is defined as follows:

1. $\delta(q_0, a, Z_0) = (q_0, aZ_0)$ // *Push*
2. $\delta(q_0, b, Z_0) = (q_0, bZ_0)$ // *Push*
3. $\delta(q_0, a, a) = (q_0, aa)$ // *Push*
4. $\delta(q_0, b, b) = (q_0, bb)$ // *Push*
5. $\delta(q_0, a, b) = (q_0, \varepsilon)$ // *Pop*
6. $\delta(q_0, b, a) = (q_0, \varepsilon)$ // *Pop*

For this problem, pushing of *a*s and *b*s after a pop operation is allowed. That is why the state does not change.

7. $\delta(q_0, \varepsilon, Z_0) = (q_f, Z_0)$

- With respect to empty stack, Transition 7 would be: $\delta(q_0, \varepsilon, Z_0) = (q_0, \varepsilon)$.
- Due to the presence of Transition 1: $\delta(q_0, a, Z_0)$ and Transition 7: $\delta(q_0, \varepsilon, Z_0)$, the second condition for DPDA fails. Hence, it is an NPDA. However, in the case of $\delta(q_0, \varepsilon, Z_0) = (q_f, Z_0)$, it may be assumed that the machine guesses that there is no input to be scanned.

ID of the proposed model

Acceptance of string: Let w = *abaabb.* The instantaneous description of the proposed model for the string using the move relation: ⊢ is presented below. Here, the RHS of ⊢ states that the output configuration is obtained by applying a specific transition.

$(q_0$, abaabb, $Z_0) \vdash (q_0$, baabb, $aZ_0)$ by Transition rule 1

⊢ $(q_0$, aabb, $Z_0)$ by Transition rule 5
⊢ $(q_0$, abb, $aZ_0)$ by Transition rule 1
⊢ $(q_0$, bb, $aaZ_0)$ by Transition rule 1
⊢ $(q_0$, b, $aZ_0)$ by Transition rule 5
⊢ (q_0, ε, Z_0) by Transition rule 5
⊢ (q_f, Z_0) by Transition rule 7

Thus, $(q_0$, abaabb, $Z_0) \overset{*}{\vdash} (q_f, Z_0)$, where $\overset{*}{\vdash}$ denotes zero or more move relations.

Rejection of string: Let w = *abaabba.*

$(q_0$, abaabba, $Z_0) \vdash (q_0$, baabba, $aZ_0)$ by Transition rule 1
⊢ $(q_0$, aabba, $Z_0)$ by Transition rule 5
⊢ $(q_0$, abba, $aZ_0)$ by Transition rule 1
⊢ $(q_0$, bba, $aaZ_0)$ by Transition rule 1
⊢ $(q_0$, ba, $aZ_0)$ by Transition rule 5
⊢ $(q_0$, a, $Z_0)$ by Transition rule 5

However, for $(q_0$, a, $Z_0)$, no transition rule is specified, so the given string is not valid.

(b) Design a PDA to accept strings in which the number of *a*s is twice that of *b*s over Σ = {a, b}.

Solution Let M = (Q, Σ, Γ, δ, q_0, Z_0, F), where Q = $\{q_0, q_1, q_f\}$, Σ = {*a*, *b*}, Γ = {*b*, Z_0} (as symbol '*b*' for input symbol '*b*' only is to be shifted to the stack). q_0 represents the start state of the machine, Z_0 stands for the initial stack symbol and F = $\{q_f\}$. Here, two *b*s are to be shifted for input '*b*'. The transition function δ is defined as follows:

1. $\delta(q_0, b, Z_0) = (q_0, bbZ_0)$
2. $\delta(q_0, b, b) = (q_0, bbb)$
3. $\delta(q_0, a, Z_0) = (q_0, aZ_0)$
4. $\delta(q_0, a, a) = (q_0, aa)$
5. $\delta(q_0, b, a) = (q_1, \varepsilon)$
6. $\delta(q_1, \varepsilon, a) = (q_0, \varepsilon)$ (Pop without consuming any input)

7. $\delta(q_0, \text{a}, \text{b}) = (q_0, \varepsilon)$
8. $\delta(q_0, \varepsilon, Z_0) = (q_f, Z_0)$

Variation: The number of occurrences of a pattern (p_1) in a valid string is as many times the number of occurrences of another pattern (p_2).

KEY POINTS

- If an FA is allowed to access one stack, it is equivalent to a PDA.
- If we allow an FA to access two stacks (instead of one), we obtain a powerful machine that is, indeed, equivalent (in terms of power) to a TM.
- A linear bounded automaton (LBA) is a device that accepts context-sensitive grammar. It is more powerful than a PDA but less powerful than a TM.
- Every FA (DFA or NFA) can be converted to an equivalent PDA.

5.2.2 Non-Deterministic Pushdown Automata (NPDA)

We have defined a transition function (δ) for PDA as: $Q \times (\Sigma \cup \{\varepsilon\}) \times \Gamma \rightarrow Q \times \Gamma^*$. Undoubtedly, it is by specification the transition function for non-deterministic pushdown automata (NPDA), as the ε move is permitted and it allows the machine to move to multiple states from a single state on reading the same input and seeing the same stack-top symbol.

However, the transition function for NPDA may be more generalised as: $Q \times (\Sigma \cup \{\varepsilon\}) \times \Gamma \rightarrow 2^Q \times \Gamma^*$. This clearly states that for the same configuration, such as (q, a, b), the machine may move to multiple states or perform multiple operations with different state moves. Either way, the designed PDA may be deterministic (as a deterministic automaton of any type is a special case of non-deterministic automata of that type, that is, the deterministic type is the subset of the non-deterministic type). Importantly, the deterministic behaviour of any PDA can be verified by the two conditions mentioned in Section 5.2.1.

Some examples for NPDA are presented below.

EXAMPLE 5.5

Design a PDA for $L = \{ww^R \mid w \in \{\text{a, b}\}^*\}$.

Solution We present two NPDAs here for the given problem based on the two different transition functions stated above.

(i) The transition diagram based on the function $Q \times (\Sigma \cup \{\varepsilon\}) \times \Gamma \rightarrow 2^Q \times \Gamma^*$ is shown in Fig. 5.5 (a).
(ii) The transition diagram based on the function $Q \times (\Sigma \cup \{\varepsilon\}) \times \Gamma \rightarrow Q \times \Gamma^*$ is shown in Fig. 5.5 (b).

The mid-point (centre) of each valid string must be non-deterministically decided in each of these two cases.

Identifying the mid-point non-deterministically: The centre may be non-deterministically identified when the present input symbol and the stack-top symbol are the same (that is, (a, a) or (b, b)).

Case I: Transition function $Q \times (\Sigma \cup \{\varepsilon\}) \times \Gamma \rightarrow 2^Q \times \Gamma$.

- Push operations are performed if the present input symbol and the stack-top symbols are different $((a, b), (b, a), (a, Z_0), (b, Z_0))$. However, push operations for (a, a), (b, b) are also executed before the machine guesses the mid-point of the string.
- As soon as the mid-point is guessed, the pop operation starts and continues for (a, a) and (b, b).

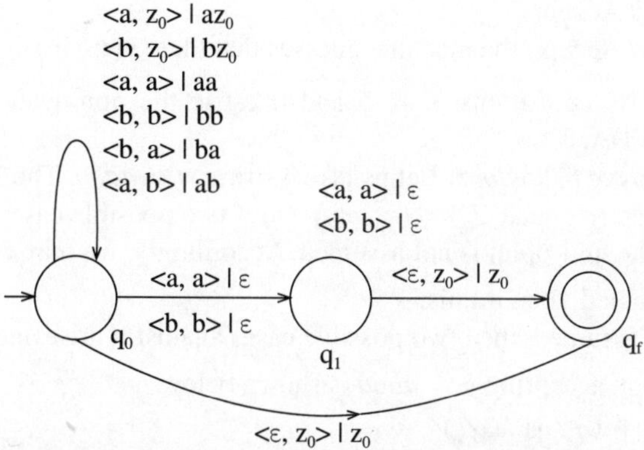

FIGURE 5.5 (a) An NPDA for $L = \{ww^R \mid w \in \{a, b\}^*\}$

Note that for the same configuration: (q_0, a, a), the machine moves to different states, q_0 and q_1, while performing two different operations: (i) push and (ii) pop. This is true for configuration: (q_0, b, b) as well. Thus, it shows non-deterministic behaviour. Hence, the designed PDA is an NPDA.

Case II: Transition function $Q \times (\Sigma \cup \{\varepsilon\}) \times \Gamma \to Q \times \Gamma^*$

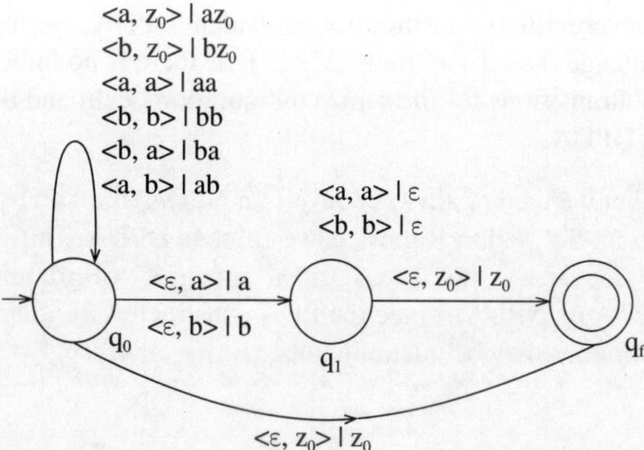

FIGURE 5.5 (b) An NPDA for $L = \{ww^R \mid w \in \{a, b\}^*\}$

The transitions of the suggested PDA are shown below:

1. $\delta(q_0, a, Z_0) = (q_0, aZ_0)$ // Push
2. $\delta(q_0, b, Z_0) = (q_0, bZ_0)$ // Push
3. $\delta(q_0, a, a) = (q_0, aa)$ // Push
4. $\delta(q_0, b, b) = (q_0, bb)$ // Push
5. $\delta(q_0, a, b) = (q_0, ab)$ // Push
6. $\delta(q_0, b, a) = (q_0, ba)$ // Push
7. $\delta(q_0, \varepsilon, a) = (q_1, a)$ // Read ε by guessing the mid-point of the string, as concatenation of ε with any string does not change the string
8. $\delta(q_0, \varepsilon, b) = (q_1, b)$ // Read ε by guessing the mid-point of the string
9. $\delta(q_1, a, a) = (q_1, \varepsilon)$ // Pop
10. $\delta(q_1, b, b) = (q_1, \varepsilon)$ // Pop

11. $\delta(q_1, \varepsilon, Z_0) = (q_f, Z_0)$ // Accept
12. $\delta(q_0, \varepsilon, Z_0) = (q_f, Z_0)$ // Accept; the machine guesses that there is no input to be scanned.

This is surely an NPDA. Transitions 3, 4, 5 and 6 cause the non-deterministic behaviour with Transitions 7 and 8 in the PDA.

Evidence for non-deterministic behaviour: Let us take a string $w = aaaa$. The ID for w is: $(q_0$, aaaa, $Z_0) = (q_0$, $az_0)$ and $(q_0$, aaaa, $Z_0) = (q_0$, $az_0)$. Now, two possible cases may occur: (i) the mid-point is assumed, or (ii) the mid-point is not assumed. Accordingly, we may consider two PDAs:

- If the mid-point is assumed, then it fails.
- If the mid-point is not assumed, then two possible cases against it arise and proceed accordingly.

The sequence of moves in accepting $w = aaaa$ is shown below.

$(q_0$, aaaa, $Z_0) \vdash (q_0$, $az_0) \vdash (q_0$, aa, $aaZ_0)$
$\vdash (q_1$, a, $aZ_0)$ // With respect to Fig. 5.5 (a)
$\vdash (q_1, \varepsilon, Z_0)$
$\vdash (q_f, Z_0)$

The non-deterministic alternative for locating the mid-point of the string is taken at the third move. In this stage, the PDA has the ID $(q_0$, aa, $aaZ_0)$ and two choices for its next move. One is shift: $\delta(q_0$, a, a$) = (q_0$, aa) and the other is pop: $\delta(q_0$, a, a$) = (q_1, \varepsilon)$. Only the latter leads to acceptance.

Importantly, a DPDA can be obtained from an NPDA if the NPDA never encounters a situation in which more than one instruction can be executed. For certain problems, there occur some transitions in NPDA that can never be converted into transitions of equivalent DPDA. For instance, we cannot have a DPDA to recognise the language: $L = \{ww^R \mid w \in \{a, b\}^*\}$, as there is no indication of when the w part is over; that is why multiple transitions for the same configuration occur and these cannot be converted to transitions of equivalent DPDA.

> It should also be noted that we cannot always convert an NPDA to a DPDA. There is no standard method for such conversion. So, NPDA is more powerful than DPDA. Thus, there exists an NPDA for each DPDA, but the converse is not true. Further, ambiguity of grammar is not related with PDA (irrespective of DPDA or NPDA), since the PDA constructed for a language may be NPDA but the corresponding grammar may be unambiguous.

KEY POINTS

- In DPDA, we can easily instruct the machine when to push and when to pop, but this is not possible in NPDA. For clarity, for string: $w = aaaa$ of the language $L = \{ww^R \mid w \in \{a, b\}^*\}$, we cannot instruct the machine when to push 'a' or pop 'a' after the first push; that is, there is no definite way to decide when to push or pop.
- In DPDA, there is only one move in every situation, whereas in NPDA, there are multiple moves with multiple operations in every situation. For example, $\delta(q_i$, a, a$) = \{(q_i$, aa$), (q_j, \varepsilon), (q_j$, a$)\}$. This transition states that the machine may move to multiple states performing multiple operations; that is, the machine may move to states q_i and q_j from q_i on reading input 'a' and viewing stack-top symbol 'a'. Push, pop and read operations are performed in the same transition.

EXAMPLE 5.6

Design a PDA for $L = \{a^i b^j c^k \mid i, j, k \geq 0, i = j \text{ or } j = k\}$ over {a, b, c}.

Solution The transitions of the PDA for solving the given problem are shown below.

// If the suggested PDA assumes a^ib^j with $i = j$, it performs Transitions 1 to 8.

1. $\delta(q_0, \varepsilon, Z_0) = (q_1, Z_0)$ // Reading without consuming any input, the PDA moves to q_1.
2. $\delta(q_1, a, Z_0) = (q_1, aZ_0)$ // Push
3. $\delta(q_1, a, a) = (q_1, aa)$ // Push
4. $\delta(q_1, b, a) = (q_2, \varepsilon)$ // Pop
5. $\delta(q_2, b, a) = (q_2, \varepsilon)$ // Pop
6. $\delta(q_2, c, Z_0) = (q_3, Z_0)$ // Read operation for '*c*'. If the state is not changed, invalid strings like *aabcb* may also be accepted using the subsequent transitions, as *c*s are not pushed onto the stack.
7. $\delta(q_3, c, Z_0) = (q_3, Z_0)$ // Reading '*c*' only
8. $\delta(q_3, \varepsilon, Z_0) = (q_4, Z_0)$ // Accept, where q_4 is the final state.

// When the PDA assumes b^jc^k with $j = k$, it performs Transitions 9 to 15

9. $\delta(q_0, \varepsilon, Z_0) = (q_5, Z_0)$ // Reading without consuming any input, the PDA moves to q_5.
10. $\delta(q_5, a, Z_0) = (q_5, Z_0)$ // Reading '*a*' only
11. $\delta(q_5, b, Z_0) = (q_5, bZ_0)$ // Push
12. $\delta(q_5, b, b) = (q_5, bb)$ // Push
13. $(q_5, b, c) = (q_6, \varepsilon)$ // Pop operation. State change is necessary here, otherwise invalid strings like *ababcc* will be accepted, as *a*s are not pushed.
14. $\delta(q_6, b, c) = (q_6, \varepsilon)$ // Pop
15. $\delta(q_6, \varepsilon, Z_0) = (q_7, Z_0)$ // Accept, and q_7 is the final state

Note: Here, the strict case: $i = j = k$ is explicitly ignored, as CSG (equivalently, LBA) is necessary to satisfy this constraint. However, any of the considerations: a^ib^j with $i = j$ or b^jc^k with $j = k$ may satisfy $i = j = k$, but there is no guarantee

EXAMPLE 5.7

(a) Can we design a PDA for the language $L = \{a^mb^na^nb^m \mid m, n \geq 1\}$? If yes, present your idea.

Solution Yes, we can design a PDA to solve the given problem. An idea is presented in brief below.

- We know that PDA remembers the last pushed symbol. More specifically, matching always takes place with the last stored symbol in the stack.

With this point in mind, all the symbols of a^mb^n are first pushed onto the stack. As soon as the first '*a*' is encountered after a^mb^n, pop operations are carried out until the stack marker (Z_0) is encountered to validate the string.

(b) Can we design a PDA for the language $L = \{a^mb^na^mb^n \mid m, n \geq 1\}$? If yes, present your idea.

Solution We cannot design any PDA (NPDA or DPDA) to solve this problem. This is because the PDA remembers the last pushed symbol. In other words, matching always takes place with the last stored symbol. However, if we use two stacks, the problem can be solved, but this set-up behaves like an LBA or higher.

THEOREM 5.1

Every context-free grammar can be converted into an equivalent pushdown automaton.

Proof The theorem states that for every CFG, there is an equivalent PDA. Let $G = (V, \Sigma, P, S)$ be a CFG. Now, the aim is to design an equivalent PDA for G. For this purpose, we will use the tuples of the grammar. Let the machine be M = (Q, Σ, Γ, δ, q_0, Z_0, {q_f}), where Q = {q_0, q_1, q_f} such that q_0, q_1 and q_f are, respectively, the start state, non-final state and accepting state of the machine. Σ is the Σ of CFG, Γ = V ∪ Σ ∪ {Z_0} (Z_0 is the marker in the stack). The PDA will use the following transitions to process each string generated by the CFG:

1. $\delta(q_0, \varepsilon, Z_0) = (q_1, SZ_0)$. This states that if the machine (M) at state q_0 observes Z_0 as the stack-top, it moves to state q_1 without consuming any input and shifts the start symbol (S) of the grammar to the stack.
2. $\delta(q_1, \varepsilon, A) = (q_1, \alpha)$, that is, the machine replaces A with α if $A \rightarrow \alpha$ exists. More explicitly, if the machine at state q_1 observes a non-terminal symbol A at the top of the stack, it first removes A from the stack and then pushes the RHS of the production associated with $A(A \rightarrow \alpha)$ in such a way that the first symbol of the RHS appears at the top, the second one next and so on. Again, no input is consumed in this action. Also, the state remains unchanged.
3. $\delta(q_1, a, a) = (q_1, \varepsilon)$, that is, the machine performs a pop operation. In state q_1, if the stack-top symbol (say, '*a*') matches the current input symbol: '*a*', a pop operation is performed. The state of the machine remains unchanged.
4. $\delta(q_1, \varepsilon, Z_0) = (q_f, Z_0)$, that is, the machine accepts the string when no input is left to be processed but the stack-top symbol is Z_0. The machine moves to the final state (q_f).

Mathematically, these operations have the ability to recognise the strings generated by the CFG.

To understand the idea, consider a CFG (G) with a rule set (P) = {S → aSb | b}. Let *w* = *abb*. This string is generated by the rule set, P. Now, recognition of *w* by a PDA based on Theorem 5.1 is partially shown in Fig. 5.6.

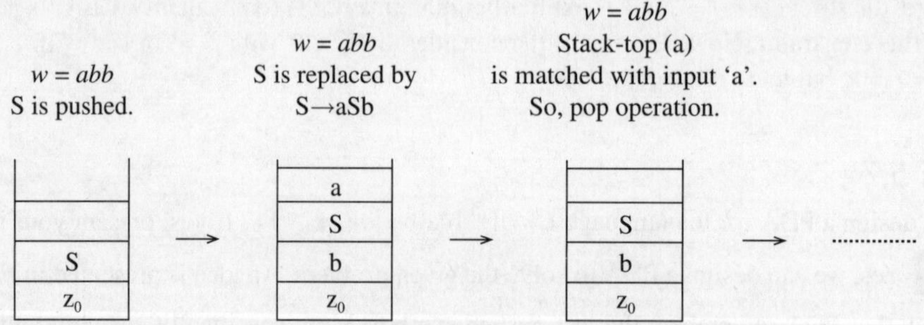

FIGURE 5.6 Validation of strings using stack

Likewise, we may convert PDA to an equivalent CFG. The conversion procedure is presented below.

Application: The idea behind construction of PDA from a CFG is to obtain a PDA that simulates the sequence of left-sentential forms that the grammar uses to generate a given terminal string w. More specifically, the approach for CFG to PDA is used to design an LL(*k*) parser ($k \geq 1$) (an efficient version of the top-down parser) during compiler construction. On the other hand, the reverse process (PDA to CFG) is useful for eliminating useless symbols from the grammar.

Procedure to convert PDA to CFG: We have learnt how to convert CFG to PDA. Now, we will learn how to convert a given PDA to CFG. The general procedure for the conversion is given below:

Step 1: The input symbols of PDA will be the terminals of CFG.

Step 2: If the PDA moves from state q_i to q_j on consuming the input $a \in \Sigma$ when Z_0 is the top of the stack, the non-terminals of CFG are triplets of the form: $(q_i Z_0 q_j)$.

Step 3: If q_0 is the start state and q_f is the final state, then $(q_0Z_0q_f)$ is the start symbol of CFG.
Step 4: The production of CFG can be obtained from the transitions of PDA:

- For each transition of the form: $\delta(q_i, a, Z_0) = (q_j, AB)$, introduce a production of the form: $(q_iZ_0q_k) \rightarrow a(q_jAq_m)(q_mBq_k)$, where q_k and q_m take all possible values of Q (set of states of the proposed PDA).
- For each transition of the form: $\delta(q_i, a, Z_0) = (q_j, \varepsilon)$ introduce a production of the form: $(q_iZ_0q_j) \rightarrow a$.

EXAMPLE 5.8

Obtain a CFG that generates the language accepted by PDA, $M = (\{q_0, q_1\}, \{a, b\}, \{A, Z\}, \delta, q_0, Z_0, \{q_1\})$, with the transitions

(i) $\delta(q_0, a, Z_0) = (q_0, AZ_0)$ (ii) $\delta(q_0, b, A) = (q_0, AA)$ (iii) $\delta(q_0, a, A) = (q_1, \varepsilon)$

Solution Now, the transition (iii): $\delta(q_0, a, A) = (q_1, \varepsilon)$ can be converted into a production, as shown below.

For δ of the form: $\delta(q_i, a, Z_0) = (q_j, \varepsilon)$	Resulting production: $(q_iZ_0q_j) \rightarrow a$
$\delta(q_0, a, A) = (q_1, \varepsilon)$	$(q_0Aq_1) \rightarrow a$

Now, the transitions:

(i) $\delta(q_0, a, Z_0) = (q_0, AZ_0)$ and (ii) $\delta(q_0, b, A) = (q_0, AA)$ can be converted into productions using Step 4 of the procedure, as shown below.

For δ of the form: $\delta(q_i, a, Z_0) = (q_j, AB)$	Resulting production: $(q_iZ_0q_k) \rightarrow a(q_jAq_m)(q_mBq_k)$
$\delta(q_0, a, Z_0) = (q_0, AZ_0)$	$(q_0Z_0q_0) \rightarrow a(q_0Aq_0)(q_0Z_0q_0) \mid a(q_0Aq_1)(q_1Z_0q_0)$, $(q_0Z_0q_1) \rightarrow a(q_0Aq_0)(q_0Z_0q_1) \mid a(q_0Aq_1)(q_1Z_0q_1)$
$\delta(q_0, b, A) = (q_0, AA)$	$(q_0Aq_0) \rightarrow b(q_0Aq_0)(q_0Aq_0) \mid b(q_0Aq_1)(q_1Aq_0)$, $(q_0Aq_1) \rightarrow b(q_0Aq_0)(q_0Aq_1) \mid b(q_0Aq_1)(q_1Aq_1)$

Obviously, the start symbol of the grammar will be $q_0Z_0q_1$.

5.3 CLOSURE PROPERTIES OF CONTEXT-FREE LANGUAGES

Certain operations performed on CFLs are guaranteed to produce new CFLs. In other words, if applying any operation over existing CFL(s) always results in new CFL, the said operation is an example of a closure operation. On the other hand, if the operation does not result in a CFL, the said operation is not an example of closure. Some such operations that produce CFLs from existing CFLs are discussed below.

Closure Under Union

THEOREM 5.2

If L_1 and L_2 are two CFLs, their union is a CFL.

Proof Let $G_1 = (V_1, \Sigma, P_1, S_1)$ and $G_2 = (V_2, \Sigma, P_2, S_2)$ be two CFGs. Suppose $L_1 = L(G_1)$ and $L_2 = L(G_2)$ are their respective languages. The languages L_1 and L_2 are CFLs as their grammars are CFG. Now, if the new language: $L = L_1 \cup L_2$ possesses strings that are guaranteed to be derived by a new CFG (resulting from existing CFGs), then the CFLs are closed under union.

Constructing a new grammar: For the union of languages, we take the union of their grammars to produce a new grammar G = (V, Σ, P, S), where $V = V_1 \cup V_2 \cup \{S\}$; S is the new non-terminal symbol (as the start symbol) introduced for G. Assume that $V_1 \cap V_2 = \Phi$. If not, then rename the common non-terminal symbols of G_1 and G_2 to avoid conflict during derivation. More specifically, if renaming of variables is not carried out, it will be problematic to decide the non-terminal symbol to be selected for deriving the string. Further, such a conflict situation may produce a string that does not belong to $L(G_1)$ and $L(G_2)$. Now, $P = P_1 \cup P_2 \cup \{S \rightarrow S_1 \mid S_2\}$, where $S \rightarrow S_1 \mid S_2$ are the new rules introduced for G. Actually, there must exist rules in G to move to S_1 (start symbol of G_1) or S_2 (start symbol of G_2), respectively, from S (start symbol of G) to generate strings, each originally belonging to either $L(G_1)$ or $L(G_2)$ or both. In other words, any string $x \in L(G)$ if and only if $x \in L(G_1)$ or $x \in L(G_2)$ or both. All the strings of L(G) come from $L(G_1)$ and $L(G_2)$, and these can be ultimately derived from G, that is, the subset: $S \rightarrow S_1 \mid S_2$ assists in generating any string of L_1 or L_2 or both.

However, both G_1 and G_2 are assumed as CFG. Hence, G is also a CFG.

For example, let $L_1 = \{a^n b^n \mid n \geq 0\}$. The rule set ($P_1$) for this language is: $\{S_1 \rightarrow aS_1b \mid \varepsilon\}$. Also, let $L_2 = \{ww^R \mid w \in \{a, b\}^*\}$. The rule set ($P_2$) for this language is: $\{S_2 \rightarrow aS_2a \mid bS_2b \mid \varepsilon\}$.

Now, for union of these two languages, $L = \{a^n b^n \mid n \geq 0\} \cup \{WW^R\} = \{\varepsilon, ab, a^2b^2, \ldots, aa, bb, abba, \ldots\}$, the new rule set will be: $P = P_1 \cup P_2 \cup \{S \rightarrow S_1 \mid S_2\}$.

EXAMPLE 5.9

By applying the concept of union of CFLs, check whether a language: $L = \{a^n b^m c^k \mid k \neq n + m\}$ is CFL or not.

Solution To solve this problem, we may first consider the following two languages: $L_1 = \{a^n b^m c^k \mid k < n + m\}$ and $L_2 = \{a^n b^m c^k \mid k > n + m\}$. Both the languages are CFL as CFG exists for each. Let G_1 and G_2 be the grammars for L_1 and L_2, respectively. Now, according to the union property of two CFLs, we may write, $L(G) = L = L_1 \cup L_2 = L = \{a^n b^m c^k \mid k \neq n + m\}$. Hence, L(G) is also CFL.

Closure Under Concatenation

THEOREM 5.3

If L_1 and L_2 are two CFLs, their concatenation is also a CFL.

Proof Let $G_1 = (V_1, \Sigma, P_1, S_1)$ and $G_2 = (V_2, \Sigma, P_2, S_2)$ be two CFGs. Suppose $L_1 = L(G_1)$ and $L_2 = L(G_2)$ are their respective languages. The languages L_1 and L_2 are CFLs as their grammars are CFG. Now, if the new language $L = L_1.L_2$ (obtained by applying concatenation operation) possesses strings that are guaranteed to be derived by a new CFG (resulting from the existing grammars), then the CFLs are closed under concatenation.

Constructing the new grammar: For the purpose of language concatenation, we may construct a new grammar G = (V, Σ, P, S) from the existing grammars: G_1 and G_2, where $V = V_1 \cup V_2 \cup \{S\}$; S is the new non-terminal symbol (as start symbol) introduced for G. Assume that $V_1 \cap V_2 = \Phi$. If not, then rename the common non-terminal symbols of G_1 and G_2 to avoid conflict during parsing. In particular, if renaming is not carried out, it will be problematic to decide whose non-terminal symbol is to be selected for deriving the string, and it may produce a string that does not belong to both $L(G_1)$ and $L(G_2)$.

Thus, $P = P_1 \cup P_2 \cup \{S \rightarrow S_1S_2\}$, where $S \rightarrow S_1S_2$ is the new rule introduced for G for $L = L_1.L_2$. Actually, there must exist a rule in G to move to S_1S_2 (S_1, start symbol of G_1 and S_2, start symbol of G_2) from S. The reason is to generate strings, each concatenated by a string originally belonging to $L(G_1)$ with a string belonging to $L(G_2)$. In other words, any string $xy \in L(G)$ if and only if $x \in L(G_1)$ and

$y \in L(G_2)$. Precisely, all the strings of L(G) can be derived by concatenating strings of $L(G_1)$ and $L(G_2)$, that is, the strings can be derived by G by applying the starting rule: $S \rightarrow S_1S_2$. However, both G_1 and G_2 are assumed here to be CFGs. Hence, G is also CFG.

Note that one may opt for a rule like: $S \rightarrow S_2S_1$ for generating strings of the form: $yx \in L(G)$, where $y \in L(G_1)$ and $x \in L(G_2)$.

Examples for concatenation of CFLs

- The concatenation of G_1 for $L(G_1) = \{a^nb^n \mid n \geq 0\}$ and G_2 for $L(G_2) = \{c^md^m \mid n \geq 0\}$ results a new grammar G, such that $L(G) = \{a^nb^n\ c^md^m \mid m, n \geq 0\}$, which is a CFL, where both $L(G_1)$ and $L(G_2)$ are CFLs.
- Let $L_1 = \{a^nb^n \mid n \geq 0\}$. The rule set (P_1) for this language is: $\{S_1 \rightarrow aS_1b \mid \varepsilon\}$. Also, let $L_2 = \{ww^R \mid w \in \{a, b\}^*\}$, and the rule set (P_2) for this language is: $\{S_2 \rightarrow aS_2a \mid bS_2b \mid \varepsilon\}$. Now, to concatenate these two languages, $L = \{a^nb^n \mid n \geq 0\}.\{ww^R\} = \{\varepsilon, aa, bb, abba, .., ab, abaa, abbb, \ldots a^2b^2, \ldots, \}$, the new rule set $P = P_1 \cup P_2 \cup \{S \rightarrow S_1S_2\}$ is necessary.

Note: Let $L = \{a^ib^nc^n \mid i, n \geq 0\}$. Applying concatenation property for CFL, we can show that the language L is CFL. For this purpose, let $L_1 = \{a^i \mid i \geq 0\}$ and $L_2 = \{b^nc^n \mid n \geq 0\}$. Now, $L_1L_2 = \{a^ib^nc^n \mid i, n \geq 0\}$ is CFL, since L_1 and L_2 are CFL. Actually, we perform concatenation operation over a^i and b^nc^n, since the constraint of a^i $(i \geq 0)$ does not depend on the constraint of a^nb^n.

However, $L = \{ww \mid w \in \{a, b\}^*\}$ cannot be shown as CFL by concatenating the two languages: $L_1 = \{w \mid w \in \{a, b\}^*\}$ and $L_2 = \{w \mid w \in \{a, b\}^*\}$, since w for L_1 and w for L_2 are dependent on each other to satisfy the specified property.

THEOREM 5.4

If L is a CFL, then L* is also a CFL.

Proof Let $G = (V, \Sigma, P, S)$ be a CFG. Suppose $L = L(G)$ is the language for G. Now, if the new language $L_1 = L^*$ (obtained by applying closure operation over L, that is, zero or more concatenations of each string of L) gives the strings that are guaranteed to be derived by a new CFG (constructed from the existing grammar), then the CFL is closed under concatenation.

Constructing the new grammar: For the purpose of language closure, let us try to construct a new grammar $G_1 = (V_1, \Sigma, P_1, S_1)$ from G, where $V_1 = V \cup \{S_1\}$; S_1 is a new non-terminal symbol introduced for G_1. It is, indeed, the start symbol of G_1.

Further, $P_1 = P \cup \{S_1 \rightarrow SS_1 \mid \varepsilon\}$, where $S_1 \rightarrow SS_1 \mid \varepsilon$ are the new rules added to P_1 of G_1. The recursive rule $S_1 \rightarrow SS_1$ is essential to concatenate the strings of L(G) through S (the start symbol of G). On the other hand, the rule: $S_1 \rightarrow \varepsilon$ is added to generate the ε string, which is one outcome of the closure operation. All the strings of $L(G_1)$ are derived by concatenating zero or more times each string of L(G), that is, all these strings can be derived from G.

Here, G is assumed to be a CFG. Hence, G_1 is also a CFG.

For example, let $L = \{a^nb^n \mid n \geq 0\}$. Clearly, $L = \{\varepsilon, ab, a^2b^2, a^3b^3, \ldots.\}$. Now, $L^* = \{a^nb^n \mid n \geq 0\}^* = L^0 \cup L^1 \cup L^2 \cup \ldots.. = \{\varepsilon\} \cup \{\varepsilon, ab, a^2b^2, a^3b^3, \ldots.\} \cup \{\varepsilon, ab, a^2b^2, a^3b^3, \ldots.\}.\{\varepsilon, ab, a^2b^2, a^3b^3, \ldots.\} \cup \ldots\ldots$

This implies that L* contains any combination of (a^kb^k), $k \geq 0$. Now, to derive strings of such a combination, we need the rule set $(P_1) = \{S \rightarrow aSb, S_1 \rightarrow SS_1 \mid \varepsilon\}$.

Note: We usually prefer $S_1 \rightarrow SS_1$ to $S_1 \rightarrow S_1S$ for closure to avoid immediate left-recursion. Likewise, we may prove that L^+ is also a CFL.

THEOREM 5.5

The intersection of two CFLs is not necessarily a CFL.

Proof Let us take two CFLs: $L_1 = \{a^n b^n c^m \mid n, m \geq 0\}$ and $L_2 = \{a^n b^m c^m \mid n, m \geq 0\}$ to prove the statement. Obviously, each string in L_1 has a number of *as* followed by an equal number of *bs*, whereas each string in L_2 has a number of *bs* followed by an equal number of *cs*.

If we take their intersection as: $L = L_1 \cap L_2$, then only the strings common to both L_1 and L_2 will belong to L. More specifically, each string of L requires an equal number of *as*, *bs* and *cs*. For example, string: $a^2b^2c^3 \in L_1$ and string: $a^3b^2c^2 \in L_2$. However, these are not common to both L_1 and L_2. On the other hand, string: $a^2b^2c^2$ belongs to both L_1 and L_2, that is, it is common to both L_1 and L_2 and a valid string of L. Thus, the resulting language L formally looks like: $\{a^n b^n c^n \mid n \geq 0\}$.

Undoubtedly, strings of the resulting language: L cannot be generated by any CFG. In particular, it requires a CSG (instead of CFG) for this L. The example shows that, though both the languages (L_1 and L_2) are CFL, their intersection does not result in a CFL. Hence, the statement is true.

Let us present some more examples on intersection of CFLs.

INTERSECTION OF CFLS MAY BE A CFL

Let us take two CFLs: $L_1 = \{a^{2n}b^{2n} \mid n \geq 0\} = \{\varepsilon, a^2b^2, a^4b^4, a^6b^6, \ldots.\}$ and $L_2 = \{a^m b^m \mid m \geq 0\}\} = \{\varepsilon,$ ab, $a^2b^2, a^3b^3, \ldots.\}$. Now, perform intersection on L_1 and L_2 to obtain $L = L_1 \cap L_2 = \{a^{2i}b^{2i} \mid i \geq 0\} = \{\varepsilon, a^2b^2, a^4b^4, a^6b^6, \ldots.\}$. Clearly, L can be generated by the CFG. Thus, the intersection of two CFLs here gives a CFL, but it is not always true.

INTERSECTION OF TWO CFLS MAY BE REGULAR

Let us take two CFLs, $L_1 = \{a^n b^n \mid n \geq 0\}$and $L_2 = \{b^m a^m \mid m \geq 0\}$. Now, $L_1 \cap L_2 = \{\varepsilon\}$, which is regular. Thus, the intersection of two CFLs here gives a regular language, but it is not always true.

INTERSECTION OF A CFL AND A REGULAR LANGUAGE IS A CFL

The intersection of a CFL and a regular language may be regular or a CFL. However, it is true that every regular language can be recognised by CFG (as it is the subset of CFG). Thus, we may say that intersection of a CFL and a regular language is a CFL.

Let us take a CFL, $L_1 = \{a^n b^n \mid n \geq 0\} = \{\varepsilon, ab, a^2b^2, a^3b^3, \ldots.\}$ and a regular language (L_2) as: $L_2 = \{a^* b^*\} = \{\varepsilon$, a, b, aa, aaa,....., b, bb,..., ab,, aab, ..., a^2b^2, aabbb, ... $a^3b^3, \ldots.\}$. Now, $L = L_1 \cap L_2 = \{a^n b^n \mid n \geq 0\} = \{\varepsilon, ab, a^2b^2, a^3b^3, \ldots.\}$ is a CFL. But, some may claim that it is only regular as regular languages are a subset of CFLs. However, the correct answer is CFL. Here is the explanation. L_2 is a regular language and it includes strings that contain any number of *as* and *bs* but no *a* appears after a *b*. On the other hand, L_1 includes strings that contain an equal number of *as* followed by an equal number of *bs* and no *a* appears after a *b*. Obviously, L_2 (the regular language) includes L_1 (the CFL). Thus, the intersection of L_1 and L_2 gives L_1, which is a CFL.

Actually, regular languages (RL) are less powerful than CFLs. In other words, RLs have less control than CFLs, that is, the generation of strings that cannot be controlled by RLs may be controlled by CFLs. In real-life, "more controlled" is a subset of "less controlled". Therefore, every RL is a CFL, but the converse is not true.

Let us take an example: $L_1 = \{a^n b^n \mid n \geq 0\}$ and $L_2 = \{a^m b^m \mid m \leq 100\}$. The intersection of L_1 and L_2 gives L_2 itself, and it is regular. However, every regular language is also a CFL as it is a subset of CFL. This proposition can be proved mathematically.

Proof Let P be a PDA that accepts language L, and let M be a DFA that accepts language R. A new PDA P′ will simulate P and M simultaneously for the same input and accept it if both accept. Then, P′ accepts L ∩ R.

- The stack of P′ is the stack of P.
- The state of P′ at any time is the pair (state of P, state of M). The pair determines the transition function of P′.
- The final states of P′ are those in which both the state of P and the state of M are accepting.

More formally, let $M = (Q_1, \Sigma, \delta_1, q_1, F_1)$ be a DFA such that L(M) = R, and $P = (Q_2, \Sigma, \Gamma, \delta_2, q_2, F_2)$ be a PDA such that L(P) = L. Then, consider P′ such that

- $Q = Q_1 \times Q_2$
- $q_0 = (q_1, q_2)$
- $F = F_1 \times F_2$
- $\delta((p, q), x, a) = \{ ((p', q'), b) \mid p' = \delta_1(p, x) \text{ and } (q', b) \in \delta_2(q, x, a)\}$

THEOREM 5.6

The complement of a CFL is not necessarily a CFL.

Proof Let L_1 and L_2 be two CFLs. We may express $L_1 \cap L_2 = (L_1' \cup L_2')'$ (by De Morgan's laws). If the complement of CFLs were always CFL, then $L_1 \cap L_2$ would be a CFL, since the union of two CFLs is also a CFL. However, $L_1 \cap L_2$ is not necessarily a CFL. Hence, the complement of a CFL is not necessarily a CFL.

Note: The complement of a regular language is also regular. Regular languages are CFLs. Hence, the complement of such regular languages (that are CFLs) is also CFL.

THEOREM 5.7

If L_1 and L_2 are two CFLs, then $L_1 - L_2$ is not necessarily a CFL.

Proof Let $(L_1 - L_2)$ be a CFL for both L_1 and L_2. Then $\Sigma^* - L$ is also a CFL, where L is a CFL and Σ^* is a universal language and also a CFL. However, $\Sigma^* - L$ represents the complement of L, and we have seen that the complement of a CFL is not necessarily a CFL. Hence, a contradiction occurs. Therefore, $L_1 - L_2$ is not necessarily a CFL.

EXAMPLE 5.10

(a) Prove that $L = \{a^n b^n \mid n \neq 100, n \geq 0\}$ is a CFL.

Proof Here, $\Sigma = \{a, b\}$. We know that $L_1 = \{a^n b^n \mid n \geq 0\}$ is a CFL. Also, $L_2 = \{a^{100} b^{100}\}$ is regular (as it is a finite language). Now, $\overline{L_2}$ (complement of L_2) = $\{(a + b)^* - \{a^{100} b^{100}\}\}$ is also regular. Further, we know that the intersection of a CFL and a regular language is also a CFL. Thus, the intersection of $L_1 = \{a^n b^n \mid n \geq 0\}$ (a CFL) and $\overline{L_2} = \{(a + b)^* - \{a^{100} b^{100}\}$ (a regular language) is also a CFL, and it results in $L = \{a^n b^n \mid n \neq 100, n \geq 0\}$.

It is important to note that the PDA to be designed for $L = \{a^n b^n \mid n \neq 100, n \geq 0\}$ accepts all the strings of the form: $a^n b^n$, $n \geq 0$ except $n = 100$. In fact, for the string: $a^{100} b^{100}$, it moves to the non-final state (but not to a dead state).

(b) Prove that $L = \{w \mid n_a(w) = n_b(w) = n_c(w)\}$ is not a CFL. Here, $n_a(w)$ = the number of *a*s in string *w*, $n_b(w)$ = the number of *b*s in string *w*, and so on.

Proof Assume that L = $\{w \mid n_a(w) = n_b(w) = n_c(w)\}$ is a CFL, where the order of appearance of *as*, *bs* and *cs* is immaterial. Let us take a regular language, $L_1 = \{a^*b^*c^*\}$. Now, $L \cap \{a^*b^*c^*\}$ will be a CFL as the intersection of a CFL and a regular language is a CFL. However, $L \cap \{a^*b^*c^*\}$ results in $\{a^nb^nc^n \mid n \geq 0\}$, which is not a CFL (it is, indeed, a CSL). Here, the order of appearance of *as*, *bs* and *cs* does matter. But, we assumed that L is a CFL. So, a contradiction occurs. Hence, L is not a CFL. So, $L = \{w \mid n_a(w) = n_b(w) = n_c(w)\}$ is not a CFL.

Limitation of PDA: For PDA, the counting criteria (of valid strings) for infinite languages must again be linear, that is, a language like a^{n^2} *or* a^{2^n} (that is, counting criteria: n^2, 2^n, are non-linear) cannot be recognised by PDA.

5.4 PUSHDOWN AUTOMATA AND PARSING

PDA is used to check the acceptability of strings. More specifically, the compiler uses stack here to check whether or not a string is syntactically correct. In general, a parser takes the input and builds a parse tree.

A parser can be of two types:

- **Top-down parser:** Parsing starts from the top with the start-symbol, and the string is derived using a parse tree.
- **Bottom-up parser:** Parsing starts from the bottom with the string, and moves towards the start symbol using a parse tree.

Design of Top-Down Parser: For top-down parsing, a PDA has the following four types of transitions:

1. Push the start symbol 'S' of the grammar (G) onto the stack.
2. Pop the non-terminal on the LHS of the production to the top of the stack and push its RHS string.
3. If the top symbol of the stack matches the input symbol being read, pop it.
4. If the input string is fully read and the stack is empty, go to the final state 'F' to validate the string.

Design of Bottom-Up Parser: For bottom-up parsing, a PDA has the following four types of transitions:

1. Push the current input symbol onto the stack.
2. Replace the RHS of a production at the top section of the stack with its LHS.
3. If the top of the stack element matches the current input symbol, pop it.
4. If the input string is fully read and if the start symbol 'S' remains in the stack, pop it and go to the final state 'F' to validate the string.

EXAMPLE 5.11

Design a simple set of CFG rules for checking the validity of C declaration statement and show how such statements are parsed (validated) by a bottom-up parser using PDA.

Solution 1. A simple left-recursive CFG is given below.

D → T L; // D stands for declaration, T stands for type, L is used to denote variables. All are non-terminal symbols.

T → char| int | float| double

L → L, id

L → id //Examples of ids are x, y, ..

Example of a C declaration statement: int x, y;

Validation of the given statement is shown in Table 5.1 based on bottom-up parsing using the stack.

TABLE 5.1 Stack operations

Input	Stack content	Operation on stack
int x, y; $	$	No operation
x, y; $	int$	Push int
x, y; $	T$	Pop int and reduce it by T→ int (that is, replace it by T)
,y;$	id TS	Push *x* (id)
,y;$	L T$	Pop id and reduce it by L → id
y;$	, LT$	Push comma (,)
;$	id, L T$	Push, *y* (id)
;$	LT$	Pop id, L and reduce it by L → L, id
$	; LT$	Push semi-colon (;)
$	D$	Pop ; LT and reduce it by D →TL;
$	D$	Processing of string is complete and the stack contains only the start symbol: D of the grammar. So, the string is accepted.

Solution 2. A simple right-recursive CFG is given below.

D → T L; // D stands for declaration, T stands for type, L is used to denote variables. All are non-terminal symbols.

T → char| int | float| double

L → id, L

L → id //Examples of ids are x, y, ...

Example of a C declaration statement: int x, y;

TABLE 5.2 Stack operations

Input	Stack content	Operation on stack
int x, y; $	$	No operation
x, y; $	int$	Push int
x, y; $	T$	Pop int and reduce it by T → int (that is, replace it by T)
, y;$	id T$	Push *x* (id)
y;$	, id T$	Push comma (,) assuming that comma (,) has higher precedence than the stack-top id
;$	id ,id T$	Push *x* (id)
;$	L, id T $	Pop id and reduce it by L → id
;$	LT$	Pop L, id and reduce it by L → id, L
$	; LT$	Push semi-colon (;)
$	D$	Pop ; LT and reduce it by D → TL;
$	D$	Processing of the string is complete and the stack contains only the start symbol: D of the grammar. So, the string is accepted.

Parsing of the string/sentence: int x, y; *using the top-down approach:*

D ⇒TL; ⇒int L; ⇒ int id, L; ⇒ int id(x), id(y); (based on left-most derivation)

Note: LL(k), $k \geq 1$ parser (a predictive parser under TDP, more specifically a recursive descent parser without backtracking) also uses the stack. Here, the operations: shift and replacement (expansion) are performed in the stack. LL parsers are table-based parsers, similar to LR (shift-reduce) parsers.

Deterministic parsing: During parsing, if only one rule in the case of LL parsing is chosen or only one action: shift/reduce/accept/error in LR parsing is executed at any instance, then it is called deterministic parsing. Deterministic LR parsing can decide whether to shift/reduce/accept/error action at any instance during parsing. The LR parser is a type of S(shift)-R(reduce) parser that efficiently handles deterministic context-free grammar (or equivalently DPDA). Let us highlight some important points:

- An efficient LR parser simulates DPDA.
- Some PDA cannot be simulated efficiently by a computer program if they are NPDA.
- A parsing algorithm that corresponds to DPDA can be more efficient than a parsing algorithm that corresponds to NPDA. In fact, the non-determinism of NPDAs can make parsing slow (due to backtracking).
- DPDA → DCFG → efficient LR parsing algorithm may exist.
- NPDA → DCFG → efficient LR parsing algorithm may not exist.

Note: In LL parsing, the first symbol of the RHS of the production will be at the top of the stack, and the next one below it, and so on, whereas in S-R parsing, the last symbol of the RHS of the production will be at the top of the stack. More specifically, the stack used in LL parsing mainly assists in expanding the rule; that is why the first symbol of the RHS of the rule appears at the top of the stack; whereas the stack used in bottom-up parsing mainly assists in reducing the production, and that is why, the last symbol of the RHS of the rule appears at the top of the stack.

Applications of PDA

- Designing compiler for programming languages
- Designing tools for artificial intelligence

5.5 ADDITIONAL SOLVED PROBLEMS

PROBLEM 1

Justify: DPDA accepts a class of languages that is between regular languages and context-free languages.

Solution FA accepts regular languages. FA has no auxiliary memory (stack), so its power is less than that of PDA (which recognises CFL). Further, DPDA does not allow the machine to guess the solution to a problem. However, NPDA does so, and this capability increases its power. That is why problems that are solvable by NPDA may not be solved by DPDA. In this respect, there is a very simple question: NDFA also allows the machine to guess the solution to problems but why is its power still equivalent to that of DFA? The reason is that decision based on a guess is not possible in NDFA.

Thus, DPDA accepts a class of languages that is between regular languages and context-free languages. In fact, it accepts deterministic context-free languages – a proper subset of context-free languages.

In short, DPDA is not as powerful as NPDA. This machine accepts a class of languages somewhere between regular languages and context-free languages. For this reason, DPDA is often skipped as a topic. In practice, it can be useful as determinism is much easier to implement. Parsers such as YACC are implemented using DPDA.

PROBLEM 2

Present your idea to design a PDA to accept the language, $L = \{a^n b^n \mid n \geq 0 \text{ and } n \neq 10\}$.

Solution For each valid string of the language $L = \{a^n b^n \mid n \neq 10\}$, we may specify distinct transitions for PDA up to $n \neq 10$ using distinct states; that is, it behaves like an FA. The transitions for $n = 11$ and onwards will be similar to that of operating an infinite language, $L = \{a^n b^n \mid n \geq 0\}$, as discussed earlier.

More specifically, the PDA to be designed for $L = \{a^n b^n \mid n \neq 10, n \geq 0\}$ accepts strings of the form: $a^n b^n$, $n \geq 0$, except $n = 10$. In fact, for string: $a^{10}b^{10}$, it moves to the non-final state (but not a dead state).

PROBLEM 3

Let $L = \{a^n b^n \mid n \geq 0\}$. Prove that: (a) L^2 is a CFL. (b) L^c (complement of L) and L^* are CFLs.

Solution (a) The given language $L = \{a^n b^n \mid n \geq 0\}$ is a CFL. We know that the concatenation property for CFL gives a CFL. Hence, $L^2 = L.L = \{a^m b^m\ a^n b^n \mid m, n \geq 0\}$ is also a CFL.
(b) Here, $L = \{a^n b^n \mid n \geq 0\}$ is a CFL over $\Sigma = \{a, b\}$. Now, $L^c = \{a^*b^*\} - \{a^n b^n \mid n \geq 0\} = \{a^m b^n \mid m \neq n\}$, which is a CFL. Further, $L^* = L^0 \cup L^1 \cup L^2 \cup \ldots$, where each L^i $(i \geq 0)$ is obtained from i-times concatenation with L. However, L is a CFL and the concatenation property for CFLs results in another CFL. Also, the union of two CFLs gives a CFL. Hence, L^* is a CFL.

PROBLEM 4

Design a PDA to accept $L = \{0^m 1^n \mid (m + n) \text{ is even}\}$.

Solution To solve this problem using PDA, only read operations are to be performed. However, accepting strings depends on moving to accepting states. The necessary transitions are given below, where q_0, q_2 and q_4 are the accepting states with q_0 as the start state as well.

- $\delta(q_0, \varepsilon, Z_0) = (q_0, Z_0)$ *// Read operation; null string is also accepted.*
- $\delta(q_0, 1, Z_0) = (q_1, Z_0)$ *// Read operation*
- $\delta(q_1, 1, Z_0) = (q_2, Z_0)$ *// Read operation*
- $\delta(q_2, 1, Z_0) = (q_1, Z_0)$ *// Read operation*
- $\delta(q_0, 0, Z_0) = (q_3, Z_0)$ *// Read operation*
- $\delta(q_3, 0, Z_0) = (q_0, Z_0)$ *// Read operation*
- $\delta(q_3, 1, Z_0) = (q_4, Z_0)$ *// Read operation; from q_3, on reading* 1. *It moves to q_4 (not q_1), since moving to q_1 gives invalid string reaching q_2 via input* 1.
- $\delta(q_4, 1, Z_0) = (q_5, Z_0)$ *// Read operation*
- $\delta(q_5, 1, Z_0) = (q_4, Z_0)$ *// Read operation*

PROBLEM 5

Is the language $L = \{a^m b^n \mid m, n \geq 0, n = m^2\}$ a CFL?

Solution No, the given language is not a CFL. Firstly, the language is not finite. As it is infinite, it is not regular since we may not insert varying values (inputs) inside the loop of the designed FA. For CFLs, no fixed number of *a*s may be pushed for tracking and matching with *b*s (through pop operation), as the value of n varies from string to string in the language. For example, let us take some values of m and n for the corresponding values of m as follows:

$m = 1, n = 1$
$m = 2, n = 4$
$m = 3, n = 8$

$m = 4, n = 16$

....

Hence, the given language is not a CFL.

Note: If $n = km$, where k is a positive constant, we may be able to design a PDA. However, if $k = m$, then k depends on m, and we will be unable to design a PDA.

PROBLEM 6

Is the language $L = \{a^nb^nc^n \mid n \geq 1\}$ a CFL?

Solution The given language is not CFL. Further, the language is not finite. As it is infinite, it is not regular. Regarding CFLs, let us attempt to design a PDA to solve this problem. First, push all the *as* onto the stack for counting. Next, match and track the *bs* with the *as* through pop operation. When all the *bs* have been consumed, nothing remains in the stack to compare with the *cs*. This is why it is not a CFL.

However, one may claim that two *as* may be shifted to the stack for each push operation in order to track the number of *bs* and *cs*. But, if it is performed thus, incorrect strings such as *aabc, aaabbc*, etc. may also be validated.

Note: Based on the same logic, we may easily prove that languages $L = \{a^nb2^nc^{3n} \mid n \geq 1\}$, $L = \{a^nb^nc^m \mid n > m\}$, $L = \{a^nb^mc^nd^m \mid m, n \geq 1\}$ are not CFLs but languages such as $L = \{a^nb^{2n}c^{3n} \mid n \leq 10\}$, $L = \{a^nb^nc^n \mid n \leq 10\}$, $L = \{a^nb^mc^nd^m \mid m, n \leq 10\}$ are all CFLs as they are finite and all finite languages are regular. Further, each regular language is a CFL as well.

PROBLEM 7

Can we design a DPDA for the language resulting from the union of two CFLs, $L_1 = \{a^nb^n \mid n \geq 1\}$ and $L_2 = \{a^nb^{2n} \mid n \geq 1\}$?

Solution The resulting language $L = L_1 \cup L_2$ is a CFL, but we may not design any DPDA for this language. No fixed number of *as* may be pushed for tracking and matching with *bs* (through a pop operation), as the value of *n* varies from string to string. However, we may design an NPDA for the resulting language. An incomplete simple NPDA is shown in Fig. 5.7.

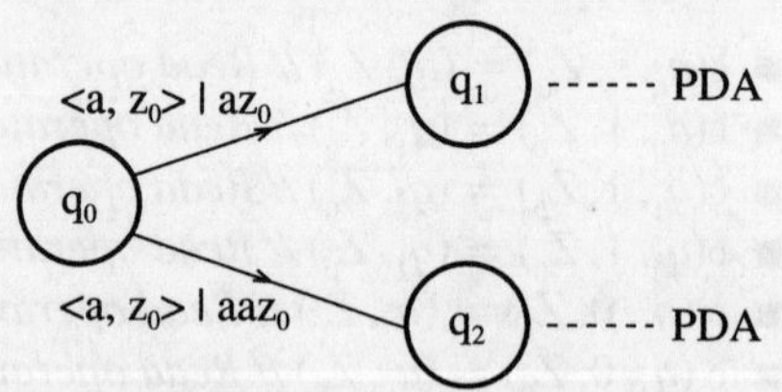

FIGURE 5.7 An NPDA for $L_1 = \{a^nb^n \mid n \geq 1\} \cup L_2 = \{a^nb^{2n} \mid n \geq 1\}$

It is clear that for the same configuration: (q_0, a, Z_0), the machine moves two different states, namely, q_1 and q_2, with stack contents (az_0) and (aaz_0), respectively. Hence, the PDA is an NPDA.

PROBLEM 8

Is the language $L = \{a^kb^lc^md^n \mid k, l, m, n \geq 1\}$ a CFL?

Solution Here, no constraint is imposed on *k, l, m* and *n*. The language is thus regular and the RE is $(a^+b^+c^+d^+)$. Hence, the language is a CFL too.

PROBLEM 9

Is $L = \{a^{n^n} \mid n \geq 1\}$ a CFL?

Solution Here, n^n is not in arithmetic progression (AP), that is, a machine such as FA and PDA cannot change the value of n in the loop. So, we may not design any FA or PDA for this language. Therefore, this language is neither regular nor CFL. Likewise, languages $L_1 = \{a^{n!} \mid n \geq 1\}$, $L_2 = \{a^{2^n} \mid n \geq 1\}$, $L_2 = \{a^m \mid m \text{ is prime}\}$ is neither regular nor CFL, since the powers of the string structures in these languages are not in AP. However, $L = \{a^k \mid k \text{ is even}\}$ is regular, as k is in AP.

PROBLEM 10

Is $L = \{a^i b^j c^k \mid i > k > j\}$ a CFL?

Solution This language is not a CFL as comparison for $a^i b^j$, $i > j$ through push and pop operations is possible but the remaining contents of the stack may not ensure a comparison of $j > k$ for the substring: $b^j c^k$.

PROBLEM 11

Is $L = \{a^i b^j c^k \mid j = i + k\}$ a CFL?

Solution This language is a CFL. The constraint: $j = i + k$ is satisfied as follows: for *as*, perform push; for *bs*, perform pop. These two operations help to match *as* and *bs*. The remaining *bs* are pushed onto the stack and, finally, the *cs* are popped. Thus, $j = i + k$ is verified.

PROBLEM 12

Is $L = \{a^i b^j c^k d^l \mid i = k \text{ or } j = l\}$ a CFL?

Solution Here, the conditions $i = k$ and $j = l$ are connected by the logical OR. So, we may construct a PDA for $a^i c^k$, where the number of *bs* and *ds* do not matter, and another PDA for $b^j d^l$, where the number of *as* and *cs* do not matter. Finally, we may combine these two PDAs by taking their union through ε move to satisfy either or both of the conditions. The resulting PDA will be an NPDA.

However, $L = \{a^i b^j c^k d^l \mid i = k \text{ and } j = l\}$ is not a CFL. We may not be able to design any NPDA for this language by taking the union of the two PDAs because both the conditions need to be satisfied.

PROBLEM 13

Is $L = \{a^n b^n c^m \mid n > m\}$ a CFL?

Solution This is not a CFL. The reason is that we may not be able to design any PDA for this language, as nothing will remain in the stack to check if $n > m$ ($n_b(w) > n_c(w)$) for $a^n b^n c^m$ after matching the number of *as* and *bs*.

PROBLEM 14

Is $L = \{w \mid w \in \{a,b\}^* \text{ and } |w| \geq 100\}$ a CFL?

Solution The given language starts with strings of finite size: 100 but there is no restriction on the number of *as* or *bs*. So, it is regular. Hence, it is a CFL too.

PROBLEM 15

Is $L = \{w \mid w \in \{a,b,c\}^* \text{ and } n_a(w) = n_b(w) = n_c(w)\}$ a CFL?

Solution The given language is not a CFL. We have already seen that the language $L_1 = \{a^n b^n c^n \mid n \geq 0\}$ is not a CFL; it is, indeed, a CSL. Again, L_1 is a subset of $L = \{w \mid w \in \{a,b,c\}^* \text{ and } n_a(w) = n_b(w) = n_c(w)\}$. As L_1 is not a CFL, L is also not a CFL.

PROBLEM 16

Is L = {$a^n b^m$ | $m \neq n$} a CFL?

Solution Here, either of the following possible cases in valid strings may occur: (i) $n_a(w) > n_b(w)$ OR (ii) $n_b(w) > n_a(w)$. To design a PDA to accept strings of this language, we may first construct an ordinary PDA for L = {$a^n b^n$ | $n \geq 1$}. Then, we modify this PDA as follows:

- $n_a(w) > n_b(w)$: If no input remains to be processed but '*a*' remains in the stack, then accept; the desired transition is: $\delta(q_i, \varepsilon, a) = (q_f, a)$.
- $n_b(w) > n_a(w)$: If the input is '*b*' but the stack-top is Z_0, then accept; $\delta(q_i, b, Z_0) = (q_f, Z_0)$.

Note: In the same way, we can design a PDA for L = {$a^n b^m$ | $n \neq 2m$}.

PROBLEM 17

Is wxw^R: $w, x \in$ {a, b}* a regular language or a context-free language?

Solution It is regular, as *x* can be extended (if allowed) beyond the limit. Thus, the regular expression can be viewed as: *r* = a*x*a + b*x*b, where $x \in$ {a, b}*.

PROBLEM 18

Is $xww^R y$: $w, x, y \in$ {a, b}* a regular language or a context-free language?

Solution It is regular, as *x* and y can be extended beyond the limit. Thus, the regular expression can be viewed as: *r* = *x*aa*y* + *x*bb*y*, where *x*, y $\in$ {a, b}*.

PROBLEM 19

Is xww^R: $w, x \in$ {a, b}* a regular language or a context-free language?

Solution It is not regular, as *x* cannot be extended beyond the limit. Take an example, where *w* = *aba*, w^R = *aba* and *x* = *aa*; that is, the entire string is (*aba*)*aa*(*aba*) and it cannot be expressed as: xww^R ((*abaaaa*)(*ba*)) by expanding *x* equal to *abaaaa*.

PROBLEM 20

Consider the following languages:

{$a^m b^n c^p d^q$ | $m + p = n + q$, where $m, n, p, q \geq 0$}
{$a^m b^n c^p d^q$ | $m = n$ and $p = q$, where $m, n, p, q \geq 0$}
{$a^m b^n c^p d^q$ | $m = n = p$ and $p \neq q$, where $m, n, p, q \geq 0$}
{$a^m b^n c^p d^q$ | $mn = p + q$, where $m, n, p, q \geq 0$}
Which of the above languages is context-free?

Solution
- Language (L) = {$a^m b^n c^p d^q$ | $m + p = n + q$, where $m, n, p, q \geq 0$} is a CFL as we can frame the equation as: $m - n + p - q = 0$, which can be accomplished by push (for *a*s), pop (for *b*s), push (for *c*s) and pop (for *d*s), and check that the stack is empty at the end.
- Language (L) = {$a^m b^n c^p d^q$ | $m = n$ and $p = q$, where $m, n, p, q \geq 0$} is a CFL as the sequence of operations for *a, b, c* and *d* is (push, pop), (push, pop).
- Language (L) = {$a^m b^n c^p d^q$ | $m = n = p$ and $p \neq q$, where $m, n, p, q \geq 0$} is not a CFL as $m = n = p$ requires double comparisons, which cannot be performed by PDA.
- Language (L) = {$a^m b^n c^p d^q$ | $mn = p + q$, where $m, n, p, q \geq 0$} is not a CFL as *mn* involves multiplying the number of *a*s and the number of *b*s, which cannot be performed by PDA.

Tip: If the constraint for a language (L) can be expressed in the form of a linear equation, it would be possible to construct a PDA for L. However, the sequence of push and pop must be decided based on analysis. Examples: $i = j + k$ in L = $\{a^ib^jc^k \mid i = j + k \text{ and } i, j, k > 0\}$ can be expressed as: $i - j - k = 0$; $j = i + k$ in L = $\{a^ib^jc^k \mid j = i + k \text{ and } i, j, k > 0\}$ can be expressed as: $j - i - k = 0$; $m + p = n + q$ in L = $\{a^mb^nc^pd^q \mid m + p = n + q, \text{ where } m, n, p, q \geq 0\}$ can be expressed as: $m - n + p - q = 0$, etc.

PROBLEM 21

Design a PDA for HTML, ensuring that all opening tags have matching closing tags within the <BODY> </BODY> tag.

Solution HTML is a language used for designing web pages. It is an interpreted language and produces no error. The language has a basic structure of the form:

<HTML>
<HEAD>
<TITLE>
....
</TITLE>
</HEAD>
</HTML>

Here, the tag <HTML> indicates the beginning of a HTML program, and </HTML> indicates the end of the program.

<TITLE>
Introduction to Turing machine. // *This is the title of the topic to be covered.*
</TITLE>

From the structure of the tags presented above, it is clear that any HTML programme can easily be parsed using PDA by visualising <tag-name> [beginning of the tag] and </tag-name> [end of the tag]

A few basic transitions in this regard are presented below.

1. δ (q_0, <HTML>, Z_0) = (q_1,<HTML> Z_0) // Push
2. δ (q_1, <HEAD>, <HTML>) = (q_2,<HEAD><HTML>) // Push
3. δ (q_2, <TITLE>, <HEAD>) = (q_3, <TITLE><HEAD>) // Push
4. δ (q_3, </TITLE>, <TITLE>) = (q_3, ε) // Pop

....

PROBLEM 22

Design a PDA to convert infix expressions to postfix expressions.

Solution To convert infix expression to postfix expression, we may extend the transition relation as follows:

δ (q_0, N, Z_0) = {(q_1, Z_0), print (N)} // Read N (number) and then print it.
δ (q_1, op, Z_0) = {(q_2, op Z_0)} // Push the operator (op)
δ (q_2, N, Z_0) = {(q_3, Z_0), print (N)} // Read N (number) and then print it.
δ (q_3, ε, op) = {print (*op*), (q_0, ε) } //Read no input, print the stack-top and pop it.
δ (q_0, ε, Z_0) = {(h, Z_0)} // The machine moves to the halt state (h).

Note: Here, the PDA works as an expression converter. It may also work for postfix evaluation, that is, as the evaluator. However, all these jobs are built into the parser part of the compiler.

PROBLEM 23

(a) Compare the pros and cons of the three notations: infix, postfix, and prefix.

Solution An infix expression is more natural and the operator acts like a separator but is difficult to compute using a stack; whereas the postfix expression is easier for computation with a stack. The prefix expression is easier for computation with a stack, just like postfix; more like a tree structure and more functional.

(b) Is deterministic PDA equivalent to non-deterministic PDA?

Solution No. Some CFG can be recognised by non-deterministic PDA but not by deterministic PDA. For example, $\{ww^R \mid w \in \{a, b\}^*\}$.

5.6 RESEARCH DIRECTIONS

By now, readers must have visualised that PDA is inherently recursive in nature. There can be several interesting applications of this property.

XML files are frequently used in software for storing various types of configurations. We can easily use PDA to determine the syntactical correctness of an XML file.

Consider the file below:

```
<?xml version="1.0" encoding="UTF-8"?>
<note>
 <to>Tove</to>
 <from>Jani</from>
 <heading>Reminder</heading>
 <body>Don't forget me this weekend!</body>
 </note>
```

This file is valid. We need to check that every opening tag has a valid closing tag. Given below is another XML file.

```
<?xml version="1.0" encoding="UTF-8"?>
<!—-This is a comment-->
<note>
 <to>Tove</to>
 <from>Jani</from>
 <heading>Reminder</heading>
 <body>Don't forget me this weekend!</body>
 </note>
```

Now, there is a need to handle the comments in a special way. We need to think along the lines of the HTML parsing problem discussed just before this section, and attempt to design a PDA for parsing an XML file.

Let us consider another interesting problem. Can we use a PDA to translate an English sentence to some other language, say French? Consider this example: The English sentence "*I do not speak French*", when translated to French, reads as "*Je ne parle pas français.*"

The word-by-word translation is as follows: I = Je, speak = parle, French = français...

The difficult part is the "*ne … pas*". In French, to imply negation of a verb, we need to add "*ne*" before the verb and "*pas*" after the verb. Can you visualise this negation as a PDA? In several inter-language translations, similar problems will be encountered.

Project-based problems

- Design a parser to check the validity of an HTML file. You must consider embedded Javascript within HTML as well.
- Design a translator to translate C++ program to Java. Both programs must behave identically on successful compilation.
- Design an HTML parser.

EXERCISES

1. Construct a PDA to recognise a language of matched parenthesis.
2. Which of the following languages over {a, b, c} is accepted by deterministic pushdown automata?
   (a) $\{w \subseteq w^R \mid w \in \{a, b\}^*\}$ (b) $\{ww^R \mid w \in \{a, b, c\}^*\}$
   (c) $\{a^nb^nc^n \mid n \geq 0$ (d) {w | w is a palindrome over {a, b, c}}
   Note: w^R is the string obtained by reversing w. Give a proper explanation for the correct answer. **[GATE 1997]**
3. Let L_D be the set of all languages accepted by a PDA in the final state and L_E the set of all languages accepted by an empty stack. Which of the following is true?
   (a) $L_D = L_E$ (b) $L_D \supset L_E$ (c) $L_E \subset L_D$ (d) None of the above
   Give a proper explanation for the correct answer. **[GATE 1999]**
4. The language accepted by a PDA in which the stack is limited to 10 items is best described as
   (a) Context-free (b) Regular (c) Deterministic context-free (d) Recursive
   Give a proper explanation for the correct answer. **[GATE 2002]**
5. Let G = ({S},{a, b}, R, S) be a context-free grammar where the rule set R is $S \rightarrow a\,S\,b \mid S\,S \mid \varepsilon$. Which of the following statements is true?
   (a) G is not ambiguous (b) There exist x, y ∈ L(G) such that xy ∉ L(G)
   (c) There is a deterministic pushdown automaton that accepts L(G)
   (d) We can find a deterministic finite state automaton that accepts L(G)
   Give a proper explanation for the correct answer. **[GATE 2003]**
6. Let Nf and Np denotes the classes of languages accepted by NFA and NPDA, respectively. Let Df and Dp denote the classes of languages accepted by DFA and DPDA, respectively. Which one of the following is true?
   (a) Df ⊂ Nf and Dp ⊂ Np (b) Df ⊂ Nf and Dp = Np
   (c) Df = Nf and Dp = Np (d) Df = Nf and Dp ⊂ Np
   Give a proper explanation for the correct answer. **[GATE 2005, Maharashtra SET 2016]**
7. Consider the NPDA < Q = {q0, q1, q2}, Σ = {0, 1}, Γ = {0, 1, ⊥ }, δ, q0, ⊥ , F = {q2}>, where (as per usual convention) Q is the set of states, Σ is the input alphabet, Γ is the stack alphabet, δ is the state transition function, q0 is the initial state, ⊥ is the initial stack symbol, and F is the set of accepting states, The state transition is shown below.

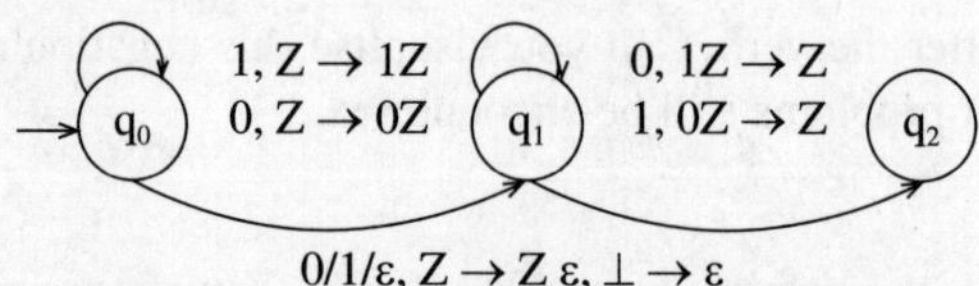

Which one of the following sequences must follow the string 101100 so that the entire string is accepted by the automaton?

(a) 10110 (b) 10010 (c) 01010 (d) 0100 **[GATE 2015]**

8. Consider the transition diagram of the PDA given below, with input alphabet $\Sigma = \{a, b\}$ and stack alphabet $\Gamma = \{X, Z\}$. Z is the initial stack symbol. Let L denote the language accepted by the PDA. Which one of the following is true?

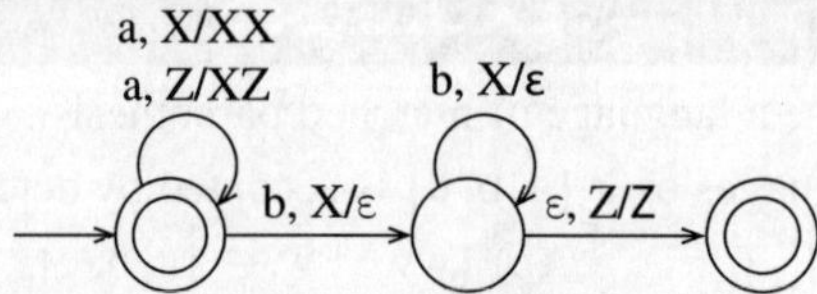

(a) $L = \{a^n b^n \mid n \geq 0\}$ and is not accepted by any finite automata.
(b) $L = \{a^n \mid n \geq 0\} \cup \{a^n b^n \mid n \geq 0\}$ and is not accepted by any deterministic PDA
(c) L is not accepted by any TM that halts on every input
(d) $L = \{a^n \mid n \geq 0\} \cup \{a^n b^n \mid n \geq 0\}$ and is deterministic context-free.

Give a proper explanation for the correct answer. **[GATE 2016]**

9. The language accepted by the non-deterministic pushdown automaton M = ({q0, q1, q2}, {a, b}, {a, b, z}, δ, q0, z, {q2}) with transitions δ (q0 a, z) = {(q1 a), (q2 ε)}; δ (q1, b, a) = {(q1, b)}; δ (q1, b, b) = {(q1 b)}, δ (q1, a, b) = {(q2, ε)} is:

(a) L(abb*a) (b) {a} ∪ L(abb*a) (c) L(ab*a) (d) {a} ∪ L(ab*a)

Give a proper explanation for the correct answer. **[UGC NET 2013]**

10. A non-deterministic pushdown automaton that accepts the language generated by the grammar: S → aSS | ab is

(a) δ(q0, ε, z) = {(q1, z)}; δ(q0, a, S) = {(q1, SS)}, (q1, B)}
δ(q0, b, B) = {(q1, ε)}, δ(q1, Λ, z) = {(qf, Λ)}

(b) δ(q0, ε, z) = {(q1, Sz)}; δ(q0, a, S) = {(q1, SS)}, (q1, B)}
δ(q0, b, B) = {(q1, ε)}, δ(q1, ε, z) = {(qf, ε)}

(c) δ(q0, ε, z) = {(q1, Sz)}; δ(q0, a, S) = {(q1, S)}, (q1, B)}
δ(q0, b, ε) = {(q1, B)}, δ(q1, ε, z) = {(qf, ε)}

(d) δ(q0, ε, z) = {(q1, z)}; δ(q0, a, S) = {(q1, SS)}, (q1, B)}
δ(q0, b, ε) = {(q1, B)}, δ(q1, ε, z) = {(qf, ε)} **[UGC NET 2013]**

11. The pushdown automation M = ({q0, q1, q2}, {a, b}, {0, 1}, δ, q0, 0, {q0}) with δ(q0, a, 0) = {(q1, 10)}, δ(q1, a, 1) = {(q1, 11)}, δ(q1, b, 1) = {(q2, ε)}, δ(q2, b, 1) = {(q2, ε)} and δ(q2, ε , 0) = {(q0, ε)}, accepts the language

(a) $L = \{a^n b^m \mid n, m \geq 0\}$ (b) $L = \{a^n b^n \mid n \geq 0\}$
(c) $L = \{a^n b^m \mid n, m > 0\}$ (d) $L = \{a^n b^n \mid n > 0\}$

Give a proper explanation for the correct answer. **[UGC NET 2014]**

12. Which of the following machines recognises even-length palindrome strings?
    (a) Deterministic finite state machine (b) Non-deterministic finite state nachine
    (c) Deterministic PDA (d) Non-deterministic PDA
    Give a proper explanation for the correct answer. **[APSET 2014]**
13. Which of these statements is true about PDA?
    (a) PDAs recognise context-sensitive languages (b) PDAs recognise context-free languages
    (c) PDAs recognise Type 1 languages (d) PDAs recognise Type 0 languages
    Give a proper explanation for the correct answer. **[Gujarat SET 2004, APSET 2014]**
14. Which of these is false regarding formal languages?
    (a) There is a unique minimal DFA for every regular language
    (b) Every NFA can be converted to an equivalent PDA
    (c) The complement of every context-free language is recursive
    (d) Every non-deterministic PDA can be converted into an equivalent deterministic PDA
    [Gujarat SET 2013, Maharashtra SET 2016]
15. Linear bounded automata (LBA):
    (a) Accept the set of regular but not context-free languages
    (b) Accept the set of context-free but not context-sensitive languages
    (c) Accept the set of context-sensitive languages (d) Accept the set of recursive languages
    Give a proper explanation for the correct answer. **[Gujarat SET 2013]**
16. Pushdown automata __________ context-free languages.
    (a) Are a superset of (b) Are the same thing as
    (c) Generate (d) Accept **[HP SET 2014]**
17. A pushdown automaton is to a __________ as a finite automaton is to a regular expression.
    (a) Arithmetic expression (b) Context-free grammar
    (c) Entity-relationship diagram (d) Wild card character **[J&K SET 2013]**
18. Context-free grammar (CFG) can be recognised by:
    (a) FA (b) Two-way LBA (c) PDA (d) both (b) and (c)
    [UGC NET 2009, HP SET 2013, Karnataka SET 2014]
19. Palindromes cannot be recognised by a finite state machine because:
    (a) It cannot remember large amounts of information (b) It cannot fix the mid-point
    (c) Even for a known mid-point, it cannot check if the second half matches the first
    (d) All of these
    Give a proper explanation for the correct answer. **[Rajasthan SET 2011]**
20. Which of the following statements is true?
    (a) If a language is context-free, it can always be accepted by a DPDA
    (b) The union of two context-free languages is context-free
    (c) The intersection of two context-free languages is context-free
    (d) The complement of a context-free language is context-free
    [HCU PhD Entrance 2010, TN SET 2012]

21. Mr. Das has a farm. On that farm, he has some chickens, some sheep, some horses and some dogs. On his farm, there must always be twice as many chickens as sheep; twice as many sheep as horses; and twice as many horses as dogs. If we represent the animals on this farm using the alphabet $\Sigma = \{c, s, h, d\}$ for chicken, sheep, horse and dog, then a string such as *ccccccccsssshhd* represents 8 chickens, 4 sheep, 2 horses and 1 dog (always in that order). Consider the formal language of all such strings with the above constraints.
   (a) What class of formal language does this language belong to if there can be at most 1000 animals on the farm? Prove your answer. *Hint:* As the size is finite, this problem can be modelled by regular grammar.
   (b) What class does it belong to if there is no limit on the number of animals? Prove your answer. (if you claim that the language is context -free, you must prove that it is context-free AND you must prove that it is not regular). *Hint:* Both the order and the number of occurrences with no limit may not be satisfied by a regular language (equivalently FA); however, it may be solved by CFG (equivalently PDA). It may be proved by applying pumping lemma for CFL.
22. (a) Construct a context-free grammar for $L = \{a^n b^m c^k \mid 2m = n \text{ and } k \geq 2\}$. *Hint:* $S \rightarrow S_1 A$, $S_1 \rightarrow aS_1 bb \mid \varepsilon$, $A \rightarrow cB$, $B \rightarrow cB \mid c$
   (b) Construct a deterministic pushdown automaton for the above language.
23. We know that the concatenation of two regular languages results in a regular language. However, consider the language L of all even palindromes over {0, 1}; L is not regular. Now consider the languages $L_1 = \{w \mid w.w^R \in L\}$ and $L_2 = \{w^R \mid w.w^R \in L\}$. Show that L_1 and L_2 are regular. Explain why although L_1 and L_2 are regular, L is not regular. *Hint*: $w \in L_1$ and $w^R \in L_2$, where $w \in \{0, 1\}^*$. Obviously, L_1 and L_2 are regular languages, where each of these languages contains any string (w and w^R: a form of w) generated over {0, 1}. But strings of the format: $w.w^R$ may not be generated by a regular language.
24. Let L consist of all binary strings beginning with 1 such that its value when converted to decimal is divisible by 5. Which of the following is true?
   (a) L can be recognised by a DFA (b) L can be recognised by an NDFA but not by a DFA
   (c) L can be recognised by a DPDA but not by an NDFA
   (d) L can be recognised by an NPDA but not by a deterministic PDA
   (e) L cannot be recognised by any PDA
   Justify your answer with proper reasoning. **[TIFR PhD Entrance 2010]**
25. Consider the following grammar G with terminals {[,]}, start symbol S, and non-terminals {A, B, C}: $S \rightarrow AC \mid SS \mid AB$, $C \rightarrow SB$, $A \rightarrow [$, $B \rightarrow]$. A language L is called prefix-closed if for every $x \rightarrow L$, every prefix of x is also in L. Which of the following is false?
   (a) L(G) is context-free (b) L(G) is infinite
   (c) L(G) can be recognised by a DPDA (d) L(G) is prefix-closed (e) L(G) is recursive
   Justify your answer with proper reasoning. **[TIFR PhD Entrance 2017]**
26. Which of the following is true for language $L = \{0^m 1^m \mid m > 0\}$?
   (a) It can be accepted by an NPDA but not by a DPDA
   (b) It can be accepted by an NPDA as well as by a DPDA
   (c) It can be accepted neither by an NPDA, nor a DPDA
   (d) It can be accepted by a DPDA but not by an NPDA
   Justify your answer with proper reasoning. **[Gujarat SET 2017]**

27. Let M = ({q_0, q_1}, {0, 1}, {z_0, X}, d, q_0, z_0, j) be a PDA, where d is given by:
    $\delta(q_o, 1, z_o) = \{(q_o, XZ_o)\}$
    $\delta(q_o, \in, z_o) = \{(q_o, \in)\}$
    $\delta(q_o, 1, X) = \{(q_o, XX)\}$
    $\delta(q_1, 1, X) = \{(q_1, \in)\}$
    $\delta(q_o, O, X) = \{(q_1, X)\}$
    $\delta(q_o, O, z_o) = \{(q_o, z_o)\}$
    (a) What is the language accepted by this PDA by empty stack?
    (b) Describe informally the working of the PDA. **[GATE 1998]**
28. Convert the following context-free grammar, where ε is the null string to Chomsky Normal Form (CNF) and then obtain a PDA for it: S → BBAC, A → aB | a, B → aBb | ε, C → aCa | bCb | ε. **[JNU PhD Fellowship Exam 2016]**
29. A PDA is shown below using the extended notation of a finite state diagram:

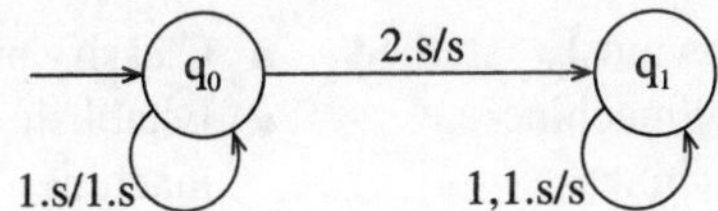

The nodes denote the states, while the edges denote the moves of the PDA. The edge labels are of the form d, s/s′ where d is the input symbol read and s, s′ are the stack contents before and after the move. For example, the edge labelled 1, s/1. s denotes the move from state q0 to q0 in which the input symbol 1 is read and pushed to the stack.
    (a) Introduce two edges with appropriate labels in the above diagram so that the resulting PDA accepts the language $\{x2x^R \mid x \in \{0, 1\}^*$, x^R denotes the reverse of x} by the empty stack.
    (b) Describe a non-deterministic PDA with three states in the above notation that accepts the language $\{0^n1^m \mid n \leq m \leq 2n\}$ by the empty stack. **[GATE 2000]**
30. Give a deterministic PDA for the language $L = \{a^ncb^{2n} \mid n \geq 1\}$ over the alphabet Σ = {a, b, c}. Specify the acceptance state. **[GATE 2001]**
31. Design a PDA for the language $L = \{a^nb^ma^mb^n \mid m > 0, n > 0\}$ over the alphabet Σ = {a, b}. Also provide a context-free grammar that generates this language. **[HCU PhD Entrance 2011]**
32. If L_1 and L_2 are context-free languages, prove that $L_1 \cup L_2$, $L_1.L_2$ and $L_1{}^*$ are also CFLs. **[UGC NET 2011]**
33. Consider the following languages: $L_1 = \{a^nb^{2n}c^m \mid m, n \geq 0\}$, $L_2 = \{a^nb^mc^{2m} \mid m, n \geq 0\}$. Determine if the sets $L_1 \cup L_2$ and $L_1 \cap L_2$ are context-free languages. **[UGC NET 2009]**

6 Turing Machine

OBJECTIVES

At the end of this chapter, you will be able to

- Provide generalisation of previously studied computation models as a Turing machine
- Explain the types of Turing machines
- Classify problems as solvable and unsolvable
- Establish a correlation between Turing machines and various classes of formal languages

6.1 INTRODUCTION

Thus far, we have studied simple classes of languages and the way in which they can be used to solve relatively constrained problems, such as for searching text, generating languages, and so on. Now, we look at methods to solve numerical problems like addition, subtraction, multiplication, division and others. Accordingly, we must study the languages for computational problems and the corresponding machines.

Such jobs can be satisfied by the Turing machine (TM), which was first modelled by Alan Turing in 1930. Turing, in effect, visualised a computer before its existence. TM is an extremely abstract way of simulating the logic of a computer algorithm (as we understand it). In fact, TM is the standard mathematical computational model of a real (digital) computer; it is an abstraction of a computer with unbounded resources. It is thus accepted that if TM is able to solve a problem, then the real computer can also solve it. Expressed in another way, if we can design a systematic approach for solving a problem, then TM exists for solving that problem. A brief discussion on the structure of TM is presented below.

Structure of TM

- TM, in its simplest form, is composed of a **tape** (memory in the form of cells) of infinite length. The tape is used to store data. It contains **input symbols** that end with a **marker**. Apart from input symbols, a few capital letters from towards the end of the English alphabet are used as **tape symbols** (mainly used during replacement of input symbols). However, the symbols B and b are generally used to represent the **blank symbol**, which denotes the end of the string or acts as a separator between two strings. Input is stored in the tape followed by the end marker (B), and the rest of the tape initially contains blanks (again, B). So, it is better to say that the tape is of infinite size but the non-blank portion is finite.
- There is a **head** (the microprocessor in the context of a real computer) that points to the tape. At any instance, the head reads a symbol (placed in a cell of the input tape) and replaces it. In fact, the replaced symbol may be a new one or the same. Next, the head moves either left or right. The

head is known as the **read/write (R/W) head**. Four basic operations (in sequence) are performed by the head of a TM at each transition: reading, erasing, writing and shifting. Erasing and writing may together be treated as *replacing*. Importantly, TM is a device that is the generalisation of an FA from the perspective that it allows two-way movement of the head and reading as well as writing of symbols over a tape of infinite size; this is why it can solve any type of feasible algorithm (even the ones that cannot be solved by pushdown automata [PDA] and linear bounded automata [LBA]).

- The **controller** part (finite control box [FCB]) of TM is the theoretical counterpart of the central processing unit (CPU) in modern computers. It possesses a finite number of states. The FCB with the help of the movable R/W head can read a symbol written at any position on the tape and replace it with some other symbol (if required). In reality, the control part consults the transition table to move from one state to another and carries out the tasks. The transitions (actions/functions) are formally placed in the transition table.

Note that each TM performs a number of transitions (a set of instructions/operations/actions). Given a current state and a symbol under the head, a transition specifies

- What should be written on the tape
- To which state the machine should move
- Whether the head should move left or right

It is important to note that TM is not only the language recogniser/acceptor but it is also a type of computational machine. In this context, each numerical computation performed by TM (via a set of transitions) is treated as a sub-routine, and the routine may be reused whenever required.

The basic model of a TM is shown in Fig. 6.1.

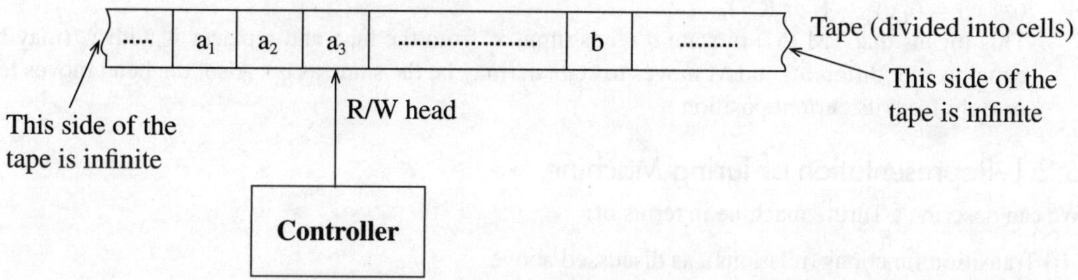

FIGURE 6.1 Basic model of a Turing machine

In summary, a Turing machine is a device with a finite number of states (memory) in the controller and an unbounded amount of read/write tape memory. The R/W head of the machine receives input from the tape, replaces the symbol in it and moves forward or back. Simultaneously, the machine may change state. Due to the presence of infinite memory (in reality in different forms) and the use of a R/W head with unrestricted movement, complex problems can be solved by TM.

Note: The tape shown here is a two-way infinite tape. However, many authors consider only a one-way infinite tape; the two-way tape is taken as a variant of TM.

6.2 FORMAL DEFINITION OF SINGLE-TAPE TURING MACHINE

A Turing machine is formally defined as a 7-tuple M = $(Q, \Sigma, \Gamma, \delta, q_0, B, F)$, where:

(i) Q is a non-empty finite set of states
(ii) Σ is a non-empty finite set of symbols, called the alphabet of the language, accepted by the machine
(iii) B (blank) $\in \Gamma$ (but it does not belong to Σ)
(iv) Γ (a Greek letter) is a non-empty finite set of tape symbols and $\Sigma \subset \Gamma$ (as $B \notin \Sigma$ but $B \in \Gamma$)

Note: Blank (B or b) does not represent null (string). If a blank is encountered at any instance, it simply indicates that no symbol is left to be processed, and so, processing of input is complete. Besides, if blank or any symbol other than the input symbols (*, #, and so on, that represent the end of input) appears at the beginning, it simply means that no string is present – it is a null string.

(v) q_0 is the start state ($q_0 \in$ Q)
(vi) F $\subseteq$ Q (subset of Q) is the set of final states that the machine accepts. F may be a null set (Φ)[1] too.
(vii) δ is the transition function, formally defined as δ: Q x $\Gamma \rightarrow$ Q x Γ x D, where D (direction) of head = {L, R}. The function states that the R/W head of the machine reads one input symbol from the tape in state q_i ($q_i \in$ Q). The machine then moves to another state (or remains in the same state) by replacing the tape symbol (Γ), if necessary. Simultaneously, the head may move left (L) or right (R) (L, R $\in$ D) depending on the characteristics of the given problem. It may be noted here that the head may sometimes remain static (S) too. Hence, D includes S. When a TM supports S, it is called a **TM with stay option**. In general, the stay option in TM is adopted in the final state in case of language acceptability or in the halt state (denoted as h: a special state), considered to indicate the end of the desired computation. However, the stay option of the head may often be used by TM while processing input (even not reaching the end of input).
The general form of the transition is:
$(q_i, x_k) = (q_j, y_k,$ <L *or* R>).
This means that TM (M) in state q_i reads input x_k from the tape and replaces it with y_k (may be equal to x_k or different) and M moves to state q_j (may be the same as q_i). Also, the head moves left or right from its current position.

6.2.1 Representation of Turing Machine

We can describe a Turing machine in terms of:

(i) Transition functions (relations), as discussed above
(ii) Transition table
(iii) Transition diagram, where each edge may be labelled as: <input symbol | replaced symbol, movement of head> (for example, <a | x, R>)

Instantaneous description (ID) of TM: To display the ID, we use the transition relations of automata. As in the case of PDA, $\delta(q, x)$ induces a change in the ID of the TM; we call this change in ID, a **move**.

Suppose the input to be processed is $x_1x_2 \ldots x_n$ and the present symbol under the R/W head is x_i. Assume $\delta(q, x_i) =$ (p, y, L). So before processing x_i, the ID is $x_{i-1}qx_i \ldots x_n$. After processing x_i, the resulting ID is $x_1x_2 .. x_{i-2}\, px_{i-1}yx_{i+1} \ldots. x_n$. This change in ID is represented by

$x_1x_2 \ldots x_{i-1}qx_i .. x_n \vdash x_1x_2 \ldots x_{i-1}ypx_{i+1} .. x_n$. Here, the symbol $\vdash$ takes the usual meaning as stated earlier.

Definition of FA in terms of TM: An FA is a TM of finite tape length with unidirectional head movement (no rewinding capability).

---

[1] As TM can write on its tape and leave an answer on it at the end of computation, the machine does not need to have a final state for all cases, as in FA and PDA. Accordingly, the set of final states (F) may be a null set. In such a case, the halt state h may be used to indicate the end of computation.

Note: In TM, there is no relation between head movement and state change. In fact, a head move does not necessarily imply a state change. Importantly, the present state and the present input taken in the transition play an important role in satisfying the properties of the problem while solving it using the machine.

Comparison between transition functions of PDA and TM: The transition function of PDA is δ: Q x ($\Sigma \cup \{\varepsilon\}$) x $\Gamma \rightarrow$ Q x Γ^*, whereas the transition function of TM is δ: Q x $\Gamma \rightarrow$ Q x Γ x D, where D (direction) of head = {L, R}.

From these two functions, it is clear that in PDA transitions, reading of two inputs (one from the tape and the other from the stack) is necessary and this causes complexity in computation, whereas only one input symbol (tape symbol) is necessary in TM.

Deterministic TM: The transition (p, a) has only one outcome (unique), where $p \in Q$, $a \in \Gamma$. More specifically, it does not result in multiple states, several replacements of the tape symbol or several movements of the head.

Replacement of the input symbol by another symbol in the tape: The head of a TM is allowed to move towards the left from its current position. Thus, if an input already processed by the machine is not replaced by some other symbol, the same input may be processed again. This may cause the machine to fall into an infinite loop. That is why, depending on the nature of the problem, the processed symbols may be required to be replaced by some other symbols to avoid such a situation.

Reason TM is more powerful than any other finite state machine: TM is more powerful than any other finite state machine because it has the ability to replace its own input. More explicitly, the head may perform three basic operations: reading, writing (replacing) and shifting head (back/forward). In fact, the last two operations cannot be performed by other machines.

6.3 VARIATIONS OF TURING MACHINE

TM has many variations, such as single tape with multiple tracks (each with a special symbol as a left end marker), multi-tape (namely, input, intermediate and output) with an arbitrary number of R/W heads, and so on. Note that the time complexity of single-tape TM is higher than that of multi-tape TM, but the space complexity is lower. With the aid of multi-tape TM, it is often easier to construct a machine to perform a particular function. In reality, variations of TM cannot improve the power of computation of single-tape TM, but may be able to reduce the time required to solve problems.

6.3.1 Multi-tape and Multi-track TM

Multi-tape TM and Its Characteristics

- In multi-tape TM, multiple distinct tapes ($n \geq 2$) are used. So there exist 'n' R/W heads (one for each tape).
- A move of the machine depends on the present state and 'n' symbols (one from each distinct tape).
- For each move, the machine performs the following:

1. On each tape, a new tape symbol is written on the cell scanned. So, 'n' symbols are simultaneously scanned and written on the tapes of an n-tape TM. The new symbol may be identical to the symbol that was present earlier.
2. Control moves to an output state (which may be the same as the previous state).
3. The movement of each R/W head for each tape may be left or right or stationary.

- On the basis of some assumptions to make computation easier, the tapes may be categorised as follows:
  (i) **Input tape**: We adopt the convention that the input string is placed in the input tape; this tape is used for scanning only the input. So, the replaced symbols are the same as the earlier symbols (conceptually, the operations on the symbol are read, replace and write). In practice, there is a single input tape in a TM, assuming the following characteristics:
  - No left end marker is considered, as it is assumed that the head of the input tape is initially positioned at the left end of the input.
  - The R/W head of this tape does not move left but may move right or be stationary.

  (ii) **Output tape**: It is used to store data or the result for the problem to be solved. It is assumed that there may be at most one output tape in a TM with the following characteristics:
  - The tape has a unique left end marker (an assumption; however, we may use a blank symbol as the left end marker).
  - The R/W head of this tape does not conventionally move left but may move right or be static.
  - The replaced symbols in the tape may be any tape symbol.
  - The existence of an output tape is not compulsory. In particular, only if the problem yields processed output value(s) does its existence become necessary.

  (iii) **Memory/Intermediate tapes**: Tapes other than the input and output tapes in multi-tape TM are called the intermediate or memory tapes, and these are mainly used for processing.
  - Each memory tape has a unique left end marker (an assumption).
  - The replaced symbols in the tape are tape symbols.
  - The movement of the R/W head corresponding to each tape may be left, right or static.

Note: If there are *n* tapes in a multi-tape TM, the number of memory tapes may be $(n - 2)$ or $(n - 1)$ depending on the presence or absence of the output tape. Another point is that the left end markers of the tapes may not be replaced by some other symbols.

Multi-track TM and Its Characteristics: A multi-track TM is a specific type of multi-tape TM that satisfies the following basic characteristics:

- It has a single tape with multiple tracks.
- In a standard *n*-tape TM, there exist *n* R/W heads, where the heads (one for each tape) move independently along *n* tapes. However, in *n*-track TM, there exists only one R/W head that scans *n* symbols (one from each track) and writes *n* symbols on the tracks simultenously. More explicitly, a tape position in an *n*-track TM contains '*n*' symbols taken from the tape alphabet.

Simple block diagrams of multi-track and multi-tape TMs are shown in Figs 6.2 (a) and (b), respectively.

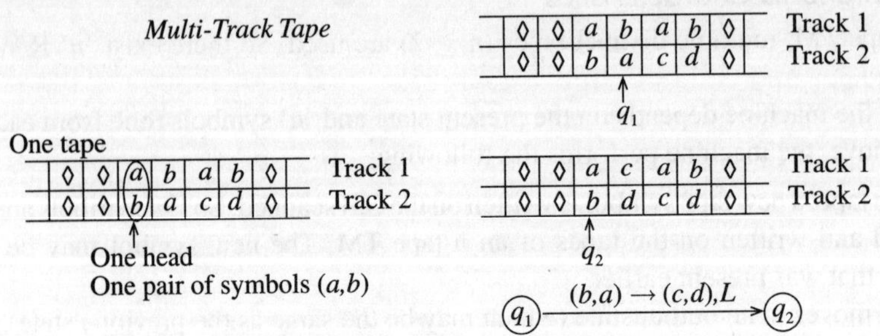

FIGURE 6.2 (a) A basic model of a two-track TM

In Fig. 6.2 (a), a two-track TM is shown at various instances. For example, in state q_1, the head of the two-track TM moves left, replacing (b, a) with (c, d) and the machine moves to state q_2. On the other hand, the positions of the heads of a multi-tape (two-tape) TM are shown in Fig. 6.2 (b).

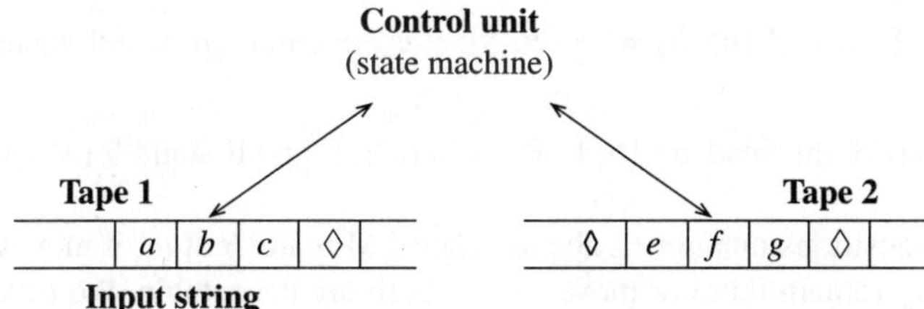

FIGURE 6.2 (b) A basic model of a two-tape TM

Transition for k-tape TM: The transition of a multi-tape (say, k-tape) TM may be expressed as δ: $Q \times \Gamma^k \rightarrow Q \times \Gamma^k \times D^k$, where $D = \{S, L, R\}$. This implies that the machine reads k input symbols from k different tapes and replaces them accordingly. It is important to note that resolving ambiguous transitions in multi-tape TM is quite easy, as more symbols (one from each tape) are read at a time. In fact, the transition depends on present state and the present inputs (each taken from one tape).

Note: Multi-tape/Multi-track TMs that retrieve and store data simultaneously are, indeed, the idea behind parallel computers. The extra tapes/tracks add no additional computing power to the model (for solving problems) as compared to single-tape TM. It only increases the speed of computation. Every multi-tape TM has an equivalent single-tape TM. Thus, the power of single-tape, multi-tape and multi-track TMs is the same.

Universal Turing Machine (UTM): A TM that can simulate any other TM is called a universal TM. UTM consists of a single tape but its working style is similar to that of the three-tape TM.

6.4 TM AS ACCEPTING AND COMPUTING DEVICE

Basic strategy to solve any problem using TM: Identify the possible inputs to be read by the R/W head of the TM while moving forward across the tape, and replace them according to the requirements.

TM as Accepting Device: TM can be used both as an accepting and as a computing device[2]. Note that it is used as an accepting device to recognise languages generated mainly by Type 0 grammars. So, it can easily recognise Type 1, 2 and 3 languages. If it is used as an accepting device, the aim is to decide whether strings are recognised or not by the TM that is designed for a language L, depending on whether the strings belong to L or not. Some languages recognised by TM are discussed below.

EXAMPLE 6.1

(a) Design a TM that accepts language $L = \{a^n \mid n \geq 1\}$.

Solution To accept the given language, we may model a TM as, $M = (Q, \Sigma, \Gamma, \delta, q_0, B, \{q_f\})$, where $Q = \{q_0, q_1, q_f\}$ such that q_0 and q_f are, respectively, the initial and final states of the machine. q_1 is the

[2] A computing device is considered as one other than the accepting or recognising device (acceptor or recogniser). For example, arithmetic computations are the primary job here.

non-final state, $\Sigma = \{a\}$, $\Gamma = \{a, x, B\}$, where B stands for blank symbol. For recognising the language, the required transitions (δ) are defined as follows:

- $\delta(q_0, a) = (q_1, x, R)$
- $\delta(q_1, a) = (q_1, x, R)$
- $\delta(q_1, B) = (q_f, B, R)$ // or $\delta(q_1, B) = (q_f, B, S)$; B is encountered, which means no input is left to scan.

Here, the movements of the head are {S, L, R}, where S, L and R stand for static (no move), left and right, respectively.

According to the transitions mentioned above, when TM reads B at q_1, it moves to the final state: q_f and the R/W head may remain there or move right – both are acceptable. We may notice here that no transition is specified at q_f for B (transition is undefined at q_f for B). The state diagram corresponding to the specified transitions is depicted in Fig. 6.3. The transitions are presented in Table 6.1.

TABLE 6.1 Transition table for Example 6.1 (a)

Present state	Tape symbols		
	a	**x**	**B**
$\rightarrow q_0$	xRq_1		
q_1	xRq_1		BRq_f
(q_f)			

FIGURE 6.3 A TM for $L = \{a^n \mid n \geq 1\}$

> **Note:** We can replace tape symbol '*a*' with '*a*' itself. Accordingly, $\Gamma = \{a, B\}$ and the transition $\delta(q_0, a) = (q_1, x, R)$ looks like $\delta(q_0, a) = (q_1, a, R)$. Likewise, the other cases are true. However, it is safe to replace input symbol '*a*' with some other symbol (say, *x*) to show that the symbols are being processed.

(b) Design a TM that accepts language $L = \{a^n \mid n \text{ is divisible by } 2\}$.

Solution To accept the given language, we may model a TM as, $M = (Q, \Sigma, \Gamma, \delta, q_0, B, \{q_f\})$, where $Q = \{q_0, q_1, q_f\}$ such that q_0 and q_f are, respectively, the initial and final states of the machine. q_1 is the non-final state, $\Sigma = \{a\}$, $\Gamma = \{a, x, B\}$, where B stands for blank symbol. The essential transitions (δ) are defined as follows:

1. $\delta(q_0, a) = (q_1, x, R)$
2. $\delta(q_1, a) = (q_0, x, R)$
3. $\delta(q_0, B) = (q_f, B, S)$ or (q_f, B, R) //when B is encountered, there is no input left to be scanned.

Here, the movements of the head are {S, L, R}, where S, L and R stand for static (no move), left and right, respectively.

ID of the TM: Let $w = aa$. For the input aaB, we obtain the following move sequence:

$q_0aaB \vdash xq_1aB$
$\vdash xxq_0B$
$\vdash (Bq_f)$ // when $\delta(q_0, B) = (q_f, B, R)$

This states that the string $w = aa$ is valid.

Note: Here too, we can replace tape symbol '*a*' with '*a*' itself. Then $\Gamma = \{a, B\}$.

(c) Design a single-tape TM that accepts language $L = \{a^n b^n \mid n \geq 1\}$.

Solution To design a TM for the given problem, we may follow the basic steps presented below:

Step 1: Replace an '*a*' with some symbol, say '*x*' and move right until a '*b*' is found.
Step 2: Replace the '*b*' with some symbol, say '*y*' and move left until an '*x*' is encountered.
Step 3: If an '*x*' is encountered, move right to search for an '*a*'.
Step 4: As soon as an '*a*' is found, go to *Step 1*; otherwise stop processing.

In order to implement these steps using transitions (operations), we may model a TM (M) as follows: $Q = \{q_0, q_1, q_2, q_3, q_f\}$, where q_0 and q_f are, respectively, the initial and final states of the machine, $\Sigma = \{a, b\}$; $\Gamma = \{a, b, x, y, B\}$, B stands for blank symbol.

D (movement of head) = {S, L, R}, where S, L and R stand for static (no move), left and right, respectively.

Transitions

- *Initial transition:*
  1. $\delta(q_0, a) = (q_1, x, R) \Rightarrow$ when '*a*' is read at q_0, the state changes to q_1 and the R/W head moves right from its current position. Also, the symbol '*a*' is replaced by '*x*'. If the state is not changed, this sub-routine replaces each '*a*' with '*x*' and the machine moves forward without keeping track of *as* and *bs*.
- *Searching for current left-most 'b' from the unprocessed part of the string*

  Case I: More as *may be found at* q_1. The essential transition to process the *as* is:

  2. $\delta(q_1, a) = (q_1, a, R)$

  Case II: At q_1, *'y' (replacement symbol for 'b') may be found between the last 'a' and the next 'b'*. (For example, *w* = *xaaybb*). The necessary transition to process the *ys* is:

  3. $\delta(q_1, y) = (q_1, y, R)$
- *Replacement of 'b' and searching for the current left-most 'a' from the unprocessed part of the string*

  Case I: *'b' is found, replacement of 'b' with 'y' is to be performed and the head moves left to search for further* as. The required transition for this purpose is:

  4. $\delta(q_1, b) = (q_2, y, L)$ [If the state is not changed, then all *bs* will be replaced with *y* using $\delta(q_1, y) = (q_1, y, R)$]

  Case II: *More* ys *may appear at* q_2. The necessary transition to process the *ys* is:

  5. $\delta(q_2, y) = (q_2, y, L)$

  Case III: More as *may appear at* q_2. The necessary transition to process the *as* is:

  6. $\delta(q_2, a) = (q_2, a, L)$ [It continues until an '*x*' is encountered. Certainly, if '*a*' is replaced with '*x*' at Transition 6, keeping track of *as* and *bs* will become complex].

  Case IV: 'x' is encountered at q_2, *the head moves right and the machine attempts to use Transition 1 again.* The necessary transition is:

  7. $\delta(q_2, x) = (q_0, x, R)$

Case V: No 'a' is found at q_0, only ys are found (that is, no 'a' is left to be processed). The required transition is:

8. δ (q_0, y) = (q_3, y, R). [The state is changed to keep track for $n \geq 1$, otherwise the machine moves to the final state from q_0. Then the transitions are δ (q_0, y) = (q_0, y, R) and δ (q_0, B) = (q_f, B, S). This implies that ε string is also valid.]

- *Attempt to get B (blank) at q_3.* The essential transition for this purpose is:

9. δ (q_3, y) = (q_3, y, R)

- *When B is encountered, the machine moves to acceptance state to validate strings.* The transition is:

10. δ (q_3, B) = (q_f, B, R) ⇒ If B is read at q_3, the machine reaches the final state, replacing B with B, and the head moves right. Accordingly, the string is accepted. Note that no transition is specified for B at q_f. However, Transition 10 may also be rewritten as: δ (q_3, B) = (q_f, B, S). This means that the head remains static, when the machine reaches q_f and no transition is specified at q_f.

The state diagram corresponding to these transitions is shown in Fig. 6.4.

Note: An important observation in the context of the given problem is that when the direction of the R/W head changes, the state of the machine also changes.

Tip to construct state diagram: First, construct the simple paths and then extend them (wherever possible) to complete the diagram.

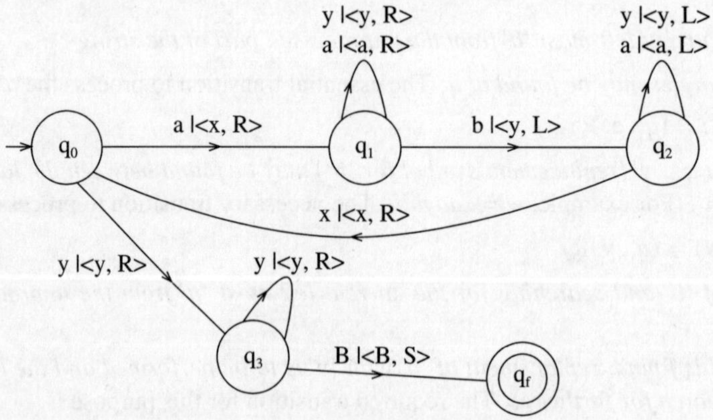

FIGURE 6.4 TM for L = $\{a^n b^n \mid n \geq 1\}$

Note: After replacing a single '*a*' and a single '*b*' (at q_0 and q_1, respectively) to solve the language L = $\{a^n b^n \mid n \geq 1\}$, if the machine does not change its state, then all *as* and *bs* in the string will be replaced by the respective replacement symbols. Accordingly, it is not possible to keep track of the number of *as* and *bs* to satisfy the given constraint. Further, if the state is not changed from q_0 for ys (when no '*a*' is left to be processed), the constraint: $n \geq 1$ may not be satisfied. In addition, if any symbol other than B (blank) is encountered at q_3 (transition is undefined), the string may not be valid. Again, if both the symbols: '*a*' and '*b*' are replaced with the same symbol (say, *x*), it is difficult to design any systematic procedure for solving the problem.

(d) Design a single-tape TM that accepts language L = {$a^nb^n c^n$ | $n \geq 1$}.

Solution Similar to the logic followed in the earlier problem, L = {a^nb^n | $n \geq 1$}, the transitions of a single-tape TM machine to recognise L = {$a^nb^n c^n$ | $n \geq 1$} are presented below. Here, *x*, *y* and *z* are the replacement symbols used for *a*, *b* and *c*, respectively.

1. δ (q_0, a) = (q_1, x, R)
2. δ (q_1, a) = (q_1, a, R)
3. δ (q_1, y) = (q_1, y, R)
4. δ (q_1, b) = (q_2 , b, R)
5. δ (q_2, b) = (q_2, b, R)
6. δ (q_2 , c) = (q_3, z, L)
7. δ (q_3, a) = (q_3, a, L)
8. δ (q_3, b) = (q_3, b, L)
9. δ (q_3, y) = (q_3, y, L)
10. δ (q_3, z) = (q_3, z, L)
11. δ (q_3, x) = (q_0, x, R)
12. δ (q_0, y) = (q_4, y, R)
13. δ (q_4, y) = (q_4, y, R)
14. δ (q_4, z) = (q_4, z, R)
15. δ (q_4, B) = (q_f, B, S)

Here, q_0 denotes the start state of the machine, whereas q_f is the final state.

(e) Give a TM to accept the language L = {a^nb^m | $n, m \geq 1; m \geq n$}.

Solution To accept the given language, we may design the following single-tape TM.

M = ({q0, q1, q2, q3, q4}, {a, b}, {a, b, B, X, Y}, δ, q0, B, {q4}), where δ is:

1. δ(q0, a) = (q1, X, R) (2) δ (q0, Y) = (q3, Y, R)
3. δ (q1, a) = (q1, a, R), (4). δ (q1, b) = (q2, Y, L)
5. δ (q1, Y) = (q1, Y, R) (6). δ (q2, a) = (q2, a, L)
7. δ (q2, X) = (q0, X, R) (8). δ (q2, Y) = (q2, Y, L)
9. δ (q3, b) = (q3, b, R) // *the number of* bs *may be greater than the number of* as
10. δ (q3, Y) = (q3, Y, R) (11). δ (q3, B) = (q4, B, S)

No transition is specified for *B* at q_4.

TM as Computing Device: The aim of the device is to compute a function. It is an example of the most generalised transducer. In fact, for numerical computation, backward movement of the R/W head and replacement of the tape symbol are usually necessary, and these are not possible in machines other than TM. That is why TM may act as a computing device. It is important to note that the theoretical model of TM assumes to accept zero or positive integers for numerical computation.

(f) Design a single-tape TM to add two positive integers (the concept of ADDER).

Solution We know that any natural number can be represented by the binary system. It is equivalently true that the same can be performed by the unary system too. In particular, the unary representation of integers was Turing's original scheme. Interestingly, the scheme works on the basis of the simple counting strategy adopted in the early stages of school education. In fact, the unary system is the simplest for arithmetic computation. However, the concept of binary representation is used later.

Now, to solve the current problem, each positive integer is assumed to be equivalent to a unary number. For example, let us take two positive integers $m = 5$ and $n = 3$. In the unary system, these are equivalent to, respectively, 5 = 11111 and 3 = 111. It is further assumed that the integers are stored in the tape, and a symbol (say, '#' as a seperator) is placed between the two integers. Clearly, $\Sigma = \{1\}$ and $\Gamma = \{1, \#, B\}$.

On the basis of the assumptions, we may consider the following transitions for adding two positive numbers:

1. $\delta(q_0, 1) = (q_0, 1, R)$
2. $\delta(q_0, \#) = (q_1, 1, R)$ [1 is inserted in place of #; state changes (from q_0 to q_1) to process two integers. If the state is not changed, the machine may process more integers (if present). Thus, finding the exact sum may be problematic. More specifically, state change here indicates that not more than two numbers are to be processed. *However, if it is assumed that only two numbers are present, then state change is not necessary*].
3. $\delta(q_1, 1) = (q_1, 1, R)$ // If more 1s are found at q_1.
4. $\delta(q_1, B) = (q_2, B, L)$ [Left move is essential to obtain the extra '1' that was replaced for #. Also, state change is essential; otherwise the machine falls into an infinite loop. The reason is that the machine reads '1' again at q_1 (if the state is not changed) and moves forward using Transition 3. It reads B and moves back, and so on].
5. $\delta(q_2, 1) = (q_3, B, R)$ or (h, B, R) // equivalently (q_3, B, S) or (h, B, S) [The extra '1' is now removed from the final outcome to obtain the actual sum. Here, q_3 is not the final state (it is just a halt state). Further, no transition is specified at q_3 or h for B. Finally, a counter machine (a form of TM) may be used to count the number of 1s to obtain the final sum. From storage point of view, a counter machine program has a finite number of counters (also called registers or cells), each holding a natural number].

A *variation* of the given problem is to design a single-tape TM to add two integers, say m and n, $m, n \geq 0$. Clearly, both m and n may be zero. To solve the problem, we may consider the following transitions, assuming that integer 0 is represented by simply B (blank). For easy understanding of the transitions, we place them in some categories (as given below).

Case I: First number is zero (0).

1. $\delta(q_0, B) = (q_1, B, R)$ // State change is necessary to indicate that the first number is zero.

   // Further, if state is not changed, then conflict occurs for Transitions 1 and 4.
2. $\delta(q_1, \#) = (q_1, B, R)$ // As the first number is 0, # need not be replaced by 1.

- *Second number may or may not be zero.*

3. $\delta(q_1, 1) = (q_1, 1, R)$
4. $\delta(q_1, B) = (h, B, S)$ // or (h, B, R)

Case II: First number is non-zero.

5. $\delta(q_0, 1) = (q_0, 1, R)$
6. $\delta(q_0, \#) = (q_2, 1, R)$ // One extra '1' is placed for '#'. The next state should not be q_1.

// If it becomes q_1, then Transitions 4 and 8 (specified later) will conflict with each other.

- *Second number may or may not be zero.*

7. $\delta(q_2, 1) = (q_2, 1, R)$
8. $\delta(q_2, B) = (q_3, B, L)$
9. $\delta(q_3, 1) = (h, B, S)$ [Irrespective of whether the second number is zero or non-zero, one '1' is replaced with B as one extra '1' is written on the tape in place of '#', assuming that the first number is non-zero].

Likewise, we may define transitions for the subtraction of two positive integers. In fact, the concept of **monus subtraction** is taken into account, that is, if '*a*' and '*b*' are two positive integers, it is defined as

$$a-b=\begin{cases}0 & \text{if } a<b\\ a-b & \text{if } a\geq b\end{cases}$$

Note: The halt state (*h*) can be used for both purposes: (*i*) *h* as a final state to accept a string, (*ii*) *h* as a state (other than the final state) to end processing of the string.

(g) Design a single-tape TM to perform: $m - n$, where both m and n are positive integers and $m \geq n$.

Solution The idea for subtraction is explained with an example. Suppose $m = 3$ and $n = 2$. These are equivalent to 111 and 11 in the unary system, and placed in a tape (as shown below).

1	1	1	#	1	1	B

The R/W head is initially set at the first '1' of the first number. The machine reads the first number by moving its head forward. When '#' is encountered, it searches for the first '1' of the second number. As soon as the first unprocessed '1' of the second number is encountered, it is replaced by symbol X and the head moves back to remove a corresponding '1' (the rightmost unprocessed '1') of the first number by replacing X. When a '1' is removed from the first number, the head moves forward to search for a '1' of the second number. This procedure continues until all the 1s of the second number are processed. A partial processing is shown below.

1	1	X	#	X	1	B

Based on this idea, the following transitions for single-tape TM to solve the given problem are presented below.

1. $\delta(q_0, 1) = (q_0, 1, R)$
2. $\delta(q_0, X) = (q_0, X, R)$ [*For the processed 1s (represented by Xs) of the first number, that is, for X appearing before marker '#'*].
3. $\delta(q_1, X) = (q_1, X, R)$ [*For all the processed 1s (represented by X) of the second number*].
4. $\delta(q_0, \#) = (q_1, \#, R)$ [*If the state is not changed, the head of the machine moves forward for all the 1s of the second number too; for X appearing after the marker '#'*].
5. $\delta(q_1, 1) = (q_2, X, L)$ [*If the state is not changed (if it remains at q_1) and the head of the machine moves left by replacing a '1' of the second number and searches for a '1' of the first number to remove, all the 1s of the second number will be replaced by Xs using Transitions 4 and 5*].
6. $\delta(q_2, X) = (q_2, X, L)$ [*For X appearing after replacement of each '1' with X*]
7. $\delta(q_2, \#) = (q_2, \#, L)$
8. $\delta(q_2, 1) = (q_0, X, R)$
9. $\delta(q_1, B) = (h, B, S)$ [*When all the 1s of the second number are consumed*].

The state diagram corresponding to these transitions is shown in Fig. 6.5.

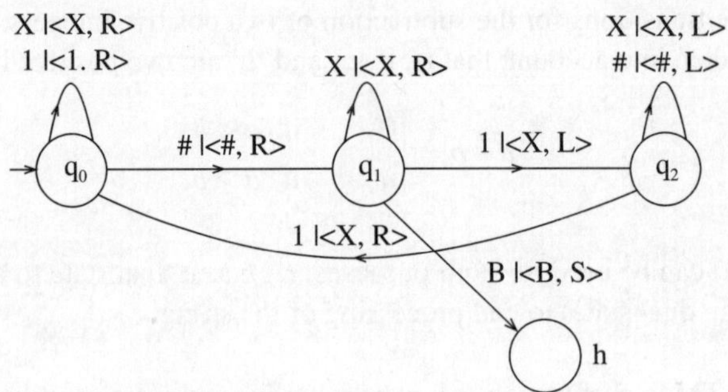

FIGURE 6.5 TM for $m - n$, $m \geq n$

Note: If $m < n$ is allowed and 0 is assumed to be returned for this case, the possibility of encountering B occurs at state q_2 and $\delta(q_2, B) = (h, 0, S)$ results in zero.

(h) Design a single-tape TM to find the larger of two positive numbers (assuming both are distinct).

Solution It is assumed that the integers (in the unary system) are stored in the tape, and the symbol (#) is placed between them as a separator. Thus, $\Sigma = \{1\}$, and $\Gamma = \{1, \#, X, Y, B\}$. Assuming that each '1' of the first number is replaced with X and each '1' of the second number is replaced with Y.

On the basis of these assumptions, we may consider the following transitions to find the larger of two numbers. For solving the problem, let us assume two notations: F for the first number and T for the second number.

1. $\delta(q_0, 1) = (q_1, X, R)$ [*X is used to replace each '1' of the first number. If the state is not changed, all the 1s will be replaced with X, but we need to replace each '1' of the second number with Y*].
2. $\delta(q_1, 1) = (q_1, 1, R)$
3. $\delta(q_1, \#) = (q_2, \#, R)$ [*If the state is not changed (if it remains at q_1), all the 1s of the second number remain unchanged via Transition 2*].
4. $\delta(q_2, Y) = (q_2, Y, R)$ [*Multiple Ys may be encountered at q_2*].
5. $\delta(q_2, 1) = (q_3, Y, L)$ [*Y is used for each '1' of the second number. If the state is not changed (if it remains at q_2), all the 1s of the second number will be replaced with Y*].
6. $\delta(q_3, Y) = (q_3, Y, L)$
7. $\delta(q_3, \#) = (q_3, \#, L)$
8. $\delta(q_3, 1) = (q_3, 1, L)$
9. $\delta(q_3, X) = (q_0, X, R)$ // This transition performs for the last replaced symbol X for the first number and right moves of the head assist in reading #.

Case I: The second number (denoted as T) is accepted as larger or equal to the first number (denoted as F).

10. $\delta(q_0, \#) = (q_4, \#, R)$
11. $\delta(q_4, Y) = (q_4, Y, R)$

For T = F (they are equal (E)): It is true if B is encountered at q_4.

12. $\delta(q_4, B) = (h, E, S)$

For T > F, it is true if 1 is encountered at q_4.

13. $\delta(q_4, 1) = (q_5, Y, R)$ [*State is changed to determine if T > F*]
14. $\delta(q_5, 1) = (q_5, 1, R)$ // For more 1s encountered at q_5.
15. $\delta(q_5, B) = (h, T, S)$

Case II: The first number is accepted as larger if B is encountered at q_2

16. $\delta(q_2, B) = (h, F, S)$

TM as Transducer: It may be noted that TM as a computing device is not limited to performing arithmetic operations over numerical problems. The machine can be used to convert one input format to another (as a coding device). The problems to be operated may be non-numerical too. For example, let $\Sigma = \{a, b\}$ and $L = \{w \mid w \in \{a, b\}^*\}$. Suppose we need each resultant string to be $f(w) = \overline{w}$, where $\overline{w}$ is the result of replacing each occurrence of '*a*' with '*b*' and vice versa. Actually, this is an example for which TM works as a coding device. Some problems in this respect are discussed below.

(i) Design a single-tape TM to find 1's complement of binary number.

Solution For this purpose, the state diagram is shown in Fig. 6.6 (a).

Here, the edge label (1 | 0, R) reads as (input symbol | replacement symbol, move of the R/W head). Likewise, (0 | 1, R), etc.

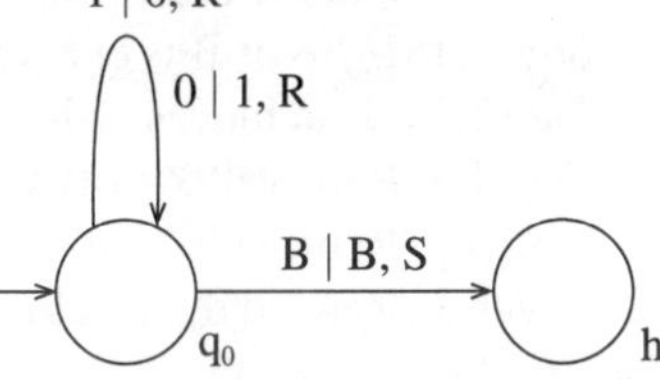

FIGURE 6.6 (a) A TM for finding 1's complement

(j) Design a single-tape TM to find 2's complement of a binary number.

Solution To compute 2's complement, a binary string is first converted into 1's complement, and then 1 is added.

Strategy to add 1 with 1's complement

- If the rightmost bit in 1's complement is 0, it is converted to 1, and we traverse left without changing any bits.
- On the other hand, if the rightmost bit is 1, it is converted to 0, and we traverse left to find 0. However, if 1 is found while searching for 0, convert it to 0 and move left. As soon as 0 is encountered, it is converted to 1, and we traverse left without changing any bits to reach the leftmost bit.

The corresponding state transitions are presented below.

- *Finding 1's complement*

1. $\delta(q_0, 0) = (q_0, 1, R)$ [0 *is converted to* 1]
2. $\delta(q_0, 1) = (q_0, 0, R)$ [1 *is converted to* 0]

- *Finding 2's complement from the 1's complement obtained*

3. $\delta(q_0, B) = (q_1, B, L)$ [*For B at 1's complement, the machine traverses left and the state changes*]
4. $\delta(q_1, 0) = (q_2, 1, L)$ [*When the rightmost bit in 1's complement is* 0]
5. $\delta(q_2, 0) = (q_2, 0, L)$
6. $\delta(q_2, 1) = (q_2, 1, L)$
7. $\delta(q_2, B) = (h, B, S)$

8. δ (q_1, 1) = (q_3, 0, L) [*If the rightmost bit in 1's complement is* 0]
9. δ (q_3, 1) = (q_3, 0, L)
10. δ (q_3, 0) = (q_4, 1, L)
11. δ (q_4, 0) = (q_4, 0, L)
12. δ (q_4, 1) = (q_4, 1, L)
13. δ (q_4, B) = (h, B, S)

(k) Design a TM to generate the successor of a given binary number.

Solution To solve this problem, it is assumed here that the binary number is given in reverse order (if not, then we may apply a sub-routine, say REVERSE, to do so). For instance, the binary representation of 47 is 101111. However, its REVERSE representation is 111101. Now, the successor of 47 is 48, whose binary representation is 110000. Its REVERSE representation is 000011. Some basic steps to find the successor of a binary number are presented below.

Assumption: Binary equivalent of the number is given in reverse order.

Step 1: If the number starts with '0', go to *Step 2*, else replace '1' with '0'. The head moves right and the following substeps are performed.

Step 1.1: If '0' or B is encountered, go to *Step 1.3*, else replace '1' with '0'.

Step 1.2: Head moves right. Go to *Step 1.1*.

Step 1.3: Replace '0' or B with '1' and stop (move to the halt state).

Step 2: Replace '0' with '1'and stop (halt).

Step 3: Rewind the resulting binary string.

The transition rules based on these steps are as follows:

1. δ (q_0, 1) = (q_1, 0, R) [Number starts with '1'. State changes to avoid replacement of B (may be found at q_0) with 1, as B does not represent any number]
2. δ (q_1, 1) = (q_1, 0, R) [Replacement of '1' with '0' for a number that starts with '1']
3. δ (q_1, 0) = (h, 1, S) [The number starts with '1', and '0' is encountered afterwards]
4. δ (q_1, B) = (h, 1, S) [The number starts with '1' and ends with '1' at q_1]
5. δ (q_0, 0) = (h, 1, S) [If the number starts with '0', then '0' is replaced with '1' and stop]
6. Rewind (Reverse the string).

(l) Design a single-tape TM to reverse binary numbers.

Solution The transitions for reversing a binary string are presented below.

1. δ (q_0, 0) = (q_0, 0, R)
2. δ (q_0, 1) = (q_0, 1, R)
3. δ (q_0, B) = (q_1, B, L) [*Moving backwards to find the rightmost unprocessed bit of the original binary string*]

Case 1: The rightmost unprocessed bit is '0'.

4. δ (q_1, 0) = (q_2, X, R) [Replacement of X for '0']
5. δ (q_2, X) = (q_2, X, R)
6. δ (q_2, Y) = (q_2, Y, R)
7. δ (q_2, 0) = (q_2, 0, R)
8. δ (q_2, 1) = (q_2, 1, R)
9. δ (q_2, B) = (q_3, 0, L) [Replacement of B with '0' for an erased '0']
10. δ (q_3, 0) = (q_3, 0, L)
11. δ (q_3, 1) = (q_3, 1, L)

12. $\delta(q_3, X) = (q_4, X, L)$ [State changes for the first X and the head moves left]
13. $\delta(q_3, Y) = (q_4, Y, L)$ [State changes for the first Y and the head moves left]

[State change for Transition 12 or 13 is necessary, otherwise transitions will be ambiguous with those specified later]

14. $\delta(q_4, X) = (q_4, X, L)$
15. $\delta(q_4, Y) = (q_4, Y, L)$
16. $\delta(q_4, 0) = (q_1, 0, S)$ [Head remains static, only state changes]
17. $\delta(q_4, 1) = (q_1, 1, S)$ [Head remains static, only state changes]

Case 2: The rightmost unprocessed bit is '1'.

18. $\delta(q_1, 1) = (q_5, Y, R)$
19. $\delta(q_5, X) = (q_5, X, R)$
20. $\delta(q_5, Y) = (q_5, Y, R)$
21. $\delta(q_5, 0) = (q_5, 0, R)$
22. $\delta(q_5, 1) = (q_5, 1, R)$
23. $\delta(q_5, B) = (q_3, 1, L)$
24. $\delta(q_4, B) = (h, B, S)$

(m) Design a TM to generate Z(x) = 0, where Z is the zero function and x is the binary number.

Solution The transition rules for this purpose are as follows:

1. $\delta(q_0, 1) = (q_0, 0, R)$
2. $\delta(q_0, 0) = (q_0, 0, R)$
3. $\delta(q_0, B) = (h, B, S)$

(n) Design a single-tape TM to generate strings of the form *ww*, where $w \in \{a\}^*$.

Solution The basic steps to solve the problem are outlined below.

Step 1: All the *a*s of *w* are first replaced with 'X' by moving the head towards the right.

Step 2: If B is encountered during the right move of the head, it moves back and searches for an 'X' (replacement of '*a*').

Step 3: If B (no X) is found on the left move of the head, the machine moves to the halt state, else if X is found, the machine replaces X with '*a*' and the head moves right to replace B with '*a*'.

Step 4: Replacing B with '*a*', the head moves back to search for B or X. If either of these is found, go to *Step 3*.

An explanation is given below. Let w = *aa*.

w	B	a	a	B	B	B	...
Stage-1	B	a | X	a | X	B | B	B	B	...
Stage-2	B	X	X | a	B | a	B	B	...
Stage-3	B	X | a	a	a	B | a	B	...
Output	B	a	a	a	a	B	....

Note: Here, the notation a | X means input symbol is '*a*' and the replaced tape symbol is X.

The state diagram based on these steps is shown in Fig. 6.6 (b).

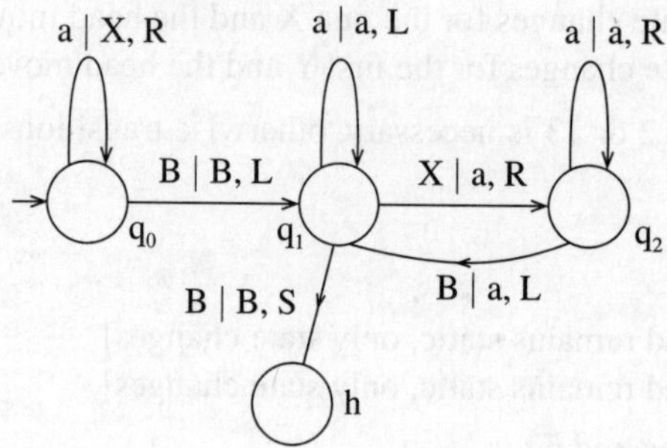

FIGURE 6.6 (b) A TM to convert *w* to *ww*

(o) Design a single-tape TM to concatenate two strings: w_1 and w_2, where each string is generated over {a, b}.

Solution Here, Σ = {a, b}. To solve the problem with a single-tape TM (M), the following transitions may be followed.

1. δ (q_0, a) = (q_0, a, R) [Right move of the head for input '*a*']
2. δ (q_0, b) = (q_0, b, R) [Right move of the head for input '*b*']
3. δ (q_0, #) = (q_1, #, R) [# is the separator between two strings; state changes here and the head moves to the right]
4. δ (q_1, a) = (q_2, #, L) [For symbol '*a*', state changes to q_2 but the head moves left by replacing '*a*' with '#'.]
5. δ (q_2, #) = (q_0, a, R) [Write '*a*' for #, as one # is replaced for '*a*']
6. δ (q_1, b) = (q_3, #, L) [For symbol '*b*', the state changes to q_3 by replacing '*b*' with #]
7. δ (q_3, #) = (q_0, b, R) [Write '*b*' for #, since one # was replaced for '*b*']
8. δ (q_1, B) = (h, B, S) [Moving to halt state for B encountered at q_1]

- **Sub-routine in TM:** Each transition of TM is considered a sub-routine. The initial sub-routine call starts from the initial state of TM.
- **Acceptance by halting**: In the case of language acceptance, another notation known as acceptance by halting (instead of moving to final state), is stated as follows: A TM halts (by accepting) if it enters a state, say q_f, scanning a tape symbol (say, X) in a state, say q_1. Importantly, there is no move in this situation. In other words, the transition: δ (q_1, X) is undefined. Here, it is understood that TM must halt if it scans X.

Note: To model a single-tape TM, considering the R/W head to be static is not essential. However, considering the head as static/stationary is essential for almost all problems solved using multi-tape TM.

A few problems solved by multi-tape TM are presented below.

EXAMPLE 6.2

(a) Design a two-tape TM that accepts language L = {$a^n b^n$ | $n \geq 1$}.

Solution To solve the given problem, we may propose a two-tape TM, as shown in Fig. 6.7.

a	a	b	b	B	

Input tape (here, B is the blank symbol)

Z_0	*	*	B		

MEMORY tape (here, Z_0 is the left end marker and B is the blank symbol)

FIGURE 6.7 Tapes of a two-tape TM for $L = \{a^n b^n \mid n \geq 1\}$

The proposed TM (M) can be modelled as follows: $Q = \{q_0, q_1, q_f\}$, where q_0 is the initial state and q_f is the final state of the machine; $\Sigma = \{a, b\}$; $\Gamma = \{a, b, *, B\}$, where B stands for the blank symbol and '*' is the replacement symbol for '*a*'; D (movement of head) = {S, L, R}, where S, L and R stand for static (no move), left and right, respectively.

Referring to the transition for the *k*-tape TM, the transition of the two-tape TM can be generalised as follows: $\delta\,(q_i, < a_i, b_k >) = (q_j, < a_i, x >, < M_1, M_2 >), M_1, M_2 \in \{S, L, R\}$.

Here, $q_i \in Q$ is the current state of the machine. $< a_i, b_k >$ denotes the two inputs, where a_i is read by the machine from the first tape (input tape) and b_k from the second tape (memory tape). $q_j \in Q$ represents the next state of the machine and $< a_i, x >$ states that the R/W head of the first tape does not replace the symbol (a_i), but the head of the second tape replaces b_k with symbol '*x*'. Finally, $< M_1, M_2 >$ specifies the movement of the heads of the tapes (first and second).

Logic: For each *a* on the first tape, the proposed machine writes a symbol (say, '*') in place of the blank symbol (*B*) on the second tape. This operation helps to match the number of *b*s to the number of *a*s in the string. The necessary transitions (operations) are presented below.

Transitions

1. $\delta\,(q_0, < a, Z_0 >) = (q_1, < a, Z_0 >, < S, R >) \Rightarrow$ Inputs '*a*' from the first tape and 'Z_0' from the second tape are read simultaneously. A state change is necessary to satisfy $n \geq 1$. The head of the first tape remains static but the head of the second tape moves right from its current position without replacing the symbol on the tape to ensure exact matching of the number of *a*s with the number of the replaced symbols.
2. $\delta\,(q_1, < a, B >) = (q_1, < a, * >, < R, R >) \Rightarrow$ Symbol of the second tape is replaced with '*'. Both the heads move right but the state remains the same.
3. $\delta\,(q_1, < b, B >) = (q_1, < b, B >, < S, L >)$ // The second head moves left to search for '*'.
4. $\delta\,(q_1, < b, \text{'*'} >) = (q_1, < b, * >, < R, L >)$
5. $\delta\,(q_1, < B, Z_0 >) = (q_f, < B, Z_0 >, < S, S >)$

Note: If $L = \{a^n b^n \mid n \geq 0\}$, ε string is also valid. So, q_0 is the initial as well as the final state and state change is not necessary. The string accepting transition will be:

$$\delta\,(q_0, < B, Z_0 >) = (q_0, < B, Z_0 >, < S, S >)$$

Comparison of complexity analysis between Examples 6.1 (c) and 6.2 (a):

- The problems stated in Examples 6.1 (c) and 6.2 (a) are identical. However, the time consumed by the single-tape TM is more than that of the two-tape TM. It is well-accepted that if $t(n)$ be a time complexity function for measuring the time to solve problems with a multi-tape TM (where $t(n) > n$ [number of tapes]), then every $t(n)$ time multi-tape TM has an equivalent $O(t^2(n))$ time single-tape TM. For each step of $(t(n))$ steps on a multi-tape TM (say, M), $O(t(n))$ steps occur for a single-tape TM (say, S). Hence, the time complexity of S is $O(t^2(n))$, whereas it is $O(t(n))$ for M.

- In addition, it is easier to design a two-tape TM than a single-tape TM. However, in the context of space, a single-tape TM is better than the two-tape TM, as only one tape is required to solve the given problem.

Hence, single-tape TM is better than multi-tape TM in terms of space and cost, but worse with respect to time.

Note: The TM to solve a problem is regarded as a mathematical formulation of the algorithm for that problem; that is, any problem can be solved if it can be formulated by TM. So, there exists a systematic approach (algorithm) to solve a problem if it can be formulated by a TM.

(b) Design a two-tape TM to add two positive integers (*m* and *n*).

Solution Assume that the two positive integers *m* and *n* (each in unary form) are stored in the input tape (say, T_1) and that they are separated by a marker, say '#'. Assume also that the final output: $m + n$ (in unary form) is to be stored in the output tape (say, T_2) with the left end marker (say, Z_0). Now, to add *m* and *n*, the following transitions may be considered.

1. $\delta (q_0, <1, Z_0>) = (q_0, <1, Z_0>, <S, R>)$
2. $\delta (q_0, <1, B>) = (q_0, <1, 1>, <R, R>)$
3. $\delta (q_0, <\#, B>) = (q_0, <\#, B>, <R, S>)$
4. $\delta (q_0, <B, B>) = (h, <B, B>, <S, S>)$

Note: A transition such as $\delta (q_0, <B, Z_0>) = (q_0, <B, Z_0>, <R, R>)$ is not defined here, so $m = 0$ is not allowed. However, to accept it, this transition is essential. Likewise, we may add transition for $n = 0$ (denoted by *B*).

Logic to solve $L = \{ww^R \mid w \in \{a, b\}^*\}$: To solve this problem (using multi-tape TM), three tapes (one input tape and two memory tapes) are necessary. The solving mechanism may be divided into two phases for easy understanding.

Phase I: Writing the symbols of the first tape simultaneously on the second and third tapes

- For this purpose, scan the inputs from the input tape and write the same content simultaneously into two memory tapes.

Phase II: At this stage, the head of the third tape is first positioned at the beginning for matching.

- Next, the R/W heads of the tapes move as follows.
  - The heads of the input tape and the first memory tape remain static but the head of the second memory tape moves left. This process continues until the R/W head of the second memory tape reaches its marker.
  - As soon as the marker is encountered, the head moves forward.
- Now, the R/W head of the input tape remains static but the heads of the first and second memory tapes move as follows.
  - Left move for the first tape but right move for the second tape; simultenously verify the corresponding symbols.
    - If they are the same, proceed in the same manner; otherwise move to the dead state.
- Proceeding in this way, if <B, Z_0, B> is finally encountered, the machine moves to the final state (to indicate that the string is valid). Here, the first B in <B, Z_0, B> refers to the blank found in the

input tape, whereas Z_0 is the left end marker in the first memory tape; and the last B denotes the blank found in the second memory tape.

Logic to solve L = {*ww* | *w* ∈ {a, b}*}: To solve this problem (using multi-tape TM), three tapes (one input tape and two memory tapes) and the following tasks are required.

- First, write the contents of the input tape in the first memory tape and the second tape.
- Next, take the first memory tape and find the starting position of the second part of the string of the form: *ww* in this tape, where the first *w* stands for the first part of the string and the second *w* denotes the second part of the string.
  - To find the starting position of the second part of the string (that is, mid-position of *ww*), perform the following steps, keeping the R/W heads of the other tapes stationary.
    1. Replace the current leftmost symbol with 'X' if it was '*a*', else with 'Y' and move forward to replace the current rightmost symbol in the same fashion.
    2. Step 1 continues till the R/W can move back.
    3. When a left move of the head is not possible, stop; this indicates the first position of the second string of the form: *ww*.
- Keep the R/W heads of the input tape and the first memory tape static, but the head of the second memory tape must move continuously back to replace '*a*' with 'X' and '*b*' with 'Y'. During this process, if the left end marker is encountered, the heads move forward.
- When the left end marker is encountered, keep the R/W head of the input tape static, but the heads of the first and second memory tapes move forward if the contents of the corresponding locations match. This process continues until B is encountered by the first memory tape. However, if B is successfully encountered by the head of the first memory tape, it indicates that the string is in the *ww* form.

THEOREM 6.1

Every language that is accepted by a multi-tape TM is also accepted by some single-tape TM (the standard TM).

Proof Suppose a language L is accepted by a *k*-tape TM (M). We may simulate M with a single-tape TM with 2*k* tracks. The second, fourth, ..., (2*k*)-th tracks hold the contents of the *k*-tapes.The first, third, ..., (2*k* – 1)-th tracks hold a head marker (a symbol, say X) to indicate the position of the respective tape head.

To simulate one iteration of a multi-tape TM, scan the tape to the left, remembering each of the *k* head markers. An implementation description of the simulation of M with a single-tape TM (M′) is given below.

Let us take the case *k* = 2. Here, the number of tracks required is 4, where the first and third tracks store the markers of the first and the second tapes, respectively, as the first symbols. The second and fourth tracks store the contents of the first and second tapes, respectively.

To simulate a move of M, the 2*k*-track TM (M_1) must visit the two head markers and store the scanned symbols in its control. After visiting the head markers, M_1 knows the tape symbols being scanned by the two heads of M. Now, M_1 revisits each of the head markers and:

(i) It changes the tape symbol in the corresponding track of M_1 based on the information regarding the move of M corresponding to the state (of M) and the tape symbol in the corresponding tape of M.

(ii) It moves the head markers to the left or right.
(iii) M_1 changes the state of M in its control.

This is the simulation of a single move of M.

Non-deterministic TM: A non-deterministic Turing machine (NTM) is formally defined as a 7-tuple, M = (Q, Σ, Γ, δ, q_0, B, F), where:

(i) Q is a non-empty finite set of states
(ii) Σ is a non-empty finite set of symbols, called the alphabet of the language, that the TM accepts
(iii) *B* (blank) ∈ Γ (but it does not belong to Σ)
(iv) Γ (a Greek letter) is a non-empty finite set of tape symbols and Σ ⊂ Γ (as $B \notin \Sigma$ but $B \in \Gamma$)
(v) q_0 is the start state ($q_0 \in Q$)
(vi) F ⊆ Q (subset of Q) is the set of final states that the machine accepts. F may be a null set (Φ) as well.
(vii) δ is the transition function formally defined as $\delta: Q \times \Gamma \rightarrow 2^{Q \times \Gamma \times D}$, where D (direction) of head = {L, R}.

Note: If $q \in Q$ and $x \in \Gamma$ and $\delta(q, x) = \{(q_1, y_1, D_1), (q_2, y_2, D_2), \dots (q_n, y_n, D_n)\}$, then the NTM can chose any one of the actions defined by (q_i, y_i, D_i) for $i = 1, 2, \dots n$.

For example, suppose a TM has a transition specified by δ (q_0, a) = {(q_1, b, R), (q_2, c, L)}. Obviously, the moves $q_0aaaB \vdash bq_1aaB$ and $q_0aaaB \vdash Bq_2caaB$ are both possible for *w* = *aaa* as per the specified transition.

Conversion of NTM to Deterministic TM: NTM can be simulated by an equivalent multi-tape deterministic Turing machine (DTM). Actually, three tapes, Tape 1, Tape 2 and Tape 3, are necessary for this purpose. Tape 1 maintains the input, Tape 2 simulates the tape of NTM and Tape 3 keeps track of the possible computations. Further, we know that for every multi-tape TM, there is an equivalent single-tape TM. So, the power of NTM and DTM is the same.

Encoding for TM: The main goal here is to devise binary code for TM. Let M = (Q, Σ, Γ, δ, q_1, B, F) be the formal model of a TM with the usual notations (discussed earlier). Here, q_1 is taken as the initial state for easy binary representation of the transition rules.

Let $\delta(q_i, x_j) = (q_k, x_l, D_m)$ be a transition rule of M, where *i, j, k, l* and *m* are the positive integers. Now, we may code this transition as $0^i10^j10^k10^l10^m$

For example, q_i is represented as 0^i, x_j is represented as 0^j …, and a single 1 is used between the two groups of 0s.

Note: This is one possible encoding scheme, but it may not necessarily be the final strategy for designing the instruction set of a real computer.

Some examples based on this idea are presented below.

For example, let M = ({q_1, q_2, q_3, q_4, q_5, q_6}, {0, 1}, {0, 1, *B*}, δ, q_1, *B*, {q_2}). Suppose the transition rules are as follows:

(i) $\delta(q_1, B) = (q_6, B, R)$. The equivalent binary code based on the above-mentioned encoding scheme is: 0100010000001000100 (assuming *B* = 000 and *R* = 00).
(ii) $\delta(q_6, 1) = (q_6, 1, R)$. Likewise, the equivalent binary code is 0000001001000000100100 (assuming 1 = 00).
(iii) $\delta(q_6, B) = (q_3, B, L)$. The equivalent binary code is 00000010001000100010 (assuming L = 0).

(iv) $\delta(q_3, 1) = (q_4, B, \text{L})$. The equivalent binary code is 0001001000010010 (assuming L = 0).
(v) $\delta(q_4, 1) = (q_3, B, \text{L})$. The equivalent binary code is 00001001000100010.
(vi) $\delta(q_4, B) = (q_5, B, \text{R})$. The equivalent binary code is 000010001000001000100.
(vii) $\delta(q_5, B) = (q_2, 1, \text{L})$. The equivalent binary code is 0000010010010010.

Here, the use of x = 0 is represented by 0, though no transition rule for input 0 is specified here. On the other hand, x = 1 is represented by 00.

Simulating a TM using a Computer: This means writing a computer program that acts like a TM performing a set of transitions.

Simulating a Computer using a TM: This means designing a TM for a computer program. Before we discuss this further, it may be good to remind readers that TM is nothing but a valid representation of what a computer can do. To simulate a real computer using a TM, the points given below may be accepted (though they still lack the details of a real computer):

- Mounting of a new hard disk in a real computer means replacing all the symbols in the disk with blank space.
- A computer program (in form of simple instructions, such as add or move, represented in machine or assembly language) is stored in some sort of memory.
- A typical computer has registers (limited in size), that is, CPU registers that may be visualised as memory in FA which has fast access.
- A number of memory tapes are required for TM to simulate a computer. One memory tape is used to store the address or contents seperated by markers such as #, *, $. The second tape holds the instruction counter (1, 2, 3, .. in binary equivalent), while the third holds the memory address or the contents of that address.
- In TM, the R/W head moves back and forth to solve problems. The equivalent task is implemented in a real computer by using the addressing scheme and by accessing data from memory.

For more details about the implementation of a real computer, readers may refer to any standard book on computer architecture.

Solvable or Decidable Problem: A problem that can be solved by a systematic (definite) algorithm is known as a solvable problem. More specifically, the upper bound complexity of the algorithm for solving the problem is fixed (known), though the complexity may be polynomial or exponential. Examples include sorting, selecting an element from a list, multiplying matrices, finding the union of two lists, checking whether a graph is Hamiltonian or not (it takes exponential time), etc. A **decidable problem** is a kind of solvable problem where the answer is always *yes* or *no* for the given instance, and the algorithm that solves the problem is called the decision procedure for that problem. Examples include searching, validity checking of a string to FA, intersection of two regular languages, etc. One may note that the solution here is definite irrespective of whether it is a solvable or decidable problem.

Unsolvable or Undecidable Problem: An unsolvable problem is one that cannot be solved by any computer algorithm. Conceptually, this means that there does not exist any systematic (definite) computer procedure to solve such a problem, and it is not known how much time will be taken to solve the problem.

Examples include ambiguity checking of CFG and post-correspondence problem (PCP). In the case of the former, we do not know for which string(s) multiple parse trees may be generated (if the grammar is ambiguous). On the other hand, the aim of PCP is to determine whether there exist identical (common) solutions (each obtained from a distinct source) following the same sequence of inputs in the sources (post-correspondence solution (PCS)). Certainly, we do not know how much time will be required to obtain the solution if it exists.

An **undecidable problem** is a decision problem for which it is known to be impossible to construct a single algorithm that always leads to a correct yes-or-no answer (here, there may be a pending situation too). The halting problem of TM (discussed below), PCP and ambiguity checking of CFG are all undecidable (and unsolvable), as the machine may continue forever until a solution with the answer 'yes' is encountered. More examples of undecidable problems are languages resulting from the intersection of CFLs (the answer is not definite; it may be regular or context-free), deciding L(G) for a given grammar G is null (Φ) (also $L_1 \cap L_2 = \Phi$, where L_1 and L_2 are two CFLs) and checking if a CFL is regular or not.

6.5 HALTING PROBLEM

Before we discuss the halting problem, let us first examine the halting status in TM.

Halting (or stopping) of a TM (T) for a string *w* (with respect to a language, say L) can be defined as follows.

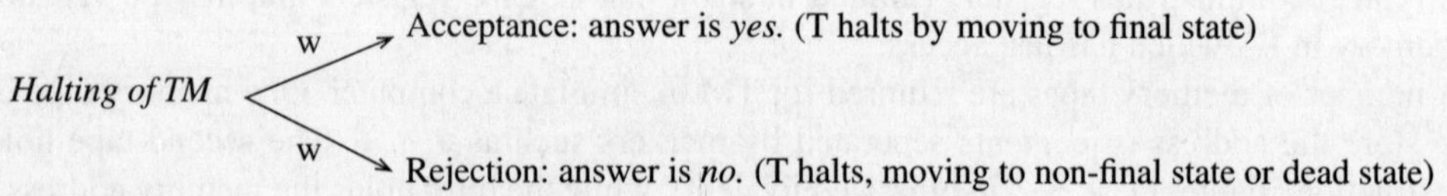

FIGURE 6.8 Halting status of TM

Therefore, the TM stops (halts) for both cases. It is interesting to note that moving to the dead/trapped state for an input does not cause any problem in halting the machine. In fact, transitions for such states are also known. In particular, if any machine moves to such a state, the machine stops and waits for further input but it cannot escape from that state.

Thus far, we have noticed that the two cases (shown in Fig. 6.8) are true for all languages of interest. However, there may still exist some languages for which consideration of only these two cases may not necessarily be true. In reality, there may exist some (one or more) strings generated over the language alphabet Σ (but not belonging to the language) such that the designed TM (T) may never halt for such string(s) or it is difficult to determine whether the machine T ever halts for such string(s). In other words, the non-halt situation of TM may occur, and it results in the machine running forever for the string (that is, no decision, accceptance or rejection, is made).

Thus, one of four outcomes is possible while executing a TM over a given input:

- The machine halts and accepts the input ($w \in L$).
- The machine halts and rejects the input ($w \notin L$).
- The machine never halts (for $w \notin L$, decision pending whether it will halt or not in the future).
- The machine crashes (it moves to the dead/trap state but halts).

Let us take an example of never-halt case. Consider the case of a TM falling into an infinite loop for an input (without waiting for the next input). Here, one may not take a decision about whether the machine stops or not by accepting or rejecting the input. In this respect, we may present a simple program segment:

while (true) continue;

Surely, this segment does not ever halt; rather, it goes on forever in an infinite loop. On the other hand, the code, *print "Hello, world!"*, halts very quickly. To better understand the never-halt case (for $w \notin L$, the decision is pending whether it will halt or not in the future), we may consider the following practical examples.

Suppose Person A visits a hotel and asks the server if he can order an item. If the item is in the menu, the server can answer yes; else the answer may be no or he may go to a nearby hotel to fulfil the demand. However, Person A will have to wait till the server returns. When he comes back, A can receive an answer to his query and be able to resume his job. However, it is possible that the server never comes back. Then A would have to wait forever. Thus, there is not enough information here to take a decision.

In another example, suppose a washing machine is designed to wash shirts. If, however, a stone is placed for washing in the machine, a correct machine will respond with a *no* or attempt to wash it. On the other hand, an incorrect machine may not return a *no* but may instead hang.

In the context of TM, the machines that are not guaranteed to halt for certain strings (not belonging to the languages) may not give us enough information to conclude that the strings are not in the languages for which they were designed. In fact, no definite algorithm exists for this purpose. Hence, the halting problem of TM is a decision problem but it is unsolvable (and undecidable).

Thus, based on the possible outcomes of TM, the halting problem is about whether a TM (T) ever halts or not for a given input string ($w \notin L$). So, the halting problem is not a problem as such. It is a question: Can we design a machine (P) and an input (w) to check whether the program will halt or not for w?

Note: In the context of a computer program, the halting problem can be stated as: "Given the description of an arbitrary computer program, decide whether the program completes running or continues to run forever".

Algorithm vs procedure: It is difficult to differentiate between these two terms. In many books, they are treated the same. However, there is a slight difference. An algorithm always terminates, resulting in an answer – yes or no – for a decision or optimisation problem; or it returns the desired output in other cases. On the other hand, a procedure gives the desired output values (if the problem is not a decision or optimisation problem), but may not necessarily show the output, yes or no, for a decision or optimisation problem. Let us attempt to illustrate the concept with examples.

Given a number n, check if it is perfect number[3] or not. For this, we first need to find out the factors of n and then check their *sum* with the number (n). This approach reports a yes or no for n, and it is treated as an algorithm. At this point, it may be noted that the complexity of an algorithm may be of polynomial or exponential measure.

For another example, given a number n, find a number greater than n to check if it is a perfect number or not. To solve this problem:

- First take the number: $n + 1$ and check if it is a perfect number based on the strategy discussed in the earlier example.
- If it is not a perfect number, take the number: $n + 2$, and so on.
- The process may fall into an infinite loop without producing a yes-or-no output. Thus, it is the case of a halting problem with respect to TM.

---

[3] If, for a number N, the sum of its factors (except itself) is equal to N itself, then N is a perfect number. For example, $6 = 1 + 2 + 3$, where 1, 2 and 3 are the factors of 6.

This approach is termed a procedure.

Thus, an algorithm always gives an output, yes or no, for a decision problem but a procedure may run *forever* without giving an output. In other words, there may occur issues to halt a procedure. If such issues occur, the problem for which the procedure is designed is called an undecidable (unsolvable) problem as well.

> **Note**: No general algorithm/program can be written that will be able to examine whether any program P will halt for a given input *w* or not. So, it is an undecidable problem.

6.6 TURING MACHINE AND LANGUAGES

Thus far we have studied the languages accepted by FA, PDA and TM. For each case, we assume the following: If a string $w \in L$, the equivalent machine accepts the language (halts). On the other hand, if $w \notin L$, the machine rejects the language (but halts).

Now, we discuss categorisation of the languages on the basis of the halting concept of TM.

6.6.1 Types of Languages

Turing Decidable Language: A language (L) is decidable iff there exists a TM that accepts strings of that language and rejects strings not in that language. Thus, there exists only the halting case of TM for a decidable language. In other words, if string $w \in L$, the corresponding TM for L moves to the accepting state. If $w \notin L$, the TM moves to the non-accepting (but halt) state to reject the string. TMs for such languages are called **total Turing machine (TTM)**.

Conversely, if L is Turing decidable, there exists a program (algorithm) P such that, for any string *w*: if $w \in L$, then P accepts *w*, else P rejects *w*. Such languages are called **recursive languages (RLs)** and they are generated by some sort of Type 0 grammar.

> **Note:** In mathematics, 'recursive' is a synonym of decidable and a recursive function (S) is defined by the following characteristics:
>
> - Values for certain arguments of S (known as base values) are necessary.
> - The successor function ($S(x) = x + 1$) is derived.
> - The function S uses itself.
>
> Thus, the value of S (for any value of argument) can be easily decided. Therefore, according to the definition of recursive function, a specification like $m*fact(m + 1)$ with base value $fact(1) = 1$ for finding the factorial value of a positive integer n and a stopping criterion such as $m = n$ is correct.

Examples of Turing decidable problems include regular languages, CFLs, CSLs, Hamiltonian cycle problem, TSP and programming languages like C, C++ and Java.

> **Note:** Programming languages are decidable, but any problem to be solved using a programming language may not be decidable, as the halting problem may occur while solving the problem.

Turing Acceptable Language: A language (L) is Turing acceptable (or recognisable or partially decidable) iff there exits a TM (say, T) that halts and accepts only for the strings in L, and rejects (but halts) or does not halt at all for the strings that do not belong to L. If string $w \in L$, the TM moves to the final state. However, if $w \notin L$, the machine moves to a non-final state (it halts) or it runs forever (it falls into an infinite loop). Hence, for $w \notin L$, there exist two possibilities: (i) T may halt in a non-final state, or (ii) T may run forever.

Conversely, if L is Turing acceptable, there exists a program P such that, for any string *w*, if w ∈ L, then P accepts *w*. If w ∉ L, there are two possibilities: (i) P may halt and will not provide "ACCEPT" or (ii) P may run forever. TMs for such languages are often called general TM or simply TM. Thus, it is applicable for both Turing decidable and acceptable languages. Note that the UTM is the general TM.

Turing acceptable languages are also called **recursively enumerable languages (RELs)**. These languages are also derived from some Type 0 grammars. Fundamentally, the term *recursively enumerable* means that a function is capable of listing all the members of a language in some order – it enumerates them. In computability theory (also called recursion theory), a set *S* of natural numbers is called recursively enumerable (or computably enumerable, semi-decidable, provable or Turing recognisable) if:

- There is an algorithm such that the set of input numbers for which the algorithm halts is exactly *S*. Or, equivalently,
- There is an algorithm that enumerates the members of *S*. This means that its output is simply a list of the members of *S*: s_1, s_2, s_3, ... , and the algorithm may run forever.

In particular, recursively enumerable languages are accepted by some sort of TM, but it may run forever for inputs not belonging to the languages. Examples include halting problem, PCP, finding the perfect number beyond a given number, and so on.

A set X is a **recursive set** if there exists an algorithm to determine whether a given element belongs to X or not. A set X is a **recursively enumerable set** if there exists a procedure to determine whether a given element is in X or not.

KEY POINTS

- For a recursively enumerable set (S), an algorithm may or may not respond for the elements (strings) not belonging to the set S.
- For a recursive set, the algorithm must state if an input is not in the set; this is not required in recursively enumerable sets.

Note: Every Turing decidable language is Turing acceptable, but the converse is not true. Importantly, we may model TM for a language irrespective of whether it is a recursive or recursively enumerable language, using the transition definition for TM.

A well-accepted relation among recursive, recursively enumerable and non-recursively enumerable languages is shown in Fig. 6.9.

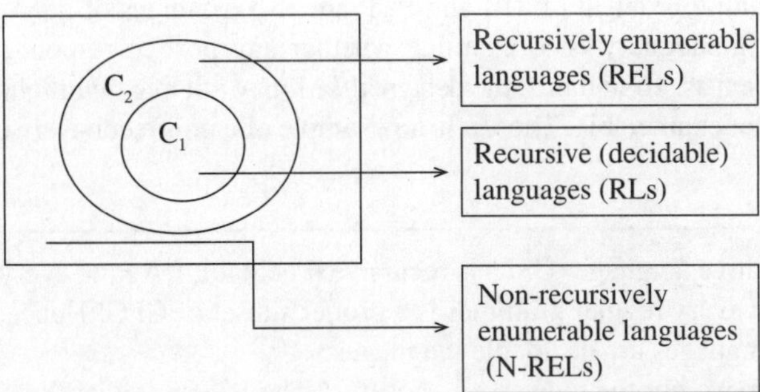

FIGURE 6.9 Venn diagram showing relationship among REL, RL and N-REL

As shown in Fig. 6.9, the region C_1 represents only recursive or decidable languages for which TMs exist, whereas the region C_2 represents all the recursively enumerable languages. Clearly, C_2 contains C_1. This means that every recursive language is also recursively enumerable but the converse is not true. Thus, the region (C_2–C_1) denotes only those languages that are recursively enumerable but not recursive, and some sort of TMs exists for such languages. Further, the region beyond C_2 represents all non-recursively enumerable languages. In particular, there are non-recursively enumerable languages that cannot be described by any Type 0 grammar. So, no TMs exist for non-recursively enumerable languages. Importantly, Type 0 grammar is capable of describing the largest number of languages and generates recursively enumerable languages; many of these are recursive.

Turing Undecidable Language: A language that is not Turing decidable is known as Turing undecidable; that is, the language is not decided by any algorithm. In other words, there can be no general and systematic (deterministic) way (algorithm) to solve undecidable languages/problems. These are, in fact, unsolvable. Obviously, the region beyond C_1 in Fig. 6.9 represents all undecidable languages. Thus, all recursively enumerable languages as well as all non-recursively enumerable languages are undecidable languages. It is important to note that the discussion on undecidable language does not concentrate on the division between recursively enumerable and non-recursively enumerable languages. A complete diagram with the different formal languages is shown in Fig.6.10 (although, a diagram in this regard is shown in Fig. 4.1).

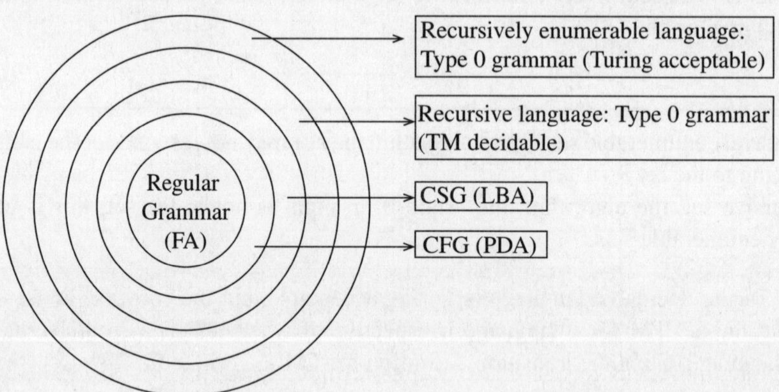

FIGURE 6.10 Venn diagram showing relationship among languages and the formal machines

The examples of the halting problem of TM and PCP are also examples of non-recursively enumerable problems. In PCP, the input/query is to examine whether any post-correspondence solution (for two different input sources) exists. To find this, the designed TM may fall into an infinite loop. Also, the power set of an infinite set is not enumerable. Thus, it is an example of a non-recursively enumerable problem.

KEY POINTS

- Every context-sensitive language (CSL) is recursive (decidable). Again, every CFG falls in a proper subset of CSG, and every regular grammar is a proper subset of CFG. Hence, all regular languages and context-free languages are decidable languages.
- Every CSL is recursive but the converse is not true. The reason is that there exist many recursive languages that are not recognised by CSG but which may be recognised by TM.

◆ Every recursive language is also an enumerable language. Hence, all regular languages, context-free languages and context-sensitive languages are recursively enumerable languages.

IS A RECURSIVELY ENUMERABLE SET FINITE OR INFINITE?

- A regular language (set) may be finite or infinite.
- Every regular language (set) is also recursive (set).
- Every recursive language (set) is also recursively enumerable (set).

Hence, a recursively enumerable set may be finite or infinite.

6.7 COMPARISON OF FA, PDA, LBA AND TM

- FA is a reading (non-writing) machine. The tape in FA is of fixed size and only input symbols are placed in the tape. No markers like B or # are explicitly placed on it. The null string (ε) cannot end processing of the string, as its explicit use causes non-deterministic behaviour in FA. Both sides (left and right) of the tape are bounded.
  FA has a read head to scan the input stored in the tape (known as input tape). The head cannot move back. The machine has no auxiliary memory apart from the finite storage capacity of its controller. As a result, FA cannot be used for comparison operations and some complex operations. Most of the combinational logic circuits of modern computers (including arithmetic and logic units, buffers, I/O handlers) are based on the concept of FA. It is used to decide whether a string (w) belongs to a language (L). In this sense, it is a language recognising tool.
- PDA is also a reading (non-writing) machine. However, the tape in PDA is infinite in size. Only one side is bounded and the other side can be extended infinitely. The machine may use the null string (ε) to indicate the end of strings, as non-deterministic behaviour caused by the use of ε move can be resolved here by the stack-top symbol.
  PDA has a read head to scan the input stored in the tape. The head cannot move backwards. The machine has auxiliary memory (known as stack) apart from the finite storage capacity of its controller. A **stack machine** is a real or emulated computer that uses a pushdown stack. A stack computer is programmed with a reverse Polish notation (RPN) instruction set. Note that Polish notation is simply pre-fix notation, whereas RPN is post-fix notation in which the operator is placed *after* the operands. As PDA has stack, validity checking of sentences (whose strings are to be stored in the stack) is carried out based on the grammar (rules).

  Advantages

  - Stack machines have much smaller instructions than other types of machines.
  - Compilers for stack machines are simpler and quicker to build than compilers for other machines.

  Disadvantages

  - Stack machines execute more data cache cycles for temporary values and local variables than register machines.

Note: These two types of machines are known as non-writing machines as they do not have the ability to modify input symbols. Further, FA cannot store past information for future processing; only a finite amount of past information may be managed by state control (state path). On the other hand, PDA can store an infinite amount of past information in the stack for future processing, but only the most recent information stored in the stack is retrieved for processing.

- TM is a writing machine. The tape in TM is not fixed in size. Both sides of the tape can be extended infinitely. It uses tape symbol B to enclose strings. TM has a read/write (R/W) head to scan the input stored in the tape. The head can move forward as well as back. That is why the computing power of TM is more than that of PDA. The tape(s) and storage of the finite controller in TM may be used as memory for storing information. The symbol-replacement strategy for the tape using the R/W head helps the machine to keep track of processed information. Furthermore, TM can write symbols on its tape, so it can leave an answer at the end of a computation. Thus, TM does not require final states for such cases, unlike FA and PDA. A TM is a general example of a CPU that controls all the data manipulation performed by a computer, with the canonical machine using sequential memory to store data. In other words, the most general form of today's computer is based on TM.
- LBA is similar to TM but its R/W head moves only to the right. The input is enclosed by two markers (left and right). The markers cannot be rewritten. The size of the tape is unbounded. Actually, the power of TM cannot be extended beyond the power of universal TM. Instead, it can be restricted and LBA is an example of such a TM in which one side of the tape is bounded (it is a semi-infinite tape).

In summary, TM = FSM with infinite tape + R/W head with unrestricted movement. It is the theoretical model of today's digital computer. In particular, adder, subtractor, multiplier, sequence detector, etc–all digital circuits are designed based on the capability of FSM, where multiple tapes (input and output) may be necessary. But overflow error occurs when the size of the input is very large, as the memory of FSM is theoretically limited. A TM is more powerful than FSM in the sense that it can execute computations that cannot be accomplished by FSM. The main reason is the ability of TM to change and write its own inputs. From a theoretical viewpoint, each finite state control unit is given access to an arbitrarily large external memory in which it executes computations, stores partial resuls, modifies and replaces input information, and finally stores the output pattern and halts. In other words, such a state control unit serves as a TM (digital computer).

KEY POINTS

- The power of the CPU of a digital computer is equivalent to the power of FSM, but the working strategy of a digital circuit may not exactly match with that of FSM, as FSM is an abstract model that hides the implementation details. Further, it may not be simulated like a full computer, as many tasks like storing the address or content and implementing the addressing scheme are out of the scope of FSM.
- The power of a stack-based machine is equivalent to that of PDA but the working strategy of a stack-based machine may not exactly match that of PDA, as PDA is an abstract model.
- The power of TM cannot be increased by increasing the number of tapes/heads, but its power can be restricted as follows:
  - TM with read-only head (R) and stack (TM with no R/W head but infinite-size tape and stack) is equivalent to PDA.
  - TM with infinite-size tape but with one side bounded is equivalent to LBA.
  - TM without an infinite-size tape (but only a finite input tape) and only a read head is equivalent to FA.
- The power of a full digital computer is equivalent to the power of TM but the working strategy of the digital computer may not exactly match with TM, as TM is an abstract model. More specifically, any problem solvable by TM is also solvable by today's digital computer.

- *Turing contribution*: It is the mathematical definition of what computers can do. In other words, physical computers are approximations of TMs.
- No computer has been built that is more powerful than TM.
- No formal proof is made that the TM model is the correct one.
- *Turing machine vs real computer*: A TM has an infinite tape, whereas real computers have finite memory. The number of states (storage space) can also be very very large but still limited.

Note: One may think that the computational problems discussed in this chapter may be solved by FSM (especially, by Moore or Mealy machines). This is partially true because FSM can solve a limited number of computational problems. However, solving problems like finding the larger number, sorting and converting a unary number to a binary number are beyond the power of FSM, as it has limited memory and the heads cannot move backwards. In particular, two tapes are used in such transducers: an Input tape and an Output tape. Accordingly, two heads are used: a Read head to read input from the Input tape and a Write head to write output on the Output tape. In fact, no R/W head (like in TM) is used to perform both read and write operations simultaneously on the same tape.

6.8 PROPERTIES OF RECURSIVE AND RECURSIVELY ENUMERABLE LANGUAGES

THEOREM 6.2

If a language L is recursive, then there exists an enumeration procedure for it.

Proof If $\Sigma = \{a, b\}$, then the procedure P can enumerate strings: a, b, aa, ab, ba, bb, aaa, ...

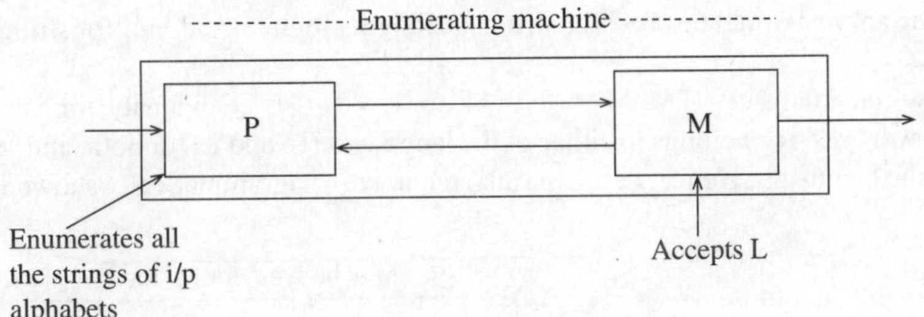

FIGURE 6.11 (a) An example of an enumerating machine

The enumeration procedure[4] is as follows. P generates string w. M checks if $w \in$ L; if yes, w is included, else w is ignored.

Let L = {a, ab, bb, aaa, ...}. Output by P = {a,b, aa, ab, ba, bb, aaa, ...};
L(M) = {a, ab, bb, aaa, . . }; enumerated output for M = a, ab, bb, aaa, . . .

- Recursive = decidable, that is, the corresponding TM always halts
- Recursively enumerable (semi-decidable) but not recursive, that is, the corresponding TM always halts if it accepts, otherwise it halts in the non-final state or in the loops.
- Non-recursively enumerable (non-RE) = there are no TMs for such problems.

[4] A procedure that generates a complete, ordered listing of all the items in a set (or language).

Complement of Recursive Languages

THEOREM 6.3

The complement of a recursive language is recursive.

Proof Let L be a recursive language and M be the TM for L. Therefore, M answers '*yes*' for strings belonging to L and '*no*' for strings not belonging to L. Hence, we may easily design (code) a TM ($\bar{M}$: complement of M) for L′ (complement of L) by interchanging the outputs: *yes* (with respect to $\bar{M}$ equivalent to *no* with respect to M) and *no* (with respect to $\bar{M}$ but *yes* with respect to M) for the strings of L (through code).

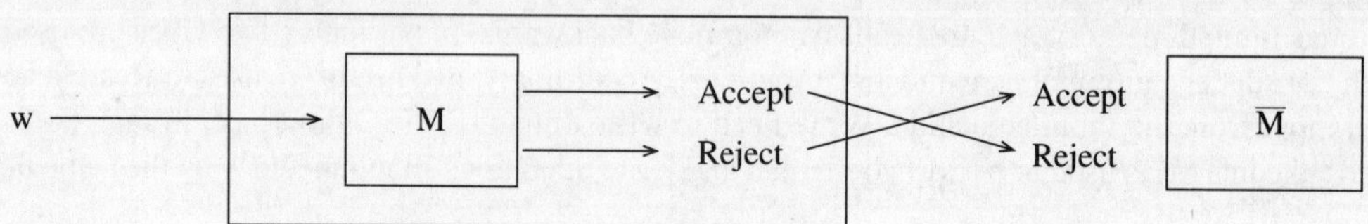

FIGURE 6.11 (b) TM ($\bar{M}$) deciding the complement of recursive language (L) decided by TM (M)

Union and Intersection of Recursive Languages

THEOREM 6.4

The union and the intersection of two recursive languages are recursive.

Proof Let L_1 and L_2 be two recursive languages decided by M_1 and M_2, respectively. As L_1 and L_2 are recursive, the answer would be '*yes*' for strings belonging to them and '*no*' for strings not belonging to them.

We may now construct a new TM, $M = M_1 \cup M_2$ for $L = L_1 \cup L_2$. Obviously, for each string (x) of L, M gives the answer '*yes*' if x belongs to either of the languages (L_1 and L_2) or both, and '*no*' if it does not belong to both the languages. Hence, $L_1 \cup L_2$ is also recursive (diagrammatically shown in Fig. 6.11 (c)).

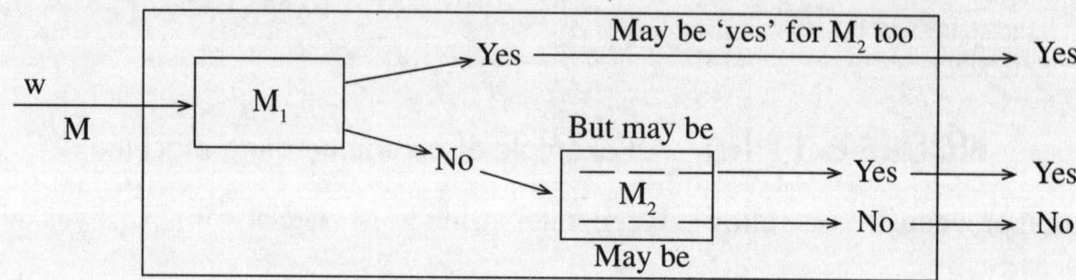

FIGURE 6.11 (c) TM (M) for the union of two recursive languages

Likewise, we may prove that $L_1 \cap L_2$ is also recursive. In this context, we may recall that valid strings in $L = L_1 \cap L_2$ satisfy a common property derived by combining two different properties (each for a language). In other words, valid strings belonging to both L_1 and L_2 implies that the answer is '*yes*'. On the other hand, the remaining strings are invalid and the answer is '*no*'; though these were either invalid with respect to L_1 or L_2 or both, or these were valid with respect to either L_1 or L_2 (but not for both). Hence, there is no halting issue for such strings (diagrammatically shown in Fig. 6.11 (d)).

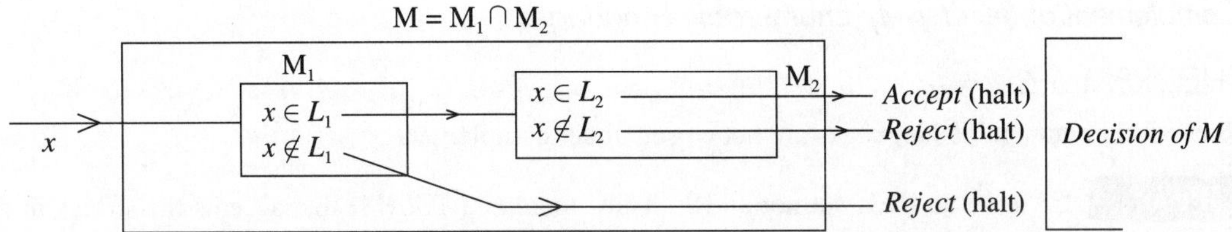

FIGURE 6.11 (d) TM (M) for intersection of two recursive languages

Note: In a similar manner, we may prove that the concatenation of two RLs is also an RL. Also, if L is RL, then L* is also an RL. Further, if L_1 and L_2 are RLs, then $L_1 - L_2$ is also an RL.

Union and Intersection of Recursively Enumerable Languages

THEOREM 6.5

The union and the intersection of two recursively enumerable languages are recursively enumerable.

Proof *Union*: Let L_1 and L_2 be two recursively enumerable languages accepted by TMs, M_1 and M_2, respectively. As L_1 and L_2 are recursively enumerable, the answer would be '*yes*' for strings belonging to them, but there may not exist any answer for strings not belonging to them. Thus, if we construct a new TM (M) for L_1 and L_2 as $M = M_1 \cup M_2$, the new machine also answers '*yes*' for strings belonging to either L_1 or L_2 or both, but no answer may result for strings not belonging to either or both. The reason is there may be a halting issue for such strings, as they are RELs.

To design (code) $M = M_1 \cup M_2$, the following cases must be taken into account:

- M accepts a string *w* if M_1 or M_2 accepts it.
- M rejects *w* and halts if both M_1 and M_2 reject and halt. In fact, if M_1 rejects the string, then it is checked by M_2, and vice versa.
- M does not stop for *w*, if neither M_1 nor M_2 stop *w*.

This is diagrammatically shown in Fig. 6.11 (e).

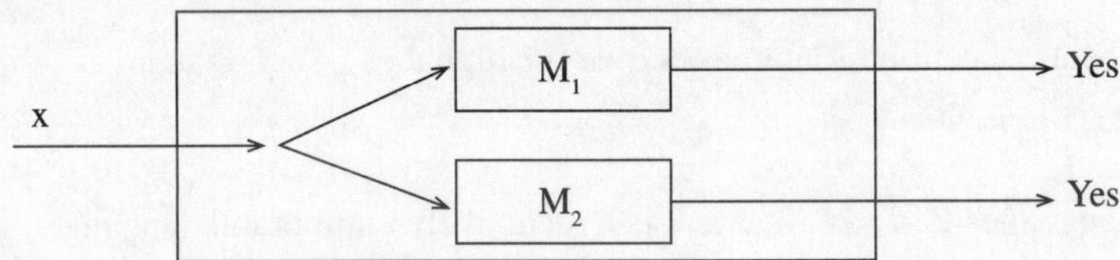

FIGURE 6.11 (e) TM (M) for the union of two recursively enumerable languages

Intersection: Let M_1 and M_2 be the TMs for the languages L_1 and L_2 (both of which are RELs). We can easily construct a new TM, $M = M_1 \cap M_2$, where M can reject (or take no decision) for string *w* if M_1 or M_2 or both reject *w* (or do not give any decision). On the other hand, it accepts *w* only when both M_1 and M_2 accept *w*.

To design (code) $M = M_1 \cap M_2$, the following cases must be taken into account:

- M accepts a string *w* iff both M_1 and M_2 accept it.
- M rejects *w* and halts if either M_1 or M_2 rejects and halts.
- M does not stop, if neither M_1 nor M_2 stop for *w*.

Complement of Recursively Enumerable Languages

THEOREM 6.6

Recursively enumerable languages are not closed under complement.

Proof Let L be an REL but not an RL. Thus, there is a TM (M) that accepts the strings in the language (L). In particular, given an input x, if x is in L, then M says '*yes*'. However, if x is not in L, the machine may not say '*no*' (because it may never halt).

Now, assume that L′ (complement of L) $= \{w \mid w \notin L\}$ is recursively enumerable. This means that there exists some other TM (say, T) that, given a string (w) of L′ (but $w \notin L$), halts and accepts w, though it may cause a halting issue for $w \notin L$ with respect to the TM (M) for L. The assumption claims that L is recursive. Hence, a contradiction occurs. Therefore, the property is proved.

> **Note:** We may also prove that the concatenation of two RELs is also an REL. Also, if L is an REL, then L* is also an REL.

THEOREM 6.7

If a language L and its complement L′ are both recursively enumerable, then L is recursive.

Proof Let T_1 and T_2 be the TMs for the languages L and L′ (both of which are RELs). Obviously, T_1 does not give any answer for strings that do not belong to L; however, it is specified that there exists a TM (say, T_2) that gives the output '*yes*' for those strings. This means that the strings not belonging to L do not cause halting issue. Hence, we can construct a two-tape TM (say, T) to process all the strings of the language (L), where one tape stores all the strings belonging to L and the other tape stores the strings not belonging to L. As it is possible to construct such a TM (M) for L, L is recursive.

Therefore, L′ is also recursive (based on Theorem 6.3).

> **Note:** Based on Theorem 6.7, we may say that if Σ is an alphabet, $L \subseteq \Sigma^*$ is a recursively enumerable language, and if $\Sigma^* - L$ is recursively enumerable, then L is recursive.

EXAMPLE 6.3

Let $L_1, L_2, \ldots, L_k$ be a collection of languages over the alphabet Σ such that:

(i) For all $i \neq j$, $L_i \cap L_j = \Phi$
(ii) $L_1 \cup L_2 \cup \ldots \cup L_k = \Sigma^*$
(iii) Each of the languages L_i, $i = 1, 2, \ldots., k$ is a recursively enumerable language.

Prove that each of these languages is therefore recursive.

Solution Given that $L_1 \cup L_2 \cup \ldots \cup L_k = \Sigma^*$. Let $L = \Sigma^*$. This means that every string (w) generated over Σ belongs to L; that is, there exists a TM (M) for L that answers '*yes*' for every w. Hence, L is a recursive language.

Further, we know that the union of two recursive languages is also a recursive language. Here, L is a recursive language and L results from the union of L_i, $i = 1, 2, \ldots, k$. Therefore, each L_i, $i = 1, 2, \ldots.,$ k must be a recursive language.

Languages with No Grammar: It is true that there exist a countable number of Turing machines. Therefore, there are a countable number of languages with grammars. However, there may exist an uncountable number of languages over any finite alphabet. As the number of languages with grammars is countable but the number of languages is uncountable, there must exist an uncountable number of languages which do not have a grammar.

Note: Pumping lemmas are not suitable for recursively enumerable languages. The main reason is that the tape heads in TMs are allowed to move backwards, and this makes creating a pumping lemma difficult.

6.9 CHURCH THESIS

TMs are the formal version of algorithms, where the operations in an algorithm are represented by the transitions of TM. Also, no computational procedure will be considered an algorithm unless it can be represented by a TM. This is known as the Church thesis or Church–Turing thesis.

TM exists only for algorithms. For each solvable problem, there exists a distinct algorithm.

6.10 POST-CORRESPONDENCE PROBLEM

The post-correspondence problem (PCP) was first introduced by Emil Post in 1946. It has many applications in the theory of formal languages. In particular, PCP consists of multiple sources of input. The aim is to determine whether there exist identical (common) solutions (each obtained from a distinct source) following the same sequence of inputs. Here, we need to determine the sequence of the solutions.

For example, let us consider two lists of non-empty strings: $X = \{x_1, x_2, \ldots, x_n\}$ and $Y = \{y_1, y_2, \ldots y_n\}$ over $\{a, b\}$. We may now attempt to find out if it is possible to construct strings such that one string (say w_1) derived from X and another string (say w_2) from Y following the same sequence are identical. The lengths of the two strings (w_1 and w_2) are certainly equal.

For instance, let $X = \{b, bab^3, ba\}$ and $Y = \{b^3, ba, a\}$. Here, $x_1 = b$, $x_2 = bab^3$, $x_3 = ba$; $y_1 = b^3$, $y_2 = ba$ and $y_3 = a$. Now, let us construct strings (one from each source) as follows: $w_1 = x_2x_1x_1x_3 = bab^3$ b b ba (taken from X) and $w_2 = y_2y_1y_1y_3 =$ ba b^3 b^3a (taken from Y). Here, $w_1 = w_2$. Thus, the PCP has a solution with the sequence (2, 1, 1, 3).

Note: PCP is undecidable (unsolvable) as the problem has no definite (systematic) algorithm to find a finite set of elements to match the two instances. More specifically, it is *harder* than NP, as repetition of permutations in the set of strings is allowed to solve the problem. Likewise, ambiguity checking of CFG is also unsolvable, as there is no systematic algorithm for detecting whether a correspondence system has a match.

EXAMPLE 6.4

Prove that a PCP with two lists, $X = \{01, 1, 1\}$ and $Y = \{01^2, 10, 1^2\}$, has no PCP solution.

Solution Here, the length of each string in X is less than the length of the correspondence string in Y. Hence, this problem may not have any PCP solution.

6.11 ADDITIONAL SOLVED PROBLEMS

PROBLEM 1

Justify that a TM like a two-way FSM can recognise only regular languages.

Solution A two-way FSM can move left and right across the input tape but cannot return to its previous position, which is possible in a regular TM. So, if a TM behaves like a two-way FSM, it cannot come back to any previous position on the tape to write a symbol. So, mere moving left or right (without returning to the previous position and writing a symbol in that position) does not make a two-way FSM more powerful than a standard FA. Equivalently, it can be seen as a read-only TM with no work tape – only a read-only input tape. Thus, a TM (M) that never writes on its tape is equivalent to an FA.

Note: Two-way DFAs (often abbreviated as 2DFAs) were introduced in a seminal 1959 paper by Rabin and Scott, who proved them to have power equivalent to that of a one-way DFA. That is, any formal language that can be recognised by a 2DFA can also be recognised by a DFA that only examines and consumes each character in order. Since DFAs are obviously a special case of 2DFAs, this implies that both kinds of machines recognise the same class of regular languages. However, the equivalent DFA for a 2DFA may require exponentially many states, making 2DFAs a much more practical representation for algorithms for some common problems.

PROBLEM 2

Design a TM to check balanced parentheses.

Solution The working of a TM for checking balanced parentheses is presented in the form of transitions:

1. δ (q_0, '(') = (q_1, '(', R) // When LP '(' is found at q_0.
2. δ (q_1, '(') = (q_1, '(', R) // When LP '(' is found at q_1.

// When the first unprocessed RP ')' is found at q_1, it is replaced by '', and the machine moves left to search for a matching '('. For this, Transitions 3 and 4 are required.*

3. δ (q_1, ')') = (q_2, *, L)
4. δ (q_2, *) = (q_2, *, L) // For the *s at q_2
5. δ (q_2, '(') = (q_1, *, R) // When the matching '(' for the replaced ')' is found, '(' is replaced with '*'.
6. δ (q_1, *) = (q_1, *, R)
7. δ (q_1, B) = (h, B, L) // The entire input string is processed.

PROBLEM 3

Design a multi-tape TM for the language L = $\{1^n \mid n > 2 \text{ is a prime number}\}$.

Solution Here, each number (n) is represented in the unary system, for example, $2 = 11 = 1^2$, $3 = 111 = 1^3, \ldots\ n = n\ 1^s = 1^n$.

Now, the straightforward logic to check the primality of a number n is to let the TM (M) try to divide n by each number (m) between 2 and $n - 1$ in turn.

For this purpose, we may design a three-tape TM, where the first (input) tape initially contains the number to be checked. The second and third tapes are taken as the processing tape with left end markers, say Z_0 and Z_1. Assume that the number $n = 1^n$ is written on the input tape as well as the first processing tape, whereas $2 = 11 = 1^2$ is initially stored in the second processing tape (the third tape). Also, assume that the heads of the first and third tapes are initially set at blank B (the end of the numbers), and the head of the second tape points to its left end marker: Z_0. The initial picture of the tapes and positions of the heads of the TM is shown in Fig. 6.12.

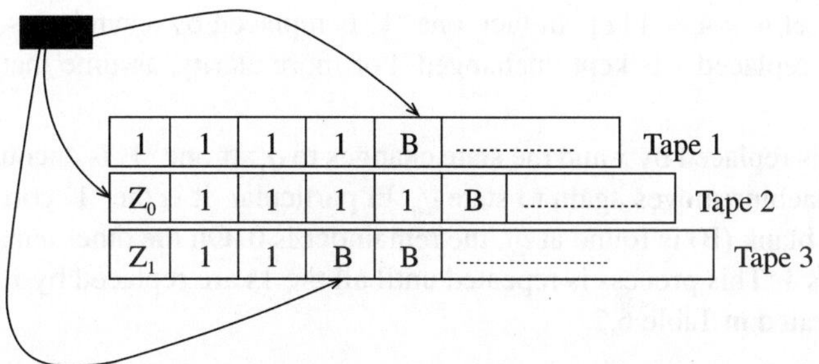

FIGURE 6.12 A snapshot of the proposed TM for the language $L = \{1^n \mid n > 2 \text{ is a prime number}\}$

Based on the assumptions, the transitions of the proposed TM are presented below. Let q_0 be the start state of the machine.

1. $\delta (q_0, < B, Z_0, B >) = (q_0, < B, Z_0, B>, < S, R, L>)$
2. $\delta (q_0, < B, 1, 1>) = (q_0, < B, 1, 1 >, < S, R, L>)$
3. $\delta (q_0, < B, B, Z_1>) = (q_Y, < B, B, Z_1>, < S, S, S >)$ // *Moving to* q_Y *(state) means that* n *is prime.*
4. $\delta (q_0, < B, 1, Z_1>) = (q_1, < B, 1, Z_1 >, < S, S, R >)$ // *When Transition 3 is not satisfied, that is, more 1s are present in Tape 2.*
5. $\delta (q_1, < B, 1, 1>) = (q_1, < B, 1, 1 >, < S, R, R >)$
6. $\delta (q_1, < B, B, B>) = (q_N, < B, B, B >, < S, S, S >)$ // n *is divisible by any number (m)* ≥ 2 *and* $\leq$ n – 1 *and the machine moves to* q_N *state; the number* n *is not prime.*
7. $\delta (q_1, < B, 1, B>) = (q_0, < B, 1, B >, < S, S, L >)$ // *When more 1s are present in Tape 2.*
8. $\delta (q_1, <B, B, 1>) = (q_2, < B, B, 1 >, < S, L, R >)$ // *Intention is to check the divisibility of* n *by the next number to be inputted on the third tape.*
   - Move the heads of the second tape towards Z_0 and the third tape towards B with the aim of increasing the content (m) of the third tape by 1
9. $\delta (q_2, < B, 1, 1>) = (q_2, <B, 1, 1 >, <S, L, R >)$
10. $\delta (q_2, < B, Z_0, 1>) = (q_2, < B, Z_0, 1 >, <S, S, R >)$ //*Contents of the first processing tape are processed but not the second processing tape.*
11. $\delta (q_2, < B, 1, B>) = (q_2, < B, 1, B >, <S, L, S >)$ // *Contents of the second processing tape are processed but not the first processing tape.*
12. $\delta (q_2, < B, Z_0, B>) = (q_3, < B, Z_0, 1 >, <S,S, R >)$ // $m = m + 1$
13. $\delta (q_3, < B, Z_0, B>) = (q_0, < B, Z_0, B >, <S, S, S >)$

Note: The complexity for primality checking of the number n can be reduced if divisibility checking is performed between 2 and $\lfloor \sqrt{n} \rfloor$.

PROBLEM 4

Design a multi-tape TM for converting a unary number to an equivalent binary number.

Solution Here, each number is represented in the unary system: $2 = 11 = 1^2$, $3 = 111 = 1^3$, …, $n = n \text{ 1s} = 1^n$.

The simple logic to covert a number n (in unary system) into its equivalent binary number is to obtain the successive remainders while dividing n by 2. The idea adopted here to find the successive remainders is illustrated below. Let $n = 4 = 1111$. In fact, one '1' is replaced by a symbol, say x, and the next '1' corresponding to the replaced x is kept unchanged. For more clarity, assume that the start state of the machine is q_0.

Logic: At q_0, one '1' is replaced by x and the state changes to q_1. If one '1' is encountered at q_1, it is kept unchanged but the machine moves again to state q_0. In particular, it is the '1' corresponding to the last replaced x. Now, if a blank (B) is found at q_0, the remainder is 0. On the other hand, if B is encountered at q_1, the remainder is 1. This process is repeated until all the 1s are replaced by x.

The logic is illustrated in Table 6.2.

TABLE 6.2 Illustration of the adopted logic

1	1	1	1	
x	1	x	1	Remainder = 0
x	x	x	1	Remainder = 0
x	x	x	x	Remainder = 1

Based on the logic, we may design a three-tape TM, where the first (input) tape initially contains the number to be converted, the second and third tapes are taken as the processing tapes with left end markers, Z_0 and Z_1, respectively. Assume that the number $n = 1^n$ is initially written on the first and second tapes. Also, assume that the head of the first tape points to the end of the number (the blank, B), whereas the heads of the second and third tapes point to the left end markers: Z_0 and Z_1. The initial picture of the tapes and positions of the heads of the TM is shown in Fig. 6.13.

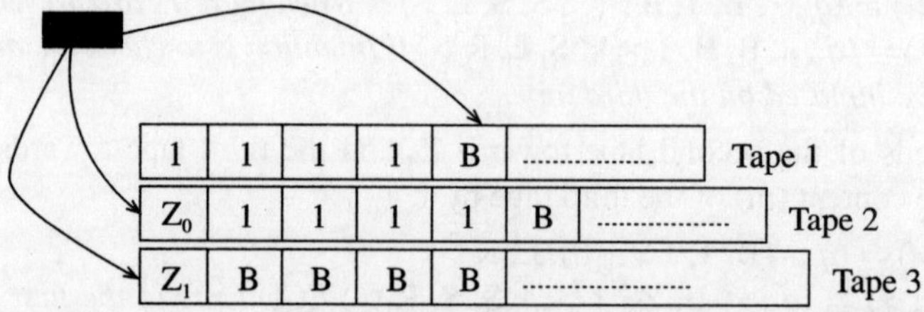

FIGURE 6.13 A snapshot of the proposed TM for converting a unary number to its equivalent binary number

Based on the assumptions and the suggested logic, the transitions of the proposed TM are presented below.

1. $\delta (q_0, < B, Z_0, Z_1 >) = (q_0, < B, Z_0, Z_1 >, < S, R, R >)$
2. $\delta (q_0, < B, 1, B >) = (q_1, < B, x, B >, < S, R, S >)$
3. $\delta (q_1, < B, x, B >) = (q_1, < B, x\ B >, < S, R, S >)$ // *Previous replaced* xs
4. $\delta (q_1, < B, 1, B >) = (q_0, < B, 1, B >, < S, R, S >)$ // *The present unprocessed '1' at* q_1.
5. $\delta (q_0, < B, B, B >) = (q_2, < B, B, 0 >, < S, L, R >)$ // *Remainder 0, and it is written on Tape 3*
6. $\delta (q_1, < B, B, B >) = (q_2, < B, B, 1 >, < S, L, R >)$ // *Remainder 1, and it is written on Tape 3*

At q_2, if the present input symbol of Tape 2 (due to left move of the head) is 1, the TM moves to q_3. The earlier inputs of this tape may be 1 or x. The corresponding transitions are given below.

7. $\delta (q_2, <B, 1, B>) = (q_3, < B, 1, B >, < S, L, S >)$
8. $\delta (q_3, <B, 1, B>) = (q_3, < B, 1, B >, < S, L, S >)$

9. $\delta(q_3, <B, x, B>) = (q_3, < B, x, B >, < S, L, S>)$
10. $\delta(q_3, <B, Z_0, B>) = (q_0, < B, Z_0, B >, < S, R, S>)$

At q_2, if the present input symbol of Tape 2 is x, the TM moves to q_4. The earlier inputs of this tape may be 1 or x. The corresponding transitions are given below.

11. $\delta(q_2, <B, x, B>) = (q_4, < B, x, B >, < S, L, S>)$
12. $\delta(q_4, <B, x, B>) = (q_4, < B, x, B >, < S, L, S>)$

If 1 is not encountered at q_4, the TM moves to the halt state (h). The corresponding transition is given below.

13. $\delta(q_4, <B, Z_0, B>) = (h, < B, Z_0, B >, < S, S, S>)$

If 1 is encountered at q_4, the TM moves to state q_5. The inputs at q_5 may be '1' or '0' and the corresponding transitions are given below.

14. $\delta(q_4, <B, 1, B>) = (q_5, < B, 1, B >, < S, L, S>)$
15. $\delta(q_5, <B, 1, B>) = (q_5, < B, 1, B >, < S, L, S>)$
16. $\delta(q_5, <B, x, B>) = (q_5, < B, x, B >, < S, L, S>)$
17. $\delta(q_5, <B, Z_0, B>) = (q_0, < B, Z_0, B >, < S, R, S>)$
18. $\delta(q_0, < B, x, B>) = (q_0, < B, x, B >, < S, R, S >)$

Lastly, the number is read in reverse order to return the binary equivalent of the given unary number, or we may take one more tape (as the output tape) to store the contents of the third tape in reverse order, and this is the final output (binary equivalent of the number n). Importantly, we may take state q_3 (instead of q_5) to reduce state if conflict in the transitions does not occur.

Note: Divisibility checking of a given binary number by any number is possible by FA but converting a number to its equivalent binary number is out of its scope. For this, we need TM.

PROBLEM 5

Design a TM for the language $L = \{0^{2^n} \mid n \geq 0\}$.

Solution The procedure to recognise the given language is shown in Fig. 6.14.

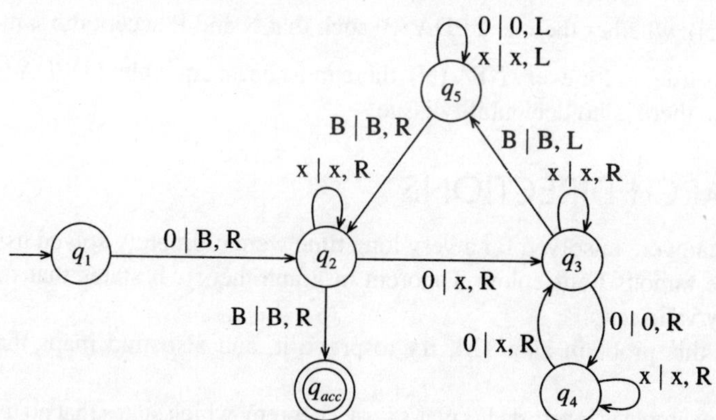

FIGURE 6.14 TM for recognising $L = \{0^{2^n} \mid n \geq 0\}$

PROBLEM 6

Can the tape alphabet Γ be equal to the input alphabet Σ?

Solution No, as Γ (a Greek letter) denotes a non-empty finite set of tape symbols and $\Sigma \subset \Gamma$ (as B (blank) $\notin \Sigma$ but $B \in \Gamma$); that is, $\Sigma \subseteq \Gamma$ may not be true.

PROBLEM 7

Justify the statement: *A language L is recognisable if it halts in an accepting state:* q_{accept} *wherever* $x \in L$.

Solution The statement is not correct because a language L is recognisable if the TM for the language halts in an accepting state: q_{accept} wherever $x \in L$; that is not a language (but the TM).

PROBLEM 8

Find out which of the options is correct for the given problem.

The set of all recursivlely enumerable languages is:

(a) Closed under complementation
(b) Closed under intersection
(c) A subset of the set of all recursive languages
(d) An uncountable set

Solution The set of RELs is closed under intersection but not closed under complementation. Further, it is not a subset of recursive languages; however, recursive languages are a subset of RELs. Again, every REL is a countable set.

PROBLEM 9

Which of the following are correct?

(a) The membership problem for unrestricted grammar is decidable.

Solution This is not true, as there is no standard derivation strategy for checking whether $x \in L$, where L is an unrestricted language. However, it is decidable for CFL and regular languages.

(b) Given two CFGs: G_1 and G_2, whether $L(G_1) = L(G_2)$ is decidable.

Solution This is incorrect, as no standard algorithm exists for solving this problem.

(c) Given an NFA (N), whether there is a PDA (P) such that N and P accept the same language.

Solution This is true, as for every NFA (N), there must be an equivalent DPDA (a superset of NFA). As the DPDA exists, there is no decidability issue.

6.12 RESEARCH DIRECTIONS

Several problems that were unsolved for a very long time were ultimately solved using computers. One such problem is the famous Four-colour Theorem in graph theory. It states that four colours are sufficient to colour any valid map.

Can you model this problem as a TM, try to prove it, and also find maps that do not obey this theorem?

Another famous example is Pierre de Fermat's Last Theorem, which states that no three positive integers a, b, c satisfy the equation $a^n + b^n = c^n$, for any integer $n > 2$. This theorem was proposed in 1637, and was successfully proved as late as 1995. Examine the proof of this theorem and try to model it as a TM.

Yet another problem for research could be to model geometrical constructions as TMs.

Project-based problems

1. Solve the Travelling Salesman problem using TM.
2. Model a system for parallel computing using TM.

EXERCISES

1. Give the general form of the transition function for a TM, and prove that the TM is more powerful than PDA. List some similarities and differences between DFAs and TMs.
2. Design a TM to accept the language L = $\{a^n \mid n \geq 0\}$.
3. Design a TM to accept the language L = $\{ww^R \mid w \in \{a, b\}^*, w^R$ is the reverse of $w\}$.
4. Design a TM to accept the language L = $\{a^n b^n c^n \mid n \geq 0\}$.
5. (a) Design a TM to accept the language L = $\{a^n b^n \mid n \geq 0\}$.
   (b) Design a TM that transforms string #w# to w# over {a, b}, where # represents the marker.
6. Design a TM to add two positive integer numbers.
7. Design a TM to subtract two positive integer numbers.
8. Design a TM to multiply two positive integer numbers.
9. Design a TM to concatenate two strings.
10. Design a TM to accept the language L = $\{w \in \{a, b\}^* \mid n_a(w) = n_b(w)\}$.
11. Design a TM for the function (a) f(x, y) = x, (b) f(x, y) = y.
12. Design a TM for computing f(n) = $n + 2$, where n is a positive integer.
13. State the implications of the halting problem.
14. What do you mean by PCP? Explain with an example.
15. Answer the following questions.
    (a) All regular languages are decidable.
    (b) All context-free languages are decidable.
    (c) All context-sensitive languages are decidable.
    (d) Give the definition of recursive languages, recursively enumerable languages and recursive functions.
    (e) Are all recursive languages also recursively enumerable?
    (f) Are there recursive languages that are not recursively enumerable?
    (g) Are there recursively enumerable languages that are not recursive?
    (h) Is the complement of a recursive language recursive?
    (i) Is the complement of a recursive language recursively enumerable?
    (j) Is the complement of a recursively enumerable language recursive?
    (k) Is the complement of a recursively enumerable language recursively enumerable?
    (l) L is Turing decidable if and only if L and L^c are both Turing recognisable.
    (m) If L is Turing decidable, then L^c is T decidable.
    (n) L is recursively enumerable if and only if L is Turing recognisable.
    (o) If L is recursively enumerable, then L is Turing recognisable.
    (p) A TM is undecidable.

(q) Halt_{TM} is not Turing decidable.
(r) Halt_{TM} is Turing recognisable.

16. In which of the cases stated below is the following statement true? "For every non-deterministic machine M1, there exists an equivalent deterministic machine M2, recognising the same language"
(a) M1 is a non-deterministic finite automaton (b) M1 is a non-deterministic PDA
(c) M1 is a non-deterministic TM (d) For no machine M1 is the above statement true
[GATE 1992]

17. Which of the following conversions is not possible (algorithmically)?
(a) Regular grammar to context-free grammar (b) Non-deterministic FSA to deterministic FSA
(c) Non-deterministic PDA to deterministic PDA (d) Non-deterministic TM to deterministic TM
[GATE 1994, Rajasthan SET 2012, Gujarat SET 2016]

18. Given $\Sigma = \{a,b\}$, which one of the following sets is not countable?
(a) Set of all strings over Σ (b) Set of all languages over Σ
(c) Set of all regular languages over Σ (d) Set of all languages over Σ accepted by TM
[GATE 1997]

19. Regarding the power of recognition of languages, which of the following statements is false?
(a) Non-deterministic FA are equivalent to deterministic FA
(b) Non-deterministic PDA are equivalent to deterministic PDA
(c) Non-deterministic TM are equivalent to deterministic PDA
(d) Non-deterministic TM are equivalent to deterministic TM
(e) Multi-tape TM are equivalent to single-tapeTM **[GATE 1998, HCU PhD Entrance 2010]**

20. The C language is:
(a) A context-free language (b) A context-sensitive language
(c) A regular language (d) Parsable fully only by a TM **[GATE 2002]**

21. A single-tape TM, M, has two states q0 and q1, of which q0 is the start state. The tape alphabet of M is {0, 1, B} and its input alphabet is {0, 1}. The symbol B is the blank symbol used to indicate the end of an input string. The transition function of M is described in the following table.

	0	1	B
q0	q1, 1, R	q1, 1, R	Halt
q1	q1, 1, R	q0, 1, R	q0, B, L

The table is interpreted as follows: The entry (q1, 1, R) in row q0 and column 1 signifies that if M is in state q0 and reads 1 on the current tape square, then it writes 1 on the same tape square, moves its tape head one position to the right and transitions to state q1. Which of the following statements is true about M?
(a) M does not halt on any string in $(0 + 1)^+$ (b) M does not halt on any string in $(00 + 1)^*$
(c) M halts on all strings ending in 0 (d) M halts on all strings ending in 1 **[GATE 2003]**

22. Define languages $L_0 = \{<M, w, 0> \mid M \text{ halts on } w\}$ and $L_1 = \{<M, w, 1> \mid M \text{ does not halt on } w\}$. Here $<M, w, i>$ is a triplet, whose first component, M, is an encoding of a TM; second component, w, is a string; and third component, i, is a bit. Let $L = L_0 \cup L_1$. Which of the following is true?
(a) L is recursively enumerable, but L′ is not (b) L′ is recursively enumerable, but L is not

(c) Both L and L′ are recursive (d) Neither L nor L′ is recursively enumerable

[GATE 2003]

23. Which of the following is true for the language $\{a^p \mid p \text{ is a prime}\}$?
    (a) It is not accepted by a TM (b) It is regular but not context-free
    (c) It is context-free but not regular
    (d) It is neither regular nor context-free, but accepted by a TM.
    Give a proper explanation for the correct answer. **[GATE 2008]**
24. Which of the following statements is false?
    (a) Every NFA can be converted to an equivalent DFA
    (b) Every non-deterministic TM can be converted to an equivalent deterministic TM
    (c) Every regular language is also a context-free language
    (d) Every subset of a recursively enumerable set is recursive **[GATE 2008]**
25. Which one of the following is false? **[GATE 2009, APSET 2013]**
    (a) There is unique minimal DFA for every regular language
    (b) Every NFA can be converted to an equivalent PDA
    (c) The complement of every context-free language is recursive
    (d) Every non-deterministic PDA can be converted to an equivalent deterministic PDA
26. The lexical analysis for a modern computer language such as Java needs the power of which one of the following machine models in a necessary and sufficient sense?
    (a) Finite state automata (b) Deterministic pushdown automata
    (c) Non-deterministic pushdown automata (d) Turing machine
    Give a proper explanation for the correct answer. **[GATE 2011]**
27. Which of the following pairs have different expressive power?
    (a) Deterministic FA and non-deterministic FA (b) Deterministic PDA and non-deterministic PDA
    (c) Deterministic single-tape TM and non-deterministic single-tape TM
    (d) Single-tape TM and multi-tape TM **[GATE 2011]**
28. Consider the languages, $L1 = \{0^p1^q \mid p, q \in N\}$, $L2 = \{0^p1^q \mid p, q \in N \text{ and } p = q\}$ and $L3 = \{0^p1^q0^r \mid p, q, r \in N \text{ and } p = q = r\}$.Which of the following statements is not true?
    (a) PDA can be used to recognize L1 and L2 (b) L1 is a regular language
    (c) All three languages are context-free (d) TM can be used to recognise all the languages
    [GATE 2011]
29. Which of the following statements is/are false?
    (1) For every non-deterministic TM, there exists an equivalent deterministic TM
    (2) Turing recognisable languages are closed under union and complementation
    (3) Turing decidable languages are closed under intersection and complementation
    (4) Turing recognisable languages are closed under union and intersection
    (a) 1 and 4 only (b) 1 and 3 only (c) 2 only (d) 3 only **[GATE 2013]**
30. Consider the languages, L1 = {<M> | M takes at least 2016 steps on some input}, L2 = {<M> | M takes at least 2016 steps on all inputs} and L3 = {<M> | M accepts ε} where for each TM, M, <M> denotes a specific encoding of M. Which one of the following is true?

(a) L1 is recursive and L2, L3 are not recursive (b) L2 is recursive and L1, L3 are not recursive
(c) L1, L2 are recursive and L3 is not recursive (d) L1, L2 and L3 are recursive **[GATE 2016]**

31. Let A and B be infinite alphabets and let # be a symbol outside A and B. Let f be a total functional from A* to B*. We say f is computable if there exists a TM, M, that given an input x in A*, always halts with f(x) on its tape. Let Lf denote the language {x #f(x) | x ∈ A *}. Which of the following statements is true?
(a) f if computable if and only if Lf is recursive
(b) f is computable if and only Lf is recursively enumerable
(c) If f is computable then Lf is recursive, but not conversely
(d) If f is computable then Lf is recursively enumerable, but not conversely
Give a proper explanation for the correct answer. **[GATE 2017]**

32. Match the following :

(i) Regular grammar	(A) Pushdown automaton
(ii) Context-free grammar	(B) Linear bounded automaton
(iii) Unrestricted grammar	(C) Deterministic finite automaton
(iv) Context-sensitive grammar	(D) Turing machine

(a) (C) (A) (B) (D) (b) (C) (A) (D) (B)
(c) (C) (B) (A) (d) (d) (C) (B) (D) (A) **[UGC NET 2012, Karnataka SET 2013, UGC NET 2014, HP SET 2014, Rajasthan CS Lecturer Recruitment 2014]**

33. Given a TM, M = ({q0, q1, q2, q3}, {a, b}, {a, b, B}, δ, B, {q3}), where δ is a transition function defined as δ(q0, a) = (q1, a, R), δ(q1, b) = (q2, b, R); δ(q2, a) = (q2, a, R), δ(q2, b) = (q3, b, R). The language L(M) accepted by the TM is given as:
(a) aa*b (b) abab (c) aba*b (d) aba*
Give a proper explanation for the correct answer. **[UGC NET 2016]**

34. Given a TM, M = ({q0, q1}, {0, 1}, {0, 1, B}, δ, B, {q1}), where δ is a transition function defined as: δ(q0, 0) = (q0, 0, R) , δ(q0, B) = (q1, B, R). The language L(M) accepted by TM is given as :
(a) 0* 1* (b) 00* (c) 10* (d) 1*0* **[UGC NET 2016]**

35. The language $\{a^m b^m c^m, m > 0\}$ can be recognised:
(a) Both by FA and PDA (b) Neither by FA nor by PDA
(c) By PDA but not by FA (d) By TM but not by PDA **[Gujarat SET 2013]**

36. The language $\{a^m b^n c^p, m > 0, n > 0, p > 0\}$ can be recognised:
(a) Both by FA and PDA (b) Neither by FA nor PDA
(c) By PDA but not by FA (d) Neither by TM nor PDA **[Gujarat SET 2014]**

37. The language $\{a^m b^m, m > 0\}$ can be recognised by:
(a) A non-deterministic FA, but not by a deterministic FA
(b) A non-deterministic PDA, but not by a deterministic PDA
(c) A deterministic PDA, but not by a non-deterministic PDA
(d) A non-deterministic TM, but not by a deterministic TM **[Gujarat SET 2014]**

38. Linear bounded automata (LBA) (i) can accept the language $L = \{a^m b^m c^m, m > 0\}$; (ii) are more powerful than deterministic Turing machines; (iii) are more powerful that non-deterministic push-down automata. Which of the above statement(s) is(are) true ? **[Gujarat SET 2014]**

(a) (i) only (b) (i) and (ii) only (c) (i) and (iii) only (d) (ii) and (iii) only

39. Consider the statements: (i) A one-tape TM can emulate a k-tape TM; (ii) A non-deterministic TM can be converted into a deterministic TM; (iii) The tape of the TM is finite; (iv) The transition table of a TM is finite. Which of the following is correct?
(a) (i), (ii), (iii) are true (b) (i), (ii), (iv) are true
(c) (i), (iii), (iv) are true (d) (ii), (iii), (iv) are true
Give a proper explanation for the correct answer. **[Gujarat SET 2014]**

40. Consider a language L for which there exists a TM, T, that accepts every word in L and either rejects or loops for every word that is not in L. The language L is:
(a) NP-hard (b) NP-complete (c) Recursive (d) Recursively enumerable
Give a proper explanation for the correct answer.
[Rajasthan CS Lecturer Recruitment 2014, Gujarat SET 2016]

41. A deterministic machine may be used to multiply two given integers. Which of the following is the correct option for the above statement:
(a) True (b) False (c) May be true (D) May be false **[HP SET 2013]**

42. The number of symbols necessary to simulate a TM with m symbols and n states is:
(a) $m + n$ (b) mn (c) $8mn + 4m$ (d) $4mn + m$ **[HP IT Lecturer Recruitment 2016]**

43. Pushdown machine behaves like a TM when the number of auxiliary storage units is
(a) 2 (b) 1 (c) 0 (d) 4
[JNU M.Phil/PhD Entrance 2014, Karnataka SET 2014]

44. Which of the following pairs have different expressive power?
P : DFA and NFA Q : DPDA and NPDA R : Single-tape TM and multi-tape TM
S : TM and two-stack PDA T : Single stack PDA and two-stack PDA
(a) Q, R, S and T (b) Q and S (c) Q and T (d) All of these
[TN SET 2012, Maharashtra SET 2013, Pondicherry University PhD Entrance 2015]

45. Identify the incorrect statement:
(a) A two-stack PDA is as powerful as TM
(b) A multi-tape TM has more computational power than a single-tape TM
(c) A deterministic FA can be used for lexical analysis
(d) PDA is not as computationally powerful as TM **[Rajasthan SET 2011]**

46. Let M1 be a single-tape TM simulating a multi-tape TM, M (k tapes). The time taken by M1 to simulate *n* moves of M, considering 'k' as constant, is given by
(a) $O(n^2)$ (b) O(n) (c) $O(n^k)$ (d) $O(k^n)$
[Rajasthan CS Lecturer Recruitment 2014]

47. A finite state machine having finite tape length without rewinding capability and unidirectional tape movement is called
(a) TM (b) PDA (c) DFA (d) All of these
[Rajasthan CS Lecturer Recruitment 2014]

48. A TM is similar to an FA, with the only difference being:
(a) Read/Write head (b) Input tape (c) Finite state control (d) All of the above
Give a proper explanation for the correct answer. **[West Bengal SET 2014]**

49. Which of the following is not true about TM?
    (a) It can accept any palindrome (b) It can accept any context-free language
    (c) It can accept $\{a^m b^n c^{mn} : m \geq 0, n \geq 0\}$ (d) One-tape TM can be converted into C++ programs.
    Give a proper explanation for the correct answer. **[JNU M.Phil/PhD Entrance 2016]**
50. Turing machine (TM) is more powerful, than FSM (Finite State Machine) because
    (a) Tape movement is confined to one direction (b) It has no finite state
    (c) It can remember arbitrarily long sequences of input symbols (d) None of the above
    Give a proper explanation for the correct answer. **[Punjab Tech University PhD Entrance 2010]**
51. Let A be defined as $A = \{<M, w> \mid$ The Turing machine M accepts the word w$\}$: and let L be some NP-complete language. Which of the following statements is false?
    (a) $L \in NP$ (b) Every problem in NP is polynomial time reducible to L
    (c) Every problem in NP is polynomial time reducible to A
    (d) Since L is NP-complete, A is polynomial time reducible to L (e) $A \in NP$
    Give a proper explanation for the correct answer. **[TIFR PhD Entrance 2011]**
52. Which of the following statements is true?
    (a) Every Turing-recognisable language is recursive
    (b) The complement of every recursively enumerable language is recursively enumerable
    (c) The complement of a recursive language is recursively enumerable
    (d) The complement of a context-free language is context-free
    (e) The set of TMs that do not halt on empty input forms a recursively enumerable set
    Give a proper explanation for the correct answer. **[TIFR PhD Entrance 2012]**
53. Which the following is false?
    (a) Complement of a recursive language is recursive
    (b) A language recognised by a non-deterministic TM can also be recognised by a deterministic TM
    (c) Complement of a context-free language can be recognised by a TM
    (d) If a language and its complement are both recursively enumerable then it is recursive
    (e) Complement of a non-recursive language can never be recognised by any TM
    Give a proper explanation for the correct answer. **[TIFR PhD Entrance 2014]**
54. Show that TMs that have a read-only input tape and constant size work tape recognise precisely the class of regular languages. **[GATE 1991]**
55. The aim of the following question is to prove that the language {M | M is the code of a TM which, irrespective of the input, halts and outputs a 1}, is undecidable. This is to be done by reducing the language {M′, x| M halts on x} which is known to be undecidable. In parts (a) and (b) describe the two main steps in the construction of M. In part (c) describe the key property which relates the behaviour of M on its input w to the behaviour of M′ on x. **[GATE 2002]**
    (a) On input w, what is the first step that M must make?
    (b) On input w, based on the outcome of the first step, what is the second step that M must make?
    (c) What key property relates the behaviour of M on w to the behaviour of M′ on x?
56. Design a TM to accept the language $L = \{a^n b^n \mid n \geq 1\}$. **[Gujarat SET 2002, UGC NET 2006, UGC NET 2008]**

57. Design a TM, T, which transforms the string bwH into the string wH, where w contains no blanks, b represents a blank and H shows the position of the head of the tape. Give the diagrammatic representation of T. **[UGC NET 2007, UGC NET 2008]**
58. Show that a pushdown stack memory machine has power that lies between FSA and TM. **[UGC NET 2008]**
59. Design a TM to compute the LCM of two given integers. **[UGC NET 2009]**
60. Design a TM to evaluate the square of an integer. **[UGC NET 2009]**
61. Design a TM to accept the language $\{a^n b^n c^n \mid n \geq 1\}$. **[UGC NET 2010]**
62. Design a TM to accept the language $\{a^n b^n c^n \mid n \geq 0\}$. **[UGC NET 2011]**
63. What are the applications of theory of formal languages in the study of natural languages? Discuss at least one. **[UGC NET 2010]**
64. Construct a TM that computes the function $f(x,y) = \begin{cases} 1, \text{ if } x < y \\ 0, \text{ otherwise} \end{cases}$ **[UGC NET 2010]**
65. Justify the need for formal languages for computation models. **[Gujarat SET 2002]**
66. Describe recursive sets and recursively enumerable sets using the formalism of Turing machines. **[Gujarat SET 2003]**
67. The Church–Turing thesis states that a TM can be built for any problem that can be solved with an effective algorithm. Explain how this is related to the Halting problem of TM. **[Gujarat SET 2006]**
68. Design a TM that can add up to four integers placed on the input tape. **[Gujarat SET 2010]**
69. Design a TM that recognises the unary language consisting of all strings of 0s whose length is a power of 2, that is, $L = \{ 0^{2^n} \mid n \geq 0\}$. **[ISI Kolkata JRF 2012]**
70. Construct a standard TM that takes as input a string $w = w_1 w_2 w_3 w_4 ... w_n$ where each w_i is *a* or *b* and *n* is an even number, and halts with the contents of the tape being $w_{n-1}\ w_{n-3} ... w_5 w_3 w_1$
    Hint. For string (*w*) ended by a marker (say B), the machine moves to some state (say q_Y or q_N) just before moving to the halt state (say *h*) depending whether on the content size is even or odd. Next, if it moves to q_Y, the machine again reads the string in reverse order and appropriately erases the even-positioned symbols and shifts to obtain the desired string. Finally, when B is encountered, the machine moves to the *h* state. On the other hand, if B is encountered at q_N, the machine stays there.
71. What is the blank tape problem for TM? Prove that it is an undecidable problem for any TM. **[HCU PhD Entrance 2009]**
72. Prove or disprove: The union of two recursive languages is recursive. **[GATE 1992]**
73. Prove that there exists a recursively enumerable language whose complement is not recursively enumerable.

7 Classes of Problems

OBJECTIVES

At the end of this chapter, you will be able to

- Explain the classes of problems: P, NP, NP-Complete and NP-Hard
- Reduce one problem to another

7.1 INTRODUCTION

Mathematically, a **problem** consists of input, output and a method to solve it in a systematic way. Problems are primarily divided into two broad categories: **numerical** (for which, operations such as addition, subtraction, multiplication, division and sorting are performed on numeric data) and **non-numerical** (for which, usual operations such as sorting, searching and editing are performed on non-numeric data).

7.2 TYPES OF PROBLEMS

We commonly speak of two types of problems (numerical or non-numerical):

- **Decision problem**: A problem that accepts an input and provides an output: *yes* or *no.* Example: Suppose there exists a weighted closed graph with *n* vertices, let us attempt to check if any cycle with k ($\leq n$) vertices and weight *w* exists.
- **Optimisation problem**: A problem that gives a feasible, optimised result for some input. Examples: Finding the minimal (optimal) cost from a vertex (say, *u*) to another vertex (say, *v*); finding the minimum spanning tree (MST) from a graph.

Note: The term *solving* includes deciding the output (yes/no) and finding fixed or optimised result; that is, it includes decision as well as computation.

Categorisation of Algorithms: In general, three types of algorithms are considered.

- **Deterministic algorithm (DA)**: An algorithm in which there is no scope for random choice. However, the input data may be generated at random. In other words, the result or outcome of every operation is uniquely defined. The performance does not vary from run to run. Examples: Sorting *n* elements, adding *n* elements.
- **Randomised algorithm (RA)**: An algorithm that makes random choices during execution to solve a problem. In other words, the outcome of each operation may not be uniquely defined. It can

exhibit different behaviours during different runs even for the same input. More specifically, it may show multiple outcomes (usually for optimisation problems, like the scheduling problem) for the same input or a single outcome with multiple paths (for example, quick sort). Randomised algorithms may give outputs with error or zero error. Its performance may vary from run to run, but it is expected that the running time improves due to random selection of data.

- **Non-deterministic algorithm**: An algorithm based on a theoretical non-deterministic Turing machine. Here too, the concept of random choice is present. Therefore, the outcome of each operation may not be uniquely defined. Hence, non-deterministic algorithm is also a kind of randomised algorithm. However, in the context of automata theory, there are some differences between randomised and non-deterministic algorithms.

  An RA can randomly explore the branches of the computation tree but a non-deterministic algorithm (based on automata) is assumed to explore the branches of the tree using the right path. It is assumed that non-deterministic algorithm follows the super-computational characteristic of choosing the optimal behaviour. A non-deterministic algorithm usually consists of the following operations:

  - Choose(), to select an unprocessed input at random or non-deterministically
  - Failure(), to stop indicating unsuccessful completion
  - Success(), to stop indicating successful completion

  Choose() is applied in RA, while Failure() and Success() may be applied in RA depending on the nature of the problem.

In the discussions that follow, the concept of non-deterministic algorithm is taken into consideration.

Examples

1. **Non-deterministic search**: Let A[1 : *n*], $n \geq 1$ be a list of elements. We need to search for an element '*x*' in A.

   Algorithm Nsearch(A, n, x)

```
begin
 j := Choose(1, n)
 if (A[j] = x) then
    begin
     write (j); Success( );
    end
 else
    begin
     write (0); Failure( );
    end
end
```

2. **Non-deterministic sort**: Let A[1 : *n*], $n \geq 1$ be a list of elements. We need to sort the list in ascending order non-deterministically.

   Algorithm Nsort(A, n)

```
begin
  for i := 1 to n do
   B[i] := 0; // initialise B[ ]
   for i := 1 to n do
    begin
```

```
j := Choose(1, n)
if (B[j] ≠ 0) then Failure( ); // Element already exists in B[j]
   else B[i] : =A[i];
   end
for i := 1 to n -1 do // Verify the order
   if (B[i] > B[i+1]) then Failure( );
 for i := 1 to n do
    write (i);
   Success( );
end
```

It takes O(n) time.

Note

- Both DA and RA follow a systematic approach. So, any algorithm for solving a problem may be deterministic or randomised. However, the non-deterministic approach may also be adopted for solving problems.
- We usually consider DA, as they always give the same output. However, we sometimes prefer RA for some problems (mainly optimisation problems) for which no known output exists.
- If no term like *deterministic* or *non-deterministic* appears explicitly before the term *algorithm*, it may be assumed to be deterministic.

Further, before discussing the classes of problems, we need to explain two terms:

- **Algorithm for performing algebraic operations**: This deals with performing algebraic operations such as matrix operations, polynomial operations, abstract algebra operations, and so on. The input size (to be executed) does not depend on the input number(s). So, we may not express input size in terms of the value of input(s). For example, consider the case of adding two matrices, A and B (each of size n x n). Here, the input size is n^2, which may not be expressed in terms of the value(s) of the matrices. So, regardless of the values provided for A and B, the complexity is bounded above by $O(n^2)$.
- **Number theoretic algorithm**: This deals with performing operations on integers and rationales. Here, the input size depends on the value of the input. So, if the size of the input increases, the running time increases. For example, primality testing of a given number, integer factorisation, and so on. Primality testing is discussed later in this section.

7.3 CLASSES OF PROBLEMS

The different complexity classes of problems, P, NP, NP-Complete and NP-Hard, define the level of hardness of the problem.

7.3.1 P Problem

This type of problem can be solved on a deterministic sequential machine in **polynomial time**, $O(n^k)$, where n is the input size and k is a constant; for example, problems such as finding the smallest integer in an array and sorting a list of elements. Note that in the context of computability theory (dealing with computational resources), P problems are decision problems. P problems are termed as easy problems.

7.3.2 NP Problem

In general, combinatorial problems are NP problems. This class usually deals with verifiable problems. So, these are also decision problems. These problems are not solvable in polynomial time and termed as hard problems too.

Definition

Definition 1 (*Verification-based*): A problem is a **non-deterministic polynomial (NP) time** problem if the validity of a given solution (found on the basis of random choice) of the problem can be verified in polynomial time, $O(n^k)$, where n is the input size and k is a constant. For example, consider the Hamiltonian cycle (HC) problem. Does a graph G with n vertices have a Hamiltonian cycle?

There is no known polynomial time algorithm for solving this problem, as there are $n!$ possible permutations (decisions) over n vertices, and it may be necessary to search all the possible permutations in the worst case. Certainly, if n increases, the value of $n!$ grows exponentially. Hence, the running time is bounded by $O(n!) \cong O(n^n)$, $n > 2$, which is exponential time. It is often called the non-polynomial time[1] that would be taken by any naïve algorithm. More specifically, the complexity cannot be polynomially bounded above – we need exponential or more time. For this reason, we often call such problems as non-polynomial–class problems (that is, NP is also used to denote *non-polynomial*). However, the term NP conceptually stands for non-deterministic polynomial time, where we focus on exploring non-deterministic solutions, and if a solution to the problem is somehow found, then that can be verified (whether the solution is correct or not) in polynomial time $O(n^k)$. In fact, the solution may be found on the basis of guess, made by random choice (in a non-deterministic way; a type of *heuristic search* to find a solution). That is why such problems are termed non-deterministic.

Thus, the term *non-deterministic* denotes using guesswork to find a solution to a problem, whereas the term *polynomial* is used strictly for verification of the guessed solution in polynomial time. The verification algorithm verifies or rejects the guess as a valid or invalid solution to the problem, and it takes polynomial time.

Therefore, non-deterministic polynomial problems consist of two phases: (i) Choosing a value/ sequence of values (solution) at random using an operation, say Choose() (non-deterministic); and (ii) Performing the desired test() operation to determine the validity of the solution (deterministic and takes polynomial time).

Definition 2 (Theoretical machine-based): NP is the set of (decision) problems that can be solved non-deterministically (by theoretical non-deterministic TMs) in polynomial time. This means that there is an accepting computation path that will be selected correctly by the non-deterministic machine.

Analysis of the two definitions: The first definition focuses on deterministic *verification* of the solution in polynomial time, whereas the second definition focuses on *solving* the problem in polynomial time using a non-deterministic TM. However, the verification-based definition tends to be more intuitive and practical in common applications than the formal machine definition.

Both the definitions are equivalent because the algorithm for the theoretical non-deterministic machine definition also consists of two phases: (i) a guess about the solution that is generated in a non-deterministic way, (ii) a deterministic algorithm that verifies (accepts or rejects) the guess as a valid solution to the problem or not.

Comment: Assuming that the non-deterministic algorithm based on Choose() gives the correct choice at each stage, we may say that every NP problem has a polynomial time non-deterministic algorithm.

---

[1] *Non-polynomial time algorithm*: If the complexity of an algorithm is bounded as $O(n^k)$, where k is a constant, the algorithm is said to be a polynomial time algorithm. Polynomial time algorithms are also known as efficient algorithms. On the other hand, an algorithm with a time complexity that cannot be bounded in this way is called a non-polynomial time algorithm. So, it is definitely exponential, like $O(2^{cn})$ or $O(n!)$. These are known as inefficient algorithms.

Polynomial time complexity can be taken by a verification algorithm for any solution (or certificate) for a Hamiltonian cycle in two ways:

Approach 1: Simply check whether the solution is a permutation of the vertices (the vertices should not repeat), and this takes $O(n^2)$ time, where n is the total number of vertices in the graph. More specifically, this strategy requires two *for* loops in cascaded (nested) manner to test the existence of each of the consecutive edges in the solution. For example,

```
for i:= 1 to n do
    for j:= i to n do
        ....................
        ....................
```

Approach 2: For the given solution, check if any vertex other than the starting vertex is repeated. This can be performed in $O(n)$ time by setting an extra one-dimensional array, say VISIT[1...*n*]. Simultaneously, test the consecutive edges in the solution to check if they exist or not.

Remarks

- If the graph is represented in adjacency matrix form, we can check if an edge is present in O(1) time; this gives $O(n)$ time in total.
- If the graph is given in adjacency list form, we need $O(n)$ time to test whether an edge is present. Since O notation gives an upper bound, $O(n^2)$ is certainly correct.

Note: Any problem of class P is also in class NP. Actually, if a problem is of class P, it is solvable in polynomial time without being supplied a certificate (solution). Truly, providing a solution generated at random is immaterial here. Besides, a polynomial is a subset of a non-polynomial (exponential). Alternatively, a non-deterministic TM can accomplish what a deterministic TM (a regular computer) can, taking less time. This means that all problems in P are also in NP.

Verification version: Examples include checking whether the given list is sorted or not, whether the given input is present in the list or not, and so on.

Search version: Examples include reading a list of numbers, X, and arranging the same list in sorted order (say, Y).

Note: If there is a quick (easily found) solution to the search version of a problem, it is said to be polynomial time (P). If the problem is confined to the verification version in a deterministic way, it is said to be non-deterministic polynomial time (NP).

Eulerian graph vs. Hamiltonian graph

- *Eulerian graph*: Given a graph, is there a closed path that visits each edge of the graph exactly once? This is a P problem. First, if there is any odd-degree vertex in the graph, it may not be a Eulerian graph. Checking for a Eulerian circuit based on Fleury's algorithm takes polynomial time.
- *Hamiltonian graph*: Given a graph, is there a closed path that visits each node of the graph exactly once? No known polynomial algorithm exists for this problem.

Primality Testing: An NP Problem: One common example of an NP problem is primality testing. Given an integer n, decide whether n is prime or not. The simple strategy for this is to check the divisibility of n by every number from 2 to $\lfloor\sqrt{n}\rfloor$. If n is not composite (if it is prime), then it must not be divisible by any number from 2 to $\lfloor\sqrt{n}\rfloor$. This is known as the **trial division approach**. It requires $\lfloor\sqrt{n}\rfloor$ iterations in the worst case. Thus, the complexity is $O(\lfloor\sqrt{n}\rfloor)$. However, it may not be treated as polynomial time complexity in the context of number theory problem.

Recall that when measuring the complexity of number-theoretic[2] algorithms having integer input, we take the input size to be the number of digits in the number (with respect to a convenient base). The number of digits in the decimal representation of n may be expressed as: $n = 10^m$. Therefore, $\sqrt{n} = n^{\frac{1}{2}} = (10^m)^{\frac{1}{2}} = 10^{\frac{m}{2}}$. This means that complexity grows faster than any polynomial, as the size of n increases; it is exponential in the size m. For example, let $n = 10^{100}$. So, $m = 100$ and $\sqrt{n} = (10^{100})^{\frac{1}{2}} = 10^{50} \ldots$ (i). However, an algorithm in this respect may take $log\, n = m$ computation time, m here is simply 100 ... (ii). Clearly, Equation (i) shows exponential behaviour, whereas Equation (ii) is bounded as: $O(m) = O(\log n)$ and it can be bounded above polynomially. In particular, $O(\sqrt{n})$ from Equation (i) results in exponential time and can be written as: $O\left(10^{\frac{m}{2}}\right)$.

To observe the growth of the exponential function obtained from Equation (i), we may consider Table 7.1.

TABLE 7.1 Running time

n	$T(n) = 10^{\frac{m}{2}}$	Execution time (when implemented)
1 000 000 = 10^6	$10^{6/2} = 10^3 = 1000$	0.906
1 000 0000 = $10^7 = 10 \times 10^6$	$(10^{1/2} \times 10^{6/2}) \approx 3.162 \times 1000$	22.407
1 000 00000 = 10^8	$10^{8/2} = 10^4 = 10000$	*Too long* !!
....	.....	.....

Note: It is an open research problem to check whether primality testing is either in P or in NP.

7.3.3 NP-Complete Problem

These problems are NP but mainly decision problems. In particular, the concept of NP-Completeness applies directly not to optimisation problems but to decision problems in which the answer is simply *yes* or *no*. Roughly speaking, these are harder than NP. For example, consider a variant of the Travelling Salesman Problem (TSP), where we want to know if there is a route of some length (say, k) that goes through all the nodes in a certain network. This is a decision problem (also an NP-C problem), as a decision needs to be made. TSP is an optimisation problem, which is also NP-C. Further, finding the largest path between two vertices in a graph is an NP-C problem, though finding the shortest path between two vertices in a graph takes polynomial time. Interestingly, by applying the reduction property (discussed later in this chapter), it may be proven that the optimisation version of the decision NP-C problem is NP-C as well.

Truly, an NP-C problem is a decision optimisation problem that cannot be solved in polynomial time. However, whether a solution belongs to a problem can be verified in polynomial time. NP-C problems

[2] In number theory, the size of the input is measured in terms of a convenient base. Here, the input size is taken from the given number (a large integer but not from many integers). More explicitly, for measuring the complexity of a number-theoretic algorithm having integer input n, we take the input size to be the number of digits in n (with respect to a convenient base). For example, the number of digits in the decimal representation of n is given by $m = \lceil \log_{10}(n) \rceil$ or $n = 10^m$. Thus, complexity $T(n) = O(10^m)$. So, if m increases, complexity may grow exponentially. From Table 7.1, we may say that the running time grows mathematically three times from the earlier input, where the present input is 10 times bigger than the earlier input. However, the execution time of the implemented program increases exponentially for these inputs. This is because modulo operations (to verify the presence of factors) are very costly and these operations increase as the size of the number increases.

are hard NP problems. To formally compare the ease/hardness of problems, we apply the concept of problem reduction (discussed below), which helps to show that one problem is no harder or no easier than another known problem.

Problem Reduction: In mathematics as in everyday life, a simple way to solve a new problem is to reduce it to a previously solved problem. An instance of the new problem is expressed completely in terms of an instance of an existing (prior) problem, and the solution of the new problem is then interpreted in terms of the solution to the existing problem. We will now discuss polynomial time reduction.

Polynomial time reduction: To better understand the idea of problem reduction, we consider first the case of reduction of easy problems. Let P_1 and P_2 be two problems (of the same infrastructure), where P_1 and P_2 are, respectively, unknown and known problems. Assume that P_1 reduces to (strictly not reduces from) P_2 in polynomial time; $P_1 \leq_P P_2$ (or $P_1 \propto P_2$) and P_2 is easy (solvable in polynomial time). This implies that there exists a computable function: f such that, for input x to problem P_1, f gives $y = f(x)$ as the input to P_2. The solution for input x of P_1 is obtained iff the output for input $y = f(x)$ to P_2 is obtained. In other words, use the solution to the instance of P_2 to obtain the solution to the instance of P_1. This is known as **Cook's reducibility**. It is true irrespective of the reduction for easy or hard cases. In case of hard problems, P_2 is taken as unknown and the solution P_2 is derived from P_1 (and the converse).

On the other hand, the **Karp reduction** applies only to decision problems, say P_1 and P_2, and it is written as: $P_1 \leq_P P_2$. Here, the solution to input x of P_1 is *'yes'* (obtained) if and only if the output for input $y = f(x)$ to P_2 is also *'yes'*. This is true irrespective of the reduction for easy or hard cases. In case of hard problems, P_2 is taken as unknown and the solution P_2 is derived from P_1 (and the converse).

The function: f in both the cases (Cook's reducibility and Karp's reducibility) is the **reduction function**, which is computed by a polynomial time algorithm (say, F). After all, the main objective of reduction is to make the problems related, and the transformation provides the means for converting any algorithm such that the converted algorithm is capable of solving the unknown problem P_1 from the known problem P_2.

In other words, for $P_1 \leq_P P_2$, we may say that P_2 is the *problem solver* and it is already a known problem (in easy case).

Now, take an instance (say, x) of P_1. Call the solver sub-routine in polynomial time and reduce P_1 to P_2 ($P_1 \leq_P P_2$) to obtain the instance (say, $y = f(x)$) of P_2. Next, P_2 gives an output, say O_2. This output is transformed to an output (say, O_1) of P_1. The symbol '$\leq$' is interpreted as 'no harder than'. For example, if $P_1 \leq_p P_2$, we say that P_1 is not harder to solve than P_2. More explicitly, P_1 is as hard to solve as P_2, and we can use P_2 to solve P_1.

Thus, the **total running time** of the entire process = Time to compute f (reduction time) + Time to apply algorithm for solution of P_2.

It may be noted here that all the instances of P_2 may not necessarily be used to solve P_1, as P_1 may be a special case of P_2. However, a reduction is blind in the sense that it is not known which instances (inputs) are true and which are false, but each instance is converted in the same manner.

KEY POINTS

- Transform unknown (new) to known (old).
- Reduction maps the solutions of one problem to the solutions of another problem.

Thus far, we have discussed $P_1 \leq_p P_2$ (easy). Let us illustrate both the cases, easy and hard, in detail. The ultimate goal of reduction is to show that one problem is no harder or no easier to solve than another problem. It is much more useful for hard problems. More specifically, reduction is taken into account if the problem is not solvable in polynomial time.

(i) **Easy**: To show that P_1 is easy, reduce it to something we already know is easy, say P_2. So, $P_1 \leq_p P_2$ (easy). Reduction will be performed in polynomial time (easily).
For example, selecting the *k*-th smallest element from a list of *n* elements is termed as P_1, whereas sorting *n* elements in a list is denoted as P_2. Here, P_2 is easy; so it is solvable in polynomial time. Further, input of P_1 is transformed into input of P_2 in polynomial time (as $P_1 \leq_p P_2$) and the solution corresponding to the input of P_2 is found in polynomial time. Now, P_1 is also easy, as the solution of the input to P_2 is found in polynomial time. Accordingly, the solution of the converted input corresponding to the input to problem P_1 too will be found in polynomial time, as the solution of P_2 gives the solution of P_1.

(ii) **Hard**: To show that P_2 is hard, reduce a known hard problem: P_1 (NP-C problem) to P_2; P_1 (hard) $\leq_p P_2$. In other words, P_2 is as hard as P_1.
Here, P_1 is hard, so it cannot be solved in polynomial time. In fact, the input of P_1 is transformed into the input of P_2 (as $P_1 \leq_p P_2$), where the solution corresponding to the input of P_1 is found in non-polynomial time (exponential). Thus, P_2 is also treated as hard, as the solution of the input to P_1 gives the solution to the converted input of the problem P_2 and the solution of the input to P_1 is found in non-polynomial time.

Note: We may think of reducibility for hard cases as: $P_1 \leq_p P_2$ (hard). In fact, if we consider it so, then P_1 (a P problem) will also be treated as hard, as any input of P_1 (P problem) is first transformed into the input of P_2, but the solution corresponding to the converted input must not be solved in polynomial time for P_2; it will take non-polynomial time, as it is hard. Hence, based on (i) (that is, the easy case) P_1 will be treated as NP, even though it may not necessarily be NP.

Let us assume that P_1 (hard) $\leq_p P_2$ is true. However, someone may think that a P problem: P_2 may be treated as an NP problem. But the claim is not valid. The solution for *w* (input) of P_1 is not found in polynomial time. Thus, the solution of *x* (converted input for *w*) for P_2 may not be obtained in polynomial time. On the other hand, if P_2 is assumed to be a P problem, any solution for P_2 can be found in polynomial time. Accordingly, the solution of P_1 (hard) can be obtained in polynomial time (in the reverse process). Thus, P_1 (hard) will be treated as easy. Hence, a contradiction arises for P_1, as it is taken as an NP problem.

Certainly, for P_1 (hard) $\leq_p P_2$ (easy), polynomial time reduction is not possible. However, if it were possible, then any hard problem could be solved in polynomial time.

The above discussion reports that if any one of the two problems, P_1 and P_2, is hard, it makes the overall problem also hard. Thus, reducibility works like an AND gate (presented in Table 7.2).

TABLE 7.2 Cases of reducibility

Level of problem: P_1	Level of problem: P_2	Conclusion
E (Easy)	E (Easy)	E
H (Hard)	H (Hard)	H
E	H	H
H	E	H

Given below is a discussion on reduction and its necessity in P and other class problems.

Reduction in P-class problem: If P_2 is a P problem and $P_1 \leq_P P_2$, P_1 is also a P problem. In other words, every input of P_1 is polynomially transformed into an input of P_2; this reduction shows the ease of solving the problem (as pointed out in (i)). The solution of P_2 is found taking polynomial time (as P_2 is a P problem). So, reduction asserts that the solution to P_1 can be obtained in polynomial time through the solutions to P_2.

Reduction for problems other than P-class problem: In most cases, when people face difficulty in solving a new problem, they classify it as hard. They try repeatedly to solve the problem but are unable to obtain a solution. In such a situation, if they attempt to show the problem as '*hard to solve*'; they stop trying to solve it and instead begin trying to solve it in the same way as other known hard problems. Essentially, investigating the consequences must take polynomial time.

Importantly, it is assumed that reduction is to be applied for non-polynomial time problems (hard problems) and each solution is a guessed solution (chosen in a non-deterministic manner).

KEY POINTS

- If P_1 is not a P problem and $P_1 \leq_P P_2$, then P_2 is also not a P problem. In other words, if P_1 cannot be solved in polynomial time, P_2 also cannot be solved in polynomial time; this reduction shows hardness (as pointed out in (ii)).
- A problem P_1 is NP-C if it is NP and any known NP-C problem (say, P_2) reduces to P_1; $P_2 \leq_P P_1$ (or all NP problems reduce to P_1).
- A problem P_1 is NP-Hard if any known NP-C problem (say, P_2) reduces to P_1; $P_2 \leq_P P_1$ (or all NP problems reduce to P_1). However, it is assumed that proving or deciding P_1 as NP is very difficult. For clarity, consider a variant of TSP – finding the shortest path with *K* cost value. This is a decision NP-C and it is verifiable for a given solution based on the given value: *K*. However, if we consider the TSP itself, there is no way to determine if a given solution is the shortest (as no shortest value is known at this moment); this level of hardness for verification may be considered as non-polynomial verification.

In summary, polynomial time reduction provides a formal means for showing that one problem is not more than polynomial-factor harder than that of the other.

Utility of Reducibility: If a new problem can be reduced to a known easy problem, it implies that the new one is also easy. On the other hand, if a known hard problem can be reduced to the unknown problem, it implies that the new one is also hard.

Both the problems, known (P_1) and unknown (P_2), are assumed to be a variant of the other; they have the same infrastructure as the inputs are to be mapped from one to the other, but the intention is to show that the problem is easy or hard based on the known problem.

The point to be noted is that if reduction (with respect to problems) is true for specific cases, they must be true for generalised cases as well. Suppose a completely new problem appears in the future. In such a case, if any existing hard problem can be reduced to the new one or the new problem be reduced to a known easy one (on the basis of input mapping), reducibility works properly to show the new one as hard or easy.

Some well-known examples of hard reducibility are the Knapsack decision problem and the Knapsack optimisation problem, Hamiltonian cycle (HC) problem and Hamiltonian path (HP) problem. For illustration, let us take the case of HC and HP.

Let G = (V, E) be a graph. Checking for the existence of the Hamiltonian cycle of a graph is hard in itself. Assume that the given G is a HC graph and it has a cycle, say C, covering the vertices. Now, delete an edge from the cycle C of G so that a new graph G′ =(V, E′) is formed. Certainly, input (I) corresponding to cycle C for G can be made for one-to-one mapping for G′. Suppose the mapped input is I′. Now, the output corresponding to I′ for G′ is 'yes'; that is, a Hamiltonian path exists because the input (I) for G corresponding to I′ results in a Hamiltonian cycle. Hence, HC $\leq_P$ HP, where HC is known as hard and the input of HC can be easily mapped to the input of HP, through a function, *f*. In the same way, it is also true for '*no*'. More such examples are given in subsequent sections.

Circuit Satisfiability as the First NP-C Problem: We should provide at least one known NP-C problem to verify if other problems are NP-C, using reducibility strategy. Circuit satisfiability (a hardware problem) is believed to be the first NP-C problem. It relies on Boolean combinational circuits connected by wires. The problem states that "Given a Boolean combinational circuit composed of AND, OR and NOT gates, is it satisfiable"[3]? In the context of formal languages, 3SAT (a specific case of satisfiability or SAT problem) may be defined as: 3SAT $= \{\langle\varphi\rangle \mid \varphi$ is a satisfiable 3-CNF formula$\}$.

In 3SAT, every clause has exactly three literals. For example: (i) (x ∨ y ∨ z) ∧ (¬x ∨ ¬y ∨ z), (ii) (x ∨ x ∨ x) ∧ (y ∨ ¬y ∨ ¬x) ∧ (x ∨ y ∨ ¬y) but not (iii) (x ∨ y ∨ z ∨ w) ∧ (x ∨ y). On the other hand, a Boolean formula is in *k*-conjunctive normal form (*k*-CNF) if it is the AND of clauses of ORs of exactly *k* variables, or their negation. For better understanding of *k*-CNF, the following points are discussed:

- **CNF**: A formula is in CNF iff it is represented as $\wedge_i^k c_i$, where c_i comprises clauses represented as $\vee l_{ij}$; $l_{i,j}$ denotes literals.
- **Disjunctive normal form (DNF)**: A formula is in DNF iff it is represented as $\vee_i^k c_i$, where c_i comprises clauses represented as $\wedge l_{ij}$; $l_{i,j}$ denotes literals.
- **3-CNF**: $(x_1 \vee \neg x_2) \wedge (\neg x_1 \vee x_3) \wedge (\neg x_2 \vee x_3)$

This 3-CNF is satisfiable for one possible case: x_1 = true, x_2 = false, x_3 = true. We can determine in polynomial time whether a 2-CNF formula is satisfiable, but the running time for satisfiability checking of a 3-CNF formula increases. Unfortunately, if the circuit has *k* inputs, we would need to check up to 2^k possible assignments.

Thus, checking or verifying a *k*-CNF problem must take $O(2^k)$ time, which is super-polynomial. In other words, it is exponential time. The reason is that we need to check all possible combinations in the worst case; otherwise a decision may not be possible. Thus, there is strong evidence that no polynomial time algorithm exists to solve this problem, and it is assumed to be the first NP-C problem. However, a polynomial time non-deterministic algorithm terminates successfully iff a given proposition formula: $E(x_1, x_2, ...x_k)$ is satisfiable. Here, one of the 2^k possible assignments is non-deterministically chosen to verify satisfiability. This takes $O(k)$ time.

Further, another valid point in favour of this problem as NP-C is that every NP problem is polynomially reducible to circuit satisfiability problem.

Note: In 1971, Cook–Levin showed that satisfiability (SAT) problem for a Boolean expression is an NP-C problem, and it is believed to be the first NP-C problem.

Significance of NP-C Problems: NP-C problems are hard NP problems. In other words, NP-C problems are defined in a precise sense as the hardest problems in NP. We believe that any NP-C problem does not have a fast algorithm. These are, in fact, decision NP-Hard problems. These problems

[3] SAT (satisfiability) is accepted as the first NP-C problem. Here, a Boolean expression of *n*-Boolean variables is considered. The value of the expression (E) is denoted as E(T), where T is the truth assignment for E. T satisfies E if E(T) = 1 or true. A Boolean expression E is said to be satisfiable (SAT) if at least E(T) = 1 is found. It is also called C-SAT (Circuit-SAT). It contains operations such as AND, OR, NOT, and so on. An example of SAT is 3-CNF-SAT, where the Boolean expression contains clauses (three variables) and conjunction operations only.

are as hard (up to polynomial factors) as every other problem in NP, and this can be verified by simply constructing a known NP-C problem. So, if any NP-C problem is solvable in polynomial time, all NP problems can be solved in polynomial time (as all NP problems reduce to that NP-C problem). However, if any NP problem is solved in polynomial time, all NP (including NP-C) problems may not necessarily be solved in polynomial time.

Further, if we are able to identify a new NP problem as NP-C, we need not waste time trying to find a solution in polynomial time. In fact, if a problem is NP-C, it may be addressed by using heuristic methods and approximation algorithms instead of by using an easy (polynomial) algorithm. In other words, identifying a problem as NP-C helps us to solve it using the heuristic or approximation approach, instead of attempting to solve it in polynomial time. So, it is necessary to verify the completeness of NP problems of interest. The easiest way for this purpose is to apply the reducibility property.

Identifying an NP-C Problem: Formally, NP-Completeness is defined in terms of reduction, meaning that one problem is easier than another. Technically, a problem (A) is said to be in NP-C if the following statements are true for A.

(i) A is in NP; that is, it is necessary to guess a solution non-deterministically and verify the solution deterministically, taking polynomial time.

(ii) Every NP problem reduces to A. However, checking the reduction of every NP problem is a complex task. That is why we may consider performing only one check:

- If A′ is a known NP-C and A′ polynomially reduces to A ($A' \leq_p A$), then A is also NP-C. In other words, it does not take more than polynomial time to reduce A′ to A. Importantly, the SAT NP-C problem may be used to prove other NP problems as NP-C. However, all NP-C problems may not be structurally the same but the transitivity property may satisfy to prove the problems as NP-C.

To be more precise, NP-C = NP + Reduction. It may be noted that the method of reducing from a single known NP-C to an unknown problem is simpler than reducing from every NP problem.

Steps for Proving NP-Completeness: Let us take a problem, A. To prove that A is NP-C, we need to apply the following steps:

1. Prove that A is NP.
2. Select a known NP-C problem, B.
3. Construct a function f that maps members of B to members of A.
4. Show that x gives a solution of B ($x \in B$) if and only if $f(x)$ gives a solution of A ($f(x) \in A$). Equivalently, if x gives a solution of B, then $f(x)$ may give a solution of A.
5. Show that f can be computed in polynomial time.

Tips to prove NP-C: The easiest way to prove that a new problem is NP-C is by first proving that it is in NP. Next, consider some known NP-C problem and polynomially reduce it to the unknown one.

To better understand the reducibility property in the context of NP-C, we use the following examples:

- Knapsack decision (KD) problem (A′) reduces to knapsack optimisation (KO) problem (A); $A' \leq_p A$, where KD is already accepted as an NP-C problem. So, KO is also NP-C. We can use the solution of KO to find the solution of KD taking an instance of KD and converting it to an instance (x) of KO by calling some sub-routine (F). Next, find the feasible solution of KO over the obtained instance by applying an objective function, say f_{obj}. This takes polynomial time; so, KO is also NP-C. Finally, we obtain the desired solution of KD for the solution obtained from KO over the instance x.

- If finding the Hamiltonian cycle is NP-C, then TSP is also NP-C; that is, the HC problem polynomially reduces to TSP. Actually, TSP appears to be similar to HC (a Hamiltonian closed path of a graph). It may be noted that solutions to both TSP and HC are chosen non-deterministically and then verified, as there is no known polynomial time algorithm to solve each of them. Hence, they are NP. We can prove that TSP is NP-C using the following two strategies:
  - *Strategy 1*: Let G = (V, E) be a graph such that |V| = *n*. Assume that Q and P denote, respectively, the HC problem and TSP with respect to the graph, G. For reduction from Q to P, first define a complete graph, K_n = (V, E′) with *n* vertices and provide weight: 1 to the edges, if these are present in G; else weight is 0. We may note that the HC problem does not require a complete graph, whereas TSP requires one. We may now attempt to map an instance of Q (instance of G) to an instance of TSP (P). Conceptually, G has a Hamiltonian cycle if and only if K_n has a tour of cost (weight) consisting of exactly *n* vertices. Further, G_{TSP} has a solution (not necessarily the optimal) iff G has an HC. In other words, if the graph G has an HC (the answer is *yes*), we may surely solve TSP (an optimisation problem). On the other hand, if there is no HC in G, then the minimum HC (TSP) will return the answer '*no*'. This means that the HC problem can be reduced to TSP; HC $\leq_p$ TSP. Now, HC is NP-C and TSP is NP. So, TSP is also NP-C. Precisely, the reduction involves the following points:
    - For every edge (*u*, *v*) in G, include (*u*, *v*) with weight 1 in K_n.
    - For every edge (*u*, *v*) not in G, include (*u*, *v*) with weight 0 in K_n.
    - There are at most n^2 edges in G; so reduction time is $O(n^2)$.
    - Transformation is expressed as follows: Let G = (V, E) be a graph with *n* nodes (HC problem). From G, construct an instance of TSP, where K_n = (V, V x V) is a complete graph on the vertices of G. The edge weight c(*u*, *v*) is 1 if the edge (*u*, *v*) is in E and 0 otherwise. The transformation may be expressed as: *f*(G) = (K_n, c, *n*).
  - *Strategy 2*: TSP is NP and HC is a special case (a subset) of TSP. Now, SAT is also NP. Further, SAT $\leq_p$ DHC (directed HC); DHC $\leq_p$ HC and HC $\leq_p$ TSP. Hence, TSP is NP-C.

Note: On the basis of edge weight (c(*u*, *v*) = 1 if the edge (*u*, *v*) is in E and 0 otherwise), a graph G has a Hamiltonian cycle if and only if H has a TSP tour of weight *n*, else the cost of the tour in H will be at least (*n* + 1). In the case of DHC, an edge between two vertices may or may not exist; these may be represented as 1 or 0. In SAT, each Boolean variable may take either 1 or 0. Therefore, relating these two problems is easy.

More NP-C problems

- **Three-colouring problem**: Given an undirected graph G, is there a legal three-colouring of its nodes? As a formal language: 3-COLOR = {⟨G⟩ | G is an undirected graph with a legal three-colouring}.
- **Vertex cover**: A set of minimum number of vertices in a graph G = (V, E), that covers all the edges of G is termed the vertex cover problem. However, finding the minimum set is itself an NP problem.
- **Subset sum**: Given a set S ⊆ N and a natural number *k*, the subset sum problem can be stated as follows: Is there a subset of S with sum exactly *k*? As a formal language:

  SUBSET SUM = {⟨S, *k*⟩| S ⊆ N, *k* ∈ N and there is a subset of S with sum of exactly *k*?}

NP vs NP-C: NP-C problems are hard NP problems. To define NP-C, we may use the statement: If any NP-C problem is solvable in polynomial time, then all NP problems are solvable in polynomial time. However, if an NP problem is solvable in polynomial time, it does not mean that all NP problems are solvable in polynomial time.

KEY POINTS

- NP-C problems are hard NP problems and we attempt to use the reducibility property to solve them. More specifically, such NP problems are supposed to be solved using only existing NP problems. One can prove that a problem is NP-C by reducing a known NP-Complete problem to it.
- If reduction of A′ (known NP-C problem) to A (unknown) requires exponential time, then statement (ii) for NP-C is not satisfied.
- If any NP-C has a polynomial time solution, every NP must have a polynomial time solution.
- It is well-accepted that NP-C problems are confined to the realm of decision problems. However, we may take advantage of the relationship between optimisation problems and decision problems. For example, given a directed graph G, vertices *u* and *v* and an integer *k*, does a path exist from *u* to *v* consisting of at most *k* edges?
- Reducibility is usually employed for NP-C, but it may be used for NP (other than NP-C) as well, to show other unknown problems as NP by applying Hard problem $\leq_p P_1$ (unknown).

The reduction of some well-known NP problems to NP-C problems is shown in Fig. 7.1.
Here, A ⇒B means that B is a known NP-C problem and A is a known NP-problem, and B reduces to A.

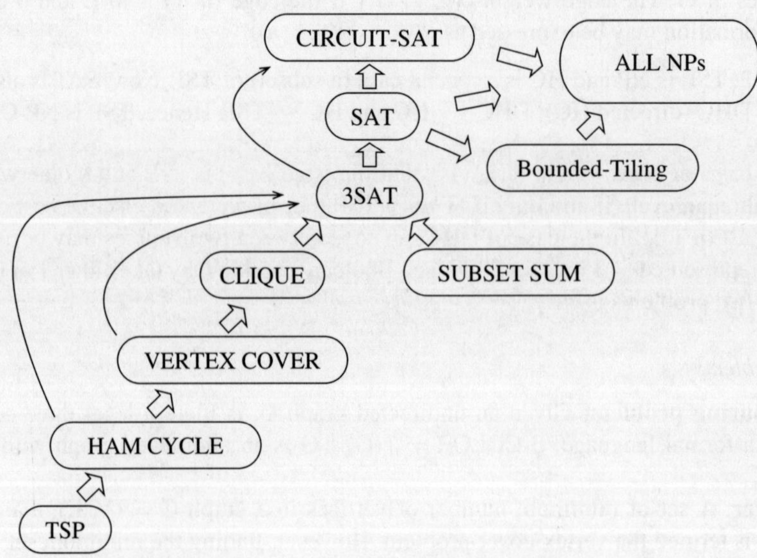

FIGURE 7.1 NP problems to NP-C problems

7.3.4 NP-Hard Problem

Intuitively, NP-Hard problems are harder than NP-C problems. These are intractable[4] problems. They are usually optimisation problems that are harder than decision problems. In particular, a problem A is NP-Hard if every problem B in NP is polynomial-time-reducible to A, but proving A as NP is very difficult. In other words, verifying the solution to A may not be possible in polynomial time. They are basically the hardest decision and non-decision problems.

The halting problem (HP) of TM is a classic NP-Hard problem. HP asks if the machine will halt for a given program (P) and input (I). Actually, for input I, the TM may run forever, and we can never be sure that the TM will halt. Hence, deciding (or verifying) whether the machine will ever halt may not be possible in polynomial time.

Examples

- Verifying whether a context-free grammar is ambiguous is an NP-Hard problem. It is often considered an unsolvable problem too.
- Finding a perfect number greater than a given number *n* is NP-Hard (it is often termed unsolvable), as it is difficult to find the factors[5] of a number. Further, some optimisation and decision problems are very hard, and so they fall under the NP-Hard category. Importantly, all NP-Hard problems are not NP-C. Only the decision NP-Hard problems can be NP-C, but optimisation NP-Hard problems may not be NP-C. However, some decision problems such as HP, ambiguity checking of CFG and PCP are assumed NP-Hard (not NP-C but harder than NP-C).
- Other examples include Hamiltonian cycle problem and graph colourability. These are NP-C as well.

It is interesting to note that some optimisation problems may reduce to decision problems.

The most common but well-accepted relationship among P, NP, NP-C and NP-Hard problems is shown in Fig. 7.2. It is also known as the **Euler diagram**. Fig. 7.2 shows that all NP-C problems are NP-Hard, but the converse is not true. Examples include HP, PCP, ambiguity checking of CFG, etc.

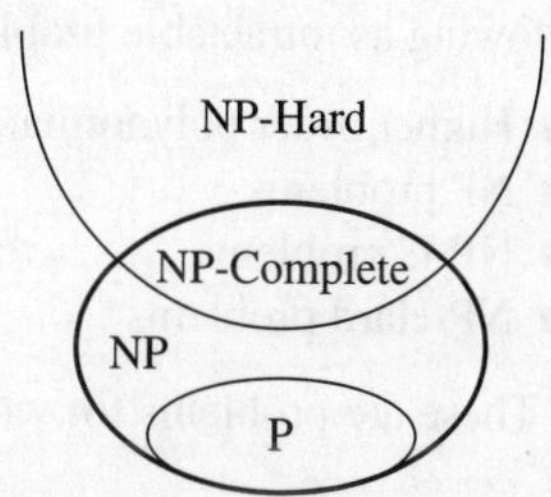

FIGURE 7.2 Commonly accepted relationship among P, NP, NP-C and NP-Hard

NP vs Unsolvable Problems: In the case of NP problems, there is no known polynomial time algorithm to solve the problem. However, there may exist exponential or super-exponential time systematic algorithms for these problems. On the other hand, there exists no systematic algorithm for unsolvable problems such as PCP, ambiguity checking of CFG, and so on. These are harder than NP problems.

---

[4] Problems that are solvable in polynomial time are known as tractable problems, whereas those that can be solved in theory (given large but finite time), but which in practice take too long (higher order polynomial such as: n^{100} or non-polynomial or super-polynomial, like 2^{cn^m}, $c > 0$ and $m \geq 1$) to obtain a solution, are known as intractable problems. In other words, a problem is intractable if there exists no efficient algorithm to solve it. The equations: n^{100} or 2^{cn}, $c > 0$ can be used to measure hardness in the context of NP-Hard problems. Surely, all NP, NP-C and NP-Hard problems are also intractable.

[5] In number theory, integer factorisation is the decomposition of a composite number into a product of smaller integers. If the integers are further restricted to prime numbers, the process is called prime factorisation. Further, when the numbers are very large, no efficient integer factorisation algorithm is known and it is hard to factor. For instance, finding the factors of an enormous number like 64^{448} can be very hard because there could be several hundred factors. It becomes harder still if we think about the product of two prime numbers; this is an NP-Hard problem.

Importantly, the size of the input is not fixed for such problems.

Co-NP-Complete Problems: In complexity theory, co-NP-Complete problems are the hardest computational NP problems. One simple example of a co-NP-complete problem is tautology – the problem of determining whether a given Boolean formula is a tautology; that is, whether every possible assignment of true/false values to variables yields a true statement. This is closely related to the Boolean satisfiability problem, which asks whether there exists at least one such assignment. Note that the tautology problem for positive Boolean formulae remains co-NP-Complete, even though the satisfiability problem is trivial, as every positive Boolean formula is satisfiable.

7.3.5 Summary of Classes of Problems

A summary of the different classes of problems and verifiability of solutions in polynomial time is given in Table 7.3.

TABLE 7.3 Problem type and verifiability

Problem type	Solution verifiable in polynomial time
P	Yes
NP	Yes
NP-C	Yes
NP-Hard	Yes or no

Intractable problems: Based on the discussion on intractable problems, we may categorise the following as intractable problems:

- Higher order polynomial, for example, n^k, $k \geq 100$
- NP problems
- NP-C problems
- NP-Hard problems

These are problems for which no efficient algorithm is known (as discussed earlier).

KEY POINTS

- Only decision NP-Hard problems are NP-C. Thus, NP-C problems are NP-Hard as well.
- There are decision problems that are NP-Hard but not NP-C; for example, the halting problem of TM. This is also an undecidable problem.

Thus far, we have not found a relation such as NP $\leq_p$ P. If any such scenario is found, then every NP problem is solvable in polynomial time (P = NP). So, this is an open research problem. However, if it is proved in the future that NP = P, problems will occur in the area of cryptography and security, where almost all cryptographic problems are NP problems that make it difficult (or quite impossible) for attackers to hack any secured data.

Note: NP – NP-C ≠ Φ but NP-C – NP = Φ, as NP-C covers the entire NP.

Now, examining NP-C problems and using the reducibility property, we highlight some important lemmas that may assist us in drawing conclusions about problems.

Lemma 1: Let P_1 and P_2 be two problems such that P_1 is NP-C and P_1 is reducible to P_2; that is, $P_1 \leq_p P_2$, then P_2 is also NP-C.

Lemma 2: If L_1 is a decision NP-C problem, L_2 is an optimisation problem and $L_1 \leq_p L_2$; then L_2 is also NP-C.

Lemma 3: Two problems, L_1 and L_2, are said to be polynomially equivalent if and only if $L_1 \leq_p L_2$ and $L_2 \leq_p L_1$.

Lemma 4: Reductions are transitive, as we can compose two polynomial transformation functions in sequence and the result is also a polynomial-time transformation function. For example, if $L_1 \leq_p L_2$ and $L_2 \leq_p L_3$, then $L_1 \leq_p L_3$.

EXAMPLE 7.1

Suppose a polynomial time algorithm is discovered that correctly computes the largest clique (largest complete graph) in a given graph. Draw the correct Venn diagram of the complexity classes P, NP and NP-C for such a scenario.

Solution The most important open question in complexity theory is whether P = NP; this asks whether polynomial time algorithms actually exist for NP-C and all NP problems (as a problem C is in NP-C iff C is in NP, and every problem in NP is reducible to C in polynomial time). According to the given question, NP $\leq_P$ P, that is, some polynomial time algorithm exists that computes the largest clique problem in the given graph. This is the known NP-C problem. Hence, P = NP = NP-C. The scenario is presented in Figs 7.3 (a) and (b). Both are conceptually equivalent, but Fig. 7.3 (b) is well-accepted for clarity.

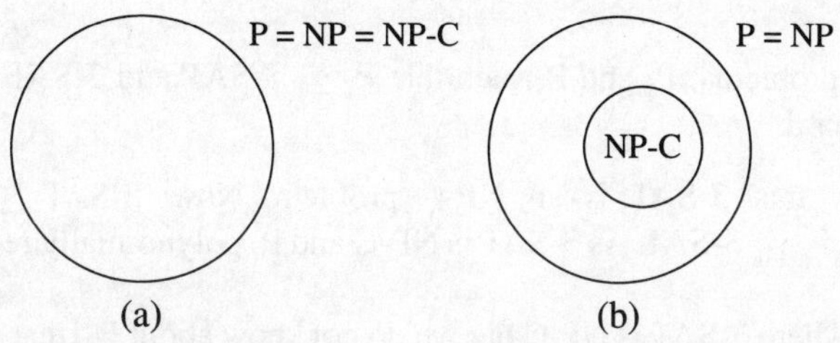

FIGURE 7.3 P = NP = NP-C

EXAMPLE 7.2

Let $L_1 \leq_P L_2$. Find which of the following is false:

(a) If L_2 is recursive, then L_1 is also recursive.

Solution True

(b) If L_1 is undecidable, then L_2 is also undecidable.

Solution True

(c) If L_2 is recursively enumerable, then L_1 is also recursively enumerable.

Solution True. Some may claim that L_1 is a recursive language (not REL). If so, L_2 will also be RE – this raises a contradiction. Hence, it is true.

(d) If L_2 is not recursively enumerable, then L_1 is also not recursively enumerable.

Solution False. The term *not recursively enumerable* is ambiguous here; L_2 may be a recursive or a non-recursively enumerable language. If it is a non-recursively enumerable language, then it is false.

EXAMPLE 7.3

(a) Let L_1 and L_2 be two languages such that L_1 is CFL and L_2 is a recursively enumerable language but not recursive. Which of the following is/are necessarily true?
(i) L'_1 (complement of L_1) is recursive.
(ii) L'_2 (complement of L_2) is recursive.
(iii) L'_1 (complement of L_1) is CFL.
(iv) $L'_1 \cup L'_2$ is recursively enumerable.
(A) Only (i) (B) Only (iii) (C) Both (iii) and (iv) (D) Both (i) and (iv)

Solution (D) is the correct option. L_1 is CFL. So, it is also recursive, and the complement of any recursive language is also recursive (*true*). However, L'_1 need not be CFL. It may be regular, which is also recursive. L_2 is a recursively enumerable language and its complement is not recursive but recursively enumerable. So, (ii) is false. L'_1 (complement of L_1) is a CFL. It is *false,* as the complement of any CFL is not necessarily CFL.

(b) Let L_1 be a recursive language and L_2 be a language that is recursively enumerable. Prove that $L'_1 \cup L'_2$ is recursively enumerable.

Solution L_1 is a recursive language. So, L'_1 is also recursive. Again, every recursive language is a recursively enumerable language. Hence, L'_1 is recursively enumerable. Further, the union of two recursively enumerable languages is also recursively enumerable.

EXAMPLE 7.4

Consider two decision problems: P_1 and P_2 such that $P_1 \leq_P$ 3-SAT and 3-SAT $\leq_P P_2$. Then, prove that P_1 is NP and P_2 is NP-Hard.

Solution We know that 3-SAT is an NP-C problem. Now, 3-SAT $\leq_P P_2$. So, P_2 is NP-C (by Lemma 1). Further, $P_1 \leq_P$ 3-SAT. As 3-SAT is NP-C and P_1 polynomially reduces to 3-SAT, we may assume that P_1 is NP.

Also, 3-SAT $\leq_P P_2$, where 3-SAT is NP-C but we do not know about P_2 (that is, it is not given as NP). So, it may be assumed that examining P_2 as NP is difficult. Hence, P_2 is NP-Hard.

EXAMPLE 7.5

Language $L_1 \leq_P L_2$, $L_3 \leq_P L_2 \leq_P L_4$. Which of the following is/are true?

(i) If $L_4 \in P$, then $L_2 \in P$.
(ii) If $L_1 \in P$ or $L_3 \in P$, then $L_2 \in P$.
(iii) $L_1 \in P$ if and only if $L_3 \in P$.
(iv) If $L_4 \in P$, then $L_1 \in P$ and $L_3 \in P$.
(A) Only (ii) (B) Only (ii) (C) Both (i) and (iv) (D) Only (i)

Solution (C) is the correct option. From $L_1 \leq_P L_2$ and $L_2 \leq_P L_4$, if L_4 is polynomial-class problem, then L_2 is also a polynomial-class problem. So, (i) is true. From $L_1 \leq_P L_2$ and $L_3 \leq_P L_2 \leq_P L_4$, if $L_4 \in P$, then $L_1 \in P$ and $L_3 \in P$. So, (iv) is also true.

EXAMPLE 7.6

Consider the following statements:

(i) The complement of every Turing decidable language is Turing decidable.
(ii) There exists some language that is in NP but is not Turing decidable.

(iii) If L is a language in NP, L is Turing decidable.
Which of the following is/are true?
(A) Only (ii) (B) Only (iii) (C) Both (i) and (ii) (D) Both (i) and (iii)

Solution (D) is the correct option. Any language (L) in NP can be recognised by a non-deterministic TM in polynomial time. This means that there is a TM that always halts for any string $w \in L$ or $w \notin L$. Therefore, any language in NP is Turing decidable. Further, the complement of any Turing-decidable language is also Turing decidable.

THEOREM 7.1

If $P_1 \leq_P P_2$, prove that P_2 is undecidable if P_1 is undecidable.

Proof Let us assume that P_2 is decidable. We know that a problem (P) is decidable if the answer for every input to P is either '*yes*' or '*no*'. However, if the answer for an input with respect to P remains pending, it is called undecidable.

Given that there exists $P_1 \leq_P P_2$, let w be an instance of P_1. As reduction: $P_1 \leq_P P_2$ exists, w can be converted polynomially to an instance (x) of P_2. As P_2 is assumed to be decidable, it answers '*yes*' for x if it is valid and '*no*' if it is invalid. Accordingly, we obtain the answer for w of P_1 as '*yes*' or '*no*'. This means that P_1 is decidable. Hence, a contradiction arises for P_1, as it is expected to be undecidable as per the given statement. Thus, P_2 is also undecidable.

Note: In this way, we can prove that B is also NP if $A \leq_P B$ and A is NP.

Longest path problem: Finding the longest path in a weighted graph is an NP-C problem. For this purpose, we may apply the dynamic programming (DP) approach to traverse all possible paths in the graph. Actually, DP performs a depth-first search for each path of the graph and it takes O(n!) running time in the worst case, where n is the number of vertices in the graph.

- *Applying reducibility property to longest path*: It is easy to conclude that the longest path (LP) is NP-C. In particular, this problem is a variant of the Hamiltonian path. So, it is NP and can be reduced as: *Hamiltonian path* $\leq_p$ *Longest path*. Hence, LP is NP-C.

DP approach: It is well-accepted that the DP approach helps us to solve an optimisation problem if any independent sub-structure can be formed to solve it, and the sub-solutions (obtained from sub-structures) can be assembled to obtain the final solution. Thus, it requires two essential features: construction of independent sub-structures and integration of the sub-solutions.

- *DP and longest path problem*: In the case of the longest path problem, both the independent sub-structure and integration of the sub-solutions on the basis of DP may not be able to solve the problem. So, the DP approach is not very useful here as compared to the brute-force method, which takes exponential time (O(n!) time in the worst case). That is why this problem is NP. However, the two features of DP mentioned above are acceptable for finding the shortest path. But if the negative edge values in the graph are allowed, then the shortest path problem also takes exponential time in the worst case.

7.4 INTRODUCTION TO RECURSION THEORY

- **Turing-computable function**: A function from numbers to numbers or from k-tuples of numbers (one tuple, that is, the value from each number) to numbers is Turing-computable if there exists a TM that computes it.

- **Turing-computable functions are μ-recursive (or primitive recursive) functions**: Primitive recursive functions are defined by three types of initial functions and two combining rules: (i) composition and (ii) recursion. The three initial functions are given below:
  - *Zero* (or initial) function, denoted as Z, is defined as Z(x) = 0.
  - *Successor* function, denoted as S, is defined as S(x) = x + 1.
  - *Projection* function, denoted as $\bigcup_1^n$, is defined as $\bigcup_i^n(x_1, x_2, x_k, \ldots x_n) = x_i$. (Here, the *i*-th element is the projected element, out of *n* elements).

The main purpose of these functions is to reuse computable functions with the help of recursion.

7.5 ADDITIONAL SOLVED PROBLEMS

PROBLEM 1

Explain the P vs NP problem.

Solution P is the class of language (problem) that can be solved in polynomial time, and NP is the class of language that can be verified in polynomial time. We know that P ⊆ NP, but it is presently not known if P = NP or P ≠ NP. It is widely believed that NP ≠ P. However, if NP = P, there would be no computationally secure cryptographic system, in the mathematical sense.

PROBLEM 2

Are there any known NP problems that are not P and NP-C?

Solution Two problems are given below:

- Graph isomorphism: Is graph G_1 isomorphic to graph G_2?
- Sub-graph isomorphism: Is graph G_1 isomorphic to any sub-graph of graph G_2?

The subgraph isomorphism problem is NP-C, as graph G_2 can use several combinations to form the sub-graph. On the other hand, the graph isomorphism problem is suspected to be neither in P nor NP-C, though it is in NP. So, this is an example of an NP problem that is thought to be hard but is not thought to be NP-C (relatively easy). In fact, the number of combinations in the case of graph isomorphism is expected to be less in comparison to the number of combinations in the case of sub-graph isomorphism.

Another example: If we can test a factorisation quickly (multiply the factors back together again), the problem is in NP. If finding the factorisation seems to be difficult, we may think that it may not be P. This is one of the (vary rare) examples of problems between N and NP-C. However, checking a perfect number beyond a given number N is NP-C (also treated as NP-Hard).

7.6 RESEARCH DIRECTIONS

As discussed earlier, NP-Hard and NP-C problems can be solved by algorithms having exponential or higher order time complexity. In many problems, the growing time complexity makes it practically infeasible to find a perfectly optimal solution. So, a trade-off is often necessary.

Consider the case of the Travelling Salesman Problem. The algorithm that generates the perfectly optimal solution has a time complexity of $O(n^2 2^n)$. Suppose the cost of an optimal tour by a travelling salesman for an *n*-vertex graph is C.

What if we can have an approximate algorithm that gives a solution whose cost is guaranteed to be between C and 2C, and a time complexity of $O(n^2 \log_2 n)$?

Table 7.4 helps us understand the impact of these numbers.

TABLE 7.4 Comparison of time complexities of optimal vs approximate solution

n	$n^2 2^n$	$n^2 \log_2 n$
2	16	4
32	4398046511104	5120
128	5.575 x 10^{42}	114688
1024	1.885 x 10^{314}	10485760
16384	3.19366 x 10^{4940}	3758096384

The higher the value of n, the more impractical the time complexity of the perfectly optimal algorithm. So, in such cases, it is acceptable if a faster algorithm gives a solution that is not worse than two times the optimal value. The time complexity benefit of obtaining a slightly sub-optimal solution is better than struggling for a long time to find the perfectly optimal solution. This is the concept behind **approximation algorithms**, a very challenging area of research. An approximation algorithm is called ***k*-approximate** if the solution generated by it is guaranteed to be in the range C to kC, where C is the value of the perfectly optimal solution. Needless to say, a 1-approximate algorithm is the optimal algorithm. Also, by definition, it is impossible to have $k < 1$.

Here is an interesting practical problem for the reader to think about.

Suppose a system (a function in some code) takes N number of parameters as its input. For the sake of simplicity, assume that each parameter has k number of possible values for testing. What would be the number of exhaustive test cases for such a system? The answer is k^n, which is an exponentially growing function. So, how many of these exponentially high number of test cases are likely to expose unique (non-duplicate) bugs in the system? Probably, not too many. In which case, is there any real value in having so many test cases for such a system? For a reasonably small system, for example, ($n = 5$, $k = 3$), ($n = 3$, $k = 2$), exhaustive testing would be acceptable. However, for larger values of n and k, it would be practically unfeasible to perform exhaustive testing.

To tackle this problem, there exists an approximation technique based on a statistical observation – "The majority of bugs are caused by the interaction of two parameters".

This gives rise to a very interesting technique, known as pairwise testing, which is a combinatorial method of software testing that, for each pair of input parameters to a system (typically, a software algorithm), tests all possible discrete combinations of those parameters.

Suppose we have written a function that takes three input parameters, x, y and z. For the sake of simplicity, let us assume that each parameter can take only two possible values, 0 or 1. This means that $n = 3$ and $k = 2$, as per the previous discussion. So, the number of exhaustive test cases $= k^n = 2^3 = 8$. Now, let us try to derive pairwise tests for this system.

Each test case is a 3-tuple. Let us generate 3-tuples using pairwise strategy. We need to ensure that we minimise the number of 3-tuples needed for this case, keeping in mind that the 3-tuples must cover all possible (x, y) pairs, (y, z) pairs and (z, x) pairs.

We start by generating all possible (x, y) pairs (Table 7.5).

TABLE 7.5 Generating all (x, y) pairs

x	y	z
0	0	
0	1	
1	0	
1	1	

Next, keeping the table undisturbed, let us try to generate all possible (*y*, *z*) pairs (Table 7.6).

TABLE 7.6 Generating all (y, z) pairs

x	*y*	*z*
0	**0**	**1**
0	**1**	**0**
1	**0**	**0**
1	**1**	**1**

It can be seen that this covers all four possible combinations: {(0, 0), (0, 1), (1, 0), (1, 1)} for (*x*, *y*) and (*y*, *z*). It should be observed that Table 7.6 has inadvertently also covered all possible (*z*, *x*) pairs.

Thus, we can see that four test cases are sufficient to cover all possible pairs in this case. This is straightforward **reduction by 50%** in the number of test cases.

Let us take another example. Suppose, the system takes four parameters, *a*, *b*, *c* and *d*, and each of these parameters can take only two possible values, 0 or 1. Here, $k = 2$ and $n = 4$, and hence, the number of exhaustive test cases is $k^n = 2^4 = 16$ (Table 7.7).

TABLE 7.7 Pairwise combinations for four parameters

a	*b*	*c*	*d*
1	1	1	1
0	0	0	0
1	0	1	0
0	0	1	1
0	1	0	0
1	0	0	1

We can verify that the six rows of Table 7.7 cover all the four possible combinations: {(0, 0), (0, 1), (1, 0), (1, 1)} for all possible pairs, (*a*, *b*), (*a*, *c*), (*a*, *d*), (*b*, *c*), (*b*, *d*) as well as (*c*, *d*). This is a **reduction by 62.5%**. Now, the question arises if this combination is really optimal. Do we need 6 tuples to achieve coverage of all the pairs? The answer is no.

Table 7.8 comprises only five rows, but they achieve the desired objective of coverage of all the pairs.

TABLE 7.8 A more optimal pairwise combination for four parameters

a	*b*	*c*	*d*
1	1	1	0
1	0	0	1
0	1	0	1
0	0	0	0
0	0	1	1

Again, we ask if five rows is the optimal solution. To verify it will take exponential time. So, pairwise testing is indeed an NP-Hard problem.

Try to come up with a 1.5-approximate or better algorithm to generate coverage of all the pairs. You can even attempt triplet-wise coverage, which covers all possible triplets.

EXERCISES

1. State which of the sentences given below are correct, false or unknown:
   (a) If a problem is in P, it must also be in NP.
   (b) If a problem is in NP, it must also be in P.
   (c) If a problem is NP-C, it must also be in NP.
   (d) If a problem is not in P, it must be NP-C.
2. (i) Prove that feature extraction from data set is NP-C.
   (ii) Finding the smallest common sub-sequence is a P class problem.
   (iii) Finding the longest common sub-sequence is NP-C.
3. Which of the following is the strongest correct statement about a finite language over some finite alphabet Σ?
   (a) It could be undecidable. (b) It is Turning machine recognisable.
   (c) It is a context-sensitive language. (d) It is a regular language.
   (e) None of these. **[GATE 1991]**
4. Which of the following statements is false?
   (a) The halting problem of TM is undecidable.
   (b) Determining whether a context-free grammar is ambiguous is undecidable.
   (c) Given two arbitrary context-free grammars G_1 and G_2, it is undecidable whether $L(G_1) = L(G_2)$.
   (d) Given two regular grammars G_1 and G_2, it is undecidable whether $L(G_1) = L(G_2)$.
   [GATE 1996]
5. Which one of the following is not decidable?
   (a) Given a TM, M, a stings *s* and an integer *k*, M accepts *s* within *k* steps
   (b) Equivalence of two given TMs
   (c) Language accepted by a given finite state machine is not empty
   (d) Language generated by a context-free grammar is non-empty **[GATE 1997]**
6. Consider the following decision problems:
   (P1) Does a given finite state machine accept a given string?
   (P2) Does a given context-free grammar generate an infinite number of stings?
   Which of the following statements is true?
   (a) Both (P1) and (P2) are decidable (b) Neither (P1) nor (P2) are decidable
   (c) Only (P1) is decidable (d) Only (P2) is decidable **[GATE 2000]**
7. Consider the following problem, X. Given a Turing machine, M, over the input alphabet Σ, any state q of M and a word w ∈ Σ*, does the computation of M on w visit the state q? Which of the following statements about X is correct?
   (a) X is decidable (b) X is undecidable but partially decidable
   (c) X is undecidable and not even partially decidable (d) X is not a decision problem
   [GATE 2001, Maharashtra SET 2016]
8. Ram and Shyam have been asked to show that a certain problem Π is NP-C. Ram shows a polynomial time reduction from the 3-SAT problem to Π, and Shyam shows a polynomial time reduction from Π to 3-SAT. Which of the following can be inferred from these reductions?

(a) Π is NP-Hard but not NP-C (b) Π is in NP, but is not NP-C
(c) Π is NP-C (d) Π is neither NP-Hard, nor in NP **[GATE 2003]**

9. Nobody knows yet if P = NP. Consider the language L defined as:

$$L = \begin{cases} (0+1)^* & \text{if } P = NP \\ j, & \text{otherwise} \end{cases}$$

Which of the following statements is true?]
(a) L is recursive (b) L is recursively enumerable but not recursive
(c) L is not recursively enumerable
(d) Whether L is recursive or not will be known after we find out if P = NP **[GATE 2003]**

10. Consider two languages L_1 and L_2, each on the alphabet Σ. Let f: Σ → Σ be a polynomial time computable bijection such that (∀x)[x ∈ L_1 iff f(x) ∈ L_2]. Further, let f^{-1} also be polynomial time computable. Which of the following cannot be true?
(a) L_1 ∈ P and L_2 is finite (b) L_1 ∈ NP and L_2 ∈ P
(c) L_1 is undecidable and L_2 is decidable (d) L_1 is recursively enumerable and L_2 is recursive **[GATE 2003]**

11. The problems 3-SAT and 2-SAT are
(a) Both in P (b) Both NP-C
(c) NP-C and in P, respectively (d) Undecidable and NP-C, respectively **[GATE 2004]**

12. Consider three problems P1, P2 and P3. It is known that P1 is decidable and P2 is undecidable. Which one of the following is true?
(a) P3 is decidable if P1 is reducible to P3.
(b) P3 is undecidable if P3 is reducible to P2.
(c) P3 is undecidable if P2 is reducible to P3.
(d) P3 is decidable if P3 is reducible to P2's complement
[GATE 2005, Maharashtra SET 2015]

13. Let S be an NP-C problem and Q and R be two other problems not known to be in NP. Q is polynomial-time reducible to S and S is polynomial-time reducible to R. Which one of the following statements is true?
(a) R is NP-C (b) R is NP-Hard
(c) Q is NP-C (d) Q is NP-Hard **[GATE 2006, APSET 2013]**

14. Which of the following problems is undecidable?
(a) Membership problem for CFGs (b) Ambiguity problem for CFGs
(c) Finiteness problem for FSAs (d) Equivalence problem for FSAs **[GATE 2007]**

15. Which of the following are decidable?
I. Whether the intersection of two regular languages is infinite
II. Whether a given context-free language is regular
III. Whether two pushdown automata accept the same language
IV. Whether a given grammar is context-free
(a) I and II (b) I and IV (c) II and III (d) II and IV **[GATE 2008]**

16. Let πA be a problem that belongs to the class NP. Which one of the following is true?

(a) There is no polynomial time algorithm for πA.

(b) If πA can be solved deterministically in polynomial time, then P = NP.

(c) If πA is NP-Hard, then it is NP-C.

(d) πA may be undecidable. **[GATE 2009, Pondicherry University PhD Entrance 2015]**

17. Which of the following problems are decidable?

(1) Does a given program ever produce an output?

(2) If L is context-free language, then, is L′ also context-free?

(3) If L is regular language, then, is L′ also regular?

(4) If L is recursive language, then, is L′ also recursive?

(a) 1,2,3,4 (b) 1,2 (c) 2,3,4 (d) 3, 4 **[GATE 2012]**

18. Which of the following is/are undecidable?

1. G is a CFG. Is L(G) = Φ?

2. G is a CFG. IS L(G) = Σ*?

3. M is a Turning machine. Is L(M) regular?

4. A is a DFA and N is a NFA. Is L(A) = L(N)?

(a) 3 only (b) 3 and 4 only (c) 1, 2 and 3 only (d) 2 and 3 only **[GATE 2013]**

19. Let A $\leq_m$ B denote that language A is mapping-reducible (or many-to-one reducible) to language B. Which one of the following is false?

(a) If A $\leq_m$ B and B is recursive then A is recursive.

(b) If A $\leq_m$ B and A is undecidable then B is undecidable.

(c) If A $\leq_m$ B and B is recursively enumerable then A is recursively enumerable.

(d) If A $\leq_m$ B and B is not recursively enumerable then A is not recursively enumerable.

[GATE 2014]

20. Let < M > be the encoding of a Turing machine as a string over Σ = {0, 1}. Let L = {<M> | M is a Turning machine that accepts a string of length 2014}. Then, L is

(a) Decidable and recursively enumerable (b) Undecidable but recursively enumerable

(c) Undecidable and not recursively enumerable (d) Decidable but not recursively enumerable

[GATE 2014]

21. Which one of the following problems is undecidable?

(a) Deciding if a given context-free grammar is ambiguous

(b) Deciding if a given string is generated by a given context-free grammar

(c) Deciding if the language generated by a given context-free grammar is empty

(d) Deciding if the language generated by a given context-free grammar is finite **[GATE 2014]**

22. Consider two decision problems Q1 and Q2 such that Q1 reduces in polynomial time to 3-SAT and 3-SAT reduces in polynomial time to Q2. Which one of the following is consistent with the above statement?

(a) Q1 is in NP, Q2 in NP-Hard (b) Q2 is in NP, Q1 is NP-Hard

(c) Both Q1 and Q2 are in NP (d) Both Q1 and Q2 are NP-Hard **[GATE 2015]**

23. Consider the following statements:

I. The complement of every Turing decidable language is Turing decidable

II. There exists some language which is in NP but is not Turing decidable

III. If L is a language in NP, L is Turing decidable

Which of the above statements is/are true?

(a) Only II (b) Only III (c) Only I and II (d) Only I and III **[GATE 2015]**

24. Language L1 is polynomial time reducible to language L2. Language L3 is polynomial time reducible to L2, which in turn is polynomial time reducible to language L4. Which of the following is/are true?

I. If L4 $\in$ P, then L2 $\in$ P II. If L1 $\in$ P or L3 $\in$ P, then L2 $\in$ P

III. L1 $\in$ P, if and only if L3 $\in$ P IV. If L4 $\in$ P, then L1 $\in$ P and L3 $\in$ P

(a) II only (b) III only (c) I and IV only (d) I only **[GATE 2015]**

25. Which of the following decision problems are undecidable?

Given NFAs N_1 and N_2, is $L(N_1) \cap L(N_2) = \phi$?

Given a CFG, $G = (N, \Sigma, P, S)$ and a string $x \in \Sigma^*$, does $x \in L(G)$?

Given CFGs G_1 and G_2, is $L(G1) = L(G_2)$?

Given a TM M, is $L(M) = \phi$?

(a) I and IV only (b) II and III only

(c) III and IV only (d) II and IV only **[GATE 2016]**

26. Let L(R) be the language represented by regular expression R. Let L(G) be the language generated by a context-free grammar, G. Let L (M) be the language accepted by a TM, M. Which of the following decision problem are undecidable?

I. Given a regular expression R and a string w, is $w \in L(R)$?

II. Given a context-free grammar G, is $L(G) = \phi$?

III. Given a context-free grammar G, is $L(G) = \Sigma^*$ for some alphabet Σ?

IV. Given a Turning machine M and a string w, is $w \in L(M)$?

(a) I and IV only (b) II and III only (c) II, III and IV only (d) III and IV only

[GATE 2017]

27. Assume statements S1 and S2. S1: Given a context-free grammar G, there exists an algorithm for determining whether L(G) is infinite. S2: There exists an algorithm to determine whether two context-free grammars generate the same language. Which of the following is true?

(a) S1 is correct and S2 is not correct. (b) Both S1 and S2 are correct.

(c) Both S1 and S2 are not correct. (d) S1 is not correct and S2 is correct. **[UGC NET 2013]**

28. Assume statements S1 and S2. S1: L2 – L1 is recursive enumerable where L1 and L2 are recursive and recursive enumerable respectively. S2: The set of all Turing machines is countable. Which of the following is true?

(a) S1 is correct and S2 is not correct. (b) Both S1 and S2 are correct.

(c) Both S1 and S2 are not correct. (d) S1 is not correct and S2 is correct. **[UGC NET 2013]**

29. Given the following statements: S1: If L1 and L2 are recursively enumerable languages over Σ, then $L1 \cup L2$ and $L1 \cap L2$ are also recursively enumerable. S2: The set of recursively enumerable languages is countable. Which of the following is correct?

(a) S1 is correct and S2 is not correct (b) S1 is not correct and S2 is correct

(c) Both S1 and S2 are not correct (d) Both S1 and S2 are correct **[UGC NET 2015]**

30. A problem is said to be NP-C:
(a) If it is as 'hard' as any other problem in NP
(b) If a non-polynomial time algorithm has been discovered
(c) If a polynomial time algorithm exists but needs a parallel computer
(d) There is a greedy solution to the problem **[HP SET 2014]**

31. Assuming P $\neq$ NP which of the following is true?
(a) NP-C = NP (b) NP-C $\cap$ P = ϕ (c) NP-Hard = NP (d) P = NP-C
[HCU PhD Entrance 2015, HP IT Lecturer Recruitment 2016]

32. If there is an NP-C language L whose complement is in NP, the complement of any language in NP is in
(a) P (b) NP (c) Both (a) and (b) (d) None of these
[Rajasthan CS Lecturer Recruitment 2014, Karnataka SET 2015]

33. If L $\in$ P, then
(a) L′ $\in$ P (b) L′ $\in$ NP (c) L′ $\in$ P space (d) None of these **[Karnataka SET 2015]**

34. Let P be the set of problems solvable in deterministic polynomial time, NP be the set of problems solvable in non-deterministic polynomial time, NPC be the set of NP complete problems. Given X = P $\cap$ NP, Y = NP – P, Z = P $\cap$ NPC, which of the following is true?
(a) X = NP, Y $\neq$ ϕ, Z = ϕ (b) X = P, Y $\supseteq$ NPC, Z $\neq$ ϕ
(c) X = P, Y $\supseteq$ NPC, Z = ϕ (d) X = NP, Y $\supseteq$ NPC, Z $\neq$ ϕ **[Maharashtra SET 2013]**

35. The solution of NP-C problem:
(a) Never solved (b) Is unknown
(d) Takes polynomial time (d) Takes non-polynomial time **[Rajasthan SET 2012]**

36. Suppose L_1 and L_2 are two problems such that L_1 converts to L_2 in polynomial time. Then, which of the following is correct?
(a) L_2 $\in$ P implies L_1 $\in$ P (b) L_1 $\in$ P implies L_2 $\in$ P
(c) Both L_1 and L_2 belong to NP (d) None of these **[Rajasthan SET 2013]**

37. In the context of problem classes, which of the following statements is not valid?
(a) NP-C $\subseteq$ NP (b) NP-Hard $\subseteq$ NP **[Rajasthan CS Lecturer Recruitment 2014]**
(c) P $\subseteq$ NP (d) If L $\in$ NP and L′ $\in$ NP, then L $\in$ co-NP

38. Let L1 be an NP-C problem, another problem L2 will also be an NP-C problem if:
(a) L2 $\in$ NP and L1 is reducible to L2 in polynomial time
(b) L2 $\notin$ NP and L1 is reducible to L2 in polynomial time
(c) L2 $\in$ NP and L2 is reducible to L1 in polynomial time
(d) L2 $\notin$ NP and L2 is reducible to L1 in polynomial time
[Rajasthan CS Lecturer Recruitment 2014]

39. A problem is NP-Complete if:
(a) Answer can be verified quickly
(b) A quick algorithm to solve this problem can be used to solve all other NP problems quickly
(c) It can be solved in polynomial time
(d) All of these **[Rajasthan CS Lecturer Recruitment 2014]**

40. Consider the following statements.
    (I) NP-C problems are those that we know we can never solve efficiently.
    (II) If we find an efficient algorithm for one NP-C problem, then we can solve all NP-C problems efficiently.
    (III) Checking whether a number is a prime is an NP-C problem.
    (a) (I) and (II) are true but (III) is false. (b) (I) and (III) are false but (II) is true.
    (c) (II) and (III) are true but (I) is false. (d) All three statements are false.
    [CMI PhD Entrance 2010]
41. Alan's task is to design an algorithm for a decision problem P. He knows that there is an algorithm A that transforms instances of P to instances of the halting problem such that *yes* instances of P map to *yes* instances of the halting problem, and *no* instances of P map to *no* instances of the halting problem. Which of the following is true?
    (a) The existence of A implies the existence of an algorithm for P.
    (b) The existence of A implies that there is no algorithm for P.
    (c) The existence of A says nothing about whether there is an algorithm for P.
    (d) The halting problem can be solved using A. **[CMI PhD Entrance 2014]**
42. Suppose we have constructed a polynomial time reduction from problem A to problem B. Which of the following can we infer from this fact?
    (a) If the best algorithm for B takes exponential time, there is no polynomial time algorithm for A.
    (b) If the best algorithm for A takes exponential time, there is no polynomial time algorithm for B.
    (c) If we have a polynomial time algorithm for A, we must also have a polynomial time algorithm for B.
    (d) If we do not know whether there is a polynomial time algorithm for B, there cannot be a polynomial time algorithm for A. **[CMI PhD Entrance 2015]**
43. Which of the following is not true?
    (a) Class of all languages is not countable. (b) Every language in P is also in NP.
    (c) Every language in NP is decidable. (d) There are some languages in NP but not in P.
    [JNU M.Phil/PhD Entrance 2016]
44. Which of the following problems is decidable?
    (a) Given a CFG G, find whether L(G) = R where R is a regular set
    (b) Given a CFG G, find whether L(G) = {}
    (c) Find whether the intersection of two CFLs is empty
    (d) Find whether the complement of a CFL is a CFL
    (e) Find whether two CFGs generate the same language **[TIFR PhD Entrance 2010]**
45. Let $\Sigma1 = \{a\}$ be a one-letter alphabet and $\Sigma2 = \{a, b\}$ be a two-letter alphabet. A language over an alphabet is a set of finite length words comprising letters of the alphabet. Let L1 and L2 be the set of languages over $\Sigma1$ and $\Sigma2$, respectively. Which of the following is true about L1 and L2 ?
    (a) Both are finite. (b) Both are countably infinite.
    (c) L1 is countable but L2 is not. (d) L2 is countable but L1 is not.
    (e) Neither of them is countable. **TIFR PhD Entrance 2015]**

46. Consider the following language: PRIMES = {A string of p number of 1s | p is prime}. Then which of the following is true?
(a) PRIMES is regular.
(b) PRIMES is undecidable.
(c) PRIMES is decidable in polynomial time.
(d) PRIMES is context-free but not regular.
(e) PRIMES is NP-complete and P $\neq$ NP **[TIFR PhD Entrance 2016]**
47. The correct way to show that a new problem is NP-C is
(a) Reduce a known NP-C problem to the new problem in P time.
(b) Reduce the new problem to a known NP-C problem in P time.
(c) Show that there exists an algorithm for the problem which is equivalent to a known NP-C problem.
(d) None of the above. **[HCU PhD Entrance 2013]**
48. Which of the following problems is solvable?
(a) Determining if a universal Turing machine and some input will halt
(b) Writing a universal Turing machine
(c) Determining if a universal Turing machine can be written in fewer than k instructions for some k
(d) Determining if an arbitrary Turing machine is a universal Turing machine?
[HCU PhD Entrance 2014]
49. Let L be a language over Σ; L $\subseteq$ Σ^*. Suppose L satisfies the two conditions: (i) L is in NP. (ii) For every n, there is exactly one string of length n that belongs to L. Let L^c be the complement of L over Σ^*. Show that L^c is also in NP. **[GATE 1995]**
50. Let a decision problem X be defined as follows: X: Given a Turing machine M over Σ and any word w $\in$ Σ, does M loop forever on w? You may assume that the halting problem of Turing machine is undecidable but partially decidable. **[GATE 2001]**
(a) Show that X is undecidable.
(b) Show that X is not even partially decidable.
51. Briefly discuss the concept of NP-completeness. How can we show that a problem is NP-C?
[Gujarat SET 2002, 2008]
52. Describe the halting problem and prove that it is undecidable. **[Gujarat SET 2003]**

Appendix: Competitive Examination Questions and Answers

Theory of Computation is a very important topic for competitive examinations on Computer Science, such as GATE, UGC NET, State-level Lectureship Recruitment Examinations and PhD/MPhil/MTech Entrance examinations of reputed universities.

Despite being a high scoring area, many candidates struggle with **objective-type problems** because of the severe time constraints of these examinations, coupled with the risk of negative marking. Typically, candidates have time ranging from about 30 seconds to a maximum of 2 minutes to solve one objective-type problem; while the negative marking can range from **–25% to –33%** of the score for the problem.

Although uncommon, some examinations do have objective-type questions that have **one or more correct alternatives**. It requires a lot of practice and mastery of handy shortcuts to quickly and accurately solve these problems.

We have put together a collection of 175 objective-type problems that have been asked in several reputed competitive examinations over the past few years. While it would be ideal to have a **one size fits all** set of techniques for solving such a vast set of problems, we have attempted here to provide hints/tips/solutions to all problems. Wherever feasible, a detailed explanation along with shortcut techniques has also been provided.

KEY POINTS FOR SUCCEEDING IN COMPETITIVE EXAMINATIONS

- Practice as many problems as you can from previous years' question papers and from as many sources as possible.
- Stick strictly to the time limit – try to complete the paper well within the prescribed time limit. For example, if your targeted examination provides an average time of 1 minute per question, then develop techniques to solve each problem in 45 seconds or less, while working out previous years' question papers.
- Try to eliminate as many incorrect options as possible in 15 seconds or less.
- Do not try to exhaustively solve the problem – use shortcuts if feasible. This is especially true for problems in the areas of DFA, NFA and regular expressions.
- Never try to guess the answer if there is any negative marking; it can cost you in terms of your final rank/result/marks. If there is no negative marking, use guesswork ONLY for the problem for which you are unable to work out the solution.
- Many reputed examinations use machine-gradable OMR sheets for evaluation; so familiarise yourself with this. A sample OMR sheet has been appended after the Hints and Solutions. You can make copies of it and use it for practice. Insufficient practice with OMR sheets can result in the consumption of a significant amount of time in the actual exam, and create unnecessary panic in candidates.

OBJECTIVE-TYPE PROBLEMS

1. Let S be an infinite set and $S_1, S_2, \ldots, S_n$ be sets such that $S_1 \cup S_2 \cup \ldots \cup S_n = S$. Then,
   (a) At least one of the sets S_i is a finite set
   (b) Not more than one of the sets S_i can be finite
   (c) At least one of the sets S_i is an infinite set
   (d) Not more than one of the sets S_i can be infinite **[GATE 1993]**
2. The less-than relation, $<$, on reals is
   (a) A partial ordering since it is asymmetric and reflexive
   (b) A partial ordering since it is antisymmetric and reflexive
   (c) Not a partial ordering because it is not asymmetric and not reflexive
   (d) Not a partial ordering because it is not antisymmetric and reflexive **[GATE 1993]**
3. The number of subsets of $\{1, 2, \ldots, n\}$ with odd cardinality is ________. **[GATE 1994]**
4. Two finite sets have m and n elements. The total number of subsets of the first set is 56 more than that of the number of subsets of the second. The values of m and n are:
   (a) 7, 6 (b) 6, 3 (c) 5, 1 (d) 8, 7 **[Karnataka SET 2015]**
5. If X and Y are sets, then the set (X – Y) union (Y – X) union (X intersection Y) is equal to
   (a) X union Y (b) X′ union Y′ (c) X intersection Y (d) X′ intersection Y′
   [TN SET 2012, Pondicherry University PhD Entrance 2013]
6. Which of the following sets can be recognised by a DFA?
   (a) The numbers 1, 2, 4, 8, …, 2^n, … written in binary
   (b) The numbers 1, 2, 4, 8, …, 2^n, … written in unary
   (c) The set of binary strings in which the number of zeros is the same as the number of ones.
   (d) The set {1, 101, 11011, 1110111, …} **[GATE 1998]**
7. Consider the regular expression (0 + 1) (0 + 1)…. n times. The minimum state finite automation that recognises the language represented by this regular expression contains
   (a) n states (b) $n + 1$ states (c) $n + 2$ states (d) None of the above
   [GATE 1999, 2010, Punjab Tech University PhD Entrance 2010, UGC NET 2012]
8. Consider the following finite state automaton:

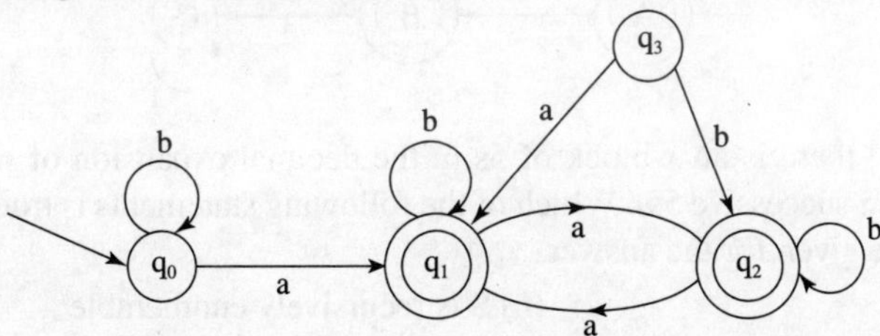

Part 1 The language accepted by this automaton is given by the regular expression
(a) b*ab*ab*ab* (b) (a+b)* (c) b*a(a+b)* (d) b*ab*ab*

Part 2 The minimum state automaton equivalent to the above FSA has the following number of states:
(a) 1 (b) 2 (c) 3 (d) 4 **[GATE 2007, Maharashtra SET 2015]**

9. A language represented by a non-deterministic finite state automaton is a:
(a) Context-free language (b) Context-sensitive language
(c) Regular language (d) Natural language
[Gujarat SET 2011, UGC NET 2017]

10. The minimum number of states required to represent L = $\{a^n b \mid n \geq 0\}$ in deterministic finite automaton is:
(a) 2 (b) 3 (c) 4 (d) 5 **[HP SET 2013]**

11. The behaviour of NFA can be simulated by DFA:
(a) Always (b) Sometimes (c) Never (d) Depends on NFA
[Punjab Tech University PhD Entrance 2010, West Bengal SET 2014]

12. Two finite state machines are said to be equivalent if they:
(a) Have the same number of edges (b) Have the same number of states
(c) Recognise the same set of strings (d) Have the same number of states and edges
[UGC NET 2018]

13. Which regular expression is equivalent to the following NFA?

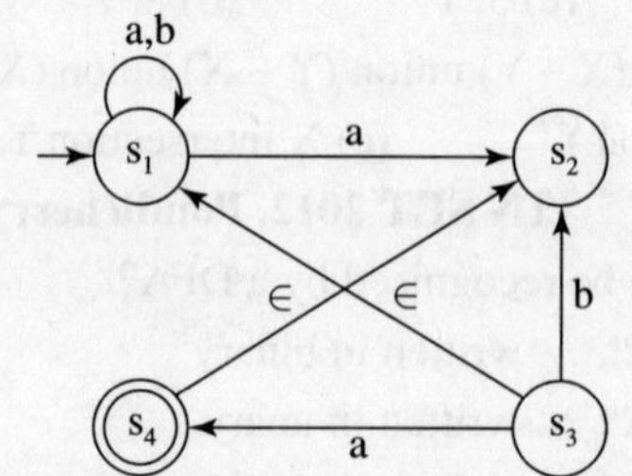

(a) (a +b)*aba (b) aba(a + b)*aba (c) (a + b)aba(b + a)* (d) aba(a + b)*
[TIFR PhD Entrance 2018]

14. The regular expression for the language recognised by the following finite automaton is ______. **[GATE 1994]**

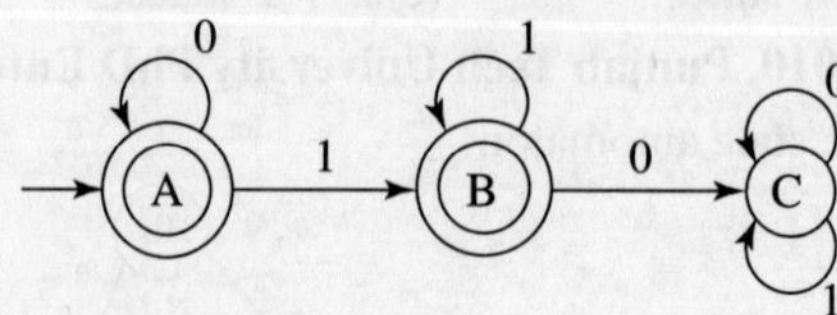

15. Given a set S {x | there is an *x*-block of 5s in the decimal expansion of π}. (Note: *x*-block is a maximal block of *x* successive 5s). Which of the following statements is true with respect to S? No reasons need to be given for the answer.
(a) S is regular (b) S is recursively enumerable
(c) S is not recursively enumerable (d) S is recursive **[GATE 1994]**

16. Which two of the following four regular expressions are equivalent? (ε is the empty string):
(i) (00)*(0 + ε) (ii) (00)* (iii) 0* (iv) 0(00)*
(a) (i) and (ii) (b) (ii) and (iii) (c) (i) and (iii) (d) (i) and (iv)
[GATE 1996, HP CS Lecturer Recruitment 2018]

17. If the regular set A is represented by A = (01 + 1)* and the regular set B is represented by B = ((01)*1*)*, which of the following is true?

(a) A ⊂ B (b) B ⊂ A (c) A and B are comparable (d) A = B

[GATE 1998, HCU PhD Entrance 2010]

18. Consider the following two statements: S1: $\{0^{2n} \mid n \geq 1\}$ is a regular language. S2: $\{0^m 1^n 0^{m+n} \mid m \geq 1 \text{ and } n \geq 1\}$ is a regular language. Which of the following statements is correct?
(a) Only S1 is correct (b) Only S2 is correct
(c) Both S1 and S2 are correct (d) Neither S1 nor S2 is correct **[GATE 2001]**

19. The regular expression 0*(10*)* denotes the same set as
(a) (1*0)*1* (b) $0^+(0^+10)^*$ (c) (0+1)*10(0+1)* (d) None of the above **[GATE 2003]**

20. Which of the following are regular sets?
I. $\{a^n b^{2m} \mid n \geq 0, m \geq 0$ II. $\{a^n b^m \mid n = 2m\}$ III. $\{a^n b^m \mid n \neq m\}$ IV. {xcy | x, y ∈ {a,b}*}
(a) I and IV only (b) I and III only (c) I only (d) IV only

[GATE 2008, Pondicherry University PhD Entrance 2016]

21. Let L = {w ∈ (0 + 1)* | w has even number of 1s}, that is, L is the set of all bit strings with even number of 1s. Which one of the regular expressions below represents L?
(a) (0*10*1)* (b) 0*(10*10*)* (c) 0*(10*1*)*0* (d) 0*1(10*1)*10* **[GATE 2010]**

22. Which of the following languages is regular?
(a) $L = \{a^n b^n \mid n \geq 0\}$ (b) $L = \{a^n \mid n \text{ is prime}\}$
(c) L = {w | w has (3k + 1) number of *b*s for some k ∈ N with Σ = {a, b}}
(d) L = {ww | w ∈ Σ* with Σ = {0, 1}} **[GATE 2014]**

23. The length of the shortest string NOT in the language (over Σ = {a, b}) of the regular expression a*b*(ba)*a* is __________. **[GATE 2014]**

24. Let Σ be a finite non-empty alphabet and let 2^{Σ^*} be the power set of Σ*. Which one of the following is true?
(a) Both 2^{Σ^*} and Σ* are countable (b) 2^{Σ^*} is countable and Σ* is uncountable
(c) 2^{Σ^*} is uncountable and Σ* is countable (d) Both 2^{Σ^*} and Σ* are uncountable **[GATE 2014]**

25. The regular expression for the language L = {w ∈ {0, 1}* | w has no pair of consecutive zeros} is:
(a) (1 + 010)* (b) (01 + 10)* (c) (1 + 010)* (0 + λ) (d) (1 + 01)* (0 + λ)

[UGC NET 2013]

26. Given the following statements, S1: If L is a regular language then the language $\{uv \mid u \in L, v \in L^R\}$ is also regular, S2 : $L = \{ww^R\}$ is regular language. Which of the following is true?
(a) S1 is not correct and S2 is not correct. (b) S1 is not correct and S2 is correct.
(c) S1 is correct and S2 is not correct. (d) S1 is correct and S2 is correct.

[UGC NET 2013, Pondicherry University PhD Entrance 2016]

27. Let L be any language. Define even(W) as the strings obtained by extracting from W the letters in the even-numbered positions and even(L) = {even (W) | W ∈ L}. We define another language Chop(L) by removing the two leftmost symbols of every string in L given by Chop(L) = {W | v W ∈ L, with |v| = 2}. If L is a regular language, then
(a) even(L) is regular and Chop(L) is not regular (b) Both even(L) and Chop(L) are regular
(c) even(L) is not regular and Chop(L) is regular (d) Both even(L) and Chop(L) are not regular

[UGC NET 2014]

28. The number of strings of length 4 that are generated by the regular expression $(0 \mid \varepsilon)1^{+}2^{*}(3 \mid \varepsilon)$, where | is an alternation character, {+, *} are quantification characters, and ε is the null string, is:
(a) 8 (b) 10 (c) 11 (d) 12 **[UGC NET 2016]**
29. The number of strings of length n over the alphabet {0, 1} is
(a) $n!$ (b) n^2 (c) n^n (d) 2^n **[APSET 2012]**
30. For which languages given below is $L^{+} = L^{*}$?
(i) Null language (ii) Empty language (iii) L = a (iv) L = aa*
(a) (i) and (ii) (b) (i) and (iii) (c) (i), (ii) and (iii) (d) All of the above
[Maharashtra SET 2013]
31. Which regular expression defined over the alphabet {a, b} recognises the language "All strings containing aba as a substring"?
(a) a*b*a (b) a*ba* (c) (ab)*a (d) (a + b)*aba(a + b)* **[Rajasthan SET 2013]**
32. The language recognised by the finite automaton is:

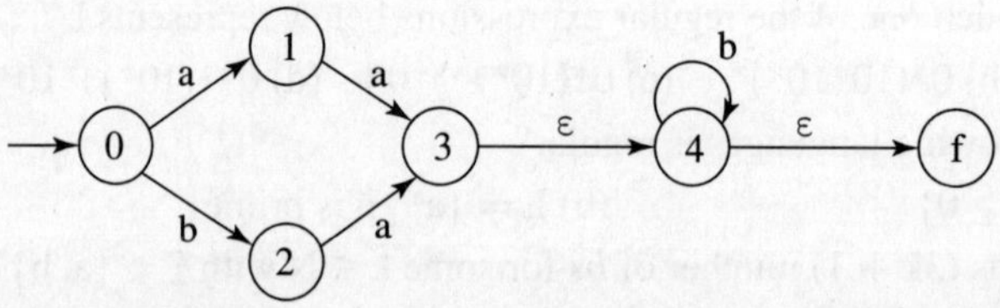

(a) aab* + bab* (b) aaba + b* (c) aa + ba + b* (d) All of these **[Rajasthan SET 2013]**
33. The set of all strings over {a, b} having exactly three *b*s is represented by the regular expression:
(a) a*bbb (b) a*ba*ba*ba* (c) a*ba*ba*b (d) ba*ba*b
[Rajasthan CS Lecturer Recruitment 2014]
34. Which of the following is NOT generated by the regular expression R = (ab|abb)*bbab?
(a) abbbab (b) ababbabbbab (c) ababab (d) ababbbbab
[Punjab Tech University PhD Entrance 2010, Uttarakhand SET 2012]
35. Let $L \subseteq \{0, 1\}^*$. Which of the following is true?
(a) If L is regular, all subsets of L are regular
(b) If all proper subsets of L are regular, then L is regular
(c) If all finite subsets of L are regular, then L is regular
(d) If a proper subset of L is not regular, then L is not regular **[CMI PhD Entrance 2012]**
36. Given $L_1 = \{a^n \mid n \text{ is not prime}\}$ and $L_2 = \{a^n \mid n \text{ is prime}\}$, then
(a) L_1 and L_2 are not regular (b) L_1 is regular and L_2 is not regular
(c) L_1 is not regular and L_2 is regular (d) L_1 and L_2 are regular
[JNU MPhil/PhD Entrance 2010]
37. Consider the following three statements:
(i) Intersection of infinitely many regular languages must be regular.
(ii) Every subset of a regular language is regular.
(iii) If L is regular and M is not regular then L . M is necessarily not regular.
Which of the following gives the correct true/false evaluation of the above?

(a) True, False, True (b) False, False, True
(c) True, False, False (d) True, True, True **[TIFR PhD Entrance 2014]**

38. Let a, b, c be regular expressions. Which of the following identities is correct?
(a) (a + b)* = a*b* (b) a(b + c) = ab + c
(c) (a + b)* = a* + b* (d) (ab + a)*a = a(ba + a)* **[TIFR PhD Entrance 2015]**

39. Which of the following regular expressions correctly accepts the set of all 0/1-strings with an even (possibly zero) number of 1s?
(a) (10*10*)* (b) (0*10*1)* (c) 0*1(10*1)*10*
(d) 0*1(0*10*10*)*10* (e) (0*10*1)*0* **[TIFR PhD Entrance 2017]**

40. A regular expression that denotes all the words with at least two 0s is
(a) (0+1)*0(0+1)*0 (b) 0(0+1)*0(0+1)*
(c) (0+1)*00(0+1)* (d) (0+1)*0(0+1)*0(0+1)* **[HCU PhD Entrance 2009]**

41. A regular expression starting with 0s or 1s and ending with '01' is equivalent to:
(a) (0 + 1)* 11 (b) (0 + 1)* 00 (c) (0 + 1)* 01 (d) (0 + 1)* 10
[Maharashtra SET 2018]

42. If G is a context-free grammar and *w* is a string of length *l* in L(G), how long is a derivation of *w* in G, if G is in Chomsky Normal Form?
(a) 2*l* (b) 2*l* + 1 (c) 2*l* – 1 (d) *l* **[GATE 1992, GATE 1994]**

43. Define a context-free language L ⊆ {0, 1}* and init(L) = {set of all prefixes of L}. If L = {w | w is non-empty and w has equal number of 0s and 1s}, then init(L) is a
(a) Set of all binary strings with unequal number of 0s and 1s
(b) Set of all binary strings including the null string
(c) Set of all binary strings with exactly one more 0 than the number of 1s OR one more 1 than the number of 0s
(d) None of these **[GATE 1996]**

44. Which of the following statements is true?
(a) If a language is context-free it can always be accepted by a deterministic pushdown automaton
(b) The union of two context-free languages is context-free
(c) The intersection of two context-free languages is context-free
(d) The complement of a context-free language is context-free
[GATE 2001, Maharashtra SET 2016]

45. S → aSa | bSb | a | b. The language generated by this grammar over the alphabet {a,b} is the set of
(a) All palindromes (b) All odd-length palindromes.
(c) Strings that begin and end with the same symbol (d) All even-length palindromes
[GATE 2009]

46. Let L1 be a recursive language. Let L2 and L3 be languages that are recursively enumerable but not recursive. Which of the following statements is not necessarily true?
(a) L2 – L1 is recursively enumerable (b) L1 – L3 is recursively enumerable
(c) L2 ∩ L1 is recursively enumerable (d) L2 ∪ L1 is recursively enumerable **[GATE 2010]**

47. Consider the DFA given below.

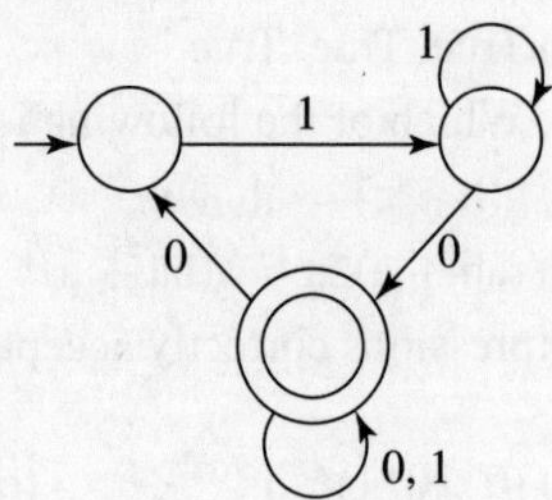

Which of the following are false?

1. Complement of L(A) is context-free 2. L(A) = ((11*0+0)(0+1)*0*1*)

3. For the language accepted by A, A is the minimal DFA

4. A accepts all strings over {0, 1} of length at least 2

(a) 1 and 3 only (b) 2 and 4 only (c) 2 and 3 only (d) 3 and 4 only **[GATE 2013]**

48. Consider the following languages: L1 = $\{0^p1^q0^r \mid p, q, r \geq 0\}$ and L2 = $\{0^p1^q0^r \mid p, q, r \geq 0, p \neq r\}$. Which one of the following statements is false?

(a) L2 is context-free (b) L1 ∩ L2 is context-free

(c) Complement of L2 is recursive (d) Complement of L1 is context-free but not regular

[GATE 2013]

49. Let X be a recursive language and Y be a recursively enumerable but not recursive language. Let W and Z be two languages such that Y′ reduces to W, and Z reduces to X′ (reduction here means the standard many-one reduction). Which one of the following statements is true?

(a) W can be recursively enumerable and Z is recursive

(b) W can be recursive and Z is recursively enumerable

(c) W is not recursively enumerable and Z is recursive

(d) W is not recursively enumerable and Z is not recursive **[GATE 2016]**

50. Language L_1 is defined by the grammar: $S_1 \to aS_1b \mid \varepsilon$. Language L_2 is defined by the grammar: $S_2 \to abS_2 \mid \varepsilon$. Consider the following statements: P: L_1 is regular. Q: L_2 is regular. Which one of the following is true?

(a) Both P and Q are true (b) P is true and Q is false

(c) P is false and Q is true (d) Both P and Q are false **[GATE 2016]**

51. Consider the grammar S → SaS | aSb | bSa | SS | ε, where S is the start variable. Which one of the following is not generated by G?

(a) abab (b) aaab (c) abbaa (d) babba **[GATE 2017]**

52. Consider the following languages over the alphabet Σ = {a, b, c}. Let $L_1 = \{a^nb^nc^m \mid m, n \geq 0\}$ and $L_2 = \{a^mb^nc^n \mid m, n \geq 0\}$. Which of the following are context-free languages?

I. $L_1 \cup L_2$ II. $L_1 \cap L_2$

(a) I only (b) II only (c) I and II (d) Neither I nor II **[GATE 2017]**

53. Which of the following regular expressions corresponds to this grammar:

S → AB | AS, A → a | aA, B → b?

(a) aa*b⁺ (b) aa*b (c) (ab)* (d) a(ab)* **[UGC NET 2006]**

54. Which one of the following statements is false?
    (a) Context-free languages are closed under union
    (b) Context-free languages are closed under concatenation
    (c) Context-free languages are closed under intersection
    (d) Context-free languages are closed under Kleene closure

[UGC NET 2011, Uttarakhand SET 2012, Gujarat SET 2014]

55. Which of the following definitions generates the same language as L, where L = $\{WW^R | W \in \{a, b\}^*\}$?
    (a) $S \to aSb|bSa|\varepsilon$ (b) $S \to aSa|bSb|\varepsilon$
    (c) $S \to aSb|bSa|aSa|bSb|\varepsilon$ (d) $S \to aSb|bSa|aSa|bSb$

[UGC NET 2012, Karnataka SET 2016]

56. Given the following productions of a grammar: $S \to aA \mid aBB$; $A \to aaA \mid \varepsilon$; $B \to bB \mid bbC \mid \varepsilon$; $C \to B$. Which of the following is true?
    (a) The language corresponding to the given grammar is a set of even number of *as*
    (b) The language corresponding to the given grammar is a set of odd number of *as*
    (c) The language corresponding to the given grammar is a set of even number of *as* followed by odd number of *bs*
    (d) The language corresponding to the given grammar is a set of odd number of *as* followed by even number of *bs* **[UGC NET 2013]**

57. Given the following statements: S1: The grammars $S \to aSb \mid bSa \mid SS \mid a$ and $S \to aSb \mid bSa \mid a$ are not equivalent. S2: The grammars $S \to SS \mid SSS \mid aSb \mid bSa \mid \varepsilon$ and $S \to SS \mid aSb \mid bSa \mid \varepsilon$ are equivalent. Which of the following is true?
    (a) S1 is correct and S2 is not correct (b) Both S1 and S2 are correct
    (c) S1 is not correct and S2 is correct (d) Both S1 and S2 are not correct **[UGC NET 2013]**

58. The Greibach Normal Form grammar for the language $L = \{a^n b^{n+1} \mid n \geq 0\}$ is:
    (a) $S \to a\,SB$, $B \to bB \mid \varepsilon$ (b) $S \to a\,SB$, $B \to bB \mid b$
    (c) $S \to a\,SB \mid b$, $B \to b$ (d) $S \to a\,Sb \mid b$ **[UGC NET 2013]**

59. If all production rules have a single non-terminal symbol on the left, the grammar defined is:
    (a) Context-free (b) Context-sensitive
    (c) Unrestricted (d) Phrase-structured **[UGC NET 2015]**

60. The language of all non-null strings of *as* can be defined by a context-free grammar as: $S \to aS \mid Sa \mid a$. The word a^3 can be generated by _________ different trees.
    (a) 2 (b) 3 (c) 4 (d) 5 **[UGC NET 2015]**

61. The context-free grammar given by $S \to XYX$, $X \to aX \mid bX \mid \varepsilon$, $Y \to bbb$ generates a language defined by the regular expression:
    (a) $(a + b)^*bbb$ (b) $abbb(a + b)^*$
    (c) $(a + b)^*(bbb)(a + b)^*$ (d) $(a + b)(bbb)(a + b)^*$ **[UGC NET 2015]**

62. Consider the following two languages: $L1 = \{0^i 1^j \mid \gcd(i, j) = 1\}$, L2 is any subset of 0^*. Which of the following is correct?
    (a) L1 is regular and L2* is not regular (b) L1 is not regular and L2* is regular
    (c) Both L1 and L2* are regular languages (d) Both L1 and L2* are not regular languages

[UGC NET 2016]

63. The grammar $S \rightarrow 0S1$, $S \rightarrow 0S$, $S \rightarrow S1$, $S \rightarrow 0$ will generate a:
(a) Context-free language (b) Context-sensitive language
(c) Regular language (d) Recursively enumerable language
[Gujarat SET 2002, HCU PhD Entrance 2012]

64. Which of the following statements regarding formal languages is false?
(a) Every regular language is a context-free language
(b) Every regular language is a context-sensitive language
(c) Every context-free language is a context-sensitive language
(d) Every context-sensitive language is a context-free language **[Gujarat SET 2013]**

65. Consider the following statements:
(i) The grammar $A \rightarrow Ab \mid a$ is left recursive.
(ii) The grammar $A \rightarrow aB$; $B \rightarrow bB \mid b$ is equivalent to the grammar in (i).
(a) Both (i) and (ii) are false (b) Both (i) and (ii) are true
(c) (i) is true and (ii) is false (d) (i) is false and (ii) is true **[Gujarat SET 2014]**

66. Consider the following language: $L = \{a^n b^n \mid n \geq 1\}$. L is:
(a) CFL but not regular (b) CSL but not CFL
(c) Regular (d) Type 0 language but not Type 1 **[Gujarat SET 2016]**

67. If the parse tree of a word W generated by a Chomsky Normal Form grammar has no path of length greater than j, then the word W is of length:
(a) j (b) 2^{j-1} (c) 2^{j+1} (d) 2^j
[JNU MPhil/PhD Entrance 2013, HP SET 2014, Rajasthan CS Lecturer Recruitment 2014]

68. If, for certain strings in the language of a grammar, there is more than one parse tree, the grammar is said to be:
(a) Context-free (b) LR(1) (c) LL(1) (d) Ambiguous **[J&K SET 2013]**

69. If L1 is a context-free language and L2 is a regular language, then $L1 \cap L2$ is:
(a) Context-free (b) Regular but not context-free
(c) Neither context-free nor regular (d) Both regular and context-free
[Rajasthan CS Lecturer Recruitment 2014]

70. The logic of pumping lemma is a good example of:
(a) Pigeon hole principle (b) Recursion (c) Divide and conquer technique (d) Iteration
[Rajasthan CS Lecturer Recruitment 2014]

71. Consider the set of all words over the alphabet {x, y, z} where the number of ys is not divisible by 2 or 7 and no x appears after a z. This language is:
(a) Regular (b) Not known to be regular
(c) Context-free but not regular (d) Recursively enumerable but not context-free
[CMI PhD Entrance 2013]

72. The production rules for the grammar $\{S \rightarrow aA, A \rightarrow bB, A \rightarrow aB, B \rightarrow a\}$ are in
(a) Chomsky Normal Form (b) Greibach Normal Form
(c) Simple form (d) None of these **[JNU MPhil/PhD Entrance 2010]**

73. Φ* (Kleene's closure of Φ, where Φ is the empty language over Σ) is equivalent to:
(a) Φ (b) ε (null string) (c) Σ (d) None of these **[JNU MPhil/PhD Entrance 2015]**

74. Let L1 = $\{a^i b^{2i} \mid i \in N\}$ and L2 = $\{a^i b^{i^2} \mid i \in N\}$ be two languages. Which of the following is correct?
(a) Both L1 and L2 are context-free languages
(b) L1 is context-free and L2 is recursive but not context-free
(c) Both L1 and L2 are recursive but not context-free
(d) L1 is regular and L2 is context-free **[TIFR PhD Entrance 2012]**

75. Let L be a given context-free language over the alphabet {a, b}. Construct L1, L2 as follows. Let L1 = L – {xyx | x, y ∈ {a, b}∗}, and L2 = L · L. Then,
(a) Both L1 and L2 are regular (b) Both L1 and L2 are context free but not necessarily regular
(c) L1 is regular and L2 is context-free (d) L1 and L2 may not be context-free
[TIFR PhD Entrance 2014]

76. Consider the following grammar (start symbol is E) for generating expressions: E → T – E | T + E | T, T → T * F | F, F → 0|1|2|3|4|5|6|7|8|9. With respect to this grammar, which of the following is the valid evaluation tree for the expression 2 * 3 * 4 – 5 * 6 + 7?
[TIFR PhD Entrance 2015]

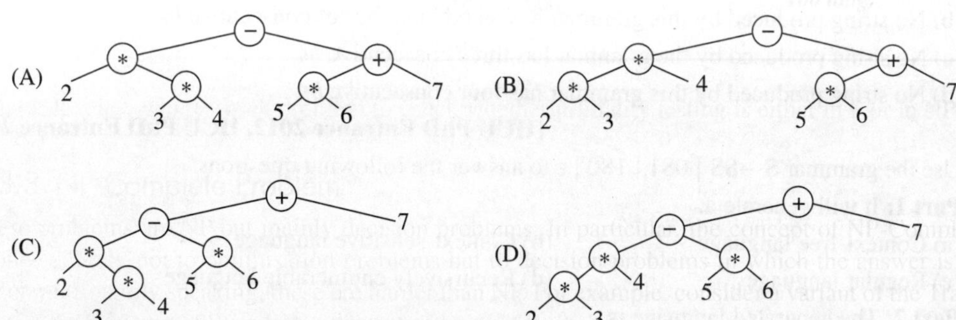

77. Which language class has the following property?
It is closed under union and intersection but not complement.
(a) Regular languages (b) Context-free languages
(c) Recursive languages (d) Recursively enumerable languages **[TIFR PhD Entrance 2016]**

78. Consider the following grammar: S → Ax | By, A → By | Cw, B → x | Bw, C → y. Which of the following regular expressions describes the same set of strings as the grammar?
(a) xw*y + xw*yx + ywx (b) xwy + xw*yx + ywx
(c) xw*y + xwyx + ywx (d) xw^+y + xw*yx + ywx **[HCU PhD Entrance 2009]**

79. The language L = $\{a^p$, where p is any non-prime, non-negative integer$\}$ defined over the alphabet Σ = {a} is:
(a) Regular (b) Non-regular and does not satisfy pumping lemma for regular languages
(c) Non-regular but satisfies pumping lemma for regular languages
(d) Context-free **[HCU PhD Entrance 2009, HCU PhD Entrance 2016]**

80. The language $L = \{a^{2k} : k \geq 0\}$ is a:
(a) Regular language (b) Non-regular language
(c) Non–context-free language (d) None of these **[HCU PhD Entrance 2010]**

81. Which of the following is(are) undecidable problem(s) ?
(i) Determination of ambiguity of context-free grammars
(ii) Given two context-free grammars G1 and G2, determining whether they generate exactly the same language
(iii) Determining whether the language accepted by Turing machine is finite or infinite
(a) (ii) only (b) (i) and (ii) only (c) (ii) and (iii) only (d) (i), (ii) and (iii)
[HCU PhD Entrance 2011]

82. Consider the following grammar where ε denotes the null string: $S \to aSb \mid bSa \mid SS \mid \varepsilon$. Which of the following best describes the language generated by the above grammar?
(a) All strings of the form $a^i b^j a^k$ where $i + j = k$ (b) All palindromes over a and b
(c) All strings with equal number of *a*s and *b*s (d) All even-length strings of *a*s and *b*s
[HCU PhD Entrance 2011, HCU PhD Entrance 2016]

83. Consider the grammar $S \to AB$, $A \to a \mid BaB$, $B \to bbA$. Which of the following statements is false?
(a) The length of every string produced by this grammar is even
(b) No string produced by this grammar has an odd number of consecutive *b*s
(c) No string produced by the grammar has three consecutive *a*s
(d) No string produced by this grammar has four consecutive *b*s
[HCU PhD Entrance 2012, HCU PhD Entrance 2013]

84. Use the grammar $S \to SS \mid 0S1 \mid 1S0 \mid \varepsilon$ to answer the following questions:
Part 1: It will generate a:
(a) Context-free language (b) Context-sensitive language
(c) Regular language (d) Recursively enumerable language
Part 2: The generated language is
(a) $\{0^n1^n$ where $n \geq 0\}$ (b) $\{0^n1^n$ where $n \geq 0\} \cup \{1^n0^n$ where $n \geq 0\}$
(c) $\{0^m1^k$ where $m, k \geq 0\} \cup \{1^m0^k$ where $m, k \geq 0\}$
(d) {W where W has equal number of 0s and 1s} **[HCU PhD Entrance 2014]**

85. Consider the following statements:
(i) The regular sets are closed under intersection.
(ii) The set of strings accepted by a finite automaton M with *n* states is finite if and only if the automaton accepts some string of length *l*, where $n \leq l \leq 2n$.
(iii) Moore machine is a finite automaton with output where the output is associated with the state.
Which of the above statements is/are false?
(a) Only (i) (b) Only (ii) (c) Both (i) and (iii) (d) Both (ii) and (iii)
[JNU M.Phil/PhD Entrance 2012]

86. Let $\Sigma = \{0, 1\}$, $L = \Sigma^*$ and $R = \{0^n1^n$ such that $n > 0\}$. Then the languages $L \cup R$ and R are, respectively:

(a) Regular, regular (b) Not regular, regular
(c) Regular, not regular (d) Not regular, not regular **[GATE 1995, Uttarakhand SET 2012]**

87. If L_1 and L_2 are context-free languages and R a regular set, one of the languages below is not necessarily a context-free language. Which one?
(a) L_1L_2 (b) $L_1 \cap L_2$ (c) $L_1 \cap R$ (d) $L_1 \cup L_2$ **[GATE 1996]**

88. Consider the alphabet $\Sigma = \{0, 1\}$, the null/empty string λ and the set of strings X_0, X_1, X_2 generated by the corresponding non-terminals of the regular grammar. X_0, X_1, X_2 are related as follows: $X_0 = 1X_1, X_1 = 0X_1 + 1X_2, X_2 = 0X_1 + \varepsilon$. Which one of the following choices precisely represents the strings in X_0?
(a) 10(0* + (10)*)1 (b) 10(0* + (10)*)*1
(c) 11(0 + 10)* (d) 10(0+ 10*)*1 + 110(0 + 10)*1 **[GATE 2015]**

89. The grammar having production rules S → AB | AS, A → aA | a, B → b generates:
(a) $(aa^*b)^+$ (b) a(ab)*b (c) (ab)* (d) aa^*b^+
[Rajasthan SET 2011, Karnataka SET 2017]

90. The language $L = \{ a^{n^2} \mid n \geq 1\}$ is __________ :
(a) Regular (b) Context-free but not regular
(c) Context-free and regular (d) Not context-free
[Pondicherry University PhD Entrance 2016]

91. Consider the following languages: I. $\{a^m b^n c^p d^q \mid m + p = n + q$, where $m, n, p, q \geq 0\}$; II. $\{a^m b^n c^p d^q \mid m = n$ and $p = q$, where $m, n, p, q \geq 0\}$; III. $\{a^m b^n c^p d^q \mid m = n = p$ and $p \neq q$, where $m, n, p, q \geq 0\}$; IV. $\{a^m b^n c^p d^q \mid mn = p + q$, where $m, n, p, q \geq 0\}$. Which of these languages are context-free?
(a) I and IV only (b) I and II only (c) II and III only (d) II and IV only **[GATE 2018]**

92. Which of the following statements is/are true?
(i) The grammar S→ SS | a is ambiguous (S is the start symbol)
(ii) The grammar S→ 0S1 | 1S0 | ε is ambiguous (ε represents the empty string, and S is the start symbol)
(iii) The grammar S→ T/U, T→ xSy/xy/ε, U→ yT generates a language consisting of the string yxxyy
(a) Only (i) and (ii) are true (b) Only (i) and (iii) are true
(c) Only (ii) and (iii) are true (d) (i), (ii) and (iii) are true **[UGC NET 2017]**

93. Consider the following languages: $L1 = \{a^m b^n \mid m \neq n\}$, $L2 = \{a^m b^n \mid m = 2n + 1\}$, $L3 = \{a^m b^n \mid m \neq 2n\}$. Which of the following statements is correct?
(a) Only L1 and L2 are context-free languages (b) Only L1 and L3 are context-free languages
(c) Only L2 and L3 are context-free languages (d) L1, L2 and L3 are context-free languages
[UGC NET 2017]

94. Consider the following two languages: $L1 = \{a^n b^l a^k : n + 1 + k > 5\}$ and $L2 = \{a^n b^l a^k : n > 5, l > 3, k \leq 1\}$. Which of the following is true?
(a) L1 is regular language and L2 is not regular language
(b) Both L1 and L2 are regular languages
(c) Both Ll and L2 are not regular languages
(d) Ll is not regular language and L2 is regular language **[JNU MPhil/PhD Entrance 2017]**

95. Consider the language L ⊆ {a, b, c}* defined as L = $\{a^p b^q c^r \mid p = q \text{ or } q = r \text{ or } r = p\}$. Which of the following statements is true for this language?
(a) L is regular but not context-free
(b) L is context-free but not regular
(c) L is decidable but not context-free
(d) L is regular, context-free and decidable
[TIFR PhD Entrance 2018]

96. Consider the following grammars: G1: A → Ab | a; G2: A → aB, B → aB | b; G3: A → aA | b. Which of the following is correct?
(a) Only G1 and G2 generate the same language
(b) Only G1 and G3 generate the same language
(c) Only G2 and G3 generate the same language
(d) G1, G2 and G3 generate the same language **[Gujarat SET 2017]**

97. Which of the following is true?
(a) Every Type 0 grammar is a Type 2 grammar
(b) Every Type 1 grammar is a Type 2 grammar
(c) Every Type 4 grammar is a Type 3 grammar
(d) Every Type 3 grammar is a Type 0 grammar
[Gujarat SET 2017]

98. Which of the following grammars is/are ambiguous? G1: A → A | ε; G2: A → aA | Aa | ε.
(a) Only G1
(b) Only G2
(c) Neither G1 nor G2
(d) Both G1 and G2
[Gujarat SET 2017]

99. Grammar G (A → A + A | a) is:
(a) Unambiguous
(b) Ambiguous as demonstrated by the derivation of string 'a + a'
(c) Ambiguous as demonstrated by the derivation of string 'a + a + a'
(d) Ambiguous for all strings generated by it **[Gujarat SET 2017]**

100. What is the language accepted by the following CFG: S → ε | 0S1S1S | 1S0S1S | 1S1S0S?
(a) Binary strings containing even number of 1s
(b) Binary strings containing odd number of 0s
(c) Odd length binary strings
(d) Binary strings with twice as many 1s as 0s
[Gujarat SET 2017]

101. The set of all palindromes over {0, 1} is:
(a) Regular but not CFL
(b) CFL but not regular
(c) Regular but not recursively enumerable
(d) Regular and CFL **[Gujarat SET 2017]**

102. Given the following two languages: $L_1 = \{0^m 1^m \mid m \geq 1\} \cup \{0\}$; $L_2 = \{w \mid w \text{ is a substring of } w^R \text{ and } w \in \{0, 1\}^*\}$. Which statement is correct?
(a) L_1 is deterministic and L_2 is non-deterministic
(b) L_1 is not deterministic and L_2 is deterministic
(c) Both L_1 and L_2 are deterministic
(d) Both L_1 and L_2 are not deterministic **[HPSET 2015, 2017]**

103. Given two grammars G1 and G2: G1: S → aSa | bSb | a | b; G2: S → 0S | 0S1 | X, X → X0 | 0. Which of these grammar(s) is(are) ambiguous?
(a) Only G1
(b) Only G2
(c) Both G1 and G2
(d) Neither G1 nor G2 **[HPSET 2015]**

104. Given the following statements: S1: Every context-sensitive language is recursive. S2: There is a recursive language which is not context-sensitive. Which of the following is correct?
(a) S1 is correct and S2 is not correct (b) S1 is not correct and S2 is correct
(c) Both S1 and S2 are correct (d) Both S1 and S2 are not correct **[HPSET 2015]**

105. The set of all nested sequences of balanced parenthesis is:
(a) String (b) Regular expression (c) Regular language (d) Context-free language **[HPSET 2015]**

106. Given the following languages: L1 is a context-free language, L2 is a regular language, L3 is a context-free language, L4 is a regular language, L5 = L1 ∩ L2, L6 = L1 ∪ L3, L7 = L1 ∪ L4. Which of the following sequences correctly describes the type of L5, L6 and L7, respectively?
(a) Regular, regular, regular (b) Context-free, context-free, regular
(c) Context-free, context-free, context-free (d) Regular, context-free, regular **[West Bengal SET 2015]**

107. A string X ∈ Σ* is said to be square-free if it cannot be written as X = uvvw for some u, v, w ∈ Σ* and |v| > 0. For example, the string *Harry* is not square-free. Suppose |Σ| > 1 and L = set of all square-free strings in Σ*. Which of the following is definitely an incorrect statement?
(a) L is regular (b) L is context-free
(c) L is context-sensitive (d) L is recursively enumerable **[West Bengal SET 2015]**

108. A finite state machine with the following state table has a single input *x* and a single output *z*

Present state	Next state, z	
	x = 1	x = 0
A	D, 0	B, 0
B	B, 1	C, 1
C	B, 0	D, 1
D	B, 1	C, 0

If the initial state is unknown, the shortest input sequence to reach the final state C is:
(a) 01 (b) 10 (c) 101 (d) 110 **[GATE 1995]**

109. Let L be the set of all binary strings whose last two symbols are the same. The number of states in the minimum state deterministic finite state automaton accepting L is:
(a) 2 (b) 5 (c) 8 (d) 3 **[GATE 1998]**

110. The finite state machine described by the following state diagram with A as starting state, where an arc label is x/y and x stands for 1-bit input and y stands for 2-bit output is:

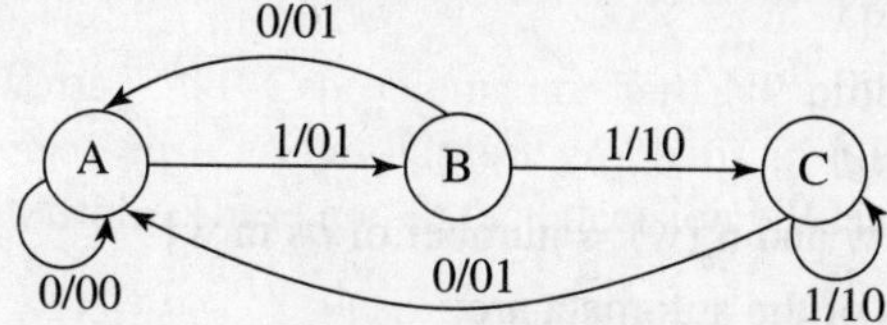

(a) Outputs the sum of the present and the previous bits of the input
(b) Outputs 01 whenever the input sequence contains 11

(c) Outputs 00 whenever the input sequence contains 10

(d) None of the above **[GATE 2002]**

111. Consider the following DFA.

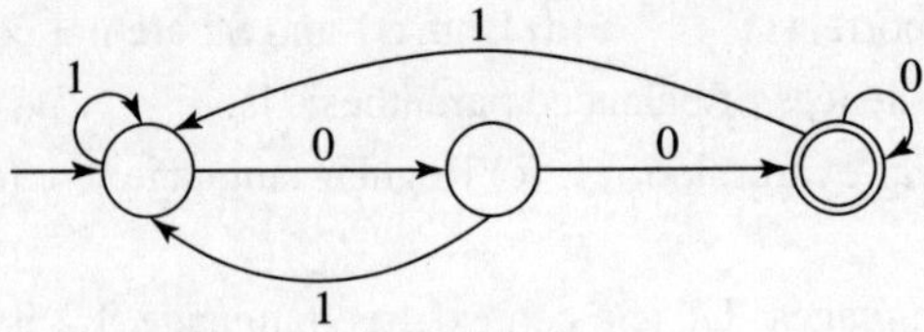

This DFA accepts the set of all strings over {0, 1} that:

(a) Begin either with 0 or 1 (b) End with 0

(c) End with 00 (d) Contain the substring 00 **[GATE 2009]**

112. Definition of a language L with alphabet {a} is given as: L = $\{a^{nk} \mid k > 0$, and n is a positive integer constant$\}$. What is the minimum number of states needed in a DFA to recognise L?

(a) $k + 1$ (b) $n + 1$ (c) 2^{n+1} (d) 2^{k+1} **[GATE 2011]**

113. Which of the regular expressions given below represent the following DFA?

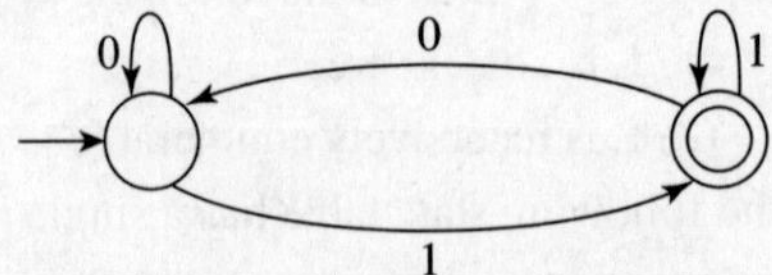

I. 0*1(1+00*1)* II. 0*1*1 + 11*0*1 III. (0+1)*1

(a) I and II only (b) I and III only

(c) II and III only (d) I, II, and III **[GATE 2014]**

114. The minimum possible number of states of a deterministic automaton that accepts the regular language $L = \{w_1aw_2 \mid w_1, w_2 \in \{a, b\}^*, |w_1| = 2, |w_2| \geq 3\}$ is ________. **[GATE 2017]**

115. "My Lafter Machine (MLM)" recognises the following strings: (i) a, (ii) aba, (iii) abaabaaba, (iv) abaabaabaabaabaabaabaabaaba. Using this information, how would you compare the following regular expressions?

(i) $(aba)^{(3^x)}$ (ii) $a.(baa)^{(3^x-1)}.ba$ (iii) $ab.(aab)^{(3^x-1)}.a$

(a) (ii) and (iii) are same, (i) is different (b) (ii) and (iii) are not the same

(c) (i), (ii) and (iii) are different. (d) (i), (ii) and (iii) are the same **[UGC NET 2010]**

116. The transition function for the language L = $\{w \mid n_a(w)$ and $n_b(w)$ are both odd$\}$ is given by:

δ(q0, a) = q1; δ(q0, b) = q2

δ(q1, a) = q0; δ(q1, b) = q3

δ(q2, a) = q3; δ(q2, b) = q0

δ(q3, a) = q2; δ(q3, b) = q1

[$n_a(w)$ = number of *a*s in w and $n_b(w)$ = number of *b*s in w]

The initial and final states of the automata are:

(a) q0 and q0, respectively (b) q0 and q1, respectively

(c) q0 and q2, respectively (d) q0 and q3, respectively **[UGC NET 2015]**

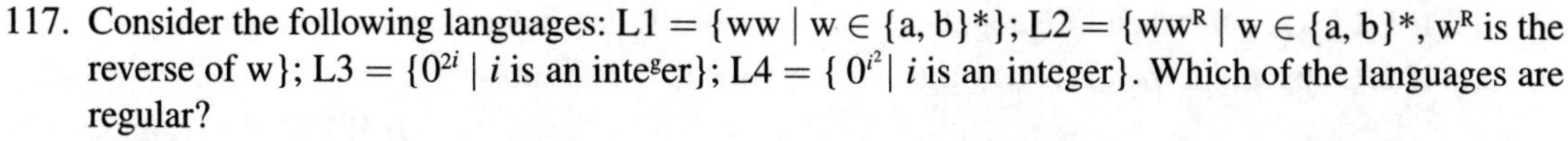

117. Consider the following languages: L1 = {ww | w ∈ {a, b}*}; L2 = {ww^R | w ∈ {a, b}*, w^R is the reverse of w}; L3 = {0^{2i} | i is an integer}; L4 = { 0^{i^2} | i is an integer}. Which of the languages are regular?

(a) Only L1 and L2 (b) Only L2, L3 and L4 (c) Only L3 and L4 (d) Only L3

[GATE 2001, Pondicherry University PhD Entrance 2013, Maharashtra SET 2016]

118. If L_1 = {a^n | $n \geq 0$} and L_2 = {b^n | $n \geq 0$}, consider the following statements: (I) $L_1.L_2$ is a regular language. (II) $L_1.L_2$ = {a^nb^n | $n \geq 0$}. Which one of the following is correct?

(a) Only (I) (b) Only (II) (c) Both (I) and (II) (d) Neither (I) nor (II) **[GATE 2014]**

119. Given two languages: L1 = {$(ab)^na^k$ | $n > k, k \geq 0$}and L2 = {a^nb^m | $n \neq m$}. Using pumping lemma for regular language, it can be shown that

(a) L1 is regular and L2 is not regular (b) L1 is not regular and L2 is regular

(c) L1 is regular and L2 is regular (d) L1 is not regular and L2 is not regular

[UGC NET 2014]

120. The regular expression corresponding to the language L where L = {x ∈ {0, 1}* | x ends with 1 and does not contain substring 00} is:

(a) (1 + 01)*(10 + 01) (b) (1 + 01)*01 (c) (1 + 01)*(1 + 01) (d) (10 + 01)*01

[UGC NET 2015]

121. Given the following two languages: L1 = {uww^{R}v | u, v, w ∈ {a, b}$^+$} and L2 = {uww^{R}v | u, v, w ∈ {a, b}$^+$, |u| ≥ |v|}. Which of the following is correct?

(a) L1 is a regular language and L2 is not a regular language

(b) L1 is not a regular language and L2 is a regular language

(c) Both L1 and L2 are regular languages

(d) Both L1 and L2 are not regular languages **[UGC NET 2016]**

122. Let A = {0, 1} and L = A*. Let R = {0^n1^n | $n > 0$}. The languages L ∪ R and R are, respectively:

(a) Regular, regular (b) Regular, not regular

(c) Not regular, regular (d) Not regular, not regular

[Rajasthan SET 2013, HP IT Lecturer Recruitment 2016]

123. Consider the language L_1 = Φ and L_2 = {a}. Which one of the following represents $L_1L_2^* \cup L_1^*$?

(a) {ε} (b) Φ (c) a* (d) {ε, a} **[Maharashtra SET 2015]**

124. If language L = {a^nb^n | $n \geq 0$}, then language L^2 is given by:

(a) {$a^{2n}b^{2n}$ | $n \geq 0$} (b) {$a^nb^na^nb^n$ | $n \geq 0$}

(c) {a^nb^n | $n \geq 0$} (d) {$a^nb^na^mb^m$ | $n \geq 0, m \geq 0$} **[JNU MPhil/PhD Entrance 2010]**

125. Which regular expression best describes the language accepted by the following FSM?

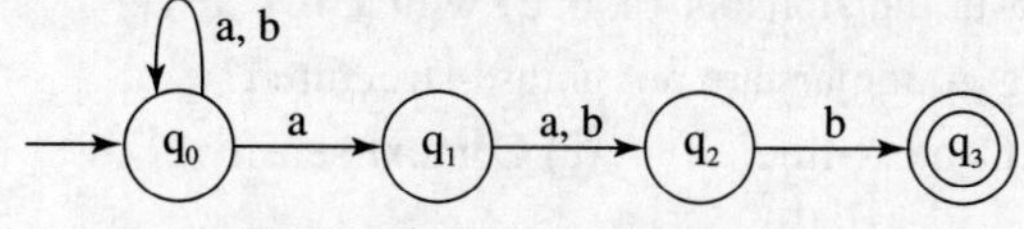

(a) (abb)* (b) (a+b)*a(a+b)*b(a+b)* (c) (a+b)* (d) (a+b)*a(a+b)b

[Pondicherry University PhD Entrance 2016]

126. Let L and L′ be languages over the alphabet Σ. The left quotient of L by L′ is defined as L/L′ = {w ∈ Σ* : wx ∈ L for some x ∈ L′}. Which of the following is true?

(a) If L/L′ is regular then L′ is regular (b) If L is regular then L/L′ is regular
(c) If L/L′ is regular then L is regular (d) L/L′ is a subset of L

[TIFR PhD Entrance 2013]

127. A regular expression for the language L over the alphabet Σ = {a, b}* consisting of strings with exactly two *b*s is:
(a) a*ba*ba* (b) a*bba* (c) bba* (d) a*bb **[HCU PhD Entrance 2015]**

128. The regular expression (a* + b)* is equivalent to which of the following regular expressions?
(a) a*b* (b) (a*b + b)* (c) (a + b*)* (d) (a*b)* **[CMI PhD Entrance 2017]**

129. Given the following languages: $L_1 = \{(ab)^n a^k \mid n > k, k \geq p\}$ and $L_2 = \{1^n \mid n \text{ is a perfect square}\}$. Which of the following is correct?
(a) L_1 is a regular language and L_2 is not a regular language
(b) L_1 is not a regular language and L_2 is a regular language
(c) Both L_1 and L_2 are regular languages
(d) Both L_1 and L_2 are not regular languages **[HPSET 2015]**

130. Consider a regular language L. Define the following: tail(L) = {y | xy ∈ L for some x ∈ Σ*}; min(L) = {w ∈ L | there is no u ∈ L, v ∈ Σ* such that w = uv}. Which of the above are regular languages?
(a) Only tail(L) (b) Only min(L)
(c) Both tail(L) and min(L) (d) Neither tail(L) nor min(L) **[HPSET 2017]**

131. Let G = (V, T, S, P) be any CFG without any unit or null productions. Let K be the maximum number of symbols on the right of any production in P. An equivalent grammar in Chomsky Normal Form has at most __________ number of production rules.
(a) (K – 1) |P| + |T| (b) K |P| + T (c) (K – 1) |P| (d) K |P| **[HPSET 2017]**

132. The number of states in a minimal DFA for accepting the language $L = \{a^n \mid n \geq 0, n \neq 5\}$ is:
(a) 5 (b) 6 (c) 7 (d) 4 **[HPSET 2017]**

133. Consider the following: CF = context-free languages; CS = context-sensitive languages; REC = recursive languages; RE = recursively enumerable languages; DCF = deterministic context-free languages. Which of the following exhibits the correct relationship among the sets of languages?
(a) DCF ⊆ CF ⊆ CS ⊆ REC ⊆ RE (b) CF ⊆ DCF ⊆ CS ⊆ REC ⊆ RE
(c) DCF ⊆ CF ⊆ CS ⊆ RE ⊆ REC (d) CF ⊆ DCF ⊆ CS ⊆ RE ⊆ REC **[HPSET 2017]**

134. The following context-free grammar generates S → TaT, T → TT | aTb | bTa | a | ε:
(a) The set of strings over the alphabet {a, b} with equal number of *a*s and *b*s
(b) The set of strings over the alphabets {a, b} with more *a*s than *b*s
(c) The set of strings over the alphabet {a, b} with more *b*s than *a*s
(d) The set of strings over the alphabet {a, b, c} with a, b, c and ε **[Karnataka SET 2017]**

135. Which of the following grammars are not phrase-structured?
(a) Regular (b) Context-free (c) Context-sensitive (d) None of these

[Karnataka SET 2017]

136. Consider the grammar G: S → abScB | ε, B → bB | b. What language does it generate?
(a) $\{(ab)^n(cb)^n \mid n = m\}$ (b) $\{a^n b^n (cb)^n \mid n \neq m\}$
(c) $\{(ab)^n(cbb^*)^n \mid n \geq 0\}$ (d) $\{(ab)^n(cb^*)^n \mid n \geq 0\}$ **[TNSET 2018]**

137. Consider the following: X: baa ∈ a*b*a*b*; Y: b*a* ∩ a*b* = a* U b*; Z: a*b* ∩ b*c* = Φ; W: abcd ∈ (a(cd)*b)*. Which of the above is(are) correct?
(a) X, W, Z (b) Y, W, Z (c) X, Y (d) Z, W
[West Bengal SET 2015]

138. All strings having an equal number of *a*s and *b*s can be recognised by:
(a) DFA (b) NDFA (c) PDA (d) All of the above
[West Bengal SET 2014]

139. A PDA behaves like an FA when the number of auxiliary memory units it has is:
(a) 0 (b) 1 (c) 2 (d) None of these
[Punjab Tech University PhD Entrance 2010]

140. Let B consist of all binary strings beginning with 1, whose value when converted to decimal is divisible by 7.
(a) B can be recognised by a DFA (b) B can be recognised by a NDFA but not by a DFA
(c) B can be recognised by a deterministic PDA but not by a NDFA
(d) B can be recognised by a non-deterministic PDA but not by a deterministic PDA
[TIFR PhD Entrance 2015]

141. Pushdown automata can recognise languages generated by__________.
(a) Only context-free grammar (b) Only regular grammar
(c) Context-free grammar or regular grammar (d) Only context-sensitive grammar
[UGC NET 2018]

142. Consider the following statements regarding languages, where w ∈ {0, 1}*, w^R is the reverse of string w: S1: x = wcw^R can be accepted by a NFA. S2: x = wcw^R can be accepted by a NPDA. S3: x = wcw^R can be accepted by a DPDA. Which of the above are correct?
(a) S1 and S2 (b) S2 and S3 (c) S1 and S3 (d) S1, S2 and S3
[Gujarat SET 2017]

143. Consider the following languages: $L_1 = \{PP^R \mid P \in \{0, 1\}^*\}$, $L_2 = \{0^n1^n \mid n \geq 1\}$, $L_3 = \{PP \mid P \in \{0, 1\}^*\}$, $L_4 = \{a^nb^mc^md^n \mid n, m \geq 0\}$and machines M_1, M_2, M_3, M_4 recognising the languages L_1, L_2, L_3, L_4, respectively. Which statement is correct?
(a) M_1, M_2 are PDAs, but M_3 and M_4 are not (b) M_2 and M_3 are PDAs, but M_1 and M_4 are not
(c) All of M_1, M_2, M_3, M_4 are PDAs (d) M_1, M_2 and M_4 are PDAs
[West Bengal SET 2017]

144. Consider the following statements: S1: There exists no algorithm for deciding if any two Turing machines M1 and M2 accept the same language. S2: The problem of determining whether a Turing machine halts on any input is undecidable. Which of the following options is correct?
(a) Both S1 and S2 are correct (b) Both S1 and S2 are not correct
(c) Only S1 is correct (d) Only S2 is correct **[UGC NET 2018]**

145. Consider the following statements regarding languages: S1: A Type 0 grammar can be accepted by a DPDA. S2: A Type 1 grammar can be accepted by an NPDA. S3: A Type 2 grammar can be accepted by a DTM. S4: A Type 3 grammar can be accepted by an NDTM. Which of the above is/are true?
(a) Only S2 (b) S2 and S3 (c) S3 and S4 (d) S1, S3 and S4 **[Gujarat SET 2017]**

146. Which of the following is false?
(a) $A = \{a^n b^n c^n \mid n \geq 0\}$ is not a CFL
(b) The shortest string not accepted by the regular expression 0*1*0*1* is 1010
(c) The regular expression (1 + 01)*0* accepts all binary strings not containing the substring 0001
(d) All regular languages are accepted by Turing machines **[Gujarat SET 2017]**

147. Which of the following conversions is not possible algorithmically?
(a) Regular grammar to context-free grammar
(b) Non-deterministic TM to deterministic TM
(c) Non-deterministic FSA to deterministic FSA
(d) Non-deterministic PDA to deterministic PDA **[Maharashtra SET 2018]**

148. Co-NP is a class of problems whose complement belongs to class NP. $X = P \cap NP$, $Y = NP \cap \text{Co-NP}$, $Z = P \cap \text{NP-Hard}$. Which one is true?
(a) $X = \Phi, Y = \Phi, Z \neq \Phi$ (b) $X \neq \Phi, Y \neq \Phi, Z = \Phi$
(c) $X = \Phi, Y = \Phi, Z = \Phi$ (d) $X \neq \Phi, Y = \Phi, Z = \Phi$ **[West Bengal SET 2014]**

149. Which of the following is true?
(a) The complement of a recursive language is recursive
(b) The complement of a recursively enumerable language is recursively enumerable
(c) The complement of a recursive language is either recursive or recursively enumerable
(d) The complement of a context-free language is context-free **[GATE 2002]**

150. If the strings of a language L can be effectively enumerated in lexicographic (alphabetical) order, which of the following statements is true?
(a) L is necessarily finite (b) L is regular but not necessarily finite
(c) L is context-free but not necessarily regular (d) L is recursive but not necessarily finite
[GATE 2003]

151. Let L_1 be a recursive language and let L_2 be a recursively enumerable but not recursive language. Which one of the following is true?
(a) L_1' is recursive and L_2' is recursively enumerable
(b) L_1' is either recursive or recursively enumerable, and L_2' is not recursively enumerable
(c) L_1' and L_2' are recursively enumerable
(d) L_1' is recursively enumerable and L_2' is recursive **[GATE 2005]**

152. Consider the following problems. L(G) denotes the language generated by a grammar G. L(M) denotes the language accepted by a machine M.
(I) For an unrestricted grammar G and a string w, whether $w \in L(G)$
(II) Given a Turing machine M, whether L(M) is regular
(III) Given two grammars G_1 and G_2, whether $L(G_1) = L(G_2)$
(IV) Given an NFA N, whether there is a deterministic PDA P, such that N and P accept the same language
Which of the following is correct?
(a) Only (I) and (II) are undecidable (b) Only (III) is undecidable
(c) Only (II) and (IV) are undecidable (d) Only (I), (II) and (III) are undecidable
[GATE 2018]

153. Which of the following statements is not correct?
(a) Every recursive language is recursively enumerable
(b) L = $\{0^n1^n0^n \mid n = 1, 2, 3, ...\}$ is recursively enumerable
(c) Recursive languages are closed under intersection
(d) Recursive languages are not closed under intersection **[UGC NET 2017]**

154. We have constructed a polynomial time reduction from problem *A* to problem *B*. Which of the following is a valid inference?
(a) If the best algorithm for *B* takes exponential time, there is no polynomial time algorithm for *A*
(b) If the best algorithm for *A* takes exponential time, there is no polynomial time algorithm for *B*
(c) If we have a polynomial time algorithm for *A*, we must also have a polynomial time algorithm for *B*
(d) If we do not know whether there is a polynomial time algorithm for *B*, there cannot be a polynomial time algorithm for *A* **[CMI PhD Entrance 2017]**

155. The halting problem is an example of:
(a) An NP-Complete problem (b) An NP-Hard problem that is not NP-Complete
(c) An undecidable problem (d) A recursively enumerable problem **[Gujarat SET 2017]**

156. Which of the following is considered the biggest open and unresolved problem in computer science?
(a) Is P = NP? (b) Is P ⊆ NP? (c) Is NP ⊆ P? (d) Is P ∩ NP = φ?
[Punjab Tech University PhD Entrance 2015]

157. The minimised DFA for the following DFA has how many states?

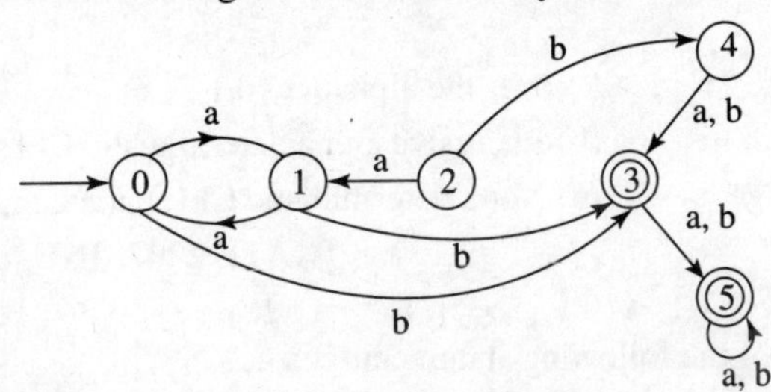

(a) 2 (b) 3 (c) 4 (d) 5 **[UGC NET 2006]**

158. Let L ⊆ {0, 1}* be a language accepted by a finite automaton. Let F be some subset of {0, 1}* containing 2011 strings. Which of the following statements is true?
(a) L ∪ F is always regular
(b) L ∪ F is regular only when L ∩ F = ∅, that is, when L and F do not have any common string
(c) L ∪ F is never regular
(d) L ∪ F is regular only if L contains ε (the empty string) **[CMI PhD Entrance 2011]**

159. Let A and B be regular languages. Define their concatenation as C = A.B = {xy | x ∈ A, y ∈ B}. Which of the following is true?
(a) C is a regular language (b) C is not a regular language
(c) C is an infinite language (d) C is recursively enumerable
[IIT Chennai PhD Entrance 2014]

160. The language L = {xcx | x ∈ {0,1}* and c is a terminal symbol, $c \neq 0$, $c \neq 1$} is:
(a) Regular (b) Context-free
(c) Regular as well as context-free (d) Neither regular nor context-free **[GATE 1999]**

161. L is a language accepted by a finite state machine. Define the following language: L^P = {s | ss′ ∈ L for some string s′.}and L^R = {s | s obtainable by reversing some string in L}. Which of the following is true?
(a) Only L^P is accepted by a finite state machine
(b) Only L^R is accepted by a finite state machine
(c) Both L^P and L^R are accepted by finite state machines
(d) Neither L^P nor L^R are accepted by finite state machines **[GATE 1997]**

162. Which one of the following languages over the alphabet {0, 1} is described by the regular expression: (0+1)*0(0+1)*0(0+1)*?
(a) The set of all strings containing the substring 00
(b) The set of all strings containing at most two 0s
(c) The set of all strings containing at least two 0s
(d) The set of all strings that begin and end with either 0 or 1 **[GATE 2009]**

163. Assume that L is a regular language. Let statements S1 and S2 be defined as:
S1: SQRT(L) = {x | for some y with $|y| = |x|^2$, xy ∈ L} is a regular language. S2: LOG(L) = {x | for some y with $|y| = 2^{|x|}$, xy ∈ L} is a regular language. Which of the following is true?
(a) S1 is correct and S2 is not correct (b) Both S1 and S2 are correct
(c) Both S1 and S2 are not correct (d) S1 is not correct and S2 is correct
[UGC NET 2013]

164. The language L = $\{0^i 2 1^i \mid i \geq 0\}$ over the alphabet {0,1,2} is:
(a) Not recursive (b) Recursive and a deterministic CFL
(c) A regular language (d) Not a deterministic CFL but a CFL
[GATE 2007, JNU M.Phil/PhD Entrance 2014]

165. Consider the languages L1 = $\{0^i 1^j \mid i \neq j\}$, L2 = $\{0^i 1^j \mid i = j\}$, L3 = $\{0^i 1^j \mid i = 2j + 1\}$, L4 = $\{0^i 1^j \mid i \neq 2j\}$. Which one of the following statements is true?
(a) Only L2 is context-free (b) Only L2 and L3 are context-free
(c) Only L1 and L2 are context-free (d) All are context-free **[GATE 2010]**

166. Which of the following is not correct?
(i) Every context-sensitive language is recursive.
(ii) There is a recursive language that is not context-sensitive.
(a) (i) is true, (ii) is false (b) (i) is true and (ii) is true
(c) (i) is false, (ii) is false (d) (i) is false and (ii) is true **[UGC NET 2012, 2013]**

167. Consider the following languages: L1 = $\{a^n b a^n \mid n > 0\}$; L2 = $\{a^n b a^n b^{n+1} \mid n > 0\}$. Which one of the following is true?
(a) Both L1 and L2 are context-free (b) L1 is context-free while L2 is not context-free
(c) L2 is context-free while L1 is not context-free (d) Neither L1 nor L2 is context-free
[UGC NET 2015]

168. Which of the following is not true?
   (a) Any language accepted by an NFA is regular
   (b) Every language is accepted by a finite state automaton
   (c) For every NFA, there is an equivalent deterministic finite state machine
   (d) Every context-free language is accepted by an NFA **[Gujarat SET 2010]**
169. A CFG is said to be in Chomsky Normal Form (CNF) if all the productions are of the form $A \to BC$ or $A \to a$. Let G be a CFG in CNF. To derive a string of terminals of length x, the number of productions to be used is:
   (a) $2x - 1$ (b) $2x$ (c) $2x + 1$ (d) None of these
   [HP IT Lecturer Recruitment 2016, UGC NET 2018]
170. Over the alphabet {0, 1}, consider the language $L = \{w \mid w$ does not contain the substring $0011\}$. Which of the following is true about L?
   (a) L is not context-free (b) L is regular
   (c) L is not regular but it is context-free (d) L is context-free but not recursively enumerable
   [CMI PhD Entrance 2010]
171. The language $L = \{a^n b^n c^n : n \geq 1\}$ is:
   (a) Not context-free but context-sensitive (b) Not regular but context-free
   (c) Not context-sensitive but regular (d) Not context-sensitive but context-free
   [Telangana SET 2017]
172. Which of the following is true?
   (a) P is a subset of NP (b) NP is a subset of P
   (c) P and NP are equal (d) NP is a subset of NP-Hard **[Telangana SET 2017]**
173. Which of the following is true about NP-Complete and NP-Hard problems?
   (a) If we want to prove that X is NP-Hard, we take a known NP-Hard problem Y and reduce X to Y
   (b) Circuit satisfiability problem was the first known NP-Complete problem
   (c) NP-Complete is a subset of NP-Hard
   (d) All of the above **[Telangana SET 2017]**
174. The string 11101 is not derived by:
   (a) $(0 + 1)^*$ (b) $1^*0^*1^*$
   (c) 111^*01 (d) $(11)^*(01)^*$ **[Telangana SET 2018]**
175. Let X be a problem belonging to class NP. Which of the following is true?
   (a) X may be decidable
   (b) If X is NP-Hard, it is not NP-Complete
   (c) If X can be solved deterministically in polynomial time, then $P \neq NP$
   (d) There is a polynomial time algorithm for X **[HP CS Lecturer Recruitment 2018]**

HINTS AND SOLUTIONS

1. $S_1 \cup S_2 \cup \ldots \cup S_n = S$. S is an infinite set. This means that at least one of the sets S_i is infinite. If any one of the sets S_i is infinite, the nature of the other sets is simply irrelevant. **Answer: (c)**
2. No element can be less that itself; hence the relation is **irreflexive.** So, it is clearly not a partial

order. We have narrowed the answers down to options (c) and (d). A relation being asymmetric (or not) has nothing to do with a partial order. Option (c) is therefore eliminated. We are only left with option (d). We do not need to check any further if it is known that only one option is correct.

Check for antisymmetry: $x < y$ and $y < x$ does not mean $x = y$, hence the relation is not antisymmetric.

Answer: (d)

3. Two cases are possible.

Case 1: If n is odd, that is, $n = 2k + 1$.

Number of subsets with 1 element $= {}^nC_1$

Number of subsets with 3 elements $= {}^nC_3$

Number of subsets with 5 elements $= {}^nC_5$

…

…

Number of subsets with $(2k + 1)$ elements $= {}^nC_{2k+1} = {}^nC_n$

Total number of subsets with odd number of elements $= {}^nC_1 + {}^nC_3 + {}^nC_5 + \dots + {}^nC_n$

$= 2^{n-1}$ (from binomial theorem)

Case 2: If n is even, that is, $n = 2k$.

Number of subsets with 1 element $= {}^nC_1$

Number of subsets with 3 elements $= {}^nC_3$

Number of subsets with 5 elements $= {}^nC_5$

…

…

Number of subsets with $(2k - 1)$ elements $= {}^nC_{2k-1} = {}^nC_{n-1}$

Total number of subsets with odd number of elements $= {}^nC_1 + {}^nC_3 + {}^nC_5 + \dots + {}^nC_{n-1}$

$= 2^{n-1}$ (from binomial theorem). **Answer: 2^{n-1}**

4. We have to try out all the options one by one.

Number of subsets of the first set $= 2^m$. Number of subsets of the second set $= 2^n$.

$2^m - 2^n = 56$

$2^7 - 2^6 = 2^6 \neq 56$

$2^6 - 2^3 = 64 - 8 = 56$. This option matches. We do not need to check any further if it is known that only one option is correct.

$2^5 - 2^1 = 30 \neq 56$

$2^8 - 2^7 = 2^7 \neq 56$

Hence, $m = 6$, $n = 3$ is the only possible match. **Answer: (b)**

5. $(X - Y) \cup (Y - X) \cup (X \cap Y)$. Draw a generic Venn diagram for this, assuming $X \cap Y \neq \varphi$. You can visualise it mentally if you have enough practice.

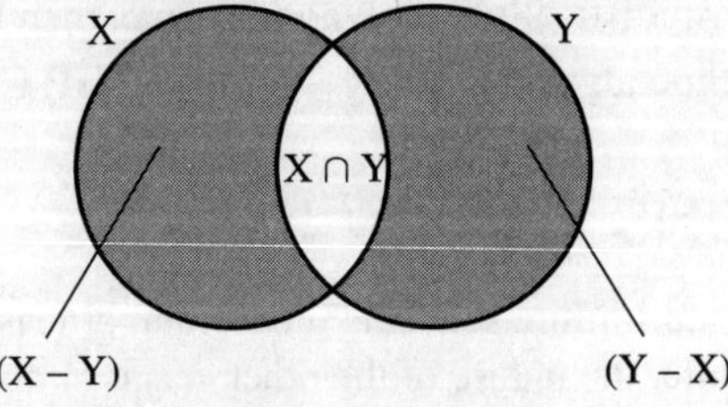

FIGURE A1 Venn diagram for the given expression

Clearly, from the above Venn diagram, $(X - Y) \cup (Y - X) \cup (X \cap Y) = (X \cup Y)$. Also, if

X ∩ Y = φ, that is, X and Y are disjoint sets, then (X – Y) = X and (Y – X) = Y. In this case also, (X – Y) ∪ (Y – X) ∪ (X ∩ Y) = (X ∪ Y).

Answer: (a)

6. The numbers 1, 2, 4, 8, …, 2^n written in binary form can be expressed by a simple regular expression 10*. Every regular expression is expressible by a DFA. Option (a) is correct. Let us analyse the other options as well.

   If the powers of 2 are written in unary form, we need a multiplier logic (or algorithm) which cannot be implemented by the O(1) memory of DFA. Therefore, option (b) is incorrect. If the number of zeros and ones is the same, we need a stack to check it, which requires expandable memory or PDA. Therefore, option (c) is also incorrect. The set {1, 101, 11011, 1110111, …} also requires a stack to check for valid/invalid strings. A stack can never be implemented by O(1) memory, and hence, this set of strings cannot be recognised by a DFA. So, option (d) is also incorrect.

   Answer: (a)

7. Observe the following automata and proceed along the lines of mathematical induction.

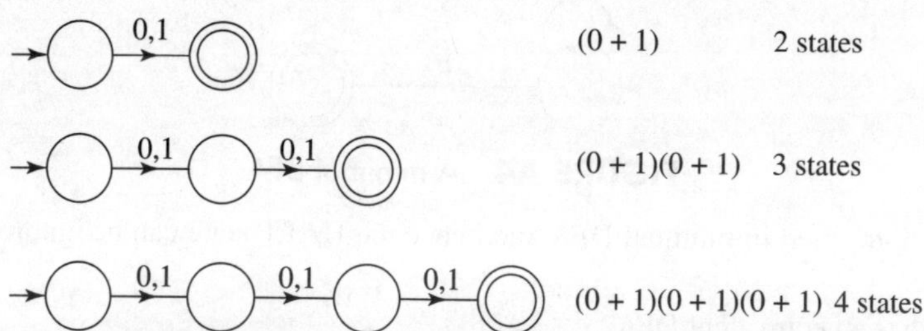

FIGURE A2 DFA and regex for various values of n

For the regular expression (0+1)(0+1)… *n* times, we need (*n* + 1) states in the DFA. **Answer: (b)**

8. Observe that state q3 is not possible. There is no path for an accepted string to reach this state. Write down the state equations and use the solution of R = Q + RP is R = QP* repeatedly.

q0 = q0b + ε and its solution is q0 = ε b* = b* (i)

q1 = q0a + q1b + q2a

⇒ q1 = b*a + q2a + q1b [using (i)]

⇒ q1 = (b*a + q2a)b*

⇒ q1 = b*ab* + q2ab* (ii)

q2 = q1a + q2b

⇒ q2 = b*ab*a + q2(ab*a + b) [using (ii)]

⇒ q2 = b*ab*a(ab*a + b)* (iii)

Using (iii) in (ii), we get q1 = b*ab* + b*ab*a(ab*a + b)*ab* (iv)

The regular expression for this automaton is q1 + q2 (as there are two final states)

= b*ab* + b*ab*a(ab*a + b)*ab* + b*ab*a(ab*a + b)*

= b*a{b* + b*a(ab*a + b)*ab* + b*a(ab*a + b)*}

= b*a{b* + b*a((ab*a)*b*)*ab* + b*a((ab*a)*b*)*}

= b*a{b* + b*a(a + b)*ab* + b*a(a + b)*}

= b*a{b* + b*a(a + b)*(ab* + ε)}

$= b^*a(a + b)^*$ **Answer for Part 1: (c)**

Minimum state FSA is given below.

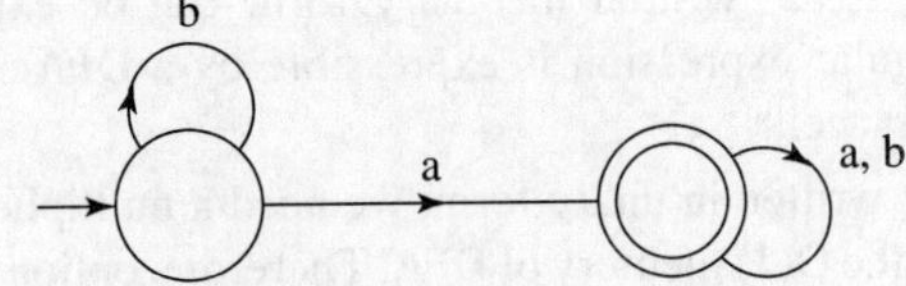

FIGURE A3 A minimal DFA

It is a DFA and needs only two states. **Answer for Part 2: (b)**

9. Every finite state automaton (irrespective of whether it is deterministic or non-deterministic) has an equivalent regular expression, which in turn, corresponds to a regular language. **Answer: (c)**
10. The DFA is given below.

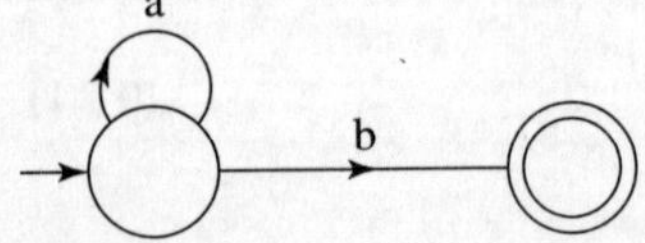

FIGURE A4 A minimal DFA

We are only interested in minimal DFA and hence the HALT state can be ignored. So, two states are adequate. **Answer: (a)**

11. Every NFA has an equivalent DFA. **Answer: (a)**
12. The number of states or edges has nothing to do with the equivalence of two finite state automata. They are equivalent if and only if they both have identical or equivalent regular expressions, that is, they accept the same set of strings. **Answer: (c)**
13. We do not need to explicitly use Arden's theorem for explicit derivation; we can use observation instead. Which string is accepted when moving from s1 to s1? $(a + b)^*$. After that, travel up to the final state through aba. Observe the null transitions – they have absolutely no impact. Try out a few of them. It is intuitively visible that the answer is $(a + b)^*aba$. **Answer: (a)**
14. Again, we do not need to explicitly use Arden's theorem. Observe that state C is the HALT state. So, no accepted string will ever reach state C. It can be intuitively seen that accepted strings are defined by the regular expression $0^*11^* = 0^*1^+$. **Answer: $0^*11^* = 0^*1^+$**
15. π is a well-known irrational number. So, we cannot derive a formula to work out whether there will be x number of consecutive 5s anywhere in the infinite decimal expansion of π. Therefore, S is not regular. Also, we cannot come up with a recursive enumeration for the same reason. **Answer: (c)**
16. $(00)^*(0 + \varepsilon) = (00)^*0 + (00)^* = \{$set of all strings with odd number of 0s$\} \cup \{$set of all strings with even (including zero) number of 0s$\} = \{$set of strings with zero or more 0s$\} = 0^*$.

    $(00)^* = \{$set of all strings with even (including zero) number of 0s$\}$; $0(00)^* = \{$set of all strings with odd number of 0s$\}$. Clearly, $(00)^*(0 + \varepsilon)$ and 0^* are equivalent, that is, expressions (i) and (iii) are equivalent. **Answer: (c)**
17. We know that for any regular expressions, P and Q, $(P + Q)^* = (P^*Q^*)^*$. Insert P = 01 and Q = 1. We get $(01 + 1)^* = ((01)^*1^*)^* \Rightarrow A = B$. **Answer: (d)**
18. The language $L_1 = \{0^{2n} \mid n \geq 1\}$ is easily expressed by the following DFA.

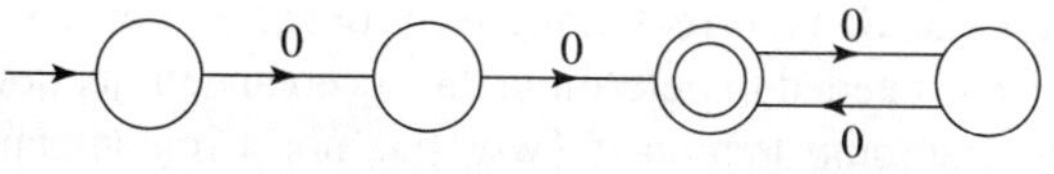

FIGURE A5 Equivalent DFA

As there is an equivalent DFA, $L_1 = \{0^{2n} \mid n \geq 1\}$ is regular $\Rightarrow$ S1 is correct. Options (b) and (d) are eliminated.

For the language $L_2 = \{0^m1^n0^{m+n} \mid m \geq 1 \text{ and } n \geq 1\}$, apply pumping lemma. Let $w = 0^m1^n0^{m+n}$. $|w| = 2(m+n) > p$, where p is some integer. Let $w = xyz$, where $|xy| \leq p$ and $|y| \geq 1$. Let $x = 0^m1^k$, $y = 1^q0^r$, $z = 0^t$, where $m + k + q = r + t$, $m \neq 0$, $k \neq 0$, $q \neq 0$, $r \neq 0$ and $t \neq 0$. Consider the string $xy^2z = 0^m1^k1^q0^r1^q0^r0^t$. Clearly, $xy^2z \notin L_2$. So, by pumping lemma, L_2 is not regular $\Rightarrow$ S2 is false. So, only S1 is correct.

Answer: (a)

19. 0*(10*)* accepts the string 010, which is clearly not accepted by (1*0)*1*. Option (a) is eliminated. 0*(10*)* also accepts the empty string ε, which is not accepted by (b) and (c). Options (b) and (c) are also eliminated. **Answer: (d)**

20. $\{a^nb^{2m} \mid n \geq 0, m \geq 0\}$ can be expressed by the regular expression a*(bb)*. Observe that there is no relation between m and n. So, I is a regular set. Option (d) is eliminated. II and III are well-known non-regular sets and can be easily proved to be non-regular by using pumping lemma. Option (b) is also eliminated. {xcy | x, y ∈{a, b}*} is easily expressed by (a + b)*c (a + b)*. Only I and IV are regular sets. **Answer: (a)**

21. Check for counterexamples. Option (a): If the string contains 1, it must end in 1, and hence, cannot generate all possible strings with an even number of 1s (for example, 1010). Option (c): Between the second and third 1, 0 is not allowed (for example, 011011). Option (d): 00 is not allowed as zero is an even number. **Answer: (b)**

22. Option (a) is not regular, and can be proved easily by pumping lemma. In fact, it can only be accepted by a PDA. Option (b) is also not regular as no known formula can generate all possible prime numbers using O(1) memory. Option (d) is also not regular and can be easily proved by using pumping lemma. In fact, it is accepted only by a PDA. Option (c) is easily expressed by a regular expression a*ba*(a*(a*ba*ba*b)*)*a*. **Answer: (c)**

23. Empty string is accepted by a*b*(ba)*a*. Size 0 is eliminated. Strings a and b are accepted. Size 1 is eliminated. Strings aa, ab, ba, bb are also accepted. Size 2 is eliminated. For size 3 strings, observe that bab is not accepted. So, the size of the smallest non-accepted string is 3. **Answer: 3**

24. A countable set implies that there exists an enumeration procedure to generate each of its elements; for a given element of a set, it takes finite steps to generate it using the enumeration procedure. Let Σ = {a, b} and let there exist an enumeration procedure to generate all the strings of the language Σ* = {∈, a, b, aa, ab, ba, bb, aaa, ...}. Here the enumeration procedure simply generates strings in alphabetical order, and in increasing order of string length. In this way, Σ* is countably infinite and 2^{Σ^*} will be an uncountable set because the power set of a countably infinite set is uncountable. **Answer: (c)**

25. (1 + 010)* can accept the string 01<u>00</u>10; hence, option (a) is eliminated. (01 + 10)* can accept the string 1<u>00</u>1; hence, option (b) is eliminated. (1 + 010)* (0 + λ) does not accept an elementary string 101; hence, option (c) is eliminated. (1 + 01)* (0 + λ) forces every zero to be immediately followed by a 1 or an empty string; however, each 1 may or may not be followed by 1 or 0. So this is valid. **Answer: (d)**

26. L is regular implies that L^R is also regular. Regular sets are closed under concatenation. So, {uv |

$u \in L, v \in L^R\}$ is also regular. S1 is correct. Languages of palindrome strings are not regular because a stack of unlimited size is needed to determine the acceptance/rejection of such languages. It can also be checked using pumping lemma if $\{ww^R\}$ is not a regular language. S2 is not correct. **Answer: (c)**

27. Regular languages are closed under homomorphism. A homomorphism is a function from strings to strings. What makes a function a homomorphism is that its output on a multi-character string is the concatenation of its outputs on each individual character in the string. Or, equivalently, $h(xy) = h(x)h(y)$ for any strings x and y. If S is a set of strings, then h(S) is $\{w : w = h(x)$ for some x in S$\}$. In this problem, even(L) and Chop(L) are both effectively homomorphisms. So, both even(L) and Chop(L) are regular. **Answer: (b)**
28. $(0 \mid \varepsilon)1 + 2^* (3 \mid \varepsilon)$ is effectively representing a union of four different regular expressions
    01+2*3 → Strings of length 4 generated by this expression are *0113, 0123*
    1+2*3 → Strings of length 4 generated by this expression are *1113, 1123, 1223*
    01+2* → Strings of length 4 generated by this expression are *0111, 0112, 0122*
    1+2* → Strings of length 4 generated by this expression are *1111, 1112, 1122, 1222*
    A total of **12** distinct strings are possible. **Answer: (d)**
29. A string of length *n* has *n* characters, and for each character, there are two possibilities: 0 or 1. So, in total, there are $2 \times 2 \times 2 \times \ldots \times 2$ [*n* times] $= 2^n$ possibilities. So, the number of possible strings $= 2^n$. **Answer: (d)**
30. Null language = Empty language = φ. By basic definitions, $\varphi^* = \varphi$, and obviously $\varphi^+ = \varphi$. (i) and (ii) clearly satisfy $L^+ = L^*$. $(a^*)^* = a^*$, and $(a^*)^+ = a^*$. (iii) clearly satisfies $L^+ = L^*$. $aa^* = a^+$. $(a^+)^* = a^*$, but $(a^+)^+ = a^+ \neq a^*$. Option (iv) clearly does not satisfy $L^+ = L^*$. **Answer: (c)**
31. The string must contain aba as a substring. This means that there can be any sequence (including empty sequence) of *as* and *bs* towards the left and right of aba. The only possible match is (a + b)*aba (a + b)*. a*b*a accepts string a, which is invalid. a*ba* accepts string b, which is also invalid. (ab)*a accepts string a, which is also invalid. **Answer: (d)**
32. The ε-moves can be eliminated to reduce the given finite state automaton to the following.

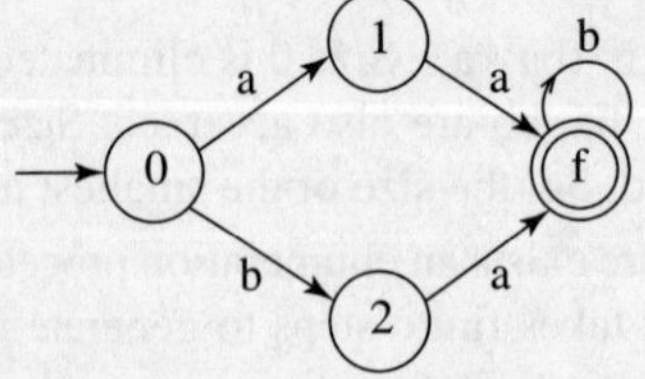

FIGURE A6 FSA without ε-moves

By basic observation, we can see that the regular expression for this automaton is (aa + ba)b* = aab* + bab*. **Answer: (a)**

33. The string needs to have exactly three *b*s: _b_b_b_. The _ in the string can be empty or contain any number of *as*. The regular expression is a*ba*ba*ba*. **Answer: (b)**
34. All strings generated by (ab|abb)*bbab must end with bbab. The only string among the given options that violates this definition is abababab. **Answer: (c)**
35. Let L = (0 + 1)*. It has one proper subset, M = {set of all binary strings whose length is a **prime** number}. M ⊆ L, but M is not regular, and L is regular. Options (a) and (d) are false. Let L = {set of all binary strings with length = prime number}. In this case, all finite subsets of L are regular,

but L is not regular. Option (c) is false. If all proper subsets of L are regular, then one proper subset of L will be some set K which has exactly 1 element less than L. If K is regular, then K union with {1 element} is also regular ⇒ L is regular. Option (b) is true. **Answer: (b)**

36. Let there be a k-state FSA accepting the language L_1. Let w = a^n such that $|w| \geq k$. Let w = xyz, such that $|xy| \leq k$ and $|y| \geq 1$. Let x = a^p, y = a^q, z = a^r. $(p + q + r)$ = prime. Consider the string xy^2z. $|xy^2z| = p + 2q + r$. If (p + q + r) is prime, then there is no guarantee that $(p + 2q + r)$ will be prime. For example, $p = 2$, $q = 3$, $r = 6$. $p + q + r = 2 + 3 + 6 = 11$ = prime. But, $p + 2q + r = 2 + 6 + 6 = 14$ = not prime. So, L_1 is not regular as it contradicts pumping lemma.

    If we apply pumping lemma for L_2, then $(p + q + r)$ is not prime does not necessarily imply that $(p + 2q + r)$ will also be not prime. For example, $p = 1, q = 2, r = 6. p + q + r = 1 + 2 + 6 = 9$ = not prime. But, $p + 2q + r = 1 + 4 + 6 = 11$ = prime. So, L_2 is not regular as it contradicts pumping lemma. **Answer: (a)**

37. Regular languages are closed under intersection. So the intersection of any number of regular languages is regular. Option (i) is true. Consider an alphabet Σ = {0, 1}. Consider the language Σ*, which is clearly regular. It has a subset S = {w | w ∈ Σ* and |w| is prime}, which is not regular. Option (ii) is false. Again, with the alphabet Σ = {0, 1}, let L = {w | w ∈ Σ* and |w| is an odd integer} and M = {w | w ∈ Σ* and |w| is prime}. Clearly, L . M = {w | w ∈ Σ* and |w| = 3, 4, 5, 6 and even numbers > 6} is regular. Option (iii) is false. **Answer: (c)**

38. (a + b)* = (a*b*)* ≠ a*b*. Option (a) is false. a(b + c) = ab + ac ≠ ab + c. Option (b) is false. (a + b)* ≠ a* + b*. Option (c) is false. (ab + a)*a = (a(b + ε))*a. Apply the rule (EF)*E = E(FE)* with E = a, F = (b + ε) to get (ab + a)*a = (a(b + ε))*a = a((b + ε)a)* = a(ba + a)*. Option (d) is true. **Answer: (d)**

39. Null/Empty string is a valid string accepted by the specified language. Options (c) and (d) are eliminated because both do not accept the null string. In fact, the smallest string accepted by both is 11. 0 is a valid string accepted by the specified language. Options (a) and (b) are also eliminated. (0∗10∗1)∗0∗ clearly accepts strings composed of only zeros, and the blank string as well. Also, whenever a 1 occurs in the string, it always occurs in even numbers. **Answer: (e)**

40. The string must contain at least two zeros. So, each string in the language must be of the form w0x0y, where w, x, y can be any sequence of zeros and ones, including empty strings. The desired regular expression is (0 + 1)*0(0 + 1)*0(0 + 1)*. **Answer is (d).**

41. The string can start with any number but must end with 01. Also, the smallest possible valid string is 01; hence, the desired regex is (0 + 1)*01. **Answer: (c)**

42. Recall the basic definition of CNF. Every rule is of the form A→BC or A → x or A → ε. So, each and every non-null terminal symbol on the RHS is derived by exactly one non-terminal symbol on the LHS. Proceed along the lines of mathematical induction.

    l = 1 ⇒ Exactly one production rule of type A → x.

    l = 2 ⇒ Only one possibility of the type A → BC → xy.

    Now, observe a pattern. If a string of length k can be derived in k number of steps, then a string of length $(k + 1)$ can be derived in exactly $(k + 1)$ number of steps. Thus, to derive a string of length l, we need exactly l number of steps. **Answer: (d)**

43. Consider a string 010101010011, which is one possible string accepted by L. Clearly, 01010101 is one possible prefix; this eliminates options (a) and (c). Now, consider the set of all possible binary strings. Recall that each string is its own prefix as well. Any binary string w can be one of three types: (i) If w has an equal number of zeros and ones, then it is clearly a prefix of strings of the

type w(01)* $\subseteq$ L. (ii) If w has k more zeros than 1s, then also it is a prefix of the string formed by appending exactly k number of 1s, and that string will be a member of L. (iii) If w has k more ones than 0s, then also it is a prefix of the string formed by appending exactly k number of 0s, and that string will be a member of L. **Answer: (b)**

44. Context-free languages can be accepted by deterministic as well as non-deterministic PDA. Option (a) is not very accurate. Context-free languages are closed only under union and Kleene closure. **Answer: (b)**

45. Even-length palindromes are not generated by this grammar. Observe along the lines of mathematical induction. All strings beginning with a end with a, and the same holds for b as well. This grammar actually generates all odd-length palindromes. **Answer: (b)**

46. (Recursively enumerable – Recursive) is recursively enumerable. So option (a) is always true. Option (b) is not always true. L1 – L3 = L1 intersection (complement L3). L1 is recursive, L3 is recursively enumerable but not recursive. Recursively enumerable languages are not closed under complement. Options (c) and (d) are always true as recursively enumerable languages are closed under intersection and union. **Answer: (b)**

47. Regular languages are closed under complement operation. As L(A) is regular, the complement of L(A) is also regular. By **Chomsky hierarchy**, every regular language is also context-free, because context-free languages are a superset of regular languages. So, statement 1 is correct. This immediately eliminates options (b), (c) and (d). However, for the sake of understanding, we present the explanation for all the statements given in the problem.

    By observation, it can be seen that L(A) = (11*0 + 0)(0 + 1)*. Statement 2 is wrong. No further minimisation of L(A) is possible. Statement 3 is correct. The smallest string accepted by the given FA is 0 – it is of length 1 – so the least length is 1 and not 2. Statement 4 is wrong. So, only statements 1 and 3 are correct. **Answer: (a)**

48. L2 is a known context-free language. So, option (a) is correct. Observe carefully that L1 $\cap$ L2 = L2, which is context-free. So, option (b) is also correct. The complement of L2 is recursive, which is true due to the fact that the complement of a CFL is a CSL. This in turn is a subset of recursive languages. So, option (c) is also correct. L1 is a regular language expressible by the regex 0*1*0*, and its complement is also a regular language as regular languages are closed under complementation. Option (d) is incorrect. **Answer: (d)**

49. X is a recursive language, so X′ is also recursive. Y is recursively enumerable, but not recursive, so Y′ is not a recursively enumerable language. A is reducible to B, that is, solving A cannot be "harder" than solving B.

    1. If A is reducible to B and B is decidable, then A is decidable.

    (i) If A is reducible to B and B is recursive, then A is recursive.

    2. If A is undecidable and reducible to B, then B is undecidable.

    (i) If B is recursively enumerable and A is reducible to B, then A is recursively enumerable.

    (ii) If A is not recursively enumerable and reducible to B, then B is not recursively enumerable.

    Y′ is not recursively enumerable and reducible to W, then W is not recursively enumerable (using 2(ii)). Z is reducible to X′ and X′ is recursive, then Z is recursive (using 1(i)). **Answer: (c)**

50. Observe the grammars, and apply mathematical induction. $L_1 = \{a^n b^n \mid n \geq 0\}$. It can be accepted only by a PDA and is a known context-free language. L_1 is not regular, implying that statement P is false. L_2 = (ab)* and is obviously regular. Hence, statement Q is true. **Answer: (c)**

51. Observe the grammar carefully – it generates strings where the number of as $\geq$ the number of bs. How? The rules S → aSb | bSa | SS | ε will generate strings with exactly equal number of as and

bs (including the empty string). The rule S → SaS introduces an extra *a* every time it is applied anywhere while expanding the derivation. So, it is impossible to have a string with more *bs* than *as*. *babba* is the only string amongst the options listed that has more *bs* than *as*. **Answer: (d)**

52. Both L_1 and L_2 are known to be context-free languages. Context-free languages are closed under union operation. So, $L_1 \cup L_2$ is context-free. $L_1 \cap L_2 = \{a^n b^n c^n \mid n \geq 0\}$ can be easily shown to be not context-free using pumping lemma. **Answer: (a)**
53. Apply the techniques for deriving regular expressions from grammar, as discussed in this book. A ≡ aa*. So, AB ≡ aa*b. So, one possible set of strings derived by S is aa*b.
Also, observe the rule S → AS, which reduces to S → AA*S when applied repeatedly, and in turn reduces to S → (aa*)(aa*)*S. Thus, the regex corresponding to S is finally expressible as (aa*)(aa*)* aa*b + aa*b, which can be simplified to aa*b. **Answer: (b)**
54. Context-free languages are closed under union, concatenation and Kleene closure only. There is an easy counterexample to show that intersection of context-free languages is not necessarily context-free. Let $L_1 = \{a^n b^n c^m \mid m, n \geq 0\}$ and $L_2 = \{a^m b^n c^n \mid m, n \geq 0\}$. Both are context-free languages. $L_1 \cap L_2 = \{a^n b^n c^n \mid n \geq 0\}$ can be easily shown to be not context-free using pumping lemma. **Answer: (c)**
55. The grammar that accepts ww^R including the null string is S → aSa | bSb | ε. You can easily check that all the other grammars can accept the string *ab*, which is invalid, as it is not a palindrome. **Answer: (b)**
56. Try to derive some minimal valid strings that can be accepted by this grammar. The smallest valid string is *a* – this is an odd number of *as and* an even number of *bs*. Another possible string is S → aBB → aB → abB → ab – this is an odd number of *as* and an even number of *bs*. So, the even/odd property of the number of *bs* is simply irrelevant. Apply expansion repeatedly to check if the number of *as* is always odd. **Answer: (b)**
57. *First set of grammars:* S → aSb | bSa | SS | a and S → aSb | bSa | a. The first grammar has an extra rule, S → SS, which cannot be derived while expanding the rules of the second grammar. So, these two grammars are not equivalent, and hence, S1 is correct.
Second set of grammars: S → SS | SSS | aSb | bSa | λ and S → SS | aSb | bSa | λ. The first grammar has an extra rule, S → SSS, which can be derived as S → SS → SSS. So, the rule S → SSS is practically redundant in the first grammar and can be removed. After removal of this redundant rule, we can see that both grammars are equivalent. S2 is correct. **Answer: (b)**
58. The straightforward form of the desired grammar for the language $L = \{a^n b^{n+1} \mid n \geq 0\}$ is given by S → aSb | b. This matches option (d), but it is not in Griebach Normal Form. We need to convert this grammar to GNF. The rule S → aSb violates the GNF format. Introduce a new rule B → b, and modify the grammar accordingly, as S → aSB | b, B → b. **Answer: (c)**
69. The broadest category of grammar that has only a singie non-terminal symbol on the left is context-free grammar. **Answer: (a)**
60. Try various leftmost derivations: S → aS → aaS → aaa; S → aS → aSa → aaa; S → Sa → Saa → aaa; S → Sa → aSa → aaa. We can apply either S → aS or S → Sa at each step, and there are 2 × 2 = 4 ways of doing so, as shown. **Answer: (c)**
61. Observe that X → aX | bX | ε is equivalent to the regular expression (a + b)*. Refer to the section which discusses the derivation of regular expressions from grammars. Also, Y → bbb. So, S → XYX is equivalent to (a + b)*bbb(a + b)*. **Answer: (c)**
62. Use pumping lemma to show that L1 is not regular. For L2, a regular expression exists for representing any subset of 0*. As there is a regular expression r for each subset of 0*, there exists a regular r* which represents the language L2*. So, L2* is regular. **Answer: (b)**

63. The regular expression for the given language is $0^+1^* = 00^*1^*$. It is clearly a regular language. **Answer:(c)**
64. CSL is a superset of CFL. Statement (d) is false. **Answer: (d)**
65. By definition, the grammar A → Ab | a is left-recursive. It generates the language defined by the regular expression ab*. The smallest string accepted by this language is *a*. Observe that the grammar A → aB; B → bB | b cannot accept the string 'a', and hence, this grammar is not equivalent to the first. (i) is true and (ii) is false. **Answer: (c)**
66. L = $\{a^nb^n \mid n \geq 1\}$is a well-known example of a language that can be accepted by PDA. So, it is CFL. It can be easily proved to be not regular using pumping lemma. **Answer: (a)**
67. All rules in a grammar of Chomsky Normal form are of the form S → AB or S → x. The parse tree is of length $\leq j$, which implies that the derivation tree (indeed a binary tree) has depth *j*. The largest possible word is of length 2^j, which is simply the maximum possible number of leaves in a binary tree of depth *j*. **Answer: (d)**
68. A grammar with more than one parse tree for at least one string is ambiguous. **Answer: (d)**
69. Every context-free language is accepted by a PDA. Every regular language is accepted by a DFA. So, the strings in the language formed by the intersection of a regular and context-free language are accepted by both DFA as well as PDA. Hence, the intersection is guaranteed to be context-free. **Answer: (a)**
70. The definition of pumping lemma is itself based on the principle of "repeatability" of substrings, which is nothing but the Pigeonhole principle. **Answer: (a)**
71. Consider these languages over the alphabet Σ = {x, y, z}. L_1 = {set of all strings over Σ where the number of *y*s is divisible by 2}; L_2= {set of all strings over Σ where the number of *y*s is divisible by 7}; L_3 = {set of all strings over Σ where no *x* appears after *z*}. The language in this problem is essentially (L1 ∪ L2)′ ∩ L3.

    Regular expressions can be easily derived for each of the three languages: L1, L2 and L3. L_1 = ((x + z)*y(x + z)y*(x + z))*; L_2 = ((x + z)*y(x + z)y*(x + z) y(x + z)y*(x + z) y(x + z) y*(x + z) y(x + z))*; L_3 = (x + y)*(y + z)*. It is known that regular languages are closed under all Boolean operations; hence the language $(L_1 \cup L_2)' \cap L_3$ is regular. **Answer: (a)**
72. Each rule is of the form A → aα, where a is a terminal symbol and α is a string of non-terminal(s). This is Greibach Normal Form. **Answer: (b)**
73. By definition, Φ* = ε (null string). **Answer: (b)**
74. Both languages are recursive as can be seen from the definitions of the languages. Also, a PDA can be easily designed for L1. However, using pumping lemma for context-free languages, L2 can be easily shown to be not context-free. **Answer: (b)**
75. Observe carefully the regular expression xyx = (a + b)*(a + b)*(a + b)*. Obviously, L1 is nothing but an empty set. It is a regular language. All strings in L2 are of the form ww, where w ∈ L. A language that accepts strings of the type ww is known to be context-free. So, L1 is regular and L2 is context-free. **Answer: (c)**
76. Apply the concept of expression grammars with proper operator precedence. **Answer: (b)**
77. Recursively enumerable languages are not closed under complement operation, but closed under union and intersection. **Answer: (d)**
78. Apply the technique of deriving regular expression from grammars. C ≡ y, B ≡ xw*, A ≡ By + Cw = xw*y + yw, S ≡ Ax + By = xw*yx + ywx + xw*y. **Answer: (a)**

79. Apply pumping lemma. Take an example string xyz where x = a^2, y = a^3, z = a^5. Clearly xyz = $a^{10} \in$ L, as 10 is non-prime. However, $xy^2z = a^{13} \notin$ L, as 13 is a prime number. So, pumping lemma for regular languages is violated. **Answer: (b)**
80. This language is expressed by a straightforward regular expression (aa)*, and hence, is a regular language. **Answer: (a)**
81. All three are well-known undecidable problems. The first two statements obviously give undecidability situation and these are already discussed in the earlier chapters. Regarding the third statement: "Determining whether the language accepted by Turing machine is finite or infinite", it makes clear that the language is Turing acceptable (not Turing decidable). So, it is REL which may be finite or infinite. But deciding the finiteness or infiniteness of REL is undecidable problem. **Answer: (d)**
82. In option (a), put $i = j = 1$ and $k = i + j = 2$. The string *abaa* cannot be derived from this grammar. In option (b), the smallest possible palindrome strings like *aa*, *bb*, *abba*, etc. cannot be derived from this grammar. In option (c), for strings with an equal number of *a*s and *b*s, we cannot derive the string *bbaaab*. All the strings generated by this grammar are of even length. This is because application of each expansion of S adds exactly zero or two non-terminal symbols. So, all possible even-length strings can be derived. **Answer: (d)**
83. As there is only one possible expansion for B, this grammar can be rewritten as S → AbbA, A → a | bbAabbA. Observe that both possible expansions of rule A have an odd number of symbols on the RHS. So, whatever expansion of A is applied at any stage, we will always end up with some expansion of S that has an even number of symbols in the finally generated string. This is because the rule for S has two occurrences of the symbol A on the RHS, and there are two terminal symbols on the RHS for S.

    Analyse carefully in terms of odd and even numbers, and it is seen that all strings generated by this grammar are of even length. Further observe that any expansion of the symbol S always has the symbol A either just before or just after bb. And, expansion of A either starts with a or bb. So, an odd number of consecutive *b*s is impossible. Three consecutive *a*s is also impossible because no possible expansion will result in the fragment AAA anywhere in the expansion. However, it is possible to end up with a string having four consecutive *b*s. S → AbbA → abbA → abbb-bAabbA → abbbbaabbA → abbbbaabba. So, option (d) is false. **Answer: (d)**
84. **Part 1:** The basic definitions of grammar show that this grammar is context-free, as all the rules have only one non-terminal symbol on the LHS. **Answer: (a)**

    Part 2: S → SS | 0S1 | 1S0 | ε. Observe that application of any expansion of the non-terminal S always adds either nothing or exactly one instance of both 0 and 1 to the currently generated sub-string. So, each generated string will have an equal number of 0s and 1s. **Answer: (d)**
85. Regular sets are closed under *all* Boolean operations. So (i) is true. If an FA with *n* states accepts any string with length $l > n$, it always accepts an infinite number of strings. The constraint mentioned in the statement is $n \le l \le 2n$, and it includes the case $l > n$. So, (ii) is false. The definition of the Moore machine is stated correctly. So, (iii) is true. (ii) is the only false statement here. **Answer: (b)**
86. Observe that L ∪ R = L = Σ* which is known to be a regular language for any alphabet Σ. So, L ∪ R is regular. R = $\{0^n1^n$ such that $n > 0\}$ can only be accepted by a PDA, and can be easily shown to be non-regular using pumping lemma. **Answer: (c)**
87. Context-free languages are closed under concatenation and union, but not under intersection. Also, intersection of a regular set and a context-free language is always context-free. **Answer: (b)**

88. This is a straightforward problem involving Arden's theorem.
Use $X_2 = 0X_1 + \varepsilon$ in the second equation to obtain
$X_1 = 0X_1 + 1(0X_1 + \varepsilon) = (0 + 10)X_1 + 1$
$\Rightarrow X_1 = 1 + (0 + 10)X_1$.
Apply Arden's theorem to get $X_1 = 1(0 + 10)^*$.
Use this in the first equation to obtain $X_0 = 11(0 + 10)^*$. **Answer: (c)**

89. Use techniques for generating regular expressions from grammars. The symbol A generates a^+. Use this to obtain the modified rule set as: $S \to a^+B \mid a^+S$, $B \to b \Rightarrow S \to a^+b \mid a^+S \Rightarrow S \to aa^*b \mid aa^*bS \Rightarrow S \to (aa^*b)^+$. **Answer: (a)**

90. Applying pumping lemma, we can easily show that this language is neither regular nor context-free. Apply pumping lemma for regular languages, as well as context-free languages. **Answer: (d)**

91. The language $\{a^m b^n c^p d^q \mid m + p = n + q$, where $m, n, p, q, \geq 0\}$ can also be expressed as $\{a^m b^n c^p d^q \mid m - n + p - q = 0$, where $m, n, p, q \geq 0\}$. This can be achieved by a sequence of push (if 'a' and stack top is empty or 'a'), pop (if 'b' and stack top is 'a'), push (if 'c' and stack top = 'a' or 'b'), pop (if 'd' and stack is not empty). Finally check if the stack is empty after the entire string has been processed. It can be accepted by PDA with empty stack, so I is context-free.

The language $\{a^m b^n c^p d^q \mid m = n$ and $p = q$, where $m, n, p, q \geq 0\}$ is essentially the concatenation of two CFLs: $\{a^m b^m$, where $m \geq 0\}$ and $\{c^p d^p$, where $p \geq 0\}$. As CFLs are closed under concatenation, II is also context-free.

The language $\{a^m b^n c^p d^q \mid m = n = p$ and $p \neq q$, where $m, n, p, q \geq 0\}$ is essentially the same as $\{a^p b^p c^p d^q \mid p \neq q$, where $p, q \geq 0\}$. It is known that the language $\{a^p b^p c^p$, where $p \geq 0\}$ is not context-free and can be easily proved by pumping lemma. So, III is not context-free.

The language $\{a^m b^n c^p d^q \mid mn = p + q$, where m $m, n, p, q \geq 0\}$ can also be proved to be not context-free by applying pumping lemma. So, IV is also not context-free. **Answer: (b)**

92. Every grammar having a rule of the form $S \to SS$ is ambiguous. So, (i) is true. The grammar $S \to 0S1 \mid 1S0 \mid \varepsilon$ is not ambiguous as every string has only one parse tree, and the same can be proved by applying mathematical induction. So, (ii) is false. Observe the grammar $S \to T/U$, $T \to xSy/xy/\varepsilon$ $U \to yT$. Use this expansion $S \to U \to yT \to yxSy \to yxTy \to yxxyy$. So, (iii) is also true. **Answer: (b)**

93. All three languages listed here are known to be context-free. **Answer: (d)**

94. Regular expressions exist for both these languages. For L2, the regular expression is aaaaaaa*bbbbb* + aaaaaaa*bbbbb*a. For L1, it is possible to work out the regular expression for the language. $L3 = \{a^n b^l a^k : n + 1 + k \leq 5\}$. Observe mathematically that there are only a finite number of non-negative integer solutions for the inequality $n + 1 + k \leq 5$. $L1 = \Sigma^* - L3$. As regular languages are closed under all Boolean operations and both Σ^* and L3 are regular languages, L1 is regular. **Answer: (b)**

95. The language L can be rewritten as the union of three languages: $\{a^m b^m c^n\} \cup \{a^m b^n c^m\} \cup \{a^m b^n c^n\}$. Each of these three languages can be easily accepted by PDA; hence, all are context-free. Also, context-free languages are closed under union. Further, using pumping lemma for regular languages, it can be shown that none of these three languages is regular. So, the given language in the problem statement is context-free but not regular. **Answer: (b)**

96. Apply the techniques of generating regular expressions from Greibach Normal Form (GNF). All three grammars are specified in GNF. G1 generates the language a*b. G2 generates the grammar aa*b. G3 generates the language a*b. [*Note: Here A is the start symbol instead of S*]. Clearly

G1 and G3 generate the same language. G2 is a different language, a*b ≠ aa*b. Another way to observe it is that G2 cannot accept the string 'b'. **Answer: (b)**

97. Recall the Chomsky hierarchy of languages. Only option (d) is correct; all other statements are invalid. Further, there is no definition of Type 4 grammar according to Chomsky hierarchy. **Answer: (d)**

98. For the grammar G1, two derivations are possible for the empty string ε. A → ε and A → A → ε. So, G1 is ambiguous. For the grammar G2, two derivations are possible for the string 'a'. A → aA → a and A → Aa → a . So, G2 is also ambiguous. **Answer: (d)**

99. Derivation of the smallest string 'a' is clearly unambiguous. The derivation of string 'a + a' is not ambiguous. Either use leftmost derivation or rightmost derivation, but not a mix of both. To derive the string 'a + a + a' using leftmost derivation, we have a choice as to which A to replace with A + A. So, there is ambiguity. **Answer: (c)**

100. Observe that each expansion of the symbol S adds two 1s and one 0 to the existing string. The smallest possible string is the empty string. So, the number of 1s is twice the number of 0s. **Answer: (d)**

> The exam in which this question was asked clearly specifies that **only one option** is correct. So, (d) is the most accurate option. However, it should be observed that option (a) is also correct, as the number of ones in the string will always be even. But, as there is only one correct option, this option does not accurately and completely describe the nature of generated strings. If the same question had stated that **one or more options** are correct, then candidates would have to mark both **(a)** and **(d)** to obtain the full marks for this question.

101. The grammar for this language is S → 0S0 | 1S1 | 0 | 1 | ε, which is clearly a context-free grammar. This language can be easily proved to be non-regular using pumping lemma for regular languages. **Answer: (b)**

102. The language X = $\{0^m1^m \mid m \geq 1\}$ is deterministic as it can be accepted by a PDA. Also, L_1 = X ∪ {0} is deterministic as it is the union of two deterministic languages. Determining whether a string is the substring of its reversed string is a decidable problem, but it cannot be solved by a deterministic automaton. This is because whenever there is a mismatch, we need to reset and start again from another symbol. The symbol to be taken as the first one to begin the matching process cannot be determined in advance in O(1) time, and hence, the language L2 is non-deterministic. **Answer: (a)**

103. G1 is the standard grammar for generating non-empty odd-length palindrome strings, and is known to be unambiguous. For G2, try the derivation of string 000. There are two ways of doing this: S → 0S → 00S → 00X → 000 and S → 0S → 0X → 0X0 → 000. As there are two parse trees for the same string, G2 is ambiguous. **Answer: (b)**

104. Recursively enumerable languages (which are recursive languages) are a superset of context-sensitive languages. So, both S1 and S2 are correct. **Answer: (c)**

105. This language cannot be regular as it involves matching, and hence, memory of past symbols is required. It can be accepted by a PDA and is therefore a context-free language. **Answer: (d)**

106. Union as well as intersection of context-free and regular languages is context-free, so L5 and L7 are context-free. Context-free languages are closed under union, so L6 is also context-free. **Answer: (c)**

107. If a string uvw ∈ L such that |v| > 0, then as per the given definition $uv^kw \notin L$ for any value $k > 1$. This means that L violates the pumping lemma for regular languages. So, L is definitely not a regular language; it can be Type 0, Type 1 or Type 2. **Answer: (a)**

108. Observe the table carefully. The final state can be reached if and only if the last input symbol is 0. Options (a) and (c) are eliminated. Observe that f(B, 0) = C and f(D, 0) = C. Further, we are interested in the shortest input sequence to reach state C from an unknown initial state. Determine if there is an input transition f(x, 1) = B or f(x, 1) = D. The answer is yes. f(B, 1) = f(C, 1) = f(D, 1) = B. So, there are multiple possible state transitions which will lead to state C through symbols 10.

For example, $B \xrightarrow{1} B \xrightarrow{0} C$. Here B is the starting state.

$D \xrightarrow{1} B \xrightarrow{0} C$. Here B is the starting state. **Answer: (b)**

109. We are considering binary strings that end with 00 or 11. The regular expression for this set of strings is (0+1)*(00+11). This problem has no possible shortcut unless and until the candidate has enough experience to quickly visualise and draw the DFA.

Step 1: The initial skeleton is as follows.

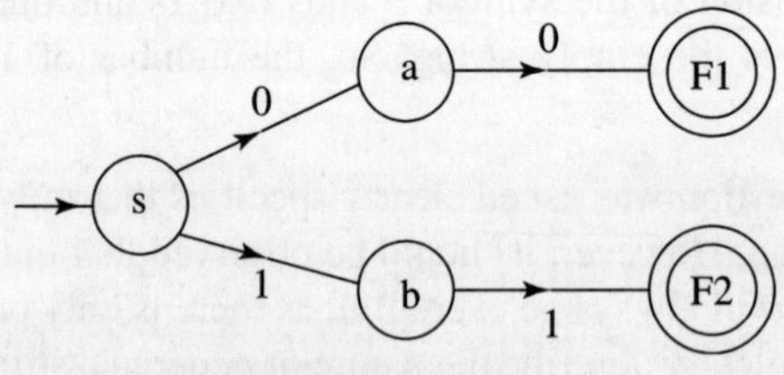

FIGURE A7 Step 1

Step 2: There are two final states, F1 and F2. We need to define state transitions for all input symbols for all states. If you encounter a 0 at F1, perform a self-loop; else go to state b. It will handle the case if the string ends with 11. Symmetrically, if you encounter a 1 at F2, perform a self-loop; else go to state a. It will handle the case if the string ends with 00. δ(F1, 0) = F1, δ(F2, 1) = F2, δ(F1, 1) = b, δ(F2, 0) = a.

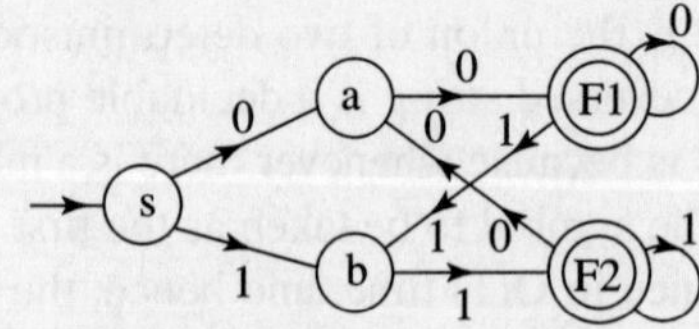

FIGURE A8 Step 2

Step 3: What is missing is input 1 for state a and input 0 for state b. Also observe that we are not handling strings 1011 and 0100. This is achieved by adding the state transitions δ(a, 1) = b and δ(b, 0) = a.

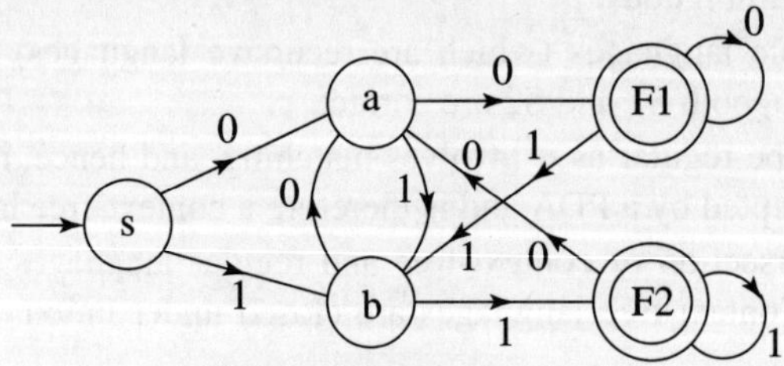

FIGURE A9 Step 3

This DFA accepts all strings accepted by (0+1)*(00+11). It has a total of 5 states. **Answer: (b)**

110. Observe the state transition $A \xrightarrow{1} B \xrightarrow{1} C$. Input sequence is 11, but output is 10 and not 01. Option (b) is eliminated. Observe the state transition $A \xrightarrow{1} B \xrightarrow{0} A$. Input sequence is 10, but output is 01 and not 00. Option (c) is also eliminated. Now, observe the case of bit addition, as per the following table.

TABLE A1 Bit addition

First bit	Second bit	Sum	Satisfied by state transitions
0	0	00	A→A→A B→A→A C→A→A
0	1	01	B→A→B A→A→B C→A→B
1	0	01	A→B→A B→C→A
1	1	10	A→B→C B→C→C C→C→C

Further observe that all possible state transitions A→A, A→B, B→A, B→C, C→A, C→C are covered by this table. **Answer: (a)**

111. Clearly, any string accepted by this DFA can start with either 0 or 1. So, option (a) is valid. Also, every string must end with at least two consecutive zeros (00), as there is no other way to reach the final state after starting from the initial state. Option (c) is also valid. As every string ends with 00, obviously 00 is also a substring of this string. Option (d) is also valid. Every string ending with 00 obviously ends with 0. Option (b) is also valid. **Answer: (a), (b), (c), (d)**

If the problem states that there is *only one* correct option, the strongest and most unambiguous statement is option (c).

112. $n > 0$ and $k > 0$.

Observe the following DFAs and proceed along the lines of mathematical induction.

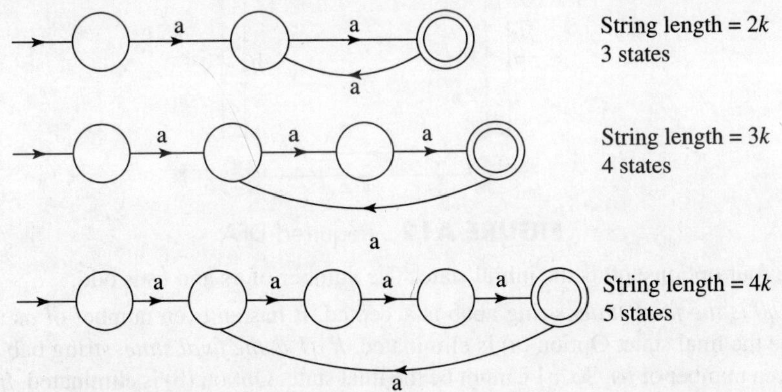

FIGURE A10 DFAs along the lines of mathematical induction

For accepting strings of length nk, we need $(n + 1)$ states. **Answer: (b)**

113. To derive the regular expression, name the states as follows.

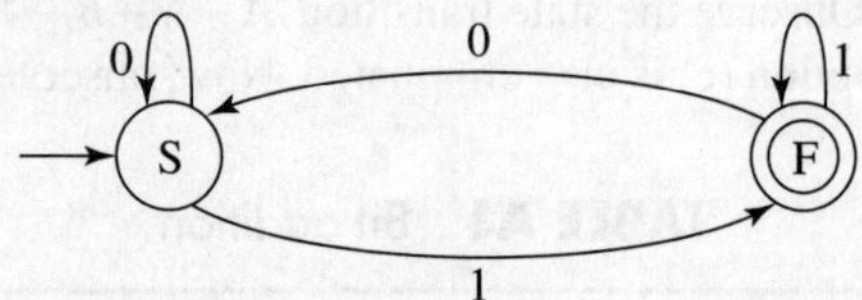

FIGURE A11 DFA with named states

The state equations are:

S = S0 + F0 + ε (i)

F = S1 + F1 (ii)

Use Arden's theorem several times. So, equation (i) gives

S = (ε + F0)0* = 0* + F00* (iii)

Using (iii) in (ii), we get F = 0*1 + F00*1 + F1 ⇒ S = 0*1(00*1 + 1)*. This matches I. Also, 0*1(00*1 + 1)* = 0*1{(00* + ε)1}* = 0*1(0*1)* = (0*1)*(0*1) = (0 + 1)*1. This matches III. II is essentially a subset of I. **Answer: (b)**

114. The minimum possible length of the strings present in L is 2 + 1 + 3 = 6. To accept a string of length 6 or more, we need at least 7 states. *[Similar results have been proved in earlier problems].* **Answer: 7**

115. Observe the string (iv) abaabaabaabaabaabaabaabaaba = **aba**aba**aba**aba**aba**aba**aba**aba**aba** = $(\text{aba})^9 = (aba)^{3^2}$

The same string can also be expressed as **abaabaabaabaabaabaabaabaaba** = $\text{a}.(baa)^3$. $\text{ba} = \text{a}.(baa)^{3^2-1}.\text{ba}$

And, the same string can also be expressed as aba**aba**aba**aba**aba**aba**aba**aba**aba = $\text{ab}.(aab)^8$. $\text{a} = \text{ab}.(aab)^{3^2-1}.\text{a}$

Apply mathematical induction to prove that all three formats are equivalent. **Answer: (d)**

116. Draw the DFA as follows.

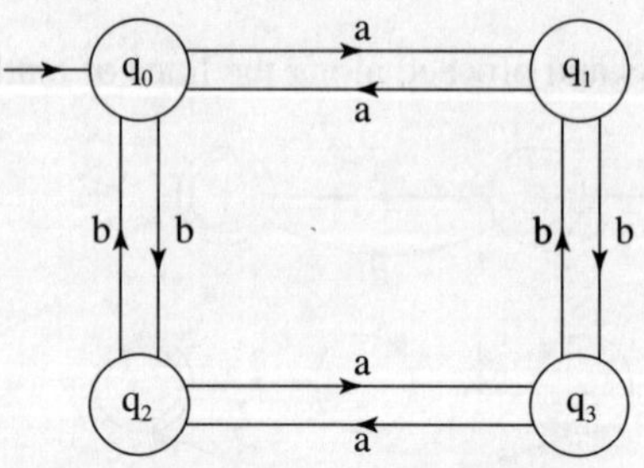

FIGURE A12 Required DFA

From the four options, q0 is the initial state. The number of *as* and *bs* is odd.

If q0 is the final state, string abab is accepted. It has an even number of *as* and *bs*. So, q0 cannot be the final state. Option (a) is eliminated. *If q1 is the final state*, string bab is accepted. It has an even number of *bs*. So, q1 cannot be the final state. Option (b) is eliminated. *If q2 is the final state*, string aba is accepted. It has an even number of *as*. So, q1 cannot be the final state. Option (c) is eliminated. *If q3 is the final state*, the number of *as* and *bs* in any accepted string will be odd.

From the initial state, you will need to traverse through an odd number of *a*s and *b*s to reach the final state, q2. **Answer: (d)**

117. L1 and L2 are known CFLs accepted by PDAs. For L3, there is a simple equivalent regular expression (00)* (assuming that i can be zero) or $(00)^+$ (assuming $i > 0$). L4 can be proved to be non-regular as well as context-free by applying the corresponding pumping lemmas. **Answer: (d)**
118. L_1 is expressed by the regular *expression* a*. L_2 is expressed by the regular expression b*. Clearly, $L_1.L_2 = a^*b^* = \{a^mb^n, m \geq 0, n \geq 0\}$. So, (I) is correct and (II) is wrong. **Answer: (a)**
119. Assume L1 is regular. Let there be an m-state DFA accepting L1. Let $w = (ab)^na^k$. Clearly, $|w| \geq m$, that is, $2n + k \geq m$. Let w = xyz, where $|xy| \leq m$ and $|y| \geq 1$. Let $x = (ab)^{(n-1)}$, y = aba, $z = a^{(k-1)}$. Clearly, $xy^2z = (ab)^{(n-1)}(aba)^2a^{(k-1)} \notin$ L1. So, L1 is not regular. Assume L2 is regular. Let there be a k-state DFA accepting L2. Let $w = a^nb^m$, $n \neq m$, and $(n + m) \geq k$. Let w = xyz, where $|xy| \leq k$ and $|y| \geq 1$. Let $x = a^{(n-p)}$, $y = a^pb^q$, $z = b^{(n-q)}$. Clearly, $xy^2z = a^nb^qa^pb^n \notin$ L2. So, L2 is not regular. **Answer: (d)**
120. Draw a DFA for the specified language.

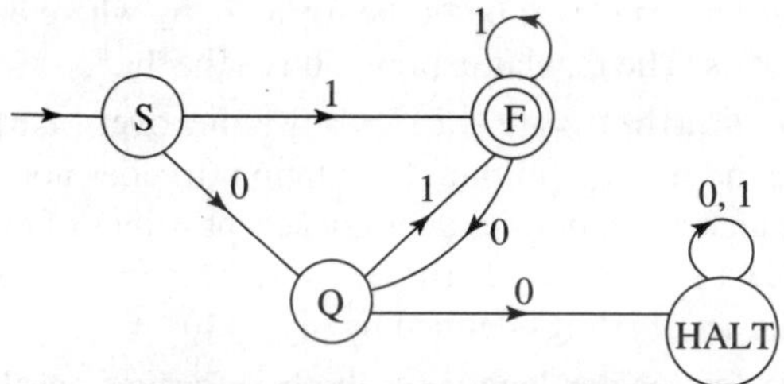

FIGURE A13 Required DFA

The state equations can be expressed as

$S = \varepsilon$ (i)

$Q = S0 + F0 = 0 + F0$ [using (i)] (ii)

$F = S1 + Q1 + F1$

$\Rightarrow F = 1 + 01 + F01 + F1$ [using (i) and (ii)]

$\Rightarrow F = (1 + 01) + F(1 + 01)$

$\Rightarrow F = (1 + 01)(1 + 01)^*$ [using Arden's theorem]

The regular expression for the above DFA is $(1 + 01)(1 + 01)^* = (1 + 01)^*(1 + 01)$. **Answer: (c)**

Shortcut: $(1 + 01)^*(10 + 01)$ accepts the string 110, which does not end with 1. Option (a) is eliminated. $(1 + 01)^*01$ does not accept the smallest possible string 1. Option (b) is also eliminated. $(10 + 01)^*01$ accepts the string 1001, which contains 00. Option (d) is also eliminated. $(1 + 01)^*(1 + 01)$ always accepts strings ending with 1, and every occurrence of 0 is followed by 1.

121. $L1 = \{uww^Rv \mid u, v, w \in \{a, b\}^+\}$.

L1 contains strings of length 4 or more. Observe that all strings of the form $(a + b)^+aa(a + b)^+$ and $(a + b)^+bb(a + b)^+$ are accepted by L1. So, L1 can be expressed by the regular expression $(a + b)^+(aa + bb)(a + b)^+$. L1 is a regular language.

L2 requires a check that $|u| \geq |\nu|$, which needs unlimited memory. L2 is not a regular language. **Answer: (a)**

122. L = (0 + 1)* is a regular language. R is a known CFL and can also be proved to be non-regular

using pumping lemma for regular languages. L ∪ R is the union of regular and context-free languages and is context-free. **Answer: (c)**

123. $L_1L_2^* \cup L_1^* = \varphi a^* \cup \varphi^* = \varphi \cup \varepsilon = \{\varepsilon\}$. **Answer: (a)**

124. L^2 is the language of concatenated pairs of strings, where each string in the pair satisfies the properties of L. L^2 is $\{x.y \mid x = a^nb^n \mid n \geq 0, y = a^mb^m \mid m \geq 0\} = \{a^nb^na^mb^m \mid n \geq 0, m \geq 0\}$. *Note that there is absolutely no correlation between m and n.* **Answer: (d)**

125. The state equations for this NFA can be written as follows: $q_0 = \varepsilon + q_0(a + b) \Rightarrow q_0 = \varepsilon(a + b)^*$ [using Arden's theorem] $\Rightarrow q_0 = (a + b)^*$.
$q_1 = q_0a = (a + b)^*a$ [using the above]. $q_2 = q_1(a + b) = (a + b)^*a(a + b)$ [using the above]. $q_3 = q_2b = (a + b)^*a(a + b)b$ [using the above]. **Answer: (d)**

126. L/L′ is regular means that the set of *w*s is regular, but there is no way to determine whether *x* is regular or not. Option (a) is false. If the set of strings $wx \in L$ can be described by a regular expression, the set of prefix strings *w* can also be described by another regular expression. Also, observe that regular sets are closed under concatenation. **Answer: (b)**

127. Strings accepted by this language will be of the form _b_b_ where each _ will be either blank or will have any number of *a*s. The regular expression is a*ba*ba*. **Answer: (a)**

128. $(a^* + b)^* = ((a^*)^*b^*)^* = (a^*b^*)^* = (a + b)^*$ = any string over {a, b}. Option (a) does not accept strings of the form aba and is hence eliminated. Option (b) does not accept strings of the type a^n, $n \geq 0$, and is hence eliminated. Option (d) does not accept strings of the type a^n, $n \geq 0$, and is hence eliminated. In option (c), $(a + b^*)^* = (a^*(b^*)^*)^* = (a^{*b*})^* = (a + b)^*$ = any string over {a, b}, which is the same as the set of strings defined by $(a^* + b)^*$. **Answer: (c)**

129. Using pumping lemma for regular languages, both languages can be shown to be not regular. **Answer: (d)**

130. If a regular expression can be defined for strings present in L, then a regular expression exists for all suffixes of strings in L as well. This means tail(L) is a regular language.

The definition of min(L) states that none of the prefixes of strings present in min(L) is a member of L, which means that min(L) is essentially a specialised subset of L. Any subset of regular sets is regular. So, both min(L) and tail(L) are regular languages. **Answer: (c)**

131. Each rule in Chomsky Normal Form is of the type A → BC or A → a, where A, B, C are non-terminal symbols and a is a terminal symbol. The largest number of symbols on the RHS of any rule is given as K. Now, analyse this conversion sequence. A rule S → ABC will be converted to $\{S \to AS_1, S_1 \to BC\}$. Three symbols on the RHS = Two rules. A rule S → ABCD will be converted to $\{S \to AS_1, S1 \to BS_2, S_2 \to CD\}$. Four symbols on the RHS = Three rules.

Apply mathematical induction along similar lines. If there are exactly K number of non-terminals on the RHS of a production rule, then at most (K – 1) rules will be generated while converting to Chomsky Normal Form. So, if the total number of rules is |P|, then at most (K – 1)|P| number of rules will be generated on converting to Chomsky Normal Form.

The number of terminal symbols is |T|. So, there will be exactly one rule of the form A → a for each terminal symbol. If this is not the case, then the grammar is incomplete/meaningless. So, the maximum possible number of rules = (K – 1) |P| + |T|. **Answer: (a)**

132. While drawing the DFA, we need to ensure that strings with five *a*s are not accepted. To accept a continuous stream of *k* number of symbols, we need $(k + 1)$ states in the DFA. So, the sixth state will correspond to a^5. Keep that state as the non-final state and add one more final state after it with a self-loop (see below).

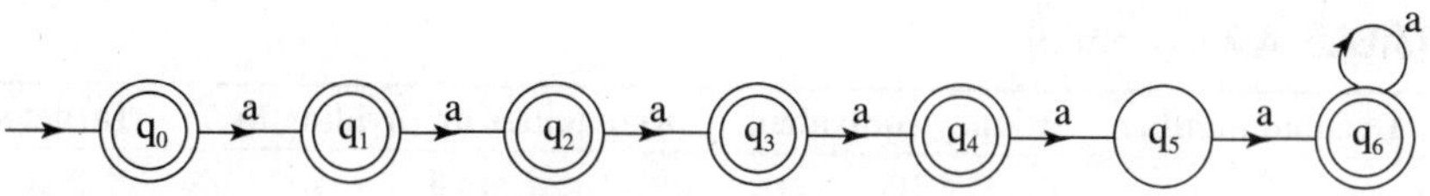

FIGURE A14 Desired DFA

State q_0 = accepting state for empty string (n = 0)
State q_1 = accepting state for string 'a' (n = 1)
State q_2 = accepting state for string 'aa' (n = 2)
State q_3 = accepting state for string 'aaa' (n = 3)
State q_4 = accepting state for string 'aaaa' (n = 4)
State q_5 = not a final state and hence 'aaaaa' is not accepted
State q_6 accepts all strings of length 6 or more.
Seven states are required. **Answer: (c)**

133. DCF $\subseteq$ CF $\subseteq$ CS $\subseteq$ REC $\subseteq$ RE is the correct hierarchy. **Answer: (a)**
134. The rule set T → TT | aTb | bTa | ε generates strings with an equal number of *a*s and *b*s. However, if we include T → a, the generated string will always have at least one more a than b. Also, the rule S → TaT will contribute to one additional a. The smallest possible string generated by this grammar is 'a'. So, the number of *a*s is always more than the number of *b*s. **Answer: (b)**
135. Phrase-structured grammar is a Type 0 grammar. CSL is Type 1, CFL is Type 2 and regular language is Type 3. Type 0 is a superset of all the other types of languages. **Answer: (d)**
136. Use the techniques of generating regex from grammars. The rule set B generates the regex bb*. Now, the rule set is modified as S → abScbb* | ε. The language generated on applying repeated expansion through the rule for S is given by $\{(ab)^n(cbb^*)^n \mid n \geq 0\}$. **Answer: (c)**
137. X is correct. Y is also correct. a*b* ∩ b*c* = b* and so Z is incorrect. W is also incorrect. **Answer: (c)**
138. The language can be easily proved to be non-regular using pumping lemma for regular languages, and is acceptable only by a PDA. **Answer: (c)**
139. FA has zero auxiliary memory. **Answer: (a)**
140. This requires a detailed explanation.

    DFA for numbers divisible by 7: When a number is divided by 7, it can give one of these values as remainder {0, 1, 2, 3, 4, 5, 6}. So, we need to have *one state for each of these remainders*. Let q_x be the state when remainder = x. Clearly q_0 is the final state. *Assumption*: Least significant digit occurs first. We begin with the following basic table.

TABLE A2 Analysis of states of DFA

Decimal number	Binary equivalent	Remainder on dividing by 7	Target state
0	0	0	q_0
1	1	1	q_1
2	10	2	q_2
3	11	3	q_3

Cont'd

TABLE A2 (Cont'd)

Decimal number	Binary equivalent	Remainder on dividing by 7	Target state
4	100	4	q_4
5	101	5	q_5
6	110	6	q_6
7	111	0	q_0

Clearly, leading zeros are insignificant. 0 is always a multiple of 7, and hence, any sequence of only zeros is also accepted by this DFA. So, insert a self-loop for input '0' on q_0. Clearly, q_0 is also the initial state. We have the following partial DFA, based on the above table.

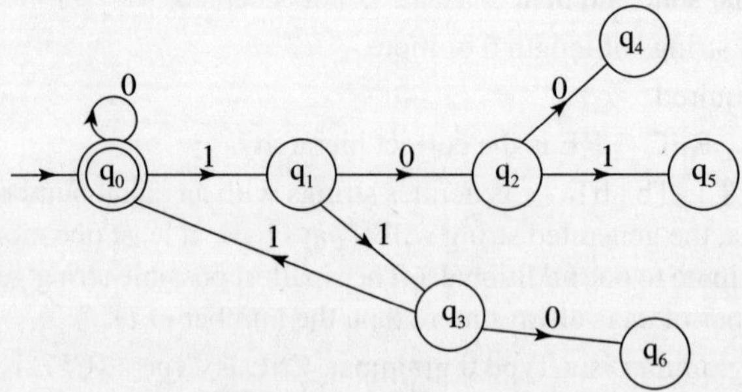

FIGURE A15 Step 1

q_0, q_1, q_2, q_3 have proper state transitions defined for inputs 0 as well as 1. The undefined transitions are given as follows:

For state q_4: 1000 = 8 and (8 mod 7) = 1. So, $\delta(q_4, 0) = q_1$. 1001 = 9 and (9 mod 7) = 2. So, $\delta(q_4, 1) = q_2$. For state q_5: 1010 = 10 and (10 mod 7) = 3. So, $\delta(q_5, 0) = q_3$. 1011 = 11 and (11 mod 7) = 4. So, $\delta(q_5, 1) = q_4$. For state q_6: 1100 = 12 and (12 mod 7) = 5. So, $\delta(q_6, 0) = q_5$. 1101 = 13 and (13 mod 7) = 6. So, $\delta(q_6, 1) = q_6$.

Add these state transitions to complete the DFA with the most significant digit first.

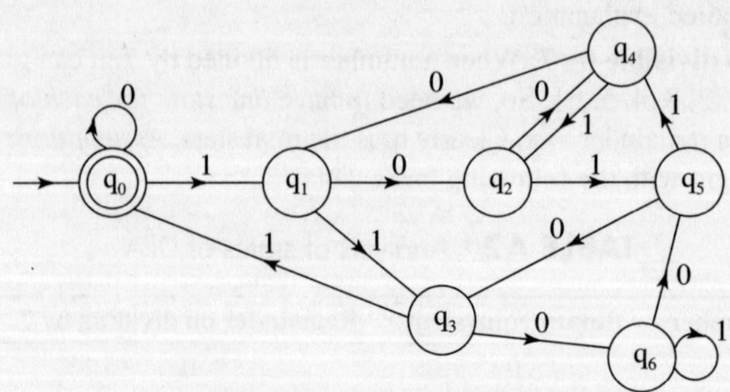

FIGURE A16 Completed DFA

So, a DFA is possible. **Answer: (a)**

141. CFG generates a context-free language, which is accepted by PDA. Also, context-free languages are a superset of regular languages, so regular languages are also accepted by PDA. **Answer: (c)**

142. Pumping lemma for regular languages can be used to prove that the language wcw^R cannot be accepted by a NFA as it is not a regular language. It can be accepted by both DPDA as well as NPDA. S1 is false, and both S2 and S3 are correct. **Answer: (b)**
143. L_3 is not context-free, and this can be proved using pumping lemma for context-free languages. All the others are CFLs. **Answer: (d)**
144. Determining the equivalence of two Turing machines is an undecidable problem. The halting problem is also undecidable. Both S1 and S2 are correct. **Answer: (a).**
145. DTM and NDTM accept all types of grammars. NPDA and DPDA can only accept Type 2 and Type 3 grammars. So, only S3 and S4 are correct. **Answer: (c)**
146. The language $\{a^n b^n c^n \mid n \geq 0\}$ is not a CFL and can be easily proved by pumping lemma for CFLs. The regular expression 1*0*1*0* accepts all binary strings of length 0, 1, 2. The smallest string not accepted by this regex is 1010. The regex $(1 + 01)^*0^*$ does not accept the string 1001, which clearly does not contain substring 0001. So, all possible binary strings not having 0001 as substring are not accepted by this regex. Turing machines accept all languages, including regular languages. **Answer: (c)**
147. There is no equivalence between deterministic and non-deterministic PDA. **Answer: (d)**
148. P is a subset of NP because any non-deterministic machine can be used as a deterministic machine. $X = P \cap NP \neq \varphi$. By definition of co-CP, $Y = NP \cap \text{Co-NP} = \varphi$. $Z = P \cap \text{NP-Hard} = \varphi$. **Answer: (d)**
149. Every recursive set is recursively enumerable, but the converse is not true. See the Venn diagram below.

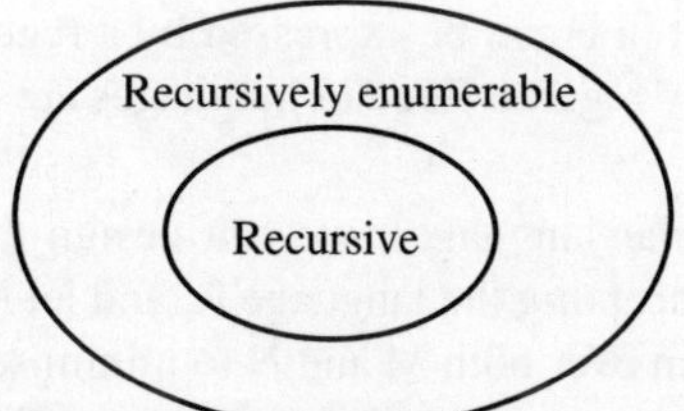

FIGURE A17 Hierarchy of recursive languages

This means that the complement of every recursive set will be either a recursive set or a recursively enumerable set. **Answer: (c)**

150. L is recursive, but nothing is mentioned about its finiteness. **Answer: (d)**
151. The complement of recursive language is either recursive or recursively enumerable. The complement of recursively enumerable language is not recursively enumerable. **Answer: (b)**
152. (I) Membership problem for TM is undecidable. (II) Regularity problem for TM is undecidable. (III) Equivalence problem for grammars is undecidable. (IV) DPDA exists for every NFA is trivially decidable. **Answer: (d)**
153. Recursive languages are closed under intersection. **Answer: (d)**
154. By definition of reducibility. **Answer: (b)**
155. The halting problem is undecidable. **Answer: (c)**
156. The biggest open problem is to determine whether P = NP. **Answer: (a)**
157. Clearly, states 2 and 4 are unreachable and are hence ignored. We write the state equations as follows:

$0 = \varepsilon + 1a$ (i)

$1 = 0a = a + 1aa$

$\Rightarrow 1 = a(aa)^*$ [since the solution of R = Q + RP is R = QP*] (ii)

Using (ii) in (i), we get $0 = \varepsilon + a(aa)^*a$ (iii)

$3 = 1b + 0b = a(aa)^*b + b + a(aa)^*ab$ [using (ii) and (iii)] (iv)

$5 = 3(a + b) + 5(a + b)$

$\Rightarrow 5 = a(aa)^*b(a + b) + b(a + b) + a(aa)^*ab(a + b) + 5(a + b)$ [using (iv)]

$\Rightarrow 5 = a(aa)^*b(a + b)(a + b)^* + b(a + b)(a + b)^* + a(aa)^*ab(a + b)(a + b)^*$ *[since the solution of R = Q + RP is R =QP*]* (v)

The regular expression for this DFA is 3 + 5 (as there are two final states) = $a(aa)^*b + a(aa)^*ab + a(aa)^*b(a + b)(a + b)^* + b(a + b)(a + b)^* a(aa)^*ab(a + b)(a + b)^* = a^*b(a + b)^*$.

The minimal DFA to accept this regular expression is given below.

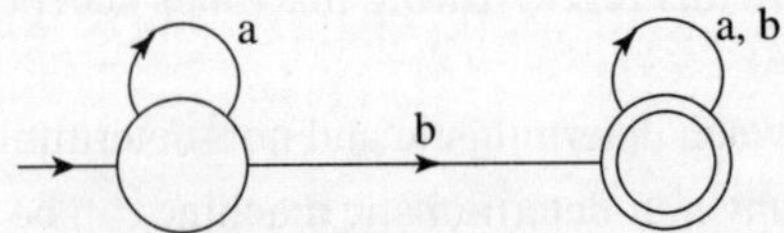

FIGURE A18 Minimal DFA

This has only two states. **Answer: (a)**

158. L is clearly a regular language because it is accepted by a finite automaton. In the worst case, the set of 2011 strings present in F can be expressed by a regular expression which is + of 2011 different terms. So, F is clearly regular. Regular languages are closed under union; hence L ∪ F is always regular. **Answer: (a)**

159. Since both A and B are regular languages, we can design a DFA for accepting each of these languages. Let M be a DFA accepting the language A, and let N be a DFA accepting the language B. Apply Hopcroft's algorithm over both M and N to minimise them, and hence, ensure that both DFAs have exactly one initial state and one final state. Now, all we need is an ε-move from the final state of M to the initial state of N to accept the language A.B = {xy x ∈ A, y ∈ B}.

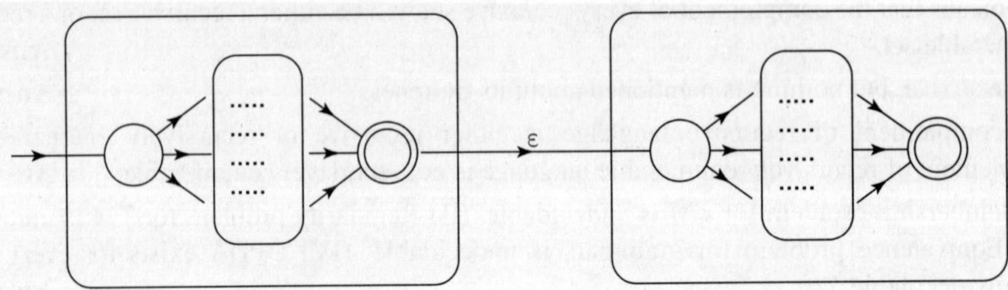

FIGURE A19 NFA for accepting A . B

So, we have defined a valid NFA to accept the language C = A.B = {xy | x ∈ A, y ∈ B}. As C is accepted by a NFA, C is regular. **Answer: (a)**

160. Apply pumping lemma for regular languages – this language is not context-free. Next, apply pumping lemma for context-free languages. This language is not context-free. **Answer: (d)**

161. Clearly, L is a regular language. Every substring of a string present in L will have a valid regular expression associated with it. So, L^P is regular. There exists a well-defined algorithm to convert an existing NFA to accept strings obtained by reversal of strings accepted by the original NFA. So, L^R is also regular. **Answer: (c)**

162. The smallest string accepted by (0+1)*0(0+1)*0(0+1)* is 00. So, this regular expression will accept all strings containing two or more zeros. **Answer: (c)**

163. $xy \in L$ and L is a regular language. So there exists a regular expression to represent the string xy. Regular languages are closed under concatenation; so, there exist regular expressions for string x as well as for string y. So, if there exists an x, such that $xy \in L$, and $|y| = |x|^2$, then the set of such xs is clearly a regular set. There is absolutely no practical significance of the constraint on string length. So, L1 is a regular language.

    By a similar argument, as there is no practical significance of the constraint on string length, L2 is also a regular language. **Answer: (b)**

164. L can be expressed by the grammar $S \rightarrow 0S1 \mid 2$, which is clearly recursive and a deterministic CFL. **Answer: (b)**

165. All four languages are context-free and PDA can be constructed for each of these. **Answer: (d)**

166. Recursive languages are a superset of context-sensitive languages. So, both (i) and (ii) are correct. **Answer: (b)**

167. L1 is easily accepted by a PDA and hence is CFL. L2 can be proved to not be a CFL by applying pumping lemma. **Answer: (b)**

168. Every language is not regular, and hence, cannot be accepted by an FSA. **Answer: (b)**

169. First, keep expanding till you have exactly x number of non-terminals on the RHS of the rule. This requires $(x - 1)$ steps. Then, apply the rules of the form $A \rightarrow a$ for each non-terminal symbol; that is a further x number of steps. The total number of productions needed $= (x - 1) + x = (2x - 1)$. **Answer: (a)**

170. A DFA can be designed for this language, implying that the language is regular. **Answer: (b)**

171. Apply pumping lemma for regular languages. Contradiction arises, showing that L is not regular. Next, apply pumping lemma for context-free languages. Contradiction arises again, showing that L is not context-free. So, L is a context-sensitive language. **Answer: (a)**

172. P is a subset of NP. All other statements are false. **Answer: (a)**

173. All statements are correct. **Answer: (d)**

174. (0 + 1)* accepts all possible strings over {0, 1}. This eliminates option (a).

    1*0*1* clearly accepts 11101. This eliminates option (b).

    111*01 also clearly accepts 11101. This eliminates option (c) as well.

    (11)*(01)* accepts strings which start with **even** number of 1s followed by any number of (01). So, 1101 and 111101 are accepted, but 11101 is not accepted by it. **Answer: (d)**

175. There is no correlation between a problem being NP and it being decidable or undecidable. So, X can be decidable. All other options are invalid statements. **Answer: (a)**

OMR SHEET

Student's Name (In Block Letters Only)

Ⓐ Ⓑ Ⓒ Ⓓ Ⓔ Ⓕ Ⓖ Ⓗ Ⓘ Ⓙ Ⓚ Ⓛ Ⓜ Ⓝ Ⓞ Ⓟ Ⓠ Ⓡ Ⓢ Ⓣ Ⓤ Ⓥ Ⓦ Ⓧ Ⓨ Ⓩ

Set

Ⓐ Ⓑ Ⓒ Ⓓ

Roll Number

⓪ ① ② ③ ④ ⑤ ⑥ ⑦ ⑧ ⑨

Subject

⓪ ① ② ③ ④ ⑤ ⑥ ⑦ ⑧ ⑨

Mobile Number

⓪ ① ② ③ ④ ⑤ ⑥ ⑦ ⑧ ⑨

Test ID

⓪ ① ② ③ ④ ⑤ ⑥ ⑦ ⑧ ⑨

No.	Options	No.	Options	No.	Options	No.	Options	No.	Options
1	Ⓐ Ⓑ Ⓒ Ⓓ	21	Ⓐ Ⓑ Ⓒ Ⓓ	41	Ⓐ Ⓑ Ⓒ Ⓓ	61	Ⓐ Ⓑ Ⓒ Ⓓ	81	Ⓐ Ⓑ Ⓒ Ⓓ
2	Ⓐ Ⓑ Ⓒ Ⓓ	22	Ⓐ Ⓑ Ⓒ Ⓓ	42	Ⓐ Ⓑ Ⓒ Ⓓ	62	Ⓐ Ⓑ Ⓒ Ⓓ	82	Ⓐ Ⓑ Ⓒ Ⓓ
3	Ⓐ Ⓑ Ⓒ Ⓓ	23	Ⓐ Ⓑ Ⓒ Ⓓ	43	Ⓐ Ⓑ Ⓒ Ⓓ	63	Ⓐ Ⓑ Ⓒ Ⓓ	83	Ⓐ Ⓑ Ⓒ Ⓓ
4	Ⓐ Ⓑ Ⓒ Ⓓ	24	Ⓐ Ⓑ Ⓒ Ⓓ	44	Ⓐ Ⓑ Ⓒ Ⓓ	64	Ⓐ Ⓑ Ⓒ Ⓓ	84	Ⓐ Ⓑ Ⓒ Ⓓ
5	Ⓐ Ⓑ Ⓒ Ⓓ	25	Ⓐ Ⓑ Ⓒ Ⓓ	45	Ⓐ Ⓑ Ⓒ Ⓓ	65	Ⓐ Ⓑ Ⓒ Ⓓ	85	Ⓐ Ⓑ Ⓒ Ⓓ
6	Ⓐ Ⓑ Ⓒ Ⓓ	26	Ⓐ Ⓑ Ⓒ Ⓓ	46	Ⓐ Ⓑ Ⓒ Ⓓ	66	Ⓐ Ⓑ Ⓒ Ⓓ	86	Ⓐ Ⓑ Ⓒ Ⓓ
7	Ⓐ Ⓑ Ⓒ Ⓓ	27	Ⓐ Ⓑ Ⓒ Ⓓ	47	Ⓐ Ⓑ Ⓒ Ⓓ	67	Ⓐ Ⓑ Ⓒ Ⓓ	87	Ⓐ Ⓑ Ⓒ Ⓓ
8	Ⓐ Ⓑ Ⓒ Ⓓ	28	Ⓐ Ⓑ Ⓒ Ⓓ	48	Ⓐ Ⓑ Ⓒ Ⓓ	68	Ⓐ Ⓑ Ⓒ Ⓓ	88	Ⓐ Ⓑ Ⓒ Ⓓ
9	Ⓐ Ⓑ Ⓒ Ⓓ	29	Ⓐ Ⓑ Ⓒ Ⓓ	49	Ⓐ Ⓑ Ⓒ Ⓓ	69	Ⓐ Ⓑ Ⓒ Ⓓ	89	Ⓐ Ⓑ Ⓒ Ⓓ
10	Ⓐ Ⓑ Ⓒ Ⓓ	30	Ⓐ Ⓑ Ⓒ Ⓓ	50	Ⓐ Ⓑ Ⓒ Ⓓ	70	Ⓐ Ⓑ Ⓒ Ⓓ	90	Ⓐ Ⓑ Ⓒ Ⓓ
11	Ⓐ Ⓑ Ⓒ Ⓓ	31	Ⓐ Ⓑ Ⓒ Ⓓ	51	Ⓐ Ⓑ Ⓒ Ⓓ	71	Ⓐ Ⓑ Ⓒ Ⓓ	91	Ⓐ Ⓑ Ⓒ Ⓓ
12	Ⓐ Ⓑ Ⓒ Ⓓ	32	Ⓐ Ⓑ Ⓒ Ⓓ	52	Ⓐ Ⓑ Ⓒ Ⓓ	72	Ⓐ Ⓑ Ⓒ Ⓓ	92	Ⓐ Ⓑ Ⓒ Ⓓ
13	Ⓐ Ⓑ Ⓒ Ⓓ	33	Ⓐ Ⓑ Ⓒ Ⓓ	53	Ⓐ Ⓑ Ⓒ Ⓓ	73	Ⓐ Ⓑ Ⓒ Ⓓ	93	Ⓐ Ⓑ Ⓒ Ⓓ
14	Ⓐ Ⓑ Ⓒ Ⓓ	34	Ⓐ Ⓑ Ⓒ Ⓓ	54	Ⓐ Ⓑ Ⓒ Ⓓ	74	Ⓐ Ⓑ Ⓒ Ⓓ	94	Ⓐ Ⓑ Ⓒ Ⓓ
15	Ⓐ Ⓑ Ⓒ Ⓓ	35	Ⓐ Ⓑ Ⓒ Ⓓ	55	Ⓐ Ⓑ Ⓒ Ⓓ	75	Ⓐ Ⓑ Ⓒ Ⓓ	95	Ⓐ Ⓑ Ⓒ Ⓓ
16	Ⓐ Ⓑ Ⓒ Ⓓ	36	Ⓐ Ⓑ Ⓒ Ⓓ	56	Ⓐ Ⓑ Ⓒ Ⓓ	76	Ⓐ Ⓑ Ⓒ Ⓓ	96	Ⓐ Ⓑ Ⓒ Ⓓ
17	Ⓐ Ⓑ Ⓒ Ⓓ	37	Ⓐ Ⓑ Ⓒ Ⓓ	57	Ⓐ Ⓑ Ⓒ Ⓓ	77	Ⓐ Ⓑ Ⓒ Ⓓ	97	Ⓐ Ⓑ Ⓒ Ⓓ
18	Ⓐ Ⓑ Ⓒ Ⓓ	38	Ⓐ Ⓑ Ⓒ Ⓓ	58	Ⓐ Ⓑ Ⓒ Ⓓ	78	Ⓐ Ⓑ Ⓒ Ⓓ	98	Ⓐ Ⓑ Ⓒ Ⓓ
19	Ⓐ Ⓑ Ⓒ Ⓓ	39	Ⓐ Ⓑ Ⓒ Ⓓ	59	Ⓐ Ⓑ Ⓒ Ⓓ	79	Ⓐ Ⓑ Ⓒ Ⓓ	99	Ⓐ Ⓑ Ⓒ Ⓓ
20	Ⓐ Ⓑ Ⓒ Ⓓ	40	Ⓐ Ⓑ Ⓒ Ⓓ	60	Ⓐ Ⓑ Ⓒ Ⓓ	80	Ⓐ Ⓑ Ⓒ Ⓓ	100	Ⓐ Ⓑ Ⓒ Ⓓ

Candidate's Signature

Invigilator's Signature

INDEX

C

G

R

U